Volcanic Processes Monitoring and Hazard Assessment Using Integration of Remote Sensing and Ground-Based Techniques

Volcanic Processes Monitoring and Hazard Assessment Using Integration of Remote Sensing and Ground-Based Techniques

Editors

Sonia Calvari
Alessandro Bonaccorso
Annalisa Cappello
Flora Giudicepietro
Eugenio Sansosti

MDPI • Basel • Beijing • Wuhan • Barcelona • Belgrade • Manchester • Tokyo • Cluj • Tianjin

Editors
Sonia Calvari
Istituto Nazionale di
Geofisica e Vulcanologia
Italy

Alessandro Bonaccorso
Istituto Nazionale di
Geofisica e Vulcanologia
Italy

Annalisa Cappello
Istituto Nazionale di
Geofisica e Vulcanologia
Italy

Flora Giudicepietro
Istituto Nazionale di
Geofisica e Vulcanologia
Italy

Eugenio Sansosti
National Research Council
(CNR) of Italy
Italy

Editorial Office
MDPI
St. Alban-Anlage 66
4052 Basel, Switzerland

This is a reprint of articles from the Special Issue published online in the open access journal *Remote Sensing* (ISSN 2072-4292) (available at: https://www.mdpi.com/journal/remotesensing/special_issues/Volcanic_Monitoring_Hazard_Assessment).

For citation purposes, cite each article independently as indicated on the article page online and as indicated below:

LastName, A.A.; LastName, B.B.; LastName, C.C. Article Title. *Journal Name* **Year**, *Volume Number*, Page Range.

ISBN 978-3-0365-5119-7 (Hbk)
ISBN 978-3-0365-5120-3 (PDF)

Cover image courtesy of Alessandro Bonaccorso

Contents

remote sensing

MDPI

Editorial

Volcanic Processes Monitoring and Hazard Assessment Using Integration of Remote Sensing and Ground-Based Techniques

Sonia Calvari [1,*], Alessandro Bonaccorso [1], Annalisa Cappello [1], Flora Giudicepietro [2] and Eugenio Sansosti [3]

1 Istituto Nazionale di Geofisica e Vulcanologia, Osservatorio Etneo, 95125 Catania, Italy; alessandro.bonaccorso@ingv.it (A.B.); annalisa.cappello@ingv.it (A.C.)
2 Istituto Nazionale di Geofisica e Vulcanologia, Osservatorio Vesuviano, 80124 Napoli, Italy; flora.giudicepietro@ingv.it
3 Istituto per il Rilevamento Elettromagnetico dell'Ambiente (IREA), National Research Council (CNR) of Italy, Via Diocleziano, 328, 80124 Naples, Italy; eugenio.sansosti@cnr.it
* Correspondence: sonia.calvari@ingv.it

Citation: Calvari, S.; Bonaccorso, A.; Cappello, A.; Giudicepietro, F.; Sansosti, E. Volcanic Processes Monitoring and Hazard Assessment Using Integration of Remote Sensing and Ground-Based Techniques. *Remote Sens.* **2022**, *14*, 3626. https://doi.org/10.3390/rs14153626

Received: 12 July 2022
Accepted: 22 July 2022
Published: 29 July 2022

The monitoring of active volcanoes is a complex task based on multidisciplinary and integrated analyses that use ground, drones, and satellite monitoring devices. Over time, and with the development of new technology and increasing frequency of acquisition, the use of remote sensing to accomplish this important task has grown enormously. This is especially so with the use of drones and satellites for classifying eruptive events, detecting the opening of new vents, the spreading of lava flows on the surface or ash plumes in the atmosphere, the fallout of tephra on the ground, the intrusion of new magma within the volcano edifice, and the deformation preceding impending eruptions, and others besides. The main challenge in using remote sensing techniques is to develop automated and reliable systems that may assist the decision-maker in volcano monitoring, hazard assessment, and risk reduction. The integration with ground-based techniques represents a valuable additional aspect that makes the proposed methods more robust and reinforces the results obtained. This collection of papers is focused on several active volcanoes, such as Stromboli, Etna, and Vulcano in Italy; the Long Valley caldera and Kilauea volcano in the USA; and Cotopaxi in Ecuador. The authors make use of several different methods to predict and forecast the volcanoes' future behavior, using insights from the available data or from new automated routines applied to the analysis of existing data. The aim is to enable rapid assessments of the state of a volcano, discovering the connection between variables apparently not related to each other and to the state of the volcano. The development of new or automated routines is an important step forward in the process of forecasting eruptive activities, and this collection comprises several such examples.

This Special Issue on the monitoring of active volcanoes using an integration of remote sensing and ground-based techniques comprises twelve papers. Three are focused on the results obtained for Stromboli volcano (Italy), where eruptive activity varies from moderate Strombolian, often accompanied by summit overflows, to highly explosive paroxysms, which are very dangerous both for the local population and for the many tourists who frequently visit the island. The first paper [1] presents the precursors of the paroxysmal and devastating explosive eruptions occurring in 2019. This paper applied an unsupervised analysis of seismic and infrasonic data, comprising a dataset of 14,289 Strombolian explosions occurring over 10 months, using a Self-Organizing Map (SOM) neural network to recognize changes in the eruptive patterns preceding the paroxysms. The SOM analysis identified three main clusters indicating a clear change in Stromboli's eruptive style before the paroxysm of 3 July 2019. The main clusters were then compared with the recordings of the fixed monitoring cameras and with the Ground-Based Interferometric Synthetic Aperture Radar measurements, showing that they were associated with different types of Strombolian explosions and different deformation patterns of the summit area.

The second paper, dealing with the Stromboli volcano [2], proposes a new classification system based on multidisciplinary volcanological and geophysical data coming from the 12 explosive events occurring at Stromboli between 25 June 2019 and 6 December 2020. The authors used images of the monitoring camera network, seismicity, and ground deformation data, to characterize, classify and distinguish paroxysms (impacting the whole island) from major explosions (that affect the summit of the volcano above 500 m elevation) and from the persistent, mild explosive activity that normally has no impact on the local population.

The third paper, dedicated to the Stromboli volcano [3], considers an eruptive period from 28 March to 1 April 2020, when the Stromboli volcano erupted overflows from the crater rim that spread along the NW slope and reached the sea. Satellite, GBInSAR, and seismic data, enabled the reconstruction of the volcanic event, which involved several small collapses of the summit cone and the generation of pyroclastic density currents (PDCs) spreading along the slope and on the sea surface. Satellite monitoring allowed for the mapping of the lava flow field and the quantification of the erupted volume, and GBInSAR continuous measurements detected the crater widening and the deflation of the summit cone caused by the last overflow. The characterization of the seismicity made it possible to identify the signals that were associated with the propagation of PDCs along the volcano flank and, for the first time, to recognize the signal that was produced by the impact of the PDCs on the coast.

The following set of three papers focuses on the results regarding the recent extraordinary sequence of lava fountains at Etna volcano (Italy). From December 2020 to February 2022, a sequence of 66 lava fountains occurred. Eruptive columns and ash plumes caused by these paroxysmal events resulted in several infrastructural problems to the urban areas and the villages around Etna's flanks and in general to the eastern part of Sicily. Moreover, they were of great concern since the ash plumes often caused the closure of the Catania international airport, and also because they expanded well beyond Italian territory.

In the first paper, Freret-Lorgeril et al. [4] investigate multi-sensor strategies for the real-time determination of eruptive source parameters of explosive eruptions, useful for accurately forecasting both tephra dispersal in the atmosphere and deposition on the ground. The authors analyze and compare data acquired by two Doppler radars, ground- and satellite-based infrared sensors, one infrasound array, visible video-monitoring cameras, as well as data from tephra-fallout deposits. A second paper by Calvari et al. [5] considers a case study by selecting and analyzing the 12 March 2021 episode, which was one of the most powerful (and best recorded) lava fountain events over the entire eruptive sequence. The authors used remote sensing data from the ground and satellite to characterize the formation and growth of the lava fountains, integrated and related with ground deformation data recorded by a high precision borehole strainmeter to infer the decompression of the source. Moreover, the authors provided an estimation and comparison of different components of the erupted volumes (pyroclasts plus lava flows) with the total erupted volume inferred from the volcano deflation recorded by the borehole strainmeter.

A further paper [6] analyzes the entire lava fountains sequence by using a new approach. This consists of a software routine able to automatically detect the start and end time of each lava fountain, the area of the hot pyroclasts, the elevation reached by the lava fountains over time, and to calculate in real-time the erupted volume of pyroclasts, giving results close to the manual analysis but more focused on the sustained portion of the lava fountain, which is also the most dangerous.

The next two papers deal with general approaches for the modeling of eruptive parameters and processes. Pulvirenti et al. [7] present a 3D finite element model that includes topography and crust heterogeneities to characterize the nature of the intrusion in the Long Valley Caldera (USA). Joint numerical inversions of uplift and Electronic Distance Measurement baseline length change data were used to infer the deformation-source size, position, and overpressure. Successively, this information was used to refine the source overpressure estimation, compute the gravity potential and infer the intrusion density from the inversion of deformation and gravity data. Pailot-Bonnétat et al. [8] use the cloud-height-from-shadow technique to model the plume emitted during the 26 October 2013 event at Mount Etna. The authors used a single Landsat-8 Operational Land Imager image to extract the cloud altitude time-series, allowing them to document the ascent and dispersion history of a plume–cloud system produced by a fountaining event. The results were validated through a comparison with the proximal plume height time-series obtained from fixed monitoring cameras, finding a good agreement.

Two papers explore methods and tools to monitor volcanic activity. Inguaggiato et al. [9] present the results obtained by the long-term monitoring of three extensive parameters measured at Vulcano Island (Italy): the SO_2 flux in the volcanic plume, the soil CO_2 flux, and the local heat flux, monitored in the mild thermal anomaly located to the east of the high-temperature fumarole. The time variations of these parameters showed a cyclical trend in the volcanic degassing and a general increase in the pattern since June 2021. Corsa et al. [10] provide a differential interferometric SAR (DInSAR) time series and integrated it with GNSS data to create a fused dataset with enhanced accuracy of 3D ground motions over Hawaii island from 2015 to 2021, giving new estimates of the spatial and temporal dynamics of the 2018 Kilauea volcanic eruption. The methodology presented can easily be repeated over any region of interest where an SAR scene overlaps with GNSS data, giving a contribution to the classification of volcanic eruption precursors and the advancement of early warning systems.

Finally, the last two papers focused on methodologies to detect and map the deposits of volcanic products, which are fundamental for hazard assessment studies. In the first paper, Andrade et al. [11] provide a detailed cartography of the lahar deposits from the 26 June 1877 event in Cotopaxi (Ecuador). The cartography was performed through a combination of geological fieldwork with the analysis and interpretation of high-definition imagery obtained by drone surveys, which produced 25 cm-pixel ortho-mosaics using Structure from Motion. These data were subsequently exploited to map the deposits with the help of remote-sensing techniques and in correlation with field data. The second paper [12] reconstructs the dynamics of the 2014 effusive eruption at Stromboli (Italy) through the main morphological changes of the entire Sciara del Fuoco area. This was constructed by integrating multisensor remote sensing data, such as lidar, photogrammetric, and bathymetric surveys coupled with SAR amplitude images collected before and after the eruption. The results highlighted the importance of integrated submarine and subaerial studies to monitor active volcanoes, providing a comprehensive view of the main processes (constructive vs. destructive) associated with eruptive dynamics.

From this brief summary, it is clear how these studies included in the Special Issue confirm the growing importance of remote sensing in the complex and multidisciplinary monitoring of active volcanoes and demonstrate how its integration with classic ground-based techniques represents an essential approach for a deeper understanding and interpretation of how volcanoes work.

Conflicts of Interest: The authors declare no conflict of interest.

References

1. Giudicepietro, F.; Calvari, S.; D'Auria, L.; Di Traglia, F.; Layer, L.; Macedonio, G.; Caputo, T.; De Cesare, W.; Ganci, G.; Martini, M.; et al. Changes in the Eruptive Style of Stromboli Volcano before the 2019 Paroxysmal Phase Discovered through SOM Clustering of Seismo-Acoustic Features Compared with Camera Images and GBInSAR Data. *Remote Sens.* **2022**, *14*, 1287. [CrossRef]
2. Calvari, S.; Giudicepietro, F.; Di Traglia, F.; Bonaccorso, A.; Macedonio, G.; Casagli, N. Variable Magnitude and Intensity of Strombolian Explosions: Focus on the Eruptive Processes for a First Classification Scheme for Stromboli Volcano (Italy). *Remote Sens.* **2021**, *13*, 944. [CrossRef]
3. Calvari, S.; Di Traglia, F.; Ganci, G.; Giudicepietro, F.; Macedonio, G.; Cappello, A.; Nolesini, T.; Pecora, E.; Bilotta, G.; Centorrino, V.; et al. Overflows and Pyroclastic Density Currents in March-April 2020 at Stromboli Volcano Detected by Remote Sensing and Seismic Monitoring Data. *Remote Sens.* **2020**, *12*, 3010. [CrossRef]
4. Freret-Lorgeril, V.; Bonadonna, C.; Corradini, S.; Donnadieu, F.; Guerrieri, L.; Lacanna, G.; Marzano, F.S.; Mereu, L.; Merucci, L.; Ripepe, M.; et al. Examples of Multi-Sensor Determination of Eruptive Source Parameters of Explosive Events at Mount Etna. *Remote Sens.* **2021**, *13*, 2097. [CrossRef]

5. Calvari, S.; Bonaccorso, A.; Ganci, G. Anatomy of a Paroxysmal Lava Fountain at Etna Volcano: The Case of the 12 March 2021, Episode. *Remote Sens.* **2021**, *13*, 3052. [CrossRef]
6. Calvari, S.; Nunnari, G. Comparison between Automated and Manual Detection of Lava Fountains from Fixed Monitoring Thermal Cameras at Etna Volcano, Italy. *Remote Sens.* **2022**, *14*, 2392. [CrossRef]
7. Pulvirenti, F.; Silverii, F.; Battaglia, M. A New Analysis of Caldera Unrest through the Integration of Geophysical Data and FEM Modeling: The Long Valley Caldera Case Study. *Remote Sens.* **2021**, *13*, 4054. [CrossRef]
8. Pailot-Bonnétat, S.; Harris, A.J.L.; Calvari, S.; De Michele, M.; Gurioli, L. Plume Height Time-Series Retrieval Using Shadow in Single Spatial Resolution Satellite Images. *Remote Sens.* **2020**, *12*, 3951. [CrossRef]
9. Inguaggiato, S.; Vita, F.; Diliberto, I.S.; Mazot, A.; Calderone, L.; Mastrolia, A.; Corrao, M. The Extensive Parameters as a Tool to Monitoring the Volcanic Activity: The Case Study of Vulcano Island (Italy). *Remote Sens.* **2022**, *14*, 1283. [CrossRef]
10. Corsa, B.; Barba-Sevilla, M.; Tiampo, K.; Meertens, C. Integration of DInSAR Time Series and GNSS Data for Continuous Volcanic Deformation Monitoring and Eruption Early Warning Applications. *Remote Sens.* **2022**, *14*, 784. [CrossRef]
11. Andrade, S.D.; Saltos, E.; Nogales, V.; Cruz, S.; Lee, G.; Barclay, J. Detailed Cartography of Cotopaxi's 1877 Primary Lahar Deposits Obtained by Drone-Imagery and Field Surveys in the Proximal Northern Drainage. *Remote Sens.* **2022**, *14*, 631. [CrossRef]
12. Casalbore, D.; Di Traglia, F.; Bosman, A.; Romagnoli, C.; Casagli, N.; Chiocci, F.L. Submarine and Subaerial Morphological Changes Associated with the 2014 Eruption at Stromboli Island. *Remote Sens.* **2021**, *13*, 2043. [CrossRef]

remote sensing

MDPI

Article

Changes in the Eruptive Style of Stromboli Volcano before the 2019 Paroxysmal Phase Discovered through SOM Clustering of Seismo-Acoustic Features Compared with Camera Images and GBInSAR Data

Flora Giudicepietro [1,*], Sonia Calvari [2], Luca D'Auria [3], Federico Di Traglia [4], Lukas Layer [5,6], Giovanni Macedonio [1], Teresa Caputo [1], Walter De Cesare [1], Gaetana Ganci [2], Marcello Martini [1], Massimo Orazi [1], Rosario Peluso [1], Giovanni Scarpato [1], Laura Spina [7], Teresa Nolesini [8], Nicola Casagli [4,9], Anna Tramelli [1] and Antonietta M. Esposito [1]

1 Istituto Nazionale di Geofisica e Vulcanologia, Osservatorio Vesuviano, 80124 Napoli, Italy; giovanni.macedonio@ingv.it (G.M.); teresa.caputo@ingv.it (T.C.); walter.decesare@ingv.it (W.D.C.); marcello.martini52@gmail.com (M.M.); massimo.orazi@ingv.it (M.O.); rosario.peluso@ingv.it (R.P.); giovanni.scarpato@ingv.it (G.S.); anna.tramelli@ingv.it (A.T.); antonietta.esposito@ingv.it (A.M.E.)
2 Istituto Nazionale di Geofisica e Vulcanologia, Osservatorio Etneo, 95125 Catania, Italy; sonia.calvari@ingv.it (S.C.); gaetana.ganci@ingv.it (G.G.)
3 Instituto Volcanológico de Canarias (INVOLCAN), 38320 San Cristobal de La Laguna, Spain; ldauria@iter.es
4 Istituto Nazionale di Oceanografia e di Geofisica Sperimentale—OGS, 34010 Sgonico, Italy; fditraglia@inogs.it (F.D.T.); nicola.casagli@unifi.it (N.C.)
5 Dipartimento di Fisica, Università degli Studi di Napoli "Federico II", 80126 Napoli, Italy; lukas.layer@cern.ch
6 Istituto Nazionale di Fisica Nucleare, Sezione di Napoli, 80126 Napoli, Italy
7 Istituto Nazionale di Geofisica e Vulcanologia, Sezione Roma 1, 00143 Roma, Italy; laura.spina@ingv.it
8 Centro per la Protezione Civile, Università degli Studi di Firenze, 50121 Firenze, Italy; teresa.nolesini@unifi.it
9 Dipartimento di Scienze della Terra, Università degli Studi di Firenze, 50121 Firenze, Italy
* Correspondence: flora.giudicepietro@ingv.it

Citation: Giudicepietro, F.; Calvari, S.; D'Auria, L.; Di Traglia, F.; Layer, L.; Macedonio, G.; Caputo, T.; De Cesare, W.; Ganci, G.; Martini, M.; et al. Changes in the Eruptive Style of Stromboli Volcano before the 2019 Paroxysmal Phase Discovered through SOM Clustering of Seismo-Acoustic Features Compared with Camera Images and GBInSAR Data. *Remote Sens.* **2022**, *14*, 1287. https://doi.org/10.3390/rs14051287

Academic Editor: Zhong Lu

Received: 7 January 2022
Accepted: 3 March 2022
Published: 6 March 2022

Abstract: Two paroxysmal explosions occurred at Stromboli on 3 July and 28 August 2019, the first of which caused the death of a young tourist. After the first paroxysm an effusive activity began from the summit vents and affected the NW flank of the island for the entire period between the two paroxysms. We carried out an unsupervised analysis of seismic and infrasonic data of Strombolian explosions over 10 months (15 November 2018–15 September 2019) using a Self-Organizing Map (SOM) neural network to recognize changes in the eruptive patterns of Stromboli that preceded the paroxysms. We used a dataset of 14,289 events. The SOM analysis identified three main clusters that showed different occurrences with time indicating a clear change in Stromboli's eruptive style before the paroxysm of 3 July 2019. We compared the main clusters with the recordings of the fixed monitoring cameras and with the Ground-Based Interferometric Synthetic Aperture Radar measurements, and found that the clusters are associated with different types of Strombolian explosions and different deformation patterns of the summit area. Our findings provide new insights into Strombolian eruptive mechanisms and new perspectives to improve the monitoring of Stromboli and other open conduit volcanoes.

Keywords: eruption precursors; neural networks; self-organizing map; seismo-acoustic signals; Stromboli volcano; volcano monitoring; ground-based visible and thermal imagery; ground deformation; volcano deformation

1. Introduction

Artificial Neural Networks (ANNs) are applied in a wide range of fields to approach classification, pattern recognition, clustering, regression analysis, and time series prediction problems. In recent years, ANNs have been successfully applied in the field of seismology and volcanology to solve geophysical signal automatic classification and clustering problems [1–4] and to perform predictive analyses [5,6]. In the field of seismology, many studies demonstrated that ANNs are powerful tools to improve the performances and the robustness of the automatic systems for seismological analyses that allow gaining critical information for people's safety in near real time [7–9]. Many applications have also been developed to automatically classify the seismicity of Stromboli [10–13] and other volcanoes [1,14], obtaining performances up to 100% correct classification [7]. Here, we focus on studying the eruptive style of the Stromboli volcano (Italy) before and during the 2019 eruptive crisis [4,15–17] through the neural analysis of seismic and infrasonic signals produced by the explosive activity.

Stromboli is a volcanic island in the Mediterranean Sea characterized by a persistent explosive activity that produces hundreds of moderate explosions per day. Three main vent regions are located in the upper part of the volcano: the North East (NE), the Central (C), and the South West (SW) vent regions (Figure 1) [18–22].

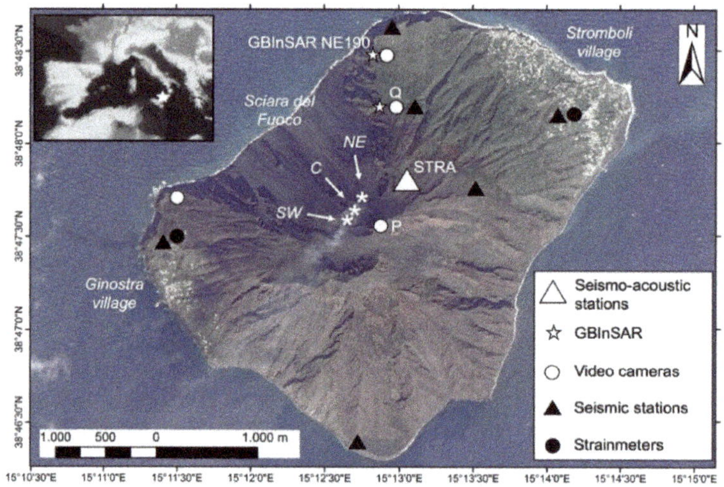

Figure 1. Map of Stromboli Island, including the location of monitoring stations. The black triangles indicate the seismic stations and the two black circles indicate strainmeters. The white triangle marks the STRA seismic–acoustic station that recorded the data used in this work. The white circles show the position of the video cameras, highlighting the SPT (P), SQT, and SQV (Q) cameras. The white stars indicate the Ground-Based Interferometric Synthetic Aperture Radar (GBInSAR) devices, and the GBInSAR NE190 is highlighted. The "Sciara del Fuoco" (SdF) depression is shown. The white asterisks indicate the main vent regions: North East (NE), Central (C) and South West (SW). The location of Stromboli in the Mediterranean Sea is reported in the inset. Maps were generated using ESRI ArcGIS CAMPUS (Università degli Studi di Firenze Licence; http://www.siaf.unifi.it/vp-1275-arcgis-licenza-campus.html, accessed on 5 March 2022). The background image is a PLÉIADES-1 image collected on 8 October 2019 (see Turchi et al. [23] for details).

In recent decades, geophysical and volcanological studies have indicated that the ordinary explosive activity of Stromboli shows a variety of eruptive mechanisms and products, whose signature is distinguishable in the geophysical signals generated by the explosions (e.g., seismic and infrasonic signals). In early studies, an association between the eruptive vent (NE or SW) and waveform of the VLP events produced by the explosions was observed [24] and led for the first time to the definition of two categories of explosions: Type 1, from the NE vent region, and Type 2, from the SW vent region. Subsequently, significant exceptions to the vent–waveform association were highlighted through a previous application of an unsupervised neural network clustering, compared with thermal camera measurements [11]. Two main types of explosions were also recognized in the ordinary activity of Stromboli through thermal camera measurements by Patrick et al. [25]: Type 1, dominated by coarse ballistic particles, and Type 2, characterized by an ash-rich plume, with (Type 2a) or without (Type 2b) ballistic particles. Leduc et al. [26] added a gas-dominated type (Type 0) to those mentioned above. More recently, Simons et al. [27], studying the explosions of the Yasur volcano (Vanuatu), defined a further type (Type 3) of explosion characterized by tephra jetting through a breccia/ash-occluded vent, followed by prolonged ash emission, belonging to the spectrum of Strombolian explosions. The rate, size, and relative occurrence of the different types of explosions characterize the eruptive style of the ordinary Strombolian activity.

Changes in the ordinary Strombolian activity are generally associated with imminent dangerous phenomena such as lava flows, major explosions, or paroxysms (e.g., [4,28,29]). Typically, an increase in Strombolian activity, in terms of the number of explosions per hour and amplitude of seismic signals associated with volcanic tremor and explosions, preceded the lava flow output [4]. They are generally caused by overflows [30] that originate from the eruptive vent regions, or by the opening of new eruptive fissures along the Sciara del Fuoco slope [31–33]. Fissure eruptions are also preceded by an intensification of landslides [8,34,35]. Major explosions are sporadic explosive events traditionally defined as explosions larger than the persistent activity, able to injure people visiting the top of the volcano. Recently, Calvari et al. [17] proposed a method based on the "VLP size" parameter of the seismic signal [4] and on the muzzle velocity by the duration of the explosive event to estimate the variable magnitude and intensity of Strombolian explosions and therefore to separate the field of the major explosions from those of paroxysms and ordinary explosions. Although to date major explosions are unpredictable, these events are most likely to occur when variations of the eruptive style happen [15,36,37]. Paroxysms are explosive eruptions that form eruptive columns of some kilometers, eject metric-sized ballistic blocks, and can generate modest pyroclastic flows [16,32,38–40].

The eruptive crisis of 2019 produced two paroxysms that occurred on July 3 and August 28, which formed eruptive columns of about 5 km and were accompanied by pyroclastic flows that traveled more than one kilometer on the sea surface. The first of these two paroxysms caused the death of a young tourist and the injury of some people, in addition to triggering extensive fires that have involved the vegetated areas of the island [23]. The effusive eruption, which began with the paroxysm on 3 July 2019, and lasted about two months, also emplaced a lava flow field on the Sciara del Fuoco slope [41,42]. It is worth pointing out that the seismic parameters that were routinely monitored, such as the seismic amplitude and the hourly occurrence of VLP events (0.05–0.5 Hz), did not show significant changes before the 2019 paroxysms (Figure 2a,b). On the contrary, the "VLP size" associated with the explosive activity and other parameters, such as the peak-to-peak amplitude of VLP events and polarization of the raw seismic signal, showed significant variations about one month before that paroxysm [4]. These medium-term seismic precursors of the paroxysmal activity (Figure 2c–e) have been interpreted as variations in the eruptive style linked to the magma–conduit interaction and to the degassing of the volcanic system, which control the Strombolian explosive mechanism.

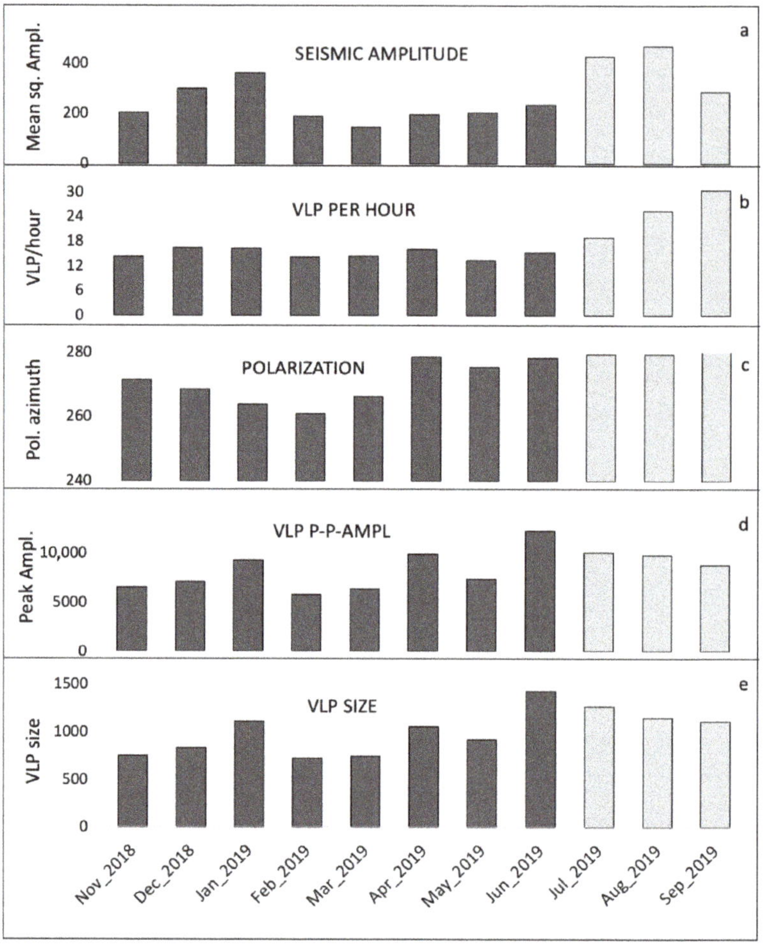

Figure 2. Monthly histograms from November 2018 to September 2019 of the raw seismic signal amplitude recorded by the STRA station (**a**); hourly occurrence of VLPs (**b**); polarization azimuth of the raw signal of the STRA station (**c**); peak-to-peak amplitude of VLP events (0.05–0.5 Hz) recorded by the STRA station, horizontal component (**d**); VLP size measured by the STRA station, horizontal component (**e**). The dark gray bars are relevant for the period before the paroxysm of 3 July 2019, the light gray bars are relevant for the period following the aforementioned paroxysm.

In addition to the geophysical studies conducted on volcanoes, analogue experiments also provided useful information to discriminate the factors that affect the degassing and eruptive style of an open conduit volcano such as Stromboli. Spina et al. [43] investigated the role of conduit surface irregularity and physical properties (e.g., viscosity and gas flux) of an analogue basaltic magma using an experimental setup [44] and produced seismo-acoustic measurements. Giudicepietro et al. [45] designed an unsupervised neural network based on a Self-Organizing Map (SOM) that was able to consistently group the artificial seismo-acoustic events generated in similar experimental conditions (conduit roughness, analogue magma viscosity, and gas flux) thanks to an appropriate strategy for extracting the seismo-acoustic features.

The aim of this work is to study the types of explosions recognizable in the persistent activity of Stromboli through unsupervised neural networks applied to seismic and infrasonic signals, which contain the fingerprints of the explosive mechanism. In particular, our target is the period that preceded and included the paroxysmal phases of 2019. For this purpose, we applied a SOM clustering of the seismic and infrasonic features to group events generated in similar conditions and we compared the result of clustering with the images recorded by the Istituto Nazionale di Geofisica e Vulcanologia (INGV) thermal and visible cameras and with the ground displacement measurements obtained by means of Ground-Based Interferometric Synthetic Aperture Radar (GBInSAR) devices, in order to gain insights into the explosive mechanisms and the pre-eruptive dynamics of the paroxysmal activity.

2. Data and Methods

2.1. Seismic and Infrasonic Data

We used the data of the STRA seismic–acoustic station (Figure 1) belonging to the Istituto Nazionale di Geofisica e Vulcanologia (INGV) monitoring network [46–48]. The seismic station is equipped with a CMG40T Guralp broadband sensor. The infrasonic sensor is a Chaparral Model 25. The signals of both sensors are acquired by a 24-bit GAIA2 digital recorder [49], optimized for low power consumption, a critical requirement for data acquisition in inaccessible areas. The sampling rate for seismic and infrasonic signals is 50 samples per second. The data are continuously transmitted to the Osservatorio Vesuviano, Osservatorio Etneo and Osservatorio Nazionale Terremoti of INGV, Italy. Figure 3 shows examples of seismic–acoustic recordings of explosive events linked to the persistent explosive activity of Stromboli. The filtered signals (red line) in frequency bands used in this work for the seismic–acoustic feature extraction are superimposed on the raw signals (gray line).

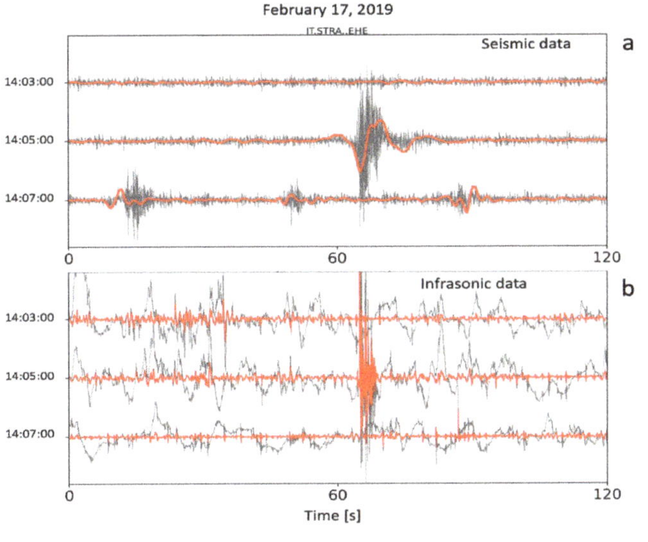

Figure 3. Example of seismic (**a**) and infrasonic (**b**) signals of a single explosion that occurred on 17 February 2019 at 14:05 UTC. The raw signals are shown in gray. In red, the seismic signal filtered in the VLP band (0.05–0.5 Hz) in panel (**a**) and the high-pass filtered infrasonic signal (>0.5 Hz) in panel (**b**).

Figure 4 shows the frequency content of the seismic and infrasonic signals of the explosive event considered in Figure 3.

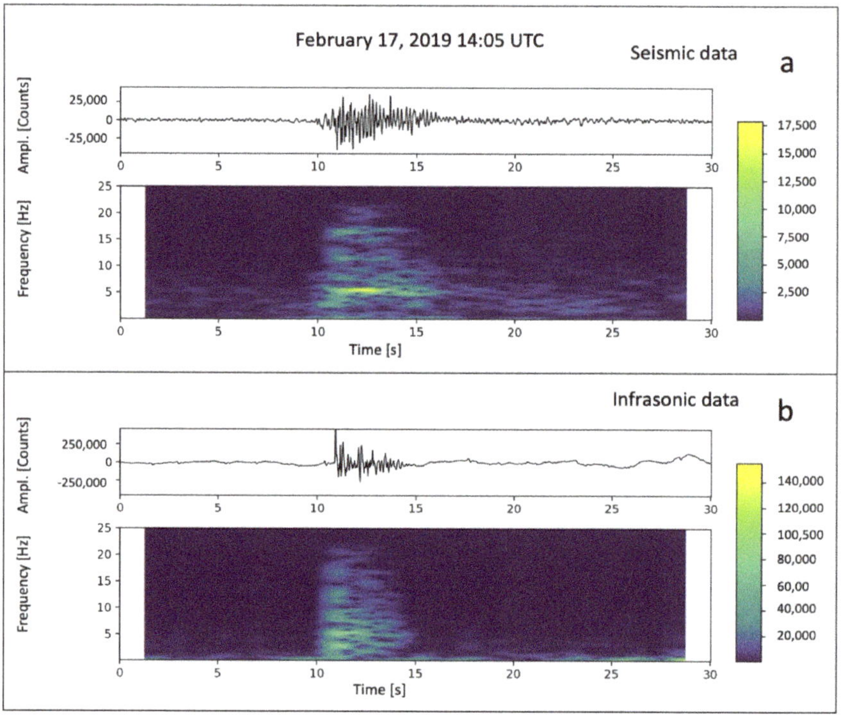

Figure 4. Example of the seismic (panel **a**) and infrasonic (panel **b**) signals of an explosion that occurred on 17 February 2019 at 14:05 UTC. The raw signals and the spectrograms for both signals are shown.

To describe the temporal evolution of the eruptive style of Stromboli before and during the 3 July–30 August 2019 eruptive crisis, we considered a wider time interval that extends from 15 November 2018 to 15 September 2019, and selected 14,289 seismic recordings (each 30 s long) and 14,179 infrasonic recordings (each 30 s long), most of which are pairs of seismo-acoustic signals linked to the same explosive event. We chose the same events related to the "VLP size" time series of Giudicepietro et al. [4], which refers to the VLPs with maximum "size" for each half-hour in the target period. We adopted this criterion for the selection of the seismic–acoustic events because the VLP size was already recognized as a parameter that effectively highlighted variations in the period preceding the eruptive crisis of July August 2019 [4]. We used the seismic data of the east–west component of the STRA station, which is the component with the maximum amplitude of the signals, with the Stromboli seismic wavefield being mainly horizontally polarized (e.g., [4,37]). In addition to the seismic data, we also selected the corresponding infrasonic data recorded by a sensor located in the same site as the STRA seismic station.

To analyze the temporal evolution of the eruptive style of Stromboli, we developed a novel preprocessing strategy suitable to extract the seismic–acoustic features representative of the fingerprints of the explosive mechanisms and to overcome the problem of the data window cutting, which cannot be based on a precise picking of seismic phases that are not recognizable in the signals of our interest (VLP events). Then, we applied the SOM algorithm to cluster the seismic–acoustic feature datasets.

2.2. Seismic–Acoustic Feature Extraction Methods

An efficient feature extraction method for seismic data typically takes into account the spectral content and the waveform of the events (e.g., [7,8,45]). Actually, these are the characteristics that the analysts visually examine to classify seismograms, for example, to distinguish a local earthquake from a regional one or a teleseism. Often, spectrograms expressed in compressed form and waveform functions calculated on sliding windows [1] are used to analyze events with impulsive onset. However, in this case, the signals of our interest are the VLPs (0.05–0.5 Hz) in which an impulsive onset cannot be recognized. Thus, we designed novel methods for seismic and infrasonic feature extraction, that are independent of the picking of a transient signal onset. The method for cutting data, which allows us to extract event records from the continuous signal, relies on an algorithm designed to detect the VLP event with the largest "size" in half-hour windows [4]. This algorithm allows cutting a 30 s signal interval from each half-hour window, but it does not return signal intervals starting precisely at the picking of a signal onset. For this reason, to encode the VLP event waveforms, we sorted the amplitude features in ascending order and, to encode the seismic signal frequency content, we used the spectrum, and not the spectrogram (Figure 5). In particular, by using the utilities of ObsPy Toolbox [50], we calculated the spectrum of every selected 30 s signal, then we applied a filter for smoothing the obtained spectrum, using the aforementioned ObsPy Toolbox, and, finally, we downsampled the smoothed spectrum by a factor of 1:3 (Figure 5d). Moreover, we encoded the VLP event waveforms by filtering the signal in the 0.05–0.5 Hz band, resampling it by a factor 1:16, and sorting the values of the filtered and resampled seismogram in ascending order (Figure 5c). We added the information of the raw seismic signal waveform using an encoding based on the peak-to-peak amplitude of a 25-sample (0.5 s) sliding window. This waveform parameterization is performed by applying the following equation:

$$WF_i = \frac{((A_{i,M}) - (A_{i,m})) * N}{\sum_{k=1}^{N}((A_{k,M}) - (A_{k,m}))} \quad (1)$$

where $A_{i,M}$ and $A_{i,m}$ are the maximum and the minimum amplitudes in a 25-sample window and N is the total number of windows. Finally, we sorted the values of the raw seismic waveform parameterization vector in descending order (Figure 5b). Figure 5 shows an example of feature extraction for one of the 14,289 seismic events in the dataset.

We also extracted the features of the infrasonic signal. To avoid the high-noise component present in the low frequencies of the infrasonic signal (Figure 3b), we high-pass filtered the data (>0.5 Hz) and applied an encoding procedure similar to that used for the raw seismic signal waveform (Equation (1)). Then, we sorted the infrasonic feature vector in ascending order to make the encoding independent from the picking of the events.

2.3. SOM Method

Once the dataset of the extracted seismic–acoustic features was prepared, we analyzed it with a SOM-based method to highlight clusters of explosive events with common characteristics.

There are different clustering techniques for the analysis of complex datasets, which can be divided mainly into two types: linear ones, such as the Principal Component Analysis (PCA) or the Multidimensional Scaling (MDS); and non-linear ones, such as the Self-Organizing Map (SOM), the Curvilinear Component Analysis (CCA), or the Curvilinear Distance Analysis (CDA). First, it has been proved that the SOM algorithm discriminates better than CCA and PCA, providing a simplified two-dimensional representation of the data and preserving the distinctive information that allows them to be separated [51,52]. Furthermore, we chose to use the SOM algorithm based on the good results obtained with SOM to analyze experimental data, proving its ability to group experimental seismo-acoustic events [45].

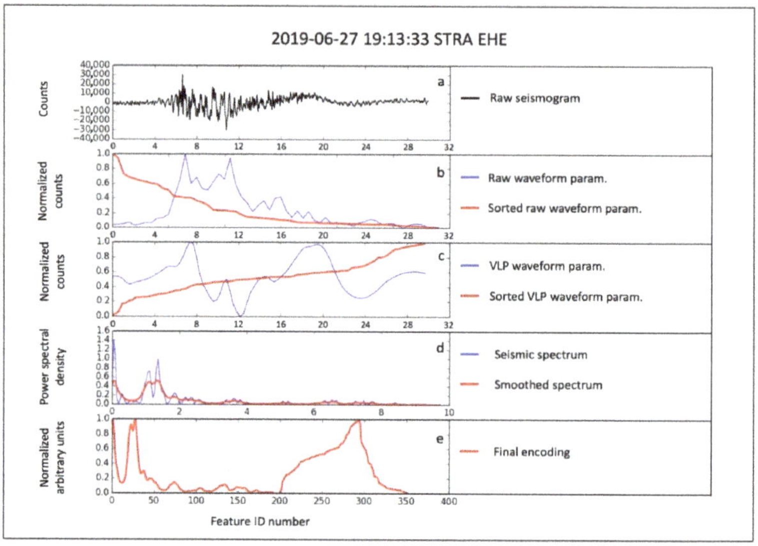

Figure 5. Feature extraction for seismic data. (**a**) Raw seismic signal of the event recorded on 27 June 2019 at 19:13:33 UTC (STRA station, east–west component; see Figure 1), time in seconds on the x axis; (**b**) normalized raw waveform parameterization (WF), time in seconds on the x axis; (**c**) normalized VLP waveform parameterization, time in seconds on the x axis; (**d**) spectrum parameterization, frequency in Hz on the x axis; (**e**) normalized seismic features (feature numbering on the x axis).

The SOM neural networks were designed by Tuevo Kohonen in 1982 [53] and inspired by brain cortex topology. In particular, he took into account the connections between neurons and how a neuron can affect its neighbors; neurons that are close to the active ones strengthen the connections, while those that are further away weaken them. The SOM network uses an unsupervised and competitive learning algorithm; this means that the process is entirely data-driven and the neurons (or nodes) compete with each other to respond to a subset of the input data. Competitive learning increases the specialization of each node in the network. The goal is to discover some hidden structures of the data so that they can be clustered.

The SOM method is used in several applications, in particular in data examination and visualization. As a clustering method, it allows the reduction of a large amount of data by grouping them. However, contrary to the classical clustering methods, being a non-parametric technique, it does not require information on the data distribution. As a projection method, it can display high-dimensional data onto an easily understandable lower-dimensional space (commonly two-dimensional), useful for improving the input pattern interpretation and classification and for finding unexpected structures in the data. Its effectiveness as a visualization technique is given by the fact that the mapping is non-linear and the resulting map preserves the topological and metric relationships of the data.

The SOM unsupervised algorithm works as follows: before the training, the prototypes are initialized with small random values to demonstrate the strong self-organizing capability of the SOM. First, a randomly extracted input vector of the dataset is presented to the network; then, the winning node (called the best-matching unit) is identified, i.e., the node whose prototype is closest to the input, in terms of Euclidean distance. The weights of the winning node and its topological neighbors are then updated or moved towards the input vector. The updating rule of the prototype vectors uses a decreasing neighborhood function of the distance between two nodes on the map grid. The most commonly used is the Gaussian. This function uses two parameters: the learning rate, which controls the intensity of the attraction of the input vector, and the neighborhood radius, which regulates the number of vectors attracted other than the winning node. Both of these parameters are time-decreasing functions and change their values during training. Then, they remain constant during the convergence phase. Thus, in the beginning, the map provides a first rough representation of the input data distribution while at the end the prototypes are settled to their final values and the final map is shown. A more exhaustive description of the SOM algorithm can be found in [11]. At the end of the iterative process, the final map consists of "ordered" prototype vectors on the grid so that similar inputs fall into topologically neighboring nodes. In this sense, the SOM is a similarity graph.

The architecture of a SOM network has two levels, one of the input nodes and one of the output nodes located on a generally two-dimensional grid. Each input node is connected to all the nodes of the grid; each output node has a vector of weights with the same dimensions as the input vectors (Figure 6a). Once the feature vectors have been processed, the final configuration of the weights will divide the input elements into the SOM nodes, which represent their clustering.

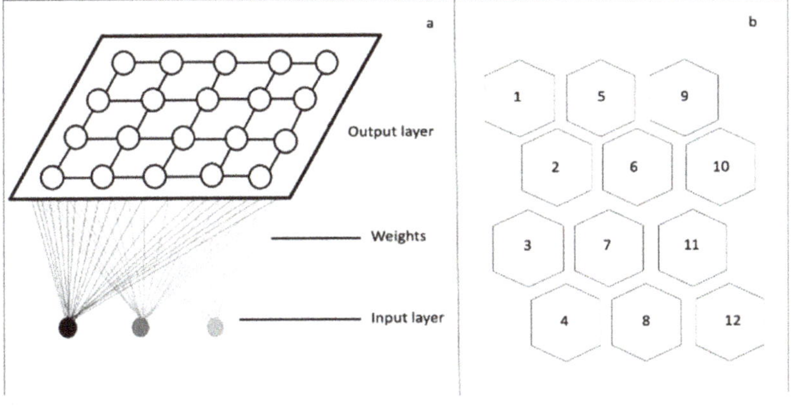

Figure 6. (**a**) Example of a SOM architecture with input layer, weights, and output layer. (**b**) Example of a two-dimensional map with a global toroid shape (visualized as a sheet) and a local hexagonal lattice structure. The size of the map is 3 × 4 = 12 (the one we adopted in our experiments) and the numbers represent the ID numbers of the nodes.

In our work, we used a SOM with a 3 × 4 = 12 local hexagonal grid (Figure 6b), and a global toroid shape displayed as a sheet to have an immediate visual cluster configuration. This means that nodes on top and bottom are neighbors as well, as are the side nodes. Figure 6b also shows the numbering of the map nodes which proceeds from top to bottom and from left to right. Finally, we fixed the SOM parameters according to Kohonen et al. [54] and the SOM toolbox for Matlab (http://www.cis.hut.fi/somtoolbox/, last accessed 4 March 2022).

2.4. Thermal and Visible Camera Data

To visually analyze the explosive activity of Stromboli and compare it with the clustering of the seismo-acoustic features, we used the recordings of the INGV monitoring fixed camera network. In particular, we used the visual and thermal images recorded by the SQV/SQT and SPV/SPT cameras (Q and P in Figure 1). These cameras acquire one image every two seconds (SQV and SQT) and two images every second (SPT), both in the visible (V) band and in the thermal (T) band. Sensors in the thermal band are particularly useful because they are not very sensitive to day/night light variations. The two cameras have different distances from the crater area and different viewing angles. This causes a different sensitivity to the detection of the explosions that occur in the three vent regions. Moreover, particular weather conditions characterized by low-lying clouds can affect the visibility of the explosions. Therefore, depending on the case, the analysis was performed by using a specific camera allowing the best view.

Based on the results of the seismo-acoustic data clustering obtained with the SOM method, we identified five days that are representative of the five main nodes, which are grouped into three clusters. Each of these days was characterized by a prevalent explosive type, according to the neural analysis of the seismo-acoustic features (see Section 3.3). Thus, we selected about 180 video recordings of explosions and characterized them on the basis of the eruptive vent, the height and shape of the ejection, and the duration of the explosive process.

2.5. GBInSAR Data

GBInSAR measurements allow retrieving ground deformation by exploiting the phase difference between pairs of Synthetic-Aperture Radar (SAR) images acquired by in situ instrumentation. They are based on the same principle as satellite-based SAR observation, with the advantage that GBInSAR has a higher rate of data acquisition and takes images short distances from the target area. At Stromboli, there are two GBInSAR devices installed in the stable area north of the Sciara del Fuoco, in order to monitor ground displacement of the unstable flank and the top of the volcano. The technical characteristics of the instruments, their settings, and the data processing methods have made this technique very important for identifying the periods of inflation/deflation of the shallow magmatic system in Stromboli (e.g., [17,30,35,41,55–57]), regardless of the weather conditions and ash content in the atmosphere. The two instruments, working in the Ku-band (17.0–17.1 mm radar), acquire with a revisiting time of 6–7 min, and then the images are averaged over 30 min in order to increase the signal-to-noise ratio. A resample operation returns images with a pixel size of about 2 m $\times$ 2 m along both range and cross-range [58] and, setting a coherence (>0.7; see Antonello et al. [59]) and a power filter (>55 dB; [30,57]), the pixel by pixel staking algorithm allows the reconstruction of the cumulative displacement maps, allowing for the tracing displacement time series of selected points (averaged over 5 $\times$ 5 pixels) with a precision in the displacement measurement of 0.5 mm [41,56]. We used GBInSAR data recorded from 15 November 2018 and 15 September 2019. The features of the two GBInSAR devices are reported in Table 1.

Table 1. Characteristics of the two GBInSAR devices.

System	Model	Band	Revisiting Time	Averaging Interval	Look Angle	Heading Angle
GBInSAR NE400	GBInSAR LiSAmobile k09	Ku	6 min	33 min	from 63.8° to 90.0°	from 143° to 217°
GBInSAR NE190	GBInSAR LiSAmobile k09	Ku	7 min	33 min	from 65.0° to 113.5°	from 115° to 245°

3. Results

Our goal is to analyze the temporal evolution of the Strombolian explosive activity in order to highlight changes in the eruptive style that preceded the paroxysmal phases of 3 July and 28 August 2019.

3.1. Seismic–Acoustic Features

We applied the novel procedures for the feature extraction from seismic and infrasonic data, which are described in the "Data and Methods" section. Starting from 30 s seismic signal recordings corresponding to 1500 samples (50 samples per second), we obtained 351-dimensional feature vectors. In particular, we encoded the seismic signal frequency content by downsampling the smoothed spectrum by a factor of 1:3 and considering the spectral features up to 10 Hz. This frequency threshold is suitable to adequately represent the signals of interest for our study [24,34,37]. Then, we encoded the VLP event waveforms by resampling the filtered signal (0.05–0.5 Hz) by a factor of 1:16 and, finally, sorting the VLP waveform features in ascending order. Finally, we extracted the raw seismic signal waveform features using an encoding based on the peak-to-peak amplitude of a 25-sample (0.5 s) sliding window (Equation (1)), sorted in descending order. Therefore, we obtained a vector of the seismic features composed of 200 coefficients for the spectral content encoding, 92 coefficients that encode the VLP waveform, and 59 coefficients for the parameterization of the raw waveform (Figures 5 and 7).

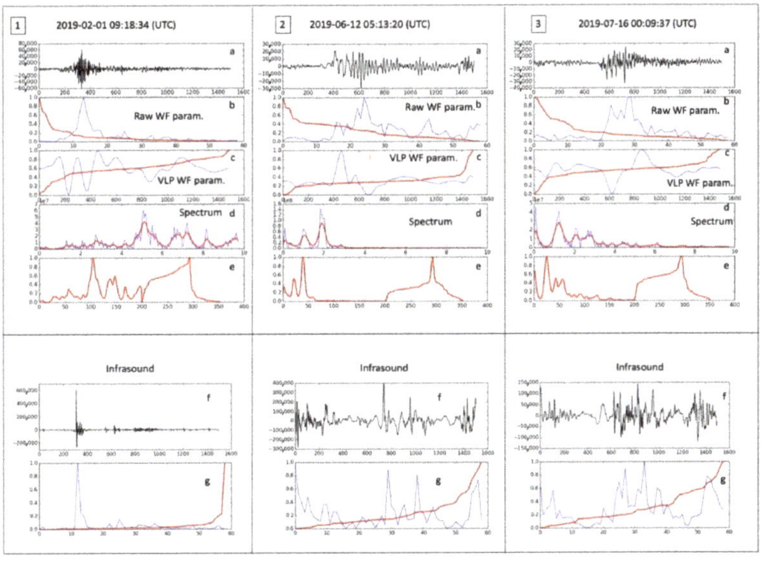

Figure 7. Parameterization of seismic and infrasound signals of three explosive events that occurred on 1 February 2019 (1), 12 June 2019 (2) and 16 July 2019 (3). In the three panels (1, 2 and 3): (**a**) The raw seismic data. (**b**) The waveform parameterization of the raw seismic data (blue line) and the same curve sorted in descending order (red line). (**c**) The seismic signal filtered in the VLP frequency band (0.05–05 Hz), resampled (blue line) and rearranged in ascending order (red line). (**d**) The spectrum up to the frequency of 10 Hz (blue line) and the smoothed and resampled spectrum (red line). (**e**) The vector of the features that encode the seismic signals given by the union of the red vectors of (**d**,**c**,**b**). At the bottom of the figure the signals (**f** panels) and the waveform parameterization (**g** panels) of the infrasonic data are shown. On the horizontal axes of panels (**a**,**c**,**f**) there is the sequential index of the sample; on the horizontal axes of panels (**b**,**g**) the sequential index of the peak-to-peak amplitude values is reported; on the horizontal axis of the panel (**d**) the frequency in Hz is reported; on the horizontal axis of the panel (**e**) there is the feature numbering.

We also extracted the infrasonic feature vectors by high-pass filtering the signal (>0.5 Hz) and applying an encoding procedure similar to that for the raw seismic signal waveform (Equation (1)). We obtained a 59-dimensional infrasonic feature vector for each infrasound record. Additionally, in this case we sorted the vector of the features in ascending order to make the encoding independent from the picking of the events (Figure 7).

3.2. SOM Analysis

We carried out three clustering experiments through the SOM analysis: (i) using only the seismic features; (ii) using only the infrasonic features; (iii) using both features of seismic and infrasonic data. Figure 8 shows the results of the three experiments. In the SOM maps, each node is shown as a yellow hexagon, whose size indicates the node density, in terms of the number of input samples associated with that node. The Euclidean distance between the prototypes of two neighboring nodes is represented according to grayscale in the spaces between the nodes. Dark gray hexagons interposed between the nodes correspond to a great distance, light gray indicates high similarity between the prototypes of neighboring nodes. We indicated the nodes of the SOM map with progressive numbers from 1 to 12 (the dimensions of the map are 3 × 4). The node numbering criterion is shown in Figure 6b.

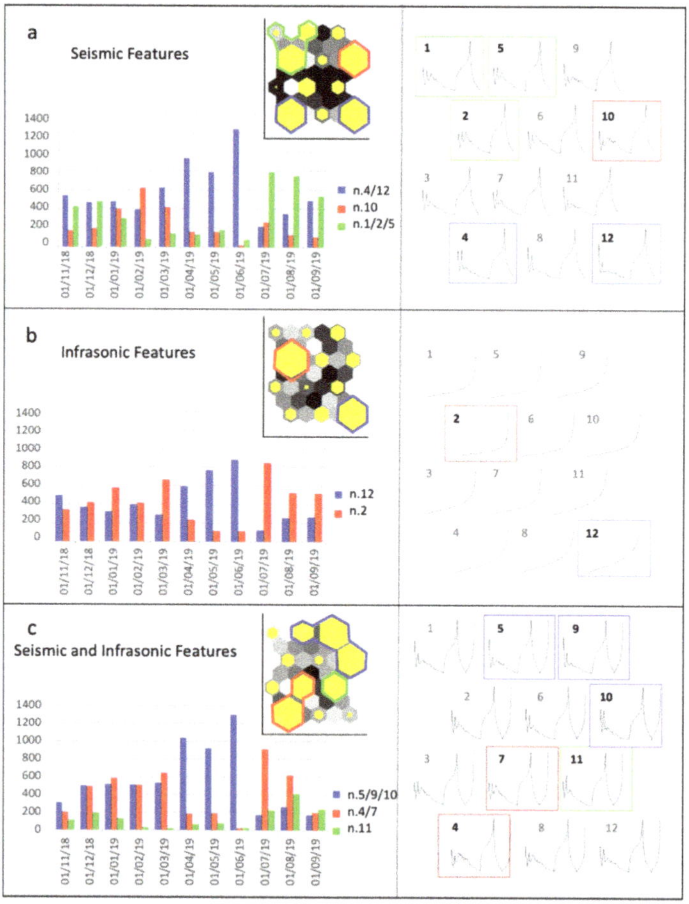

Figure 8. Summary of the experiment results. (**a**) Monthly histogram of the three main clusters obtained in the experiment based only on the seismic features. The top right inset shows the SOM map. Nodes outlined with the same color belong to the same cluster (red, blue, and green). On the right are the final prototypes of the nodes, marked with the same colors that identify the three main clusters. (**b**) Monthly histogram of the two main clusters obtained in the experiment based only on the infrasonic features. The top right inset shows the SOM map. Nodes outlined with the same color belong to the same cluster (red and blue). On the right, the final prototypes of the nodes, marked with the same colors that identify the two main clusters. (**c**) Monthly histogram of the three main clusters obtained in the experiment based both on the seismic and infrasonic features. The top right inset shows the SOM map. Nodes outlined with the same color belong to the same cluster (red, blue, and green). On the right the final prototypes of the nodes are shown, marked with the same colors that identify the three main clusters. In November 2018 and September 2019, fewer events are reported in the graph because only 15 days are considered.

The results of the experiment that was based only on the seismic features (Figure 8a) highlight three main clusters: a red cluster, formed by node 10; a blue cluster, formed by nodes 4 and 12, and a green cluster composed of nodes 1, 2, and 5.

The experiment that was based only on the infrasonic features (Figure 8b) provided only two main distinct nodes of the SOM network, in which most of the examples of the dataset are grouped. Thus, only two clusters were identified in this experiment: a red cluster, coinciding with node 2, and a blue cluster, coinciding with node 12.

The results of the experiment that was based both on seismic and infrasonic features (Figure 8c) highlight three main clusters: a red cluster, composed of nodes 4 and 7; a blue cluster, which includes nodes 5, 9, and 10; and a green cluster composed of node 11.

In all the three experiments that we carried out, the results indicate a variation in the relative occurrence of the clusters in the three months preceding the eruptive crisis, which began on 3 July 2019 (Figure 8). In particular, the experiment with only infrasonic data separates two large families of events: one characterized by large-amplitude impulsive infrasonic signals, that is marked as a red cluster in the results of our experiment, and another with infrasonic signals almost indistinguishable from background noise, marked as a blue cluster in the results of our experiment (Figure 8b). The experiment with only the features of the seismic signals identifies a greater variety of types that can be grouped into three main clusters. Finally, the experiment with the seismo-acoustic features used jointly also identifies three main clusters and more clearly emphasizes the variation in the relative occurrence of the clusters before the paroxysm of 3 July 2019 (Figure 8c).

By associating the seismic features with the infrasonic ones, the event parameterization is more complete and the SOM clustering experiment provides more significant information on the temporal evolution of the eruptive style of Stromboli in the target period. For this reason, in the following, we will focus on the results of this experiment by comparing the retrieved clusters with the camera images and GBInSAR deformation data. We will call the three clusters obtained in the seismo-acoustic SOM experiment cluster Red (in total 4539 explosions: 2950 in node 4 and 1589 in node 7), cluster Blue (in total 6332 explosions: 1183 in node 5; 2638 in node 9; and 2511 in node 10) and cluster Green (1682 explosions in node 11) (Figure 8c).

3.3. Classification of the Explosions Belonging to Clusters through the Analysis of Camera Images

To link the three seismo-acoustic clusters obtained from the SOM analysis to types of explosions, we selected a subset of seismo-acoustic events representative of Blue, Green, and Red clusters (Table 2), and compared them with the INGV camera recordings (Figure 1). We analyzed the main nodes of cluster Blue, which are 9 and 10, with node 5 being very similar to node 9. First, we selected the camera images relevant to the days when there was the highest concentration of explosions falling in the main nodes belonging to a specific cluster. These days are 17 February, 16 May, 8 June, 9 July, and 6 August 2019. Table 2 shows the distribution of the seismo-acoustic clusters on the selected days. February 17 was chosen to represent cluster Red, with a prevalence of seismo-acoustic events that fall into node 4 (43 out of 47); 16 May and 8 June were selected to represent cluster Blue, with a prevalence of seismo-acoustic events belonging to nodes 9 and 10, respectively; 9 July was again representative of cluster Red, with a prevalence of events in node 7; and August 6 represented the events of cluster Green, all of which fall into node 11. Thus, there are 182 explosions of interest, equal to the sum of the values in bold underlined in Table 2. Unfortunately, some of these explosions were not properly recorded by the INGV cameras due to poor weather conditions or technical problems. Furthermore, several explosions belonging to cluster Blue are inherently undetectable by cameras due to the types of events that this cluster groups together, namely gas explosions. In particular, thirty of the forty 16 May explosions, belonging to node 9 in our dataset, are not visible in the camera recordings. The same happens for nine of the forty explosions relevant to 8 June, belonging to node 10. In summary, the cameras recorded: 42 explosions falling into node 4 on 17 February; 32 explosions belonging to node 7 on 9 July; 10 explosions belonging to node 9 on 16 May; 31 explosions belonging to node 10 on 8 June; and 20 explosions falling into node 11 on 6 August 2019. The analyzed dataset allowed us to characterize the types of explosions corresponding to the seismo-acoustic clusters. Figure 9 shows some examples of the seismograms associated with the seismo-acoustic events that characterize the main SOM nodes.

Table 2. Days representative of the seismo-acoustic events of the three clusters. The values in bold and underlined indicate the number of events in our dataset, which, on the day indicated in the first column, falls into the prevailing node (reported in the last column). The column "Detected" gives the number of explosions that have been identified by the monitoring cameras.

Date	Red n.4	Red n.7	Blue n.9	Blue n.10	Green n.11	Tot.	Detected	Prevailing Node
17 February 2019	**<u>43</u>**	0	2	2	0	47	42	node 4
16 May 2019	0	0	**<u>40</u>**	3	3	46	10	node 9
8 June 2019	0	0	2	**<u>40</u>**	1	43	31	node 10
9 July 2019	8	**<u>33</u>**	0	0	4	45	32	node 7
6 August 2019	4	9	0	6	**<u>26</u>**	45	20	node 11

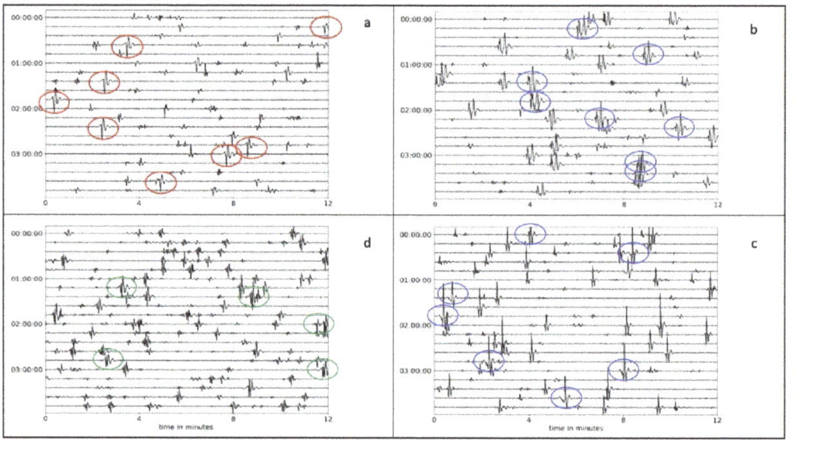

Figure 9. The first 4 h (UTC time) of the seismogram (east–west STRA channel) filtered in the VLP frequency band (0.05–0.5 Hz) of four of the five sample days representative of the main SOM nodes. (**a**) 2019-02-17, the red ovals mark the events belonging to node 4 (Red cluster). (**b**) 2019-05-16, the blue ovals mark the events belonging to node 9 (Blue cluster). (**c**) 2019-06-08, the blue ovals mark the events belonging to node 10 (Blue cluster). (**d**) 2019-08-06, the green ovals mark the events belonging to node 11 (Green cluster).

Table 2 reports in the column "Detected" the number of explosions recorded by the seismo-acoustic trace, and falling into the corresponding prevailing node (last column) that could be identified by the camera images. Some of them are shown in Figure 10. Actually, the observation of the camera images allowed us to recognize the vent (Figure 10a,f) and the eruptive style of prototypal explosions belonging to the three clusters. In particular, on 17 February, the SQV camera (Q in Figure 1) recorded 43 explosions from node 4, which belongs to cluster Red (Table 2). All of them were characterized by well-collimated jets from the N1 vent (Figure 10a), with approximately the same elevation (~200 m), and lasting on average ~5.6 s (three SQV frames). Only two events out of 43 lasted longer (8 s), whereas 12 events lasted less (4 s). These explosions ejected juvenile pyroclastic fragments showing ballistic trajectories.

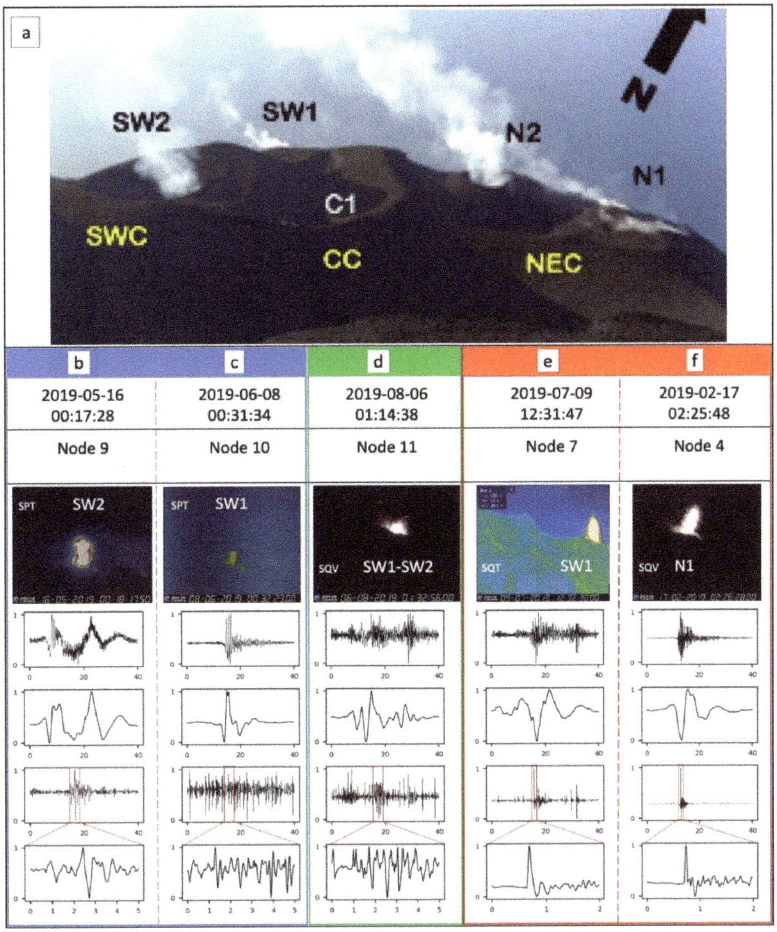

Figure 10. (**a**) Photo taken from Il Pizzo Sopra la Fossa showing the labels of the active vents within the crater terrace of Stromboli. (**b**) Example of an explosion falling into cluster Blue, node 9. (**c**) Example of an explosion falling in cluster Blue, node 10. (**d**) Example of an explosion falling in cluster Green, node 11. (**e**) Example of an explosion falling in cluster Red, node 7. (**f**) Example of an explosion falling in cluster Red, node 4. Panels (**b**) to (**f**) show from top to bottom: UTC date and time, node to which they belong, camera image, raw seismic signal, VLP seismic signal, the high-pass filtered infrasonic signal (>0.5 Hz), the zoomed-in view of the infrasonic signal.

On May 16, the SPT camera (P in Figure 1) recorded 10 of the 40 explosions that fall into node 9 (cluster Blue). Most of the SPT videos recorded on 16 May 2019 were damaged due to technical problems, probably related to the data transmission system, and could not be used, but the few available allowed observations of this activity that cannot be detected from other cameras because of too-low intensity and very short gas plume. The observed explosions were all from the SW2 vent (Figure 10a). They were mild and mostly gas-dominated (Type 0, according to Leduc et al. [26]), displaying slow bowl-shaped gas emissions with no visible ash or incandescent lapilli. The max height reached by these explosions was around 10–20 m and their duration ranged between 11 and 33 s (average 18.7 s).

On June 8, the SPT camera recorded 31 of the 40 explosions that fell into node 10 (cluster Blue). These explosions were Type 0 events occurring from SW1, SW2 (most common), or C1 vents (Figure 10a,c), having a duration between 9 and 72 s (average 22.5 s). Some of them were not visible on the surface (nine events) probably because they were too weak and occurred within the conduit.

On 9 July, the SQT camera recorded 32 powerful explosions belonging to node 7 (cluster Red) generally with well-collimated jet and ballistic ejections from the SW vent region (Figure 10a,e). We could not exactly distinguish the vent because of the inclined view offered by this camera. The duration ranged between 10 s and 28 s (18.6 average).

On 6 August, 26 explosions belonging to node 11 (cluster Green) were recorded. In this period, a lava flow descended along the SW slope of the Sciara del Fuoco (Figure 1). Given the almost continuous explosive activity that accompanied this effusive phase, it was difficult to identify the explosions associated with the seismo-acoustic signals. However, the SQV camera recorded some of these events. They may be related to explosions from multiple vents, generally the South West (SW) and Central (C) vent regions (Figure 1). These explosions were characterized by the ejection of pyroclastic fragments, most of which were incandescent spatter-like, with a wide range of ejection angles that gave the explosion an almost hemispherical shape, and the height reached a maximum of 80 m.

Figure 10 shows an example for each of the main nodes belonging to one of the three clusters. In particular, panels (b) to (f) show the images and the seismo-acoustic recordings of the event types for nodes 9 and 10 (cluster Blue), node 11 (cluster Green), and nodes 4 and 7 (cluster Red). For each of them, the date, the node they belong to, the image of one of the cameras used for the analysis, the raw seismic signal, the seismic signal filtered in the VLP band (0.05–0.5 Hz), the high-pass filtered infrasonic signal (>0.5 Hz), and a zoomed-in view of the infrasonic signal are shown from top to bottom. The SPV and SPT cameras are very close to the vents whereas the SQV and SQT cameras are further away from them (Figures 1 and 10a), therefore the explosions that produce a weak signal in the camera recordings (e.g., panels b and c) are visible only from the cameras installed at site P in Figure 1 (SPV and SPT). Figure 10b,c represents two event types of cluster Blue, which is associated with gas explosions, belonging to nodes 9 and 10, respectively. The events of this cluster show VLPs (Figure 9b,c) characterized by prolonged oscillation, especially evident in the events falling into node 9, and peak-to-peak amplitude generally higher than that of the events belonging to the other two clusters (particularly evident in node 10). The infrasonic signal associated with these events is almost indistinguishable from the background noise. Figure 10d shows an example of cluster Green, consisting of only node 11, which groups explosions with ballistic spatters and hemispherical shapes. The raw seismic signal associated with this explosion is modest in amplitude whereas the sustained VLP signal includes numerous oscillations. The infrasonic signal does not show an evident pulse linked to the explosion and is characterized by repeated minor pulses linked to spattering activities. Finally, Figure 10e,f represents two event types of cluster Red, falling on nodes 4 and 7, respectively. These events are characterized by a VLP signal with a distinct amplitude pulse and an infrasonic transient of remarkable amplitude. The raw seismic signal shows a greater contribution of the high-frequency components compared to the other types of events, in part due to the coupling of the infrasonic signal with the ground [60]. These seismo-acoustic events are associated with explosions that produce a well-collimated jet, with ejection of ballistic fragments as described above for the events of cluster Red.

3.4. Seismo-Acoustic Clusters and GBInSAR Measurements

We compared the time evolution of seismo-acoustic clusters with the ground deformations (Figure 11a) measured by the GBInSAR device in the summit area of the volcano (Figure 1). The investigated period was characterized by an oscillatory trend of deformations, with displacements towards the sensor (i.e., inflation), and displacements away from it (deflation). We observed an initial period from 15 November 2018 to 5 February 2019 (1 in Figure 11), characterized by high displacements (on average 1.8 mm/day) towards the sensor (inflation), a period of low to null displacement until March 3 (2 in Figure 11), and new inflation of about 5 mm/day toward the sensor, which lasted until March 15, 2019 (3 in Figure 11). After this inflation, there was a period characterized by small fluctuations in displacements, which in any case remained low or null until 19 July 2019 (4 in Figure 11). The following period was characterized by displacement towards the sensor (inflation), with an average rate of 2.8 mm/day and peaks that reached even more than 30 mm/day (9 August 2019 and 10 September 2019).

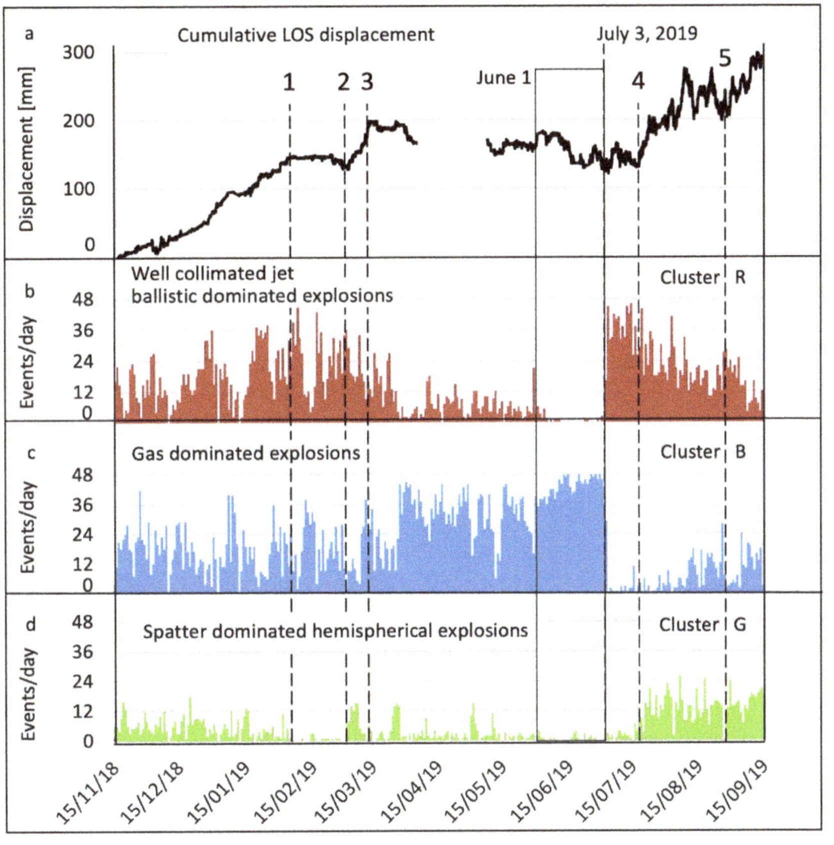

Figure 11. Comparison between the cumulative LOS displacement measured by the GBInSAR device (**a**) and the temporal evolution of the three main clusters, Red (**b**), Blue (**c**), and Green (**d**), obtained from the SOM analysis. Numbers 1 to 5 indicate the dates when the displacement rate changes. (1) 5 February 2019; (2) 3 March 2019; (3) 15 March 2019; (4) 19 July 2019; (5) 28 August 2019, when the second paroxysm occurred. The black rectangle indicates the interval between 1 June 2019 and 3 July 2019 when the first paroxysm occurred.

A very striking feature, currently never observed in the GBInSAR data from Stromboli (e.g., [56]), was the increase in the oscillations of the crater terrace, which can be deduced here from the increase in the standard deviation of the daily displacement rate. In particular, the period considered can be divided into three subperiods: (i) from 15 November 2018 to 5 April 2019, with a low standard deviation (on average 24 mm/day); (ii) from 8 May 2019 to 8 July 2019 (period preceded by the absence of data due to technical problems of the instrument), with an increase in the standard deviation (on average 45 mm/day); (iii) from 9 July 2019 to 15 September 2019, which was characterized by high standard deviation values (on average 105 mm/day), testifying the strong oscillations of the crater terrace in the time of acquisition of the GBInSAR data.

By comparing this displacement data with SOM clusters and then with the camera images, we found that the period dominated by the gas explosions of cluster Blue (Figure 11c), which begins in early April (the predominance of cluster Blue is highest in June 2019, as shown by the black rectangle in Figure 11), occurs during a stasis of ground deformation interposed between two phases of inflation of the upper part of the volcanic edifice (Figure 11a). On the contrary, the explosions of clusters Red and Green (Figure 11b,d), that are dominated by the ejection of coarse juvenile ballistic particles, occur in periods characterized by inflation of the crater area. In particular, cluster Green, erupting large spatter, seems temporally correlated with the phases of more intense inflation of the top of the volcano.

4. Discussion

In a previous study, Giudicepietro et al. [4] highlighted the precursors of the 2019 paroxysmal phase through the calculation of seismic parameters such as the polarization of the seismic signal, the peak-to-peak amplitude of VLP events, and the VLP size. The comparison of these parameters with the temporal evolution of the seismo-acoustic clusters retrieved with the SOM analysis clearly shows that the anomalies of the seismic parameters are linked to a significant change in the types of explosions before the 2019 paroxysmal phase (Figure 12).

In particular, significant variations have been recognized thanks to the definition of the VLP size parameter, which provides a value representative of the magnitude of the main VLP event for each half an hour. When the continuity of the seismic signal is satisfactory, 48 values per day relating to 48 VLP events are retrieved. The events identified by the VLP size calculation carried out in Giudicepietro et al. [4] have been selected to constitute the dataset analyzed in this work.

The time series of the VLP size in the period 15 November 2018–15 September 2019 shows a remarkable increase before the 3 July 2019 paroxysm. This increase is reflected in the time evolution of the seismo-acoustic clusters (Figure 13). Actually, about three months before the first paroxysm (3 July 2019), the occurrence of seismo-acoustic events belonging to cluster Blue (gas explosions or Type 0) increased with respect to the occurrence of seismo-acoustic events belonging to clusters Red and Green. This indicates that the gas explosions were predominant in the persistent Stromboli activity for about three months before the 3 July 2019 paroxysm. Furthermore, our findings indicate that Type 0 explosions produce large VLP events whereas they do not generate evident signals in the camera recordings, which in some cases do not record the event at all. We interpret this as the effect of large gas slugs that cause a volumetric variation in the source area of the VLP seismic signals when they rise along the conduit [24]. However, they do not generate ejection of pyroclastic fragments, or hot material, which should be detected by visible and thermal cameras, nor do they generate remarkable infrasonic signals, in frequencies greater than 0.5 Hz (Figure 10). Therefore, Type 0 explosions may not be detected at all by monitoring cameras and infrasonic networks whereas they are always clearly evident in broadband seismic signals. A low-frequency infrasonic signal, e.g., within the frequency range of the band of VLP seismic events (0.05–0.5 Hz), has been observed in some cases, but this component of the infrasonic spectrum has not been considered for the parameterization of the signals because it is generally affected by strong noise due to atmospheric weather conditions.

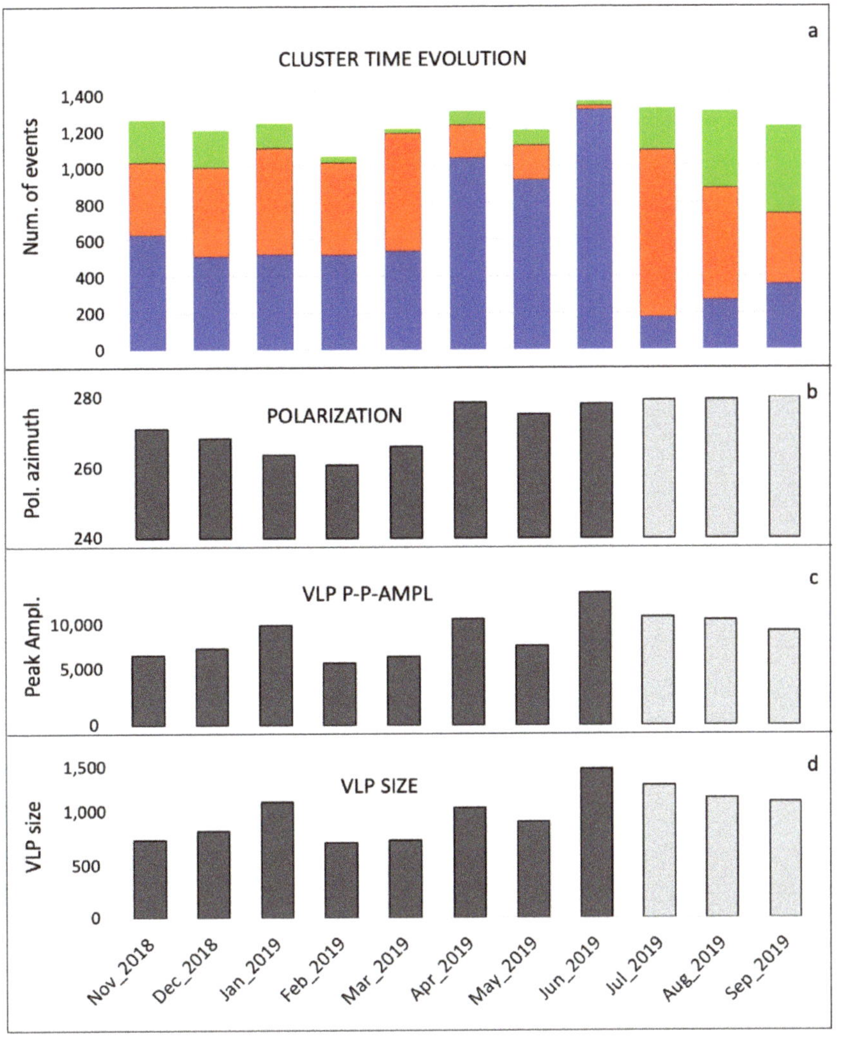

Figure 12. Comparison between: (**a**) seismo-acoustic clusters obtained with SOM analysis, (**b**) polarization of the seismic signal, (**c**) peak-to-peak amplitude of VLP events, (**d**) VLP size. The parameters shown in panels (**b**–**d**) were calculated in Giudicepietro et al. [4]. The dark gray bars are relevant for the period before the paroxysm of 3 July 2019 whereas the light gray bars are relevant for the period following that paroxysm.

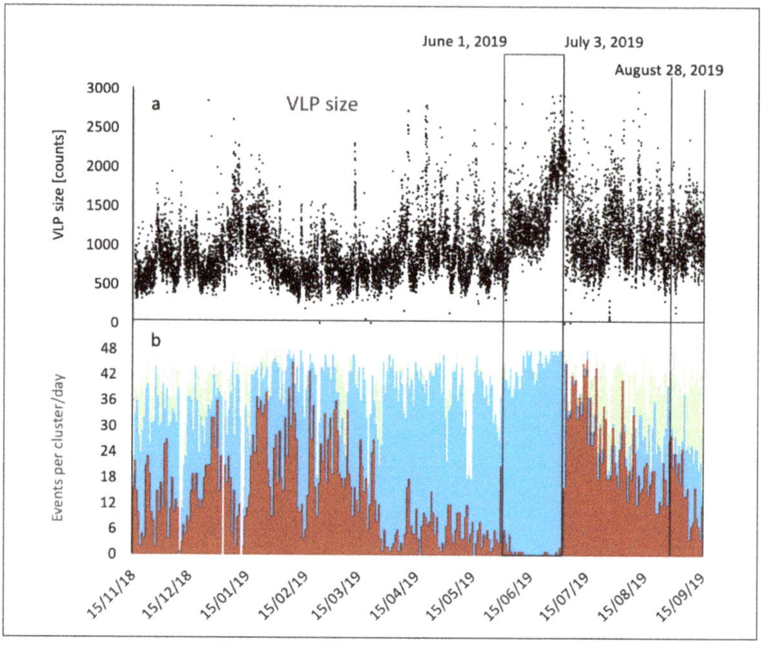

Figure 13. The time evolution of the VLP size parameter (**a**) and seismo-acoustic clusters (**b**). Relative daily occurrence of the three seismo-acoustic clusters (Red, Blue, and Green) contains one event per half hour of signal, so the maximum number of events in a day is 48. The black rectangle indicates the interval between 1 June 2019 and 3 July 2019 when the paroxysm occurred. The vertical black line indicates the 28 August 2019 paroxysm.

The comparison between the GBInSAR measures and the SOM clustering of the seismo-acoustic features highlights that, counterintuitively, the geodetic precursor of the paroxysm of 3 July 2019 was not a phase of inflation but rather an interruption of the inflation and a trend towards deflation in the last month before the paroxysm (Figure 11). The relationship between the prevailing type of explosions and the ground deformations in the crater area (Figure 11) indicates the consistency of the clustering obtained with the SOM with physical variations of the state of the volcano. In particular, the prevalence of gas explosions (cluster Blue) during a period of little or no inflation of the crater area is consistent with the fact that the gas is compressible and therefore when it passes through the final part of the conduit it produces less deformation than magma. On the other hand, the temporal correlation between the inflation phases in the crater area with the prevalence of explosions belonging to cluster Red (well-collimated jets of ballistics), and especially to cluster Green (erupting large spatter), is consistent with a condition in which the final part of a shallow conduit is filled with magma. This condition is typical of the effusive phase (3 July–30 August 2019) fed by the SW vent region, during which the occurrence of the Green cluster explosions increased.

The abrupt change in the eruptive style that arose when the paroxysm of 3 July 2019 occurred is noteworthy, suddenly determining the transition between an activity characterized by a prevalence of gas explosions with little or no emission of pyroclastic material (explosions of the Blue cluster) to an activity characterized by explosions that eject incandescent ballistic pyroclasts in conjunction with effusive activity (Figures 11 and 13). The explosions that emit incandescent ballistic fragments, which appeared immediately after the paroxysm of 3 July 2019, are distributed in two different clusters that correspond to different characteristics of the explosive mechanism whose fingerprints are recognizable in the seismo-acoustic features. In particular, the explosions of cluster Red are characterized by a well-collimated jet, which reaches a height of more than 200 m above the vent, and by a remarkable infrasonic transient. Those of cluster Green are characterized by the emission of incandescent ballistic spatter with a wide range of ejection angles and do not show an easily recognizable infrasonic transient associated with them. The latter show a hemispherical shape and reach a lower height (around 80 m). Compared to the explosions of cluster Red, this second type of explosion is linked to a greater height of the magma column in the conduit, which is completely filled with magma, as also observed in other volcanoes, for example, Etna [61]. After the paroxysm of 3 July 2019, the Green cluster explosions became frequent and probably occurred at the same SW vents that fed the lava flow. A small but significant variation in the locations of the VLP events reported in Giudicepietro et al. [4] corresponded to the sudden change in the eruptive style (3 July 2019, in Figures 11 and 13). These locations were concentrated in the SW sector of the VLP source volume before the paroxysm of 3 July 2019, and migrated slightly NE after this paroxysm, indicating a resumption of the activity in the NE vent region (see Figure 7 in Giudicepietro et al. [4]).

Information on the final part of the conduits linked to the eruptive vents is contained in the seismo-acoustic features as also highlighted in the analysis of experimental seismo-acoustic events in Giudicepietro et al. [45]. Actually, clusters Blue and Red are composed of more than one node, and the subdivision of the seismo-acoustic events into the different nodes typically corresponds to explosions with a similar mechanism emitted from different vent regions, as in the case of nodes 4 (N1 vent) and 7 (SW vents) that form cluster Red (Figure 10a,e,f).

All the three main types of explosions recognized by the SOM analysis generally manifest themselves in the persistent activity of Stromboli, each of which can occur in different vent regions (Figure 1). Therefore, the anomaly that preceded the first paroxysm of 2019 was the clear predominance, within our dataset, of gas explosions (cluster Blue), which reached 96.12% of the total in the last month before the 3 July 2019 paroxysm (Figure 13). As already specified in the Data and Methods section, our dataset does not include all explosions, which can exceed 400 per day, but only those associated with the largest VLP size of every half-hour, for a maximum of 48 events per day. This selection allowed us to extract the 14,289 and 14,179 most significant seismic and infrasonic recordings, respectively, and to prevent the dataset from reaching dimensions that are not easy to handle for analysis. In the period preceding the paroxysm of 3 July 2019, the predominance of cluster Blue in this dataset indicates a degassing activity that is not accompanied by an effective emission of juvenile material, consistently with the deflation or the absence of inflation in the crater area, therefore indicating a remarkable anomaly in the pattern of the persistent activity of Stromboli. The eruptive style change before the paroxysmal phase, which is clearly recognizable in the temporal evolution of the seismo-acoustic clusters found with the SOM analysis (Figures 11 and 13), is an important finding because it highlights hidden variations in the state of the volcano that may reveal undetected escalation of volcanic plume degassing and/or precursory leakage from deeply stored gas-rich magma (e.g., [62]). Actually, despite Stromboli being a well-monitored volcano, when the first paroxysm of the 2019 eruptive crisis occurred, it was considered to be in a state of normal activity.

The second paroxysm, which occurred on 28 August 2019, happened 56 days after the start of the effusive activity, which began immediately after the first paroxysm on 3 July 2019. Therefore, this event occurred in a different condition compared with the first one, as also indicated by the temporal evolution of the seismo-acoustic clusters (Figures 11–13). Considering the models of Stromboli paroxysm triggering proposed in the literature, the first paroxysm (3 July 2019) could be explained by an increased supply of gas and magma from the depths (e.g., [39,62,63]). However, the neural analysis of the eruptive style and its comparison with the deformation of the summit area allowed us to discover that this paroxysm was preceded by a phase of decrease in the feeding of the persistent activity, which is highlighted by the decreased emission of pyroclastic material and by the deflation of the summit area accompanied by a greater release of degassing (Type 0 explosions of the Blue cluster). For this reason, the input of gas and magma from the depths that caused the paroxysm does not seem linked only to increased activity of the deep magma system but also to a deceleration, in the period preceding the paroxysm, of persistent activity, which is partly controlled by the shallow volcanic system. On the contrary, the paroxysm of 28 August 2019 is consistent with a trigger due to the drainage of the highly porphyritic magma which is typically found in the upper part of the conduit, due to the effusive activity that began about two months earlier, which determined the rise of low-porphyritic magma capable of producing paroxysmal eruptions (e.g., [64]). In any case, the GBInSAR measurements indicate that in the medium term a deflation shortly before the event occurred for both paroxysms. Furthermore, in the short term, the strainmeter data show that similar dynamics occurred in both paroxysms, as reported in Giudicepietro et al. [4] and Di Lieto et al. [65].

5. Conclusions

The SOM analysis of the seismo-acoustic features associated with a set of about 14,200 explosions selected based on the VLP size parameter allowed us to identify three main clusters in the period 15 November 2018–15 September 2019, which contains the paroxysmal phase of July–August 2019.

The comparison of a subset of events with the visible and thermal camera images allowed us to associate distinct explosive types to the three main seismo-acoustic clusters. In particular, the cluster called Red is associated with explosions characterized by well-collimated vertical jets of ~200 m in height, which eject incandescent ballistic pyroclastic fragments and produce a remarkable infrasonic signal. Cluster Blue is associated with gas explosions with height in the range 10–20 m and with little or no ash and ballistic emission. These bursts may not be detected by the camera recordings and infrasonic signals whereas they are evident in the VLP seismic signals (filtered in the 0.05–0.5 Hz frequency band). Cluster Green groups explosions characterized by the ejection of incandescent spatter-like pyroclastic fragments, with a wide range of ejection angles and hemispherical shape. The explosions of cluster Red are mainly generated in the NE vent region whereas the explosions of clusters Blue and Green are mainly emitted from the central and SW vent regions.

Looking at the time evolution of the three main clusters, we discovered that the eruptive style of Stromboli was affected by significant changes in the three months preceding the 3 July 2019 paroxysm and that the gas explosions (Type 0; Leduc et al. [26]) falling into cluster Blue dominated the persistent Strombolian activity, especially in the last month before this paroxysm, forecasting the ascent of gas-rich magma from a depth [62].

Finally, by comparing the temporal evolution of the clusters with the deformations of the top of the volcano retrieved through GBInSAR measurements, we were able to recognize a relationship between the eruptive style and the inflation/deflation phases of the crater area. Actually, the period dominated by the gas explosions of cluster Blue (early April–late June 2019) was characterized by the absence of significant deformations whereas the effusive phase between the two paroxysms (early July–mid September 2019), dominated by explosions falling into clusters Red and Green, was characterized by inflation of the crater area, especially from July 19 until the end of our target period (15 September 2019). The explosions of clusters Red and Green are both characterized by the emission of incandescent ballistic pyroclasts but with different mechanisms: the explosions of cluster Red produce vertical jets, with a narrow ejection cone, and generate a distinct infrasonic transient associated with them; the explosions of cluster Green eject the ballistic pyroclasts according to a wide range of ejection angles assuming a hemispherical shape. The latter are linked to a high level of magma in the conduit and are often associated with spattering. Among the three main clusters, only the explosions falling in the Red cluster generate clearly recognizable infrasonic transients in the frequency band >0.5 Hz.

This study allowed us to discover variations in the pattern of the persistent activity of Stromboli that preceded the 2019 eruptive crisis and to interpret the geophysical data in terms of variations in the eruptive style and the state of activity of the volcano. The results obtained increase our ability to distinguish the different Strombolian mechanisms and suggest new opportunities for an advancement in the monitoring of Stromboli focused on the forecasting of potentially dangerous eruptive activity variations and early warning for paroxysms.

Author Contributions: Conceptualization, F.G., S.C., A.M.E., L.S., L.D., M.M. and G.M.; methodology, F.G., A.M.E., L.L., G.M., F.D.T., A.T., T.N., N.C. and S.C.; formal analysis, F.G., A.M.E., L.L., S.C., T.N., N.C. and G.M.; data curation, W.D.C., M.O., R.P., G.S., F.D.T., T.N., N.C., G.G., T.C. and S.C.; writing—original draft preparation, F.G., A.M.E., S.C., F.D.T. and G.M.; writing—review and editing, F.G., A.M.E., L.L., S.C., F.D.T., T.N., N.C., L.D., A.T., M.M. and G.M. All authors have read and agreed to the published version of the manuscript.

Funding: This work was supported by the project Progetto Strategico Dipartimentale INGV 2019 "Forecasting eruptive activity at Stromboli volcano: timing, eruptive style, size, intensity and duration" (FIRST). This work is also supported by a Marie Sklodowska-Curie Innovative Training Network Fellowship of the European Commission's Horizon 2020 Programme under Contract Number 765710 INSIGHTS. This work benefited from the EU (DG ECHO) Project EVE n. 826292. This work has been partially supported by the "Presidenza del Consiglio dei Ministri–Dipartimento della Protezione Civile" (Presidency of the Council of Ministers–Department of Civil Protection; Scientific Responsibility: N.C.). However, this publication does not necessarily represent the official opinion and policies of the department.

Institutional Review Board Statement: Not applicable.

Informed Consent Statement: Not applicable.

Data Availability Statement: Not applicable.

Acknowledgments: We wish to thank all the many colleagues who have contributed to the monitoring effort on Stromboli including technical and administrative staff. We are particularly indebted to the INGV technical staff ensuring the regular working of the multidisciplinary monitoring networks. The data used in this study were provided by the Istituto Nazionale di Geofisica e Vulcanologia (Osservatorio Vesuviano, Osservatorio Etneo). L.S. wishes to thank the project Progetto Strategico Dipartimentale INGV 2019 "UNderstanding the Ordinary to forecast the extraordinary: An integrated approach for studying and interpreting the explosive activity at Stromboli volcano" (UNO). The authors are also grateful to the Italian Presidenza del Consiglio dei Ministri, Dipartimento della Protezione Civile (DPC) for supporting the monitoring activities at Stromboli. This paper does not necessarily represent DPC official opinion and policies.

Conflicts of Interest: The authors declare no conflict of interest.

References

1. Esposito, A.M.; D'Auria, L.; Giudicepietro, F.; Caputo, T.; Martini, M. Neural analysis of seismic data: Applications to the monitoring of Mt. Vesuvius. *Ann. Geophys.* **2013**, *56*, 446. [CrossRef]
2. Esposito, A.M.; Alaia, G.; Giudicepietro, F.; Pappalardo, L.; D'Antonio, M. Unsupervised geochemical analysis of the eruptive products of Ischia, Vesuvius and Campi Flegrei. In *Progresses in Artificial Intelligence and Neural Systems*; Esposito, A., Faundez-Zanuy, M., Morabito, F.C., Pasero, E., Eds.; Smart Innovation, Systems and Technologies; Springer: Singapore, 2020; Volume 184, pp. 175–184. [CrossRef]
3. Esposito, A.M.; De Bernardo, A.; Ferrara, S.; Giudicepietro, F.; Pappalardo, L. SOM-Based analysis of volcanic rocks: An application to Somma-Vesuvius and Campi Flegrei volcanoes (Italy). In *Neural Approaches to Dynamics of Signal Exchanges*; Esposito, A., Faundez-Zanuy, M., Morabito, F.C., Pasero, E., Eds.; Smart Innovation, Systems and Technologies; Springer: Singapore, 2020; Chapter 6; Volume 151, pp. 55–60. [CrossRef]
4. Giudicepietro, F.; López, C.; Macedonio, G.; Alparone, S.; Bianco, F.; Calvari, S.; De Cesare, W.; Delle Donne, D.; Di Lieto, B.; Esposito, A.M.; et al. Geophysical precursors of the July–August 2019 paroxysmal eruptive phase and their implications for Stromboli volcano (Italy) monitoring. *Sci. Rep.* **2020**, *10*, 10296. [CrossRef] [PubMed]
5. Esposito, A.M.; D'Auria, L.; Angelillo, A.; Giudicepietro, F.; Martini, M. Predictive analysis of the seismicity level at Campi Flegrei volcano using a data-driven approach. In *Recent Advances of Neural Network Models and Applications*; Bassis, S., Esposito, A., Morabito, F.C., Eds.; Smart Innovation, Systems and Technologies; Springer: Cham, Switzerland, 2014; Volume 19, pp. 133–145. [CrossRef]

6. Ambrosino, F.; Sabbarese, C.; Roca, V.; Giudicepietro, F.; Chiodini, G. Analysis of 7-years Radon time series at Campi Flegrei area (Naples, Italy) using artificial neural network method. *Appl. Radiat. Isotopes* **2020**, *163*, 109239. [CrossRef] [PubMed]
7. Scarpetta, S.; Giudicepietro, F.; Ezin, E.C.; Petrosino, S.; Del Pezzo, E.; Martini, M.; Marinaro, M. Automatic classification of seismic signals at Mt. Vesuvius volcano, Italy, using neural networks. *BSSA* **2005**, *95*, 185–196. [CrossRef]
8. Esposito, A.M.; D'Auria, L.; Giudicepietro, F.; Peluso, R.; Martini, M. Automatic recognition of landslides based on neural network analysis of seismic signals: An Application to the monitoring of Stromboli Volcano (Southern Italy). *Pure Appl. Geophys.* **2013**, *170*, 1821–1832. [CrossRef]
9. Giudicepietro, F.; Esposito, A.M.; Ricciolino, P. Fast discrimination of local earthquakes using a neural approach. *Seismol. Res. Lett.* **2017**, *88*, 1089–1096. [CrossRef]
10. Esposito, A.M.; Giudicepietro, F.; Scarpetta, S.; D'Auria, L.; Marinaro, M.; Martini, M. Automatic discrimination among landslide, explosion-quake, and microtremor seismic signals at Stromboli Volcano using neural networks. *BSSA* **2006**, *96*, 1230–1240. [CrossRef]
11. Esposito, A.M.; Giudicepietro, F.; D'Auria, L.; Scarpetta, S.; Martini, M.; Coltelli, M.; Marinaro, M. Unsupervised neural analysis of very-long-period events at Stromboli volcano using the self-organizing maps. *BSSA* **2008**, *98*, 2449–2459. [CrossRef]
12. Esposito, A.M.; D'Auria, L.; Giudicepietro, F.; Martini, M. Waveform Variation of the explosion-quakes as a function of the eruptive activity at Stromboli volcano. In *Neural Nets and Surroundings. Smart Innovation, Systems and Technologies*; Apolloni, B., Bassis, S., Esposito, A., Martini, M., Eds.; Springer: Berlin/Heidelberg, Germany, 2013; Volume 19, pp. 111–119. [CrossRef]
13. Esposito, A.M.; Giudicepietro, F.; Scarpetta, S.; Khilnani, S. A neural approach for hybrid events discrimination at Stromboli volcano. In *Multidisciplinary Approaches to Neural Computing*; Number 69 in Smart Innovation, Systems and Technologies; Springer International Publishing: Berlin/Heidelberg, Germany, 2018; Chapter 2; pp. 11–21. [CrossRef]
14. Del Pezzo, E.; Esposito, A.; Giudicepietro, F.; Marinaro, M.; Martini, M.; Scarpetta, S. Discrimination of earthquakes and underwater explosions using neural networks. *BSSA* **2003**, *93*, 215–223. [CrossRef]
15. Andronico, D.; Del Bello, E.; D'Oriano, C.; Landi, P.; Pardini, F.; Scarlato, P.; de' Michieli Vitturi, M.; Taddeucci, J.; Cristaldi, A.; Ciancitto, F.; et al. Uncovering the eruptive patterns of the 2019 double paroxysm eruption crisis of Stromboli volcano. *Nat. Commun.* **2021**, *12*, 4213. [CrossRef]
16. Giordano, G.; De Astis, G. The summer 2019 basaltic Vulcanian eruptions (paroxysms) of Stromboli. *Bull. Volcanol.* **2021**, *83*, 1–27. [CrossRef]
17. Calvari, S.; Giudicepietro, F.; Di Traglia, F.; Bonaccorso, A.; Macedonio, G.; Casagli, N. Variable magnitude and intensity of Strombolian explosions: Focus on the eruptive processes for a first classification scheme for Stromboli volcano (Italy). *Remote Sens.* **2021**, *13*, 944. [CrossRef]
18. Washington, H.S. Persistence of vents at Stromboli and its bearing on volcanic mechanism. *Geol. Soc. Am. Bull.* **1917**, *28*, 249–278. [CrossRef]
19. Harris, A.; Ripepe, M. Temperature and dynamics of degassing at Stromboli. *J. Geophys. Res.* **2007**, *112*, B03205. [CrossRef]
20. Salvatore, V.; Silleni, A.; Corneli, D.; Taddeucci, J.; Palladino, D.M.; Sottili, G.; Bernini, D.; Andronico, D.; Cristaldi, A. Parameterizing multi-vent activity at Stromboli volcano (Aeolian Islands, Italy). *Bull. Volcanol.* **2018**, *80*, 64. [CrossRef]
21. Tioukov, V.; Alexandrov, A.; Bozza, C.; Consiglio, L.; D'Ambrosio, N.; De Lellis, G.; De Sio, C.; Giudicepietro, F.; Macedonio, G.; Miyamoto, S.; et al. First muography of Stromboli volcano. *Sci. Rep.* **2019**, *9*, 6695. [CrossRef]
22. Tioukov, V.; Giudicepietro, F.; Macedonio, G.; Calvari, S.; Di Traglia, F.; Fornaciai, A.; Favalli, M. Structure of the Shallow Supply System at Stromboli Volcano Through Integration of Muography, Digital Elevation Models, Seismicity and Ground Deformation Data. In *Muography: Exploring Earth's Subsurface with Elementary Particles*; Oláh, L., Tanaka, H.K.M., Varga, D., Eds.; AGU Books, Wiley-AGU: Hoboken, NJ, USA, 2022; Chapter 6. [CrossRef]
23. Turchi, A.; Di Traglia, F.; Luti, T.; Olori, D.; Zetti, I.; Fanti, R. Environmental aftermath of the 2019 Stromboli eruption. *Remote Sens.* **2020**, *12*, 994. [CrossRef]
24. Chouet, B.; Dawson, P.; Ohminato, T.; Martini, M.; Saccorotti, G.; Giudicepietro, F.; De Luca, G.; Milana, G.; Scarpa, R. Source mechanisms of explosions at Stromboli Volcano, Italy, determined from moment-tensor inversion of very-long period data. *J. Geophys. Res.* **2003**, *108*, 2019. [CrossRef]
25. Patrick, M.R.; Harris, A.J.L.; Ripepe, M.; Dehn, J.; Rothery, D.A.; Calvari, S. Strombolian explosive styles and source conditions: Insights from thermal (FLIR) video. *Bull. Volcanol.* **2007**, *69*, 679–784. [CrossRef]
26. Leduc, L.; Gurioli, L.; Harris, A.; Colò, L.; Rose-Koga, E.F. Types and mechanisms of strombolian explosions: Characterization of a gas-dominated explosion at Stromboli. *Bull. Volcanol.* **2015**, *77*, 8. [CrossRef]
27. Simons, B.C.; Jolly, A.D.; Eccles, J.D.; Cronin, S.J. Spatiotemporal relationships between two closely-spaced Strombolian-style vents, Yasur, Vanuatu. *Geophys. Res. Lett.* **2020**, *47*, e2019GL085687. [CrossRef]
28. Cimarelli, C.; Di Traglia, F.; Taddeucci, J. Basaltic scoria textures from a zoned conduit as precursors to violent Strombolian activity. *Geology* **2010**, *38*, 439–442. [CrossRef]
29. Inguaggiato, S.; Vita, F.; Cangemi, M.; Calderone, L. Increasing summit degassing at the Stromboli volcano and relationships with volcanic activity (2016–2018). *Geosciences* **2019**, *9*, 176. [CrossRef]
30. Calvari, S.; Di Traglia, F.; Ganci, G.; Giudicepietro, F.; Macedonio, G.; Cappello, A.; Nolesini, T.; Pecora, E.; Bilotta, G.; Centorrino, V.; et al. Overflows and hot rock avalanches in March-April 2020 at Stromboli Volcano detected by remote sensing and seismic monitoring data. *Remote Sens.* **2020**, *12*, 3010. [CrossRef]

31. Bonaccorso, A.; Calvari, S.; Garfì, G.; Lodato, L.; Patanè, D. Dynamics of the December 2002 flank failure and tsunami at Stromboli volcano inferred by volcanological and geophysical observations. *Geophys. Res. Lett.* **2003**, *30*, 1941. [CrossRef]
32. Calvari, S.; Lodato, L.; Steffke, A.; Cristaldi, A.; Harris, A.J.L.; Spampinato, L.; Boschi, E. The 2007 Stromboli eruption: Event chronology and effusion rates using thermal infrared data. *J. Geophys. Res.* **2010**, *115*, B04201. [CrossRef]
33. Casalbore, D.; Di Traglia, F.; Bosman, A.; Romagnoli, C.; Casagli, N.; Chiocci, F.L. Submarine and subaerial morphological changes associated with the 2014 eruption at Stromboli Island. *Remote Sens.* **2021**, *13*, 2043. [CrossRef]
34. Martini, M.; Giudicepietro, F.; D'Auria, L.; Esposito, A.M.; Caputo, T.; Curciotti, R.; De Cesare, W.; Orazi, M.; Scarpato, G.; Caputo, A.; et al. Seismological monitoring of the February 2007 effusive eruption of the Stromboli volcano. *Ann. Geophys.* **2007**, *50*, 775–788. [CrossRef]
35. Di Traglia, F.; Calvari, S.; D'Auria, L.; Nolesini, T.; Bonaccorso, A.; Fornaciai, A.; Esposito, A.; Cristaldi, A.; Favalli, M.; Casagli, N. The 2014 effusive eruption at Stromboli: New insights from in situ and remote-sensing measurements. *Remote Sens.* **2018**, *10*, 2035. [CrossRef]
36. Calvari, S.; Büttner, R.; Cristaldi, A.; Dellino, P.; Giudicepietro, F.; Orazi, M.; Peluso, R.; Spampinato, L.; Zimanowski, B.; Boschi, E. The 7 September 2008 Vulcanian explosion at Stromboli volcano: Multiparametric characterization of the event and quantification of the ejecta. *J. Geophys. Res.* **2012**, *117*, B05201. [CrossRef]
37. Giudicepietro, F.; Calvari, S.; Alparone, S.; Bianco, F.; Bonaccorso, A.; Bruno, V.; Caputo, T.; Cristaldi, A.; D'Auria, L.; De Cesare, W.; et al. Integration of ground-based remote-sensing and in situ multidisciplinary monitoring data to analyze the eruptive activity of Stromboli volcano in 2017–2018. *Remote Sens.* **2019**, *11*, 1813. [CrossRef]
38. D'Auria, L.; Giudicepietro, F.; Martini, M.; Peluso, R. Seismological insight into the kinematics of the 5 April 2003 vulcanian explosion at Stromboli volcano (southern Italy). *Geophys. Res. Lett.* **2006**, *33*, L08308. [CrossRef]
39. Bertagnini, A.; Di Roberto, A.; Pompilio, M. Paroxysmal activity at Stromboli: Lessons from the past. *Bull. Volcanol.* **2011**, *73*, 1229–1243. [CrossRef]
40. Bonaccorso, A.; Calvari, S.; Linde, A.; Sacks, S.; Boschi, E. Dynamics of the shallow plumbing system investigated from borehole strainmeters and cameras during the 15 March, 2007 Vulcanian paroxysm at Stromboli volcano. *Earth Planet. Sci. Lett.* **2012**, *357–358*, 249–256. [CrossRef]
41. Di Traglia, F.; De Luca, C.; Manzo, M.; Nolesini, T.; Casagli, N.; Lanari, R.; Casu, F. Joint exploitation of space-borne and ground-based multitemporal InSAR measurements for volcano monitoring: The Stromboli volcano case study. *Remote Sens. Environ.* **2021**, *260*, 112441. [CrossRef]
42. Di Traglia, F.; Fornaciai, A.; Casalbore, D.; Favalli, M.; Manzella, I.; Romagnoli, C.; Chiocci, F.L.; Cole, P.; Nolesini, T.; Casagli, N. Subaerial-submarine morphological changes at Stromboli volcano (Italy) induced by the 2019–2020 eruptive activity. *Geomorfology* **2022**, *400*, 108093. [CrossRef]
43. Spina, L.; Morgavi, D.; Cannata, A.; Campeggi, C.; Perugini, D. An experimental device for characterizing degassing processes and related elastic fingerprints: Analog volcano seismo-acoustic observations. *Rev. Sci. Instrum.* **2018**, *88*, 055102. [CrossRef]
44. Spina, L.; Cannata, A.; Morgavi, D.; Perugini, D. Degassing behaviour at basaltic volcanoes: New insights from experimental investigations of different conduit geometry and magma viscosity. *Earth Sci. Rev.* **2019**, *192*, 317–336. [CrossRef]
45. Giudicepietro, F.; Esposito, A.; Spina, L.; Cannata, A.; Morgavi, D.; Layer, L.; Macedonio, G. Clustering of experimental seismo-acoustic events using Self Organizing Maps (SOM). *Front. Earth Sci.* **2021**, *8*, 58174. [CrossRef]
46. Orazi, M.; Martini, M.; Peluso, R. Data acquisition for volcano monitoring. Eos Trans. Am. Geophys. *Union* **2006**, *87*, 385–392. [CrossRef]
47. Orazi, M.; Peluso, R.; Caputo, A.; Capello, M.; Buonocunto, C.; Martini, M. *A Multiparametric Low Power Digitizer: Project and Results*; Istituto Nazionale di Geofisica e Vulcanologia, Osservatorio Vesuviano: Naples, Italy, 2007.
48. De Cesare, W.; Orazi, M.; Peluso, R.; Scarpato, G.; Caputo, A.; D'Auria, L.; Giudicepietro, F.; Martini, M.; Buonocunto, C.; Capello, M.; et al. The broadband seismic network of Stromboli volcano, Italy. *Seismol. Res. Lett.* **2009**, *80*, 435–439. [CrossRef]
49. Salvaterra, L.; Pintore, S.; Badiali, L. *Rete Sismologica Basata su Stazioni GAIA*; Rapporti Tecnici 68; Istituto Nazionale di Geofisica e Vulcanologia: Rome, Italy, 2008.
50. Krischer, L.; Megies, T.; Barsch, R.; Beyreuther, M.; Lecocq, T.; Caudron, C.; Wassermann, J. ObsPy: A bridge for seismology into the scientific Python ecosystem. *Comput. Sci. Discov.* **2015**, *8*, 014003. [CrossRef]
51. Esposito, A.M.; Scarpetta, S.; Giudicepietro, F.; Masiello, S.; Pugliese, L.; Esposito, A. Nonlinear Exploratory Data Analysis Applied to Seismic Signals. In *Neural Nets. Neural Nets Lecture Notes in Computer Science*; Apolloni, B., Marinaro, M., Nicosia, G., Tagliaferri, R., Eds.; Springer: Berlin/Heidelberg, Germany, 2006; Volume 3931, pp. 70–77. [CrossRef]
52. Esposito, A.; Esposito, A.M.; Giudicepietro, F.; Marinaro, M.; Scarpetta, S. Models for Identifying Structures in the Data: A Performance Comparison. In *Knowledge-Based Intelligent Information and Engineering Systems. KES 2007*; Apolloni, B., Howlett, R.J., Jain, L., Eds.; Lecture Notes in Computer Science; Springer: Berlin/Heidelberg, Germany, 2007; Volume 4694. [CrossRef]
53. Kohonen, T. Self-organized formation of topologically correct feature maps. *Biol. Cybern.* **1982**, *43*, 59–69. [CrossRef]
54. Kohonen, T.; Hynninen, J.; Kangas, J.; Laaksonen, J. *SOM_PAK: The Self-Organizing Map Program Package*; Report A31; Helsinki University of Technology, Laboratory of Computer and Information Science: Espoo, Finland, 1996.
55. Di Traglia, F.; Nolesini, T.; Intrieri, E.; Mugnai, F.; Leva, D.; Rosi, M.; Casagli, N. Review of ten years of volcano deformations recorded by the ground-based InSAR monitoring system at Stromboli volcano: A tool to mitigate volcano flank dynamics and intense volcanic activity. *Earth Sci. Rev.* **2014**, *139*, 317–335. [CrossRef]

56. Di Traglia, F.; Battaglia, M.; Nolesini, T.; Lagomarsino, D.; Casagli, N. Shifts in the eruptive styles at Stromboli in 2010–2014 revealed by ground-based InSAR data. *Sci. Rep.* **2015**, *5*, 13569. [CrossRef] [PubMed]
57. Calvari, S.; Intrieri, E.; Di Traglia, F.; Bonaccorso, A.; Casagli, N.; Cristaldi, A. Monitoring crater-wall collapse at active volcanoes: A study of the 12 January 2013 event at Stromboli. *Bull. Volcanol.* **2016**, *78*, 39. [CrossRef]
58. Casagli, N.; Tibaldi, A.; Merri, A.; Del Ventisette, C.; Apuani, T.; Guerri, L.; Fortuny-Guasch, J.; Tarchi, D. Deformation of Stromboli Volcano (Italy) during the 2007 crisis by radar interferometry, numerical modeling and field structural data. *J. Volcanol. Geotherm. Res.* **2009**, *182*, 182–200. [CrossRef]
59. Antonello, G.; Casagli, N.; Farina, P.; Leva, D.; Nico, G.; Sieber, A.J.; Tarchi, D. Ground-based SAR interferometry for monitoring mass movements. *Landslides* **2004**, *1*, 21–28. [CrossRef]
60. Braun, T.; Ripepe, M. Interaction of seismic and air waves recorded at Stromboli Volcano. *Geophys. Res. Lett.* **1993**, *20*, 65–68. [CrossRef]
61. Gouhier, M.; Donnadieu, F. The geometry of Strombolian explosions: Insights from Doppler radar measurements. *Geophys. J. Int.* **2010**, *183*, 1376–1391. [CrossRef]
62. Aiuppa, A.; Bitetto, M.; Delle Donne, D.; La Monica, F.P.; Tamburello, G.; Coppola, D.; Della Schiava, M.; Innocenti, L.; Lacanna, G.; Laiolo, M.; et al. Volcanic CO_2 tracks the incubation period of basaltic paroxysms. *Sci. Adv.* **2021**, *7*, eabh0191. [CrossRef] [PubMed]
63. Barberi, F.; Rosi, M.; Sodi, A. Volcanic hazard assessment at Stromboli based on review of historical data. *Acta Vulcanol.* **1993**, *3*, 173–187.
64. Calvari, S.; Spampinato, L.; Bonaccorso, A.; Oppenheimer, C.; Rivalta, E.; Boschi, E. Lava effusion—A slow fuse for paroxysms at Stromboli volcano? *Earth Planet. Sci. Lett.* **2011**, *301*, 317–323. [CrossRef]
65. Di Lieto, B.; Romano, P.; Scarpa, R.; Linde, A.T. Strain signals before and during paroxysmal activity at Stromboli volcano, Italy. *Geophys. Res. Lett.* **2020**, *47*, e2020GL088521. [CrossRef]

remote sensing

MDPI

Article

Variable Magnitude and Intensity of Strombolian Explosions: Focus on the Eruptive Processes for a First Classification Scheme for Stromboli Volcano (Italy)

Sonia Calvari [1,*], Flora Giudicepietro [2], Federico Di Traglia [3], Alessandro Bonaccorso [1], Giovanni Macedonio [2] and Nicola Casagli [3,4]

1 Istituto Nazionale di Geofisica e Vulcanologia, Osservatorio Etneo-Sezione di Catania, 95125 Catania, Italy; alessandro.bonaccorso@ingv.it
2 Istituto Nazionale di Geofisica e Vulcanologia, Osservatorio Vesuviano, 80125 Napoli, Italy; flora.giudicepietro@ingv.it (F.G.); giovanni.macedonio@ingv.it (G.M.)
3 Dipartimento di Scienze della Terra, Universitá degli Studi di Firenze, 50121 Firenze, Italy; federico.ditraglia@unifi.it (F.D.T.); nicola.casagli@inogs.it (N.C.)
4 National Institute of Oceanography and Applied Geophysics—OGS, 30010 Trieste, Italy
* Correspondence: sonia.calvari@ingv.it

Citation: Calvari, S.; Giudicepietro, F.; Di Traglia, F.; Bonaccorso, A.; Macedonio, G.; Casagli, N. Variable Magnitude and Intensity of Strombolian Explosions: Focus on the Eruptive Processes for a First Classification Scheme for Stromboli Volcano (Italy). *Remote Sens.* **2021**, *13*, 944. https://doi.org/10.3390/rs13050944

Academic Editor: Fumio Yamazaki

Received: 9 February 2021
Accepted: 25 February 2021
Published: 3 March 2021

Abstract: Strombolian activity varies in magnitude and intensity and may evolve into a threat for the local populations living on volcanoes with persistent or semi-persistent activity. A key example comes from the activity of Stromboli volcano (Italy). The "ordinary" Strombolian activity, consisting in intermittent ejection of bombs and lapilli around the eruptive vents, is sometimes interrupted by high-energy explosive events (locally called major or paroxysmal explosions), which can affect very large areas. Recently, the 3 July 2019 explosive paroxysm at Stromboli volcano caused serious concerns in the local population and media, having killed one tourist while hiking on the volcano. Major explosions, albeit not endangering inhabited areas, often produce a fallout of bombs and lapilli in zones frequented by tourists. Despite this, the classification of Strombolian explosions on the basis of their intensity derives from measurements that are not always replicable (i.e., field surveys). Hence the need for a fast, objective and quantitative classification of explosive activity. Here, we use images of the monitoring camera network, seismicity and ground deformation data, to characterize and distinguish paroxysms, impacting the whole island, from major explosions, that affect the summit of the volcano above 500 m elevation, and from the persistent, mild explosive activity that normally has no impact on the local population. This analysis comprises 12 explosive events occurring at Stromboli after 25 June 2019 and is updated to 6 December 2020.

Keywords: Stromboli volcano; paroxysmal explosions; major explosive events; ground and remote sensing monitoring; classification of mild Strombolian events

1. Introduction

Strombolian activity is characterized by explosive transients of variable intensity, from pyroclast-free gas explosions (puffing) to intense explosions, with the formation of a few-km-high eruptive columns, ballistic ejection, and occasional generation of pyroclastic density currents [1–6]. The classification, as well as the understanding of the dynamics that trigger explosions of different intensities, is fundamental for the hazard assessment in areas characterized by Strombolian activity, both for territorial planning and for forecasting through monitoring and surveillance. A multi-parametric approach, combining geophysical and volcanological monitoring data with remote sensing techniques, is fundamental in order to find an objective as possible way to classify these transient explosions.

The best site to investigate this activity is Stromboli (Italy), known as the "Lighthouse of the Mediterranean" for its persistent explosive activity, characterized by bursts of incandescent ejecta occurring every few minutes [4,7–9], and clearly visible especially at night by sailors. Stromboli volcano is the easternmost of the islands comprising the Aeolian Archipelago volcanic arc in the south Tyrrhenian Sea (Figure 1a). It is 924 m high above sea level (a.s.l.; Figure 1b) and extends below the sea down to ~2000 m depth, reaching a total elevation of ~3000 m [10]. Explosions occur from vents located within the summit crater terrace depression at ~750 m elevation (Figure 1c), which is ~ 300 m long in a NE-SW direction, ~50 m wide and ~50 m deep (Figure 1c). Three crater areas are located within the summit depression: the NE crater zone (NEC), the Central crater zone (CC) and the SW crater zone (SWC), each of them comprising a variable number of active vents (Figure 1c).

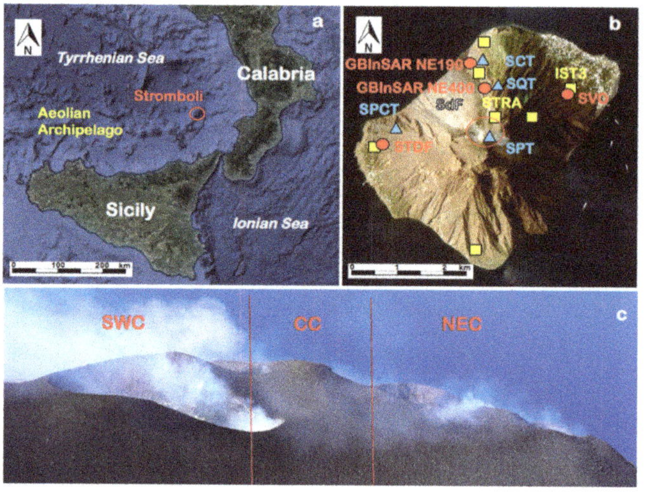

Figure 1. (**a**) Google map of southern Italy, with the red circle showing the position of Stromboli volcano, at the NE end of the Aeolian Archipelago. (**b**) Stromboli island with the position and labels of the monitoring instruments used in this study. The blue triangles are the monitoring cameras, with SPT being the thermal camera located at Il Pizzo Sopra La Fossa viewpoint. The SCV camera is located in the same place as SCT, and the SQV camera is in the same place as SQT; the red circles are the GBInSARs, the tilt (STDF) and strainmeter (SVO) stations; the yellow squares are the seismic stations. The empty red circle outlines the position of the summit craters displayed in frame c. SdF = Sciara del Fuoco slope. (**c**) View from South and from Il Pizzo Sopra La Fossa viewpoint of Stromboli summit craters; photo taken on 21 February 2020, showing the names of the active crater areas. NEC = NE Crater zone; CC = Central Crater zone; SWC = SW Crater zone. The field of view is about 300 m wide. Photo courtesy of F. Ciancitto, INGV.

The Strombolian activity of Stromboli is characterized by explosive transients of short duration (<10 s, obtained from the monitoring cameras) and small eruptive ash columns (<100 m), with variable intensity and frequency [11], which depend on the supply rate of the deeper system towards the surface [8,12–14]. This volcano gave its name to the Strombolian explosive activity, with mild explosions typical of basaltic explosive volcanism, that often feature at the summit of Yasur (Vanuatu), Piton de la Fournaise (La Réunion), Shishaldin (Alaska), Fuego (Guatemala), Nyiragongo (R.D. Congo), Masaya (Nicaragua), Turrialba (Costa Rica), Etna (Italy), Kilauea (Hawaii), and several other open conduit basaltic volcanoes [15–24].

The persistent, mild explosive activity of Stromboli is sometimes interrupted by more intense and powerful explosions. Several classifications of this "more intense" activity have been proposed over time. The first was put forward by Barberi et al. [12], who distinguished three types of explosions: in addition to the "ordinary" or persistent activity (Figure 2a), major (Figure 2b) and paroxysmal explosions (Figure 2c) were identified, depending on size, covering a broad variability in intensity and magnitude, with the latter having significantly larger intensities (>10^6 kg/s) than the former (10^4 kg/s) [25]. Major explosions (Figure 2b) normally involve more than one crater zone, may follow partial obstructions of the summit vents, and cause the rise of a mixture of spatter, bombs, ash and gas forming an eruptive column that extends > 300 m above the vents, with fallout of bombs and ash up to several hundred meters from the crater area [12,26,27]. More rarely, the volcano is the site of extremely powerful explosions, called "paroxysms", that result in eruptive columns rising a few km above the craters (Figure 2c) and causing fires and damages to the populated villages on the lower flanks of the volcano, 2.0–2.5 km away from the summit craters [12,28–33]. Paroxysms are also characterized by greater volumes of emitted materials, higher muzzle velocities, and higher mass discharge rates [12,34,35]. Besides major eruptions (Figure 2b), there is a complete range of intermediate events from the persistent "ordinary" mild Strombolian activity (Figure 2a) to the most powerful paroxysms (Figure 2c). Occasional flank fissures discharging lava flows within the Sciara del Fuoco (SdF) barren NW slope may also occur (Figure 1b). Lava fountains are generally not common and of short duration (minutes; [12,31]), whereas the periods without eruptive activity are extremely rare [12].

Figure 2. (**a**) Ordinary Strombolian explosion from the SWC producing an ash plume ~80 m high, recorded by the SPI infrared camera located at Il Pizzo Sopra La Fossa on 7 September 2008, view from South. (**b**) Image from the SQV camera showing the eruptive ash column rising ~300 m above the craters during the major explosion of 7 September 2008, view from NE. (**c**) Photo taken from helicopter by S. Calvari during the 5 April 2003 paroxysm, showing the eruptive column rising > 1 km above the summit of the volcano, view from South.

During paroxysms, and to a lower extent also during major explosions, a deep-seated, gas-rich and low porphyritic (LP) magma is erupted together with the gas-poor, high porphyritic (HP) magma residing in the upper conduit [28,36–39]. A common result of paroxysms is a deep modification of the crater area, with cinder cones and hornitos around vents being blown out and leaving a much wider and deeper crater depression [12,28–31,40,41]. Some paroxysms are associated with the emission of lava flows, and may occur both during [29,30,42] or at the start of [32,33] effusive eruptions, whereas others are not associated with lava effusion [12]. For those occurring during lava flow output, Calvari et al. [43] proposed calculating the daily erupted volume, suggesting that the drainage of degassed lava from the upper conduit could trigger the decompression and rise of the gas-rich LP magma from the source region causing the paroxysm. Paroxysms are often accompanied by the formation of hot avalanches or pyroclastic density currents (PDCs) spreading along the SdF slope and over the sea surface [29,42,44,45], thus having the potential to impact not just the island, but also boats sailing close to the coast. More rarely, PDCs may affect the other slopes of the island, such as occurred after the 1930 and 1944 paroxysms [12,46]. PDCs are caused by the opening of flank fissures [42,47,48], by the collapse of eruptive columns during paroxysms [41], by the collapse of small portions of the summit cone [49], by flank failure [50], or by the brecciation of lava flow fronts along the steep and incoherent SdF slope [29–31,42,44].

From a geophysical point of view, paroxysms (Figure 2c) at Stromboli share many common features with signals recorded during Vulcanian explosions [29,51]. They are associated with ultra-long-period (ULP) signals (having period > 100 s) starting several seconds before and ending after the event [33]. In addition, they occur with a sharp signal in the borehole strainmeters revealing a strong overpressure build-up in the uppermost conduit by the LP gas-rich magma and moving from a source located at 1.4 km b.s.l. from seconds to minutes before the blast [31,33,52]. Conversely, major explosions (Figure 2b) are not associated with ULP signals, may involve little or none of the LP gas-rich magma [37,53], and have the source region located at ~500 m a.s.l., roughly in the same place as the persistent ordinary explosions (Figure 2a; [27,32]).

A first classification scheme that ranked eruption types qualitatively in order of increasing explosivity was proposed by Lacroix at the start of the twentieth century (reported by [54]). It distinguished four types of explosions: Hawaiian, Strombolian, Vulcanian, and Peléan. However, probably the first volcanological classification of explosive activity, based on collected data rather than on similarities with previous observations of key-type eruptions, was put forward by Walker in 1973 [55]. This was based on the extent and features of the resulting deposit, namely the dispersal area and the degree of fragmentation of the erupted material [55]. Following Walker [55], the most important features of an explosive eruption are its magnitude, which can be determined by estimating the volume of erupted ejecta, and the explosive violence or intensity, which depends on the eruption rate and affects the widespread of the products and their degree of fragmentation. In turn, the dispersal area is a function of the height reached by the eruptive column, thus the greater the height of the eruptive column, the wider the dispersal area [56], although the wind speed also influences the shape and extent of the final distribution of the ash particles [43,57,58]. The main problem with Walker's [55] classification is that it cannot be used for a rapid volcanic hazard assessment, given that the time needed to collect and interpret the volume, fragmentation degree and spread of the deposit is notable. The Walker [55] classification suggested that "Strombolian" activity can be defined by pyroclastic fall deposits with dispersal areas smaller than 10 km^2 and a fragmentation index lower than 10% [59]. In addition, Walker's [55] classification does not sufficiently detail the scale of Strombolian activity, in order to allow distinguishing between persistent Strombolian explosions, major explosions or paroxysms. Newhall and Self [60] proposed using a Volcanic Explosivity Index (VEI) as a general indicator of the explosivity of an eruption, ranging from 0 (effusive) to 8 (highly destructive) on the basis of erupted volume, column height, and qualitative description of the power of the eruptive episode. However, Newhall and Self [60] rate Strombolian-type activity between VEI 1 and 2, and thus their classification is not appropriate to describe small differences like those observed at Stromboli. Houghton et al. [61] improved the VEI classification, extending the classification to negative values and expanding the limits of the Strombolian activity up to VEI -6. Barberi et al. [12] proposed a new classification just for Stromboli, defining "major explosions" as being those discrete explosions much more powerful than the persistent explosive activity and that cause fallout at Il Pizzo (Figures 1 and 2), ~250 m away from the vents and where the tourists stop to watch the activity. Conversely, paroxysms are those impacting the settled areas, located 1.5 km beyond the craters. The main problem with Barberi's et al. [12] classification is that major explosions may or not impact the summit area of Il Pizzo as a result not of the explosion magnitude and intensity, but of the vents shape [27,62–64] and/or wind speed and direction (e.g., [57,58]).

A further distinction within "more intense than ordinary" explosions has been put forward by Andronico et al. [26]. The authors considered those explosions that have greater effects than ordinary Strombolian activity, but which at the same time cannot be classified as major explosions. Andronico et al. [26] suggested several criteria, based on measurements using the Istituto Nazionale di Geofisica e Vulcanologia (INGV) monitoring cameras, to define those events whose activity is intermediate between ordinary and major Strombolian activity: (1) larger-than-ordinary eruptive jets (>300 m) and dispersal area (>250 m around the vent; potentially reaching the area at Il Pizzo Sopra La Fossa, see Figure 1) of coarser erupted products (decimetre-sized blocks and bombs); (2) multiple vents involved ("explosion sequence" instead of "single burst"); (3) longer duration of the tephra ejection (>30 s).

A more recent classification, proposed by Houghton et al. [19], distinguishes the different Strombolian explosions, as well as the Hawaiian activity from the Strombolian one, by measuring the duration and height of the column of explosive events. They [19] made a distinction on the basis of a threshold duration of 300 s, with sustained Hawaiian lava fountains displaying durations greater than 300 s, and shorter events grouping all transient Strombolian explosions. The main limitation of this classification scheme is that it does not allow distinguishing between paroxysms, major explosions and persistent Strombolian activity at Stromboli, because these all fall below the 300 s threshold. This is why Houghton et al. [19] use an additional plot considering erupted mass (kg) together with duration (s), with several fields of mass eruption rates, from 1 kg/s to 10^8 kg/s.

A new approach, here proposed for the first time and tested on twelve events occurring at Stromboli since 25 June 2019, combines different geophysical monitoring and independent data in order to obtain a straightforward classification that can be used any time an explosion occurs. This classification scheme could be easily applied to other basaltic volcanoes, provided that a suitable monitoring network exists. In this paper, we describe the twelve explosive events on the basis of observations gathered from the INGV monitoring camera network, integrated with geophysical data from the INGV seismic network, as well as the ground deformations obtained from different remote sensing and geodetic techniques. These data, working at distinct sampling frequencies, allow analyzing the ground movements associated with different phenomena.

2. Methods

In the following, the data relative to each major explosion and paroxysms were derived from the INGV monitoring weekly bulletin, as specified, integrated by a more in-depth analysis of the images recorded by all INGV monitoring cameras. The starting time of each event is expressed in UTC, to make comparison easier with other geophysical data, and is obtained from the INGV monitoring cameras images, such as the duration of each event and the height of the eruptive column, intended as the maximum vertical extension of the ash plume, when within the field of view of the instruments. The position of the INGV monitoring cameras is shown in Figure 1, and their details are listed in Table 1. The time of each image is automatically attributed by the system using the Network Time Protocol [65]. Paroxysms produce much higher eruptive columns than the field of view of the INGV monitoring cameras (extending to a maximum of 750 m above the craters for SPCT, see Table 1), and in these three cases (3 July 2019, 28 August 2019, and 19 July 2020) we refer to published data. The maximum speed of ejecta or muzzle velocity is obtained from the analysis of the SQV camera images (Table 1), the only one that detected all the 12 explosive episodes considered here. The error on the vertical measurement is estimated at 9.5 m. The resulting speed or muzzle velocity is averaged over 2 s of time lapse for each episode. The VLP size and other seismic signals are obtained from the INGV monitoring seismic network shown in Figure 1. The seismic network initially comprised 14 stations [66] and was deployed by INGV-Osservatorio Vesuviano (INGV-OV). From 2013, their number decreased because some sites became inaccessible both by land and by helicopter. The seismic stations are equipped with Guralp CMG 40T (60 s–50 Hz) velocimeters and Gilda data logger [67].

Table 1. List of the INGV monitoring cameras and of their main features. SdF = Sciara del Fuoco. The field of view is considered at the crater rim.

Label	Type	Location	Distance from the Craters (m)	Frame Rate (hz)	Field of View (m)
SPT	Thermal FLIR A310	Pizzo Sopra La Fossa 890 m a.s.l.	250	2	500 × 370
SPCT	Thermal FLIR A320	West SdF flank 85 m a.s.l.	1698	2	2150 × 1613
SCT	Thermal FLIR A655sc	East SdF flank 165 m a.s.l.	1538	2	807 × 605
SQT	Thermal FLIR A320	East SdF flank 390 m a.s.l.	1027	0.5	455 × 340
SQV	Visual Sony FCB-EX480CP	East SdF flank, 390 m a.s.l.	1027	0.5	657 × 493
SCV	Visual Mobotix M26	East SdF flank 163 m a.s.l.	1538	2	1776 × 1274

This study was also supported by the information from borehole geophysical instruments managed by INGV, in particular by the STDF tilt station and the SVO volumetric strainmeter station (Figure 1). The signals recorded at these two borehole stations are used in the official weekly reports produced by INGV to update the Italian Civil Protection Department and the local authorities on the eruptive state of the Stromboli volcanic island (http://www.ct.ingv.it/index.php/monitoraggio-e-sorveglianza/prodotti-del-monitoraggio/bollettini-settimanali-multidisciplinari (accessed on 9 February 2021)).

At Stromboli, the first shallow borehole tiltmeters were operating from 1992 by installing two stations with AGI 722 biaxial sensors with 10^{-7} rad precision at shallow depth of ~3 m at Punta Labronzo in the northern flank and at Timpone del Fuoco (STDF) in the western flank (Figure 1), respectively [68]. In order to reduce the thermo-elastic noise affecting the shallow depth installations [69], STDF was installed in 2011 at ~27 m below ground surface by using an AGI Lily sensor [70]. The data are collected with a sampling rate of 1 data/minute.

In order to improve the recording sensibility, two borehole strainmeters were installed at San Vincenzo Observatory (SVO) and at the Timpone del Fuoco (STDF) area in 2006 (Figure 1). These instruments, also called dilatometers, are Sacks-Everton types [71] which measure the volumetric strain with a nominal resolution up 10^{-11} in strain, depending on the final response of the coupling of the instruments with the surrounding rock. The devices were installed at a depth of 120 m. The data are recorded and sampled at 50 Hz using a 24-bit digital recorder and are sent to INGV via TCP/IP [52,72]. The STDF strainmeter is unfortunately located in an unconsolidated medium causing a weak coupling and a low effective sensitivity, and moreover it suffered several signal interruptions. Instead, the SVO (Figure 1) is installed in massive rock providing a reliable signal with a sensitivity of 1×10^{-11} per digital count [52].

Measuring surface deformation using the phase difference between two GBInSAR images enables recognising millimetre-scale displacements of the ground along the device line of sight (LOS) direction [73]. GBInSAR devices have the additional advantage of producing frequent SAR images (in the order of seconds to minutes), resulting in very high frequency deformation maps and time series [74]. Two GBInSAR devices are located in a stable area N of the SdF (Figure 1, Table 2). The GBInSAR devices are remote sensing imaging systems [73,75,76] that emit and receive a burst of microwave pulses, repeating this operation while the sensor is moving [75] along a rail (track), that in the case of Stromboli is 4 m long [73]. The use of GBInSAR in the Ku-band (17–17.1 mm wavelength radar), can penetrate dust clouds (abundant especially during collapse events), and works in variable light and atmospheric conditions [49,77].

The GBInSAR measures the backscattered energy (amplitude) and the phase of the received radar signal. The latter can be used to estimate ground movements (along the Line-Of-Sight, LOS) by using interferometric techniques. The interferograms are obtained from phase information of "averaged" images (i.e., by averaging the phase information derived from the different acquisitions, see Table 2 for averaging specification). Displacement cumulated maps are calculated using a staking algorithm to sum, pixel by pixel, the displacements for every consecutive pair of images, whereas displacement time series of selected points (averaged over 10 pixels) are obtained from cumulative displacement maps with a precision in the displacement measurement of 0.5 mm [73,74]. Pixel resolutions are about 2 m in range, and 2 m on average in cross range, with a measurement precision referred to the displacement of less than 1 mm [78].

Table 2. Technical features of the two GBInSAR devices installed at Stromboli volcano.

System	Model	Band	Revisiting Time [min]	Averaging Interval [min]	Look Angle [deg]	Heading Angle [deg]
GBInSAR NE400 *	GBInSAR LiSAmobile k09	Ku	11 (until November 2017) 6 (since November 2017)	33	from 63.8° to 90.0°	from 143° to 217°
GBInSAR NE190 **	GBInSAR LiSAmobile k09	Ku	2	30	from 65.0° to 113.5°	from 115° to 245°

From * Di Traglia et al. [74] and ** Schaefer et al. [79].

3. Results

We describe here the 12 explosive events analysed in this paper on the basis of information gathered from the INGV weekly reports (Table 3) integrated with the analysis of the images recorded by the INGV monitoring webcams (Table 4). We then describe the seismic trace recorded for each explosive event, as well as the ground deformation recorded by the available instruments at different rate. Finally, we compare all these data together, listed in Table 4, to select the useful parameters that can be used for the Strombolian event classification at Stromboli volcano, and possibly for other basaltic volcanoes, provided that they have a suitable monitoring system.

Table 3. List of major explosions (ME) and paroxysms (PA) at Stromboli since 25 June 2019 and updated to 6 Dec 2020. The date, time and features of the events are from the INGV weekly reports on the monitoring activity. NEC = NE crater zone; CC = central crater zone; SWC = SW crater zone.

Date	ME/PA	Time (UT)	Main Features	References and Notes
25 June 2019	ME	23:03:08	CC crater zone widened after the explosive event.	INGV weekly report 27/2019 [1]
3 July 2019	PA	14:45:43	Lava flow within the crater. Blast starting from SWC and NEC. The N rim of the crater terrace was blown away. Two PDCs along the SdF and several small lava flows.	INGV weekly report 28/2019 [1]
13 July 2019	ME	20:33	Intense event detected from seismicity.	INGV weekly report 29/2019 [1]
15 July 2019	ME	19:09	Intense event detected from seismicity.	INGV weekly report 29/2019 [1]
28 August 2019	PA	10:17:20	Paroxysm comprising 3 pulses from SWC and NEC. PDC along SdF, NEC crater rim eroded by the blast.	INGV special report 35/2019 [1]
29 August 2019	ME	20:43:18	Two fountaining during lava flow output with fallout on Ginostra.	INGV daily report n. 32, 30 August 2019 [1]

Table 3. *Cont.*

Date	ME/PA	Time (UT)	Main Features	References and Notes
19 July 2020	PA	03:00:42 03:01:11 03:01:28	Explosive sequence in 3 pulses from CC and SWC, with fallout of bombs to Il Pizzo and down to 500 m asl.	INGV weekly report 30/2020 [1]
13 August 2020	ME	14:50:28	Explosive sequence from SWC. No fallout.	INGV weekly report 34/2020 [1]
10 November 2020	ME	20:04:21 20:04:51	Explosive sequence from SWC followed by several pulses at NEC and CC.	INGV weekly report 47/2020 [1]
16 November 2020	ME	09:17:45	SWC, CC and NEC produced a blast expanding horizontally like a rose. PDC along the SdF that spread over the sea surface for ~250 m. Fallout at Il Pizzo.	INGV weekly reports 47/2020, 48/2020 [1]
21 November 2020	ME	00:33:17	Sequence of 3 explosive events from NEC and CC.	INGV weekly report 48/2020 [1]
6 December 2020	ME	05:12:44	Two pulses. Ballistics to 300 m height, ash plume, 2 PDCs along SdF.	INGV weekly report 50/2020 [1]

[1] The INGV monitoring reports can be found at http://www.ct.ingv.it/index.php/monitoraggio-e-sorveglianza/prodotti-del-monitoraggio/bollettini-settimanali-multidisciplinari (accessed on 9 February 2021).

3.1. Explosive Events Description

Table 3 shows a summary of the main features for each one of the explosive events analysed here, together with a preliminary classification of each episode taken by the INGV monitoring weekly reports. For each event, we display in Figures 3–14 a few thermal and visual frames extracted from the monitoring cameras, together with the seismic trace recorded by the IST3 INGV seismic stations (Z component, see Figure 1), in order to appreciate the size of the erupted ejecta and ash plume extension together with the seismic amplitude of the episode. Table 4 lists a number of parameters obtained for each of the explosive events from the analysis of available data.

3.1.1. The 25 June 2019 Event

The 25 June 2019 episode occurred at 23:03:08 from the CC vent of the crater area (Table 3, Figure 3a). It lasted 8 s (Table 4) and the erupted products spread laterally like a rose (Figure 3a) expanding mainly towards W and up to the crater rim. It caused a widening of the CC vent. The eruptive plume (Figure 3b) extended beyond the ~250 m of the field of view (FOV) of the SQT camera and reached ~500 m, as detected from SQV. The seismic trace recorded during the event is shown in Figure 3c. The maximum speed of the ejecta (Table 4, 54.41 m s^{-1}) was normally recorded at the start of the event and close to the vent, and rapidly declining upwards.

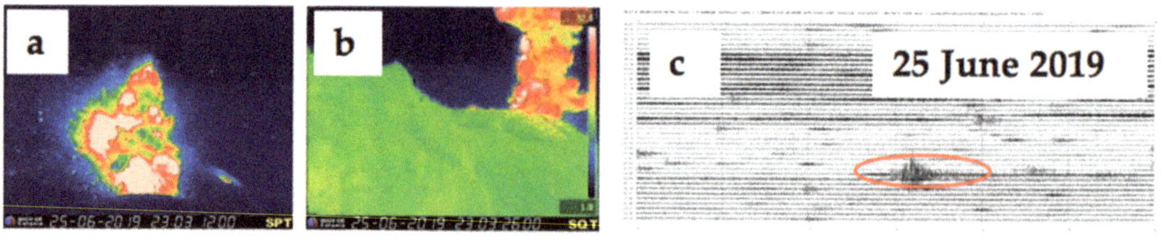

Figure 3. (**a**) Thermal image from the SPT camera of the CC vent with the start of the explosion on 25 June 2019 at 23:03:1200, view from S. Blue is cold, white is hot. (**b**) Thermal image from the SQT camera showing the eruptive plume at 23:03:2600, view from NE. Blue is cold, white is hot. (**c**) Seismic trace of the explosive event as recorded from the IST3 seismic station, Z component, with the red oval marking the explosive event here considered. See Figure 1 for station locations.

3.1.2. The 3 July 2019 Paroxysm

The 3 July episode (Figure 4) was preceded by a significant increase in the intensity of explosive activity at all the summit vents. At 13:59 a strong explosion from the SWC was followed by lava flow output along the upper part of the Sciara del Fuoco (SdF). Images of the SQT camera show a small vent opening at the base of NEC, widening and feeding a lava flow that started spreading along the SdF from 14:02:40. Several lava flows from this vent overlapped the previous flux, spreading downslope. In the meantime, at 14:43:10 another small lava flow started from the CC crater, slowly spreading S within the crater terrace (Figure 4a–d), followed by flows from the two NEC vents, and eventually joining with another flow erupted from the SWC vent. The maximum velocity of the jet, estimated at 101.92 m s^{-1} (Table 4), was recorded after 6 s of gradual increase due to the initial horizontal expansion of the jet. The main phase of the paroxysm involved the whole crater area. The column collapse produced two pyroclastic flows spreading NW along the SdF and over the sea surface for several hundred meters, and caused a significant widening of the crater terrace. The duration of the event, estimated on the basis of the images recorded by the monitoring cameras, is more than 2 min (Table 4). The end is difficult to estimate due to fallout and the ash cloud spreading for several minutes and limiting visibility from all cameras. Giordano and De Astis [41] estimated a maximum height of the eruptive column to 8.4 km above the crater (Table 4).

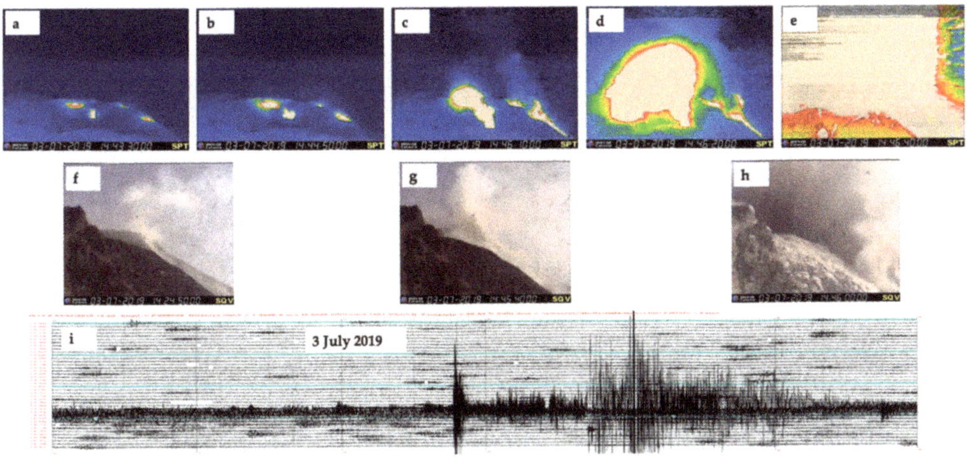

Figure 4. (**a**–**e**) Thermal images from the SPT camera showing the summit vents on 3 July 2019 at 14:43:3000 (**a**), at 14:44:5000 (**b**), at 14:46:1000 (**c**), at 14:46:2000 (**d**) and at 14:46:4000 (**e**), with (**a**–**c**) the emission of the intracrater lava flow from the CC vent and (**d**,**e**) the start of the jet explosion, view from S. Blue is cold, white is hot. (**f**–**h**) Visual images from the SQV camera showing (**f**) at 14:24:5000 the ash along the upper Sciara del Fuoco due to the lava flow output from the NEC, (**g**) at 14:45:4000 the eruptive plume, and (**h**) the fallout of bombs at 14:46:0000 along the upper Sciara del Fuoco, view from NE. (**i**) Seismic trace of the explosive event as recorded from the IST3 seismic station, Z component. See Figure 1 for station locations.

3.1.3. The 13 July 2019 Event

This strong explosion occurred while lava was flowing along the SdF slope erupted from the SWC and NEC vents [33,80,81]. This explosive event (Figure 5a,b) was considered stronger than the persistent activity just on the basis of the seismic trace, as shown in Figure 5c.

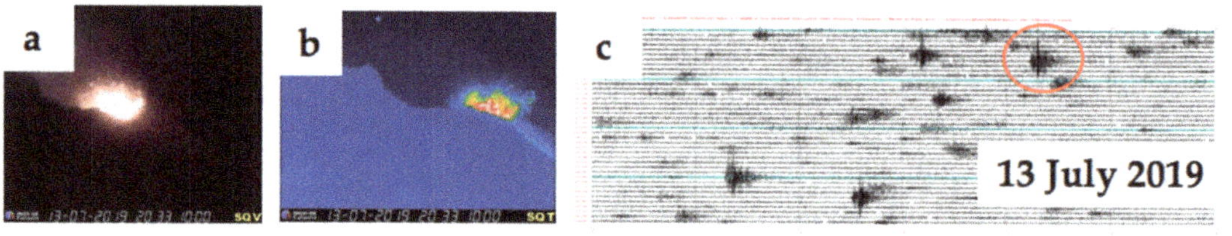

Figure 5. (**a**) Visual image from the SQV camera of the 13 July 2019 event at 20:33:10, view from NE with (**b**) corresponding thermal image taken from the SQT camera at 20:33:10. Blue is cold, white is hot. (**c**) Seismic trace with the red circle evidencing the explosive event displayed in the images above. See Figure 1 for station locations.

3.1.4. The 15 July 2019 Event

This explosion was considered stronger than the persistent ordinary explosive activity on the basis of the seismic trace. It occurred while lava flows were being erupted along the SdF [33,80,81]. The peculiar aspect of this explosion occurring from the SWC is that it took the form of a lateral jet (Figure 6a) spreading and widening upwards up to an estimated elevation of ~380 m (Figure 6b and Table 4), with incandescent bombs falling on the upper Sdf and at Il Pizzo Sopra La Fossa (Figure 6c,f, red circles). Figure 6g shows the seismic signal, slightly differing from the persistent Strombolian explosions occurring during the day.

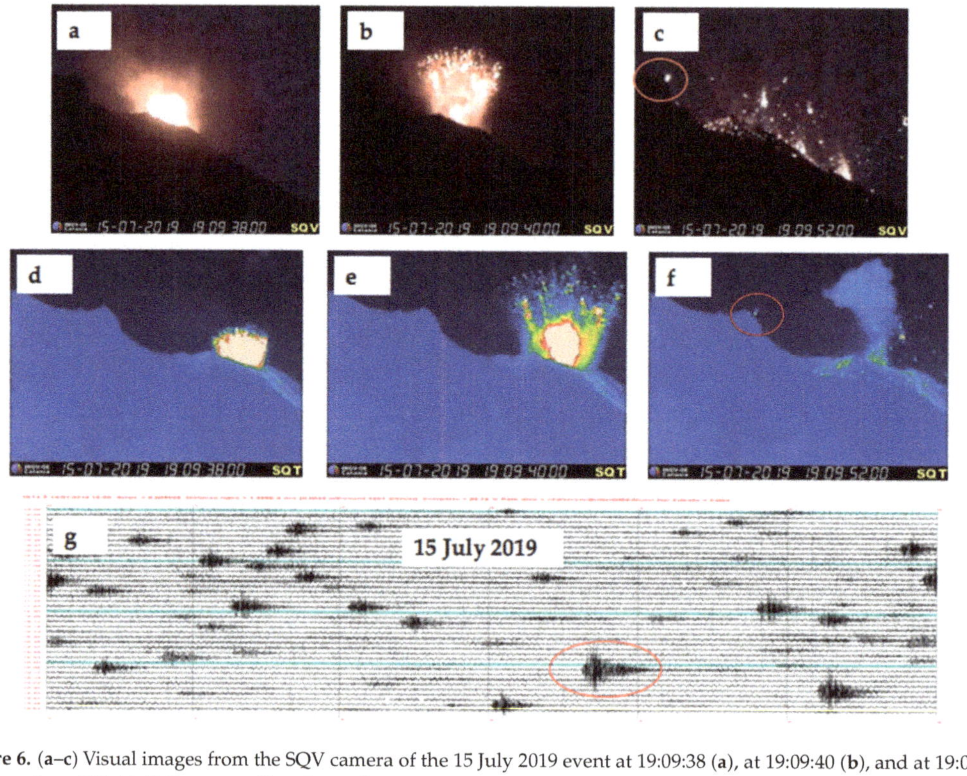

Figure 6. (**a–c**) Visual images from the SQV camera of the 15 July 2019 event at 19:09:38 (**a**), at 19:09:40 (**b**), and at 19:09:52 (**c**), view from NE. (**d–f**) Corresponding thermal images taken from the SQT camera, view from NE. Blue is cold, white is hot. The red circles in (**c**) and (**f**) indicate the incandescent block falling at Il Pizzo. (**g**) Seismic trace with the red ellipse evidencing the explosive event shown in the images above. See Figure 1 for station locations.

3.1.5. The 28 August 2019 Paroxysm

The 28 August 2019 paroxysm (Figure 7) followed a day of increased explosive intensity at the summit craters of the volcano and occurred while the volcano was still the site of an effusive eruption from the summit craters [33,80,81]. At 10:17:20 an eruptive sequence started from the SWC portion of the crater terrace, giving rise to three pulses (Figure 7a–i), of which the first two from SWC and the last one, less intense from NEC, took the form of a lateral jet. The maximum velocity of the jet was ~71.11 m s^{-1} (Table 4), and the duration of the event is almost 2 min, with the end difficult to estimate due to fallout and ash cloud spreading for several minutes and obscuring sight from all cameras (Table 4). The eruptive column rose up to 4 km above the craters (INGV report, Table 4), or as much as 6.4 km ([41]; Table 4), and the fallout from the collapsing column produced two PDCs (Figure 7e) that spread along the SdF and over the sea surface for at least 540 m (Figure 7h). The explosion deeply modified the morphology of the crater area, widening the NEC towards NW. The seismic trace of the event is shown in Figure 7j.

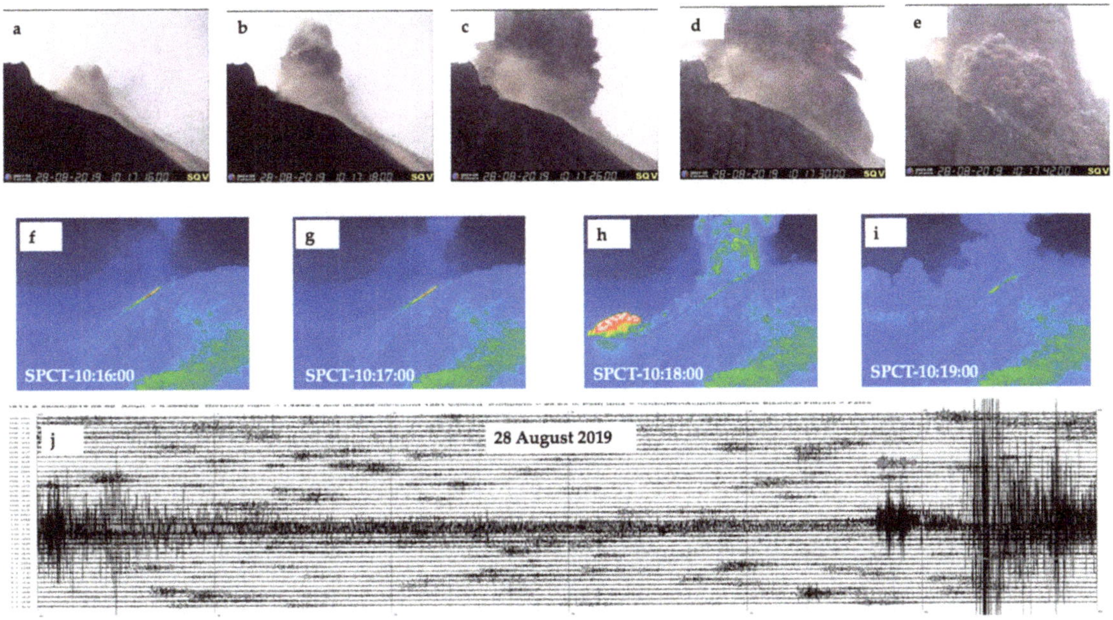

Figure 7. (**a**–**e**) Visual images from the SQV camera of the 28 August 2019 event recorded at 10:17:16 (**a**), 10:17:18 (**b**), 10:17:26 (**c**), 10:17:30 (**d**), and 10:17:42 (**e**), view from NE. (**f**,**i**) Thermal images from the SPCT camera, showing (**f**) the lava flow from the crater zone at 10:16:00 and at 10:17:00 (**g**), the eruptive column and PDC spreading over the sea surface for ~300 m at 10:18:00 (**h**), and the ash cloud at 10:19:00 (**i**), view from W. Blue is cold, white is hot. (**j**) Seismic trace of the 28 August explosive paroxysm as recorded from the IST3 seismic station, Z component. See Figure 1 for station locations.

3.1.6. The 29 August 2019 Event

Two major explosive events occurred on this date, more or less with the same maximum elevation of the ejected spatter, which reached ~350 m above the crater rim (Table 4). However, this activity was unusual, because it happened while lava was overflowing from the NEC crater rim, and because the volcanic tremor was extremely high for several hours, this activity resulting more as a continuous fountaining from SWC and CC rather than as single explosive Strombolian pulses. Figure 8 shows two frames recorded from the visible SQV camera, and the seismic trace as recorded by the IST3 seismic station.

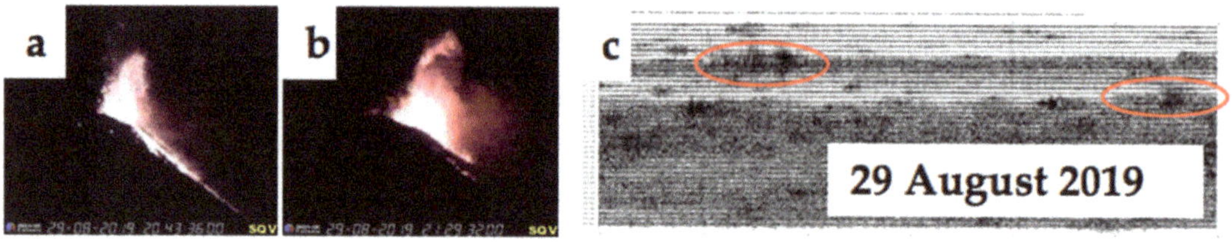

Figure 8. (**a**,**b**) Visual images from the SQV camera of the 29 August 2019 events recorded at 20:43:36 (**a**) and at 21:29:32 (**b**), view from NE. (**c**) Seismicity of 29 August 2019 as recorded from the IST3 seismic station, Z component. The two red ovals indicate the seismic trace corresponding to the two events shown above. See Figure 1 for station locations.

3.1.7. The 19 July 2020 Paroxysmal Event

This explosive event started at 03:00:42 from the CC (Figure 9a) and expanded to the SWC (Figure 9b), erupting most of the volume discharged during this event (Figure 9c). Another pulse occurred at 03:01:11 (Figure 9d) and a third pulse at 03:01:28, again from the SWC and displaying decreasing intensities, for a total explosive time of 58 s (Table 4). The maximum elevation of the ejecta, based on the images recorded by the SPCT camera, was more than 750 m above the crater rim (Figure 9k), but most of the fallout was spread horizontally all around the crater and up to the Pizzo Sopra La Fossa (Figure 9j, red circle). The fallout triggered several landslides along the SdF (Figure 9e–h), reaching the coast after 40 s.

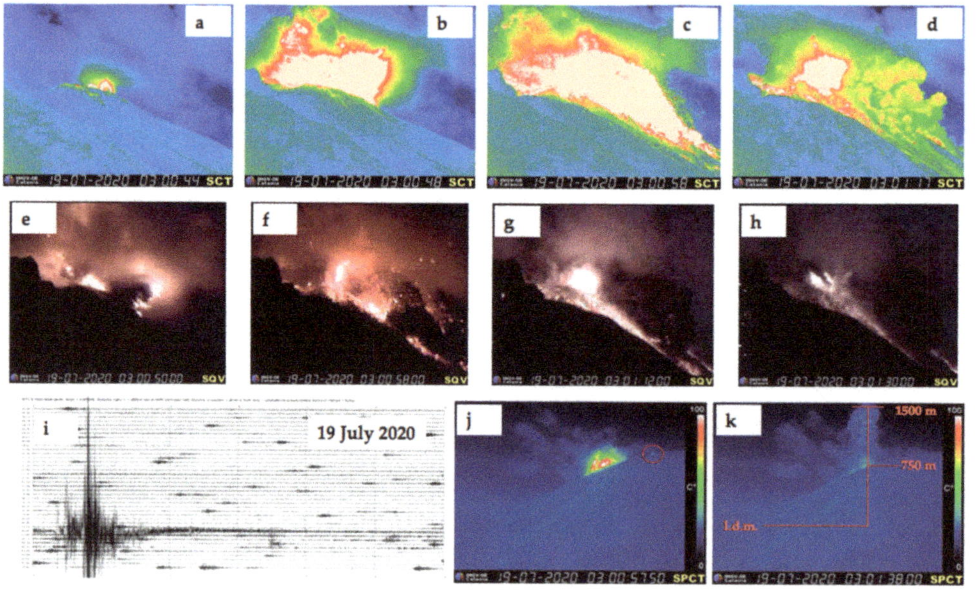

Figure 9. (**a**–**d**) Thermal images from the SCT camera of the 19 July 2020 explosive event at 03:00:44 (**a**), at 03:00:48 (**b**), at 03:00:58 (**c**), 03:01:17 (**d**), view from NE. Blue is cold, white is hot. (**e**–**h**) Visual images from the SQV camera recorded at 03:00:50 (**e**), at 03:00:58 (**f**), at 03:01:12 (**g**) and at 03:01:30 (**h**), view from NE. (**i**) Seismic trace of the 19 July 2020 paroxysm as recorded from the IST3 seismic station, Z component. (**j**,**k**) Thermal images of the 19 July 2020 explosive event from the SPCT camera, view from W, showing (**j**) the fallout of bombs on the Il Pizzo Sopra La Fossa (indicated by the red circle) at 03:00:57.50, and (**k**) the vertical extent of the eruptive plume (750 m above the craters), recorded at 03:01:38. See Figure 1 for station locations.

3.1.8. The 13 August 2020 Event

On 13 August at 14:50:27 (Figure 10), the SWC produced a stronger than average explosion, that from seismicity lasted about one minute and from the monitoring cameras 64 s (Table 4). It was characterized by a sequence of short events producing an ash plume that rose to 550 m above the craters (Table 4).

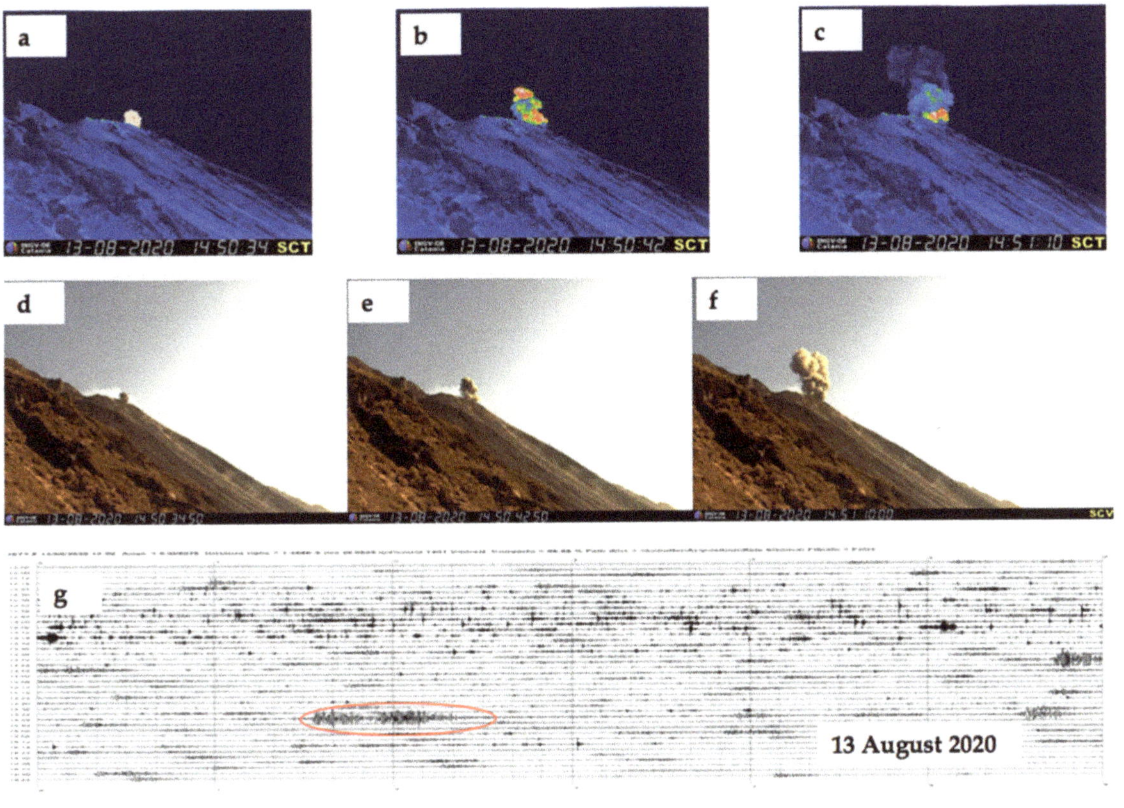

Figure 10. (**a–c**) Thermal images recorded from the SCT camera on 13 August 2020 at 14:50:34 (**a**), at 14:50:42 (**b**), and at 14:51:10 (**c**). Blue is cold, white is hot. View from NE. (**d–f**) Corresponding visual images from the SCV camera, view from NE, recorded at 14:50:3450 (**d**), at 14:50:4250 (**e**), and at 14:51:1000 (**f**). (**g**) Seismic trace of the explosive event as recorded from the IST3 seismic station, Z component. The red oval indicates the seismic trace corresponding to the event shown above. See Figure 1 for station locations.

3.1.9. The 10 November 2020 Event

The 10 November 2020 episode started from the SWC at 20:04:21 (Figure 11), forming an eruptive cloud reaching up to 600 m above the crater rim (Table 4). The explosive event then expanded to the NEC forming a jet expanding horizontally and causing a wide spatter fallout on the upper SdF, and eventually expanded to the CC producing a low fountaining with little or no fallout outside the crater. The duration of the first pulse was 20 s, and the muzzle velocity of the ejecta was 54.50 m s^{-1} (Table 4).

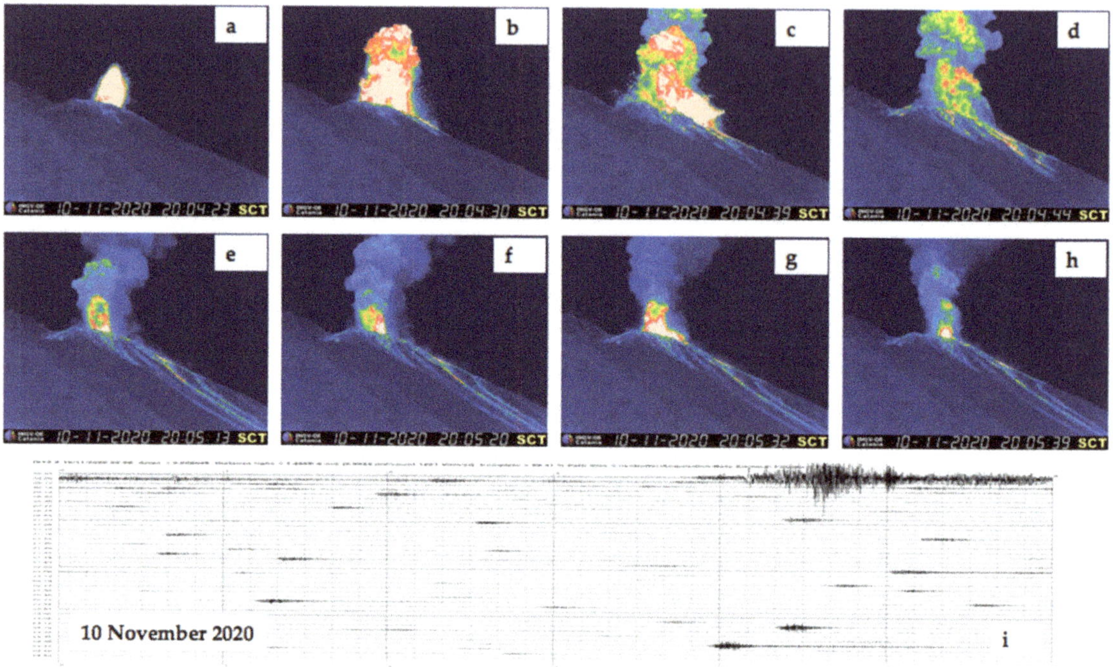

Figure 11. (**a**–**c**) Thermal images from the SCT camera recorded on 10 November 2020 at 20:04:23 (**a**), at 20:04:30 (**b**), at 20:04:39 (**c**), at 20:04:44 (**d**), at 20:05:13 (**e**), at 20:05:13 (**f**), at 20:05:32 (**g**) and at 20:05:39 (**h**), view from NE. Blue is cold, white is hot. (**i**) Seismic trace of the explosive episode, recorded by the IST3 station. See Figure 1 for station locations.

3.1.10. The 16 November 2020 Event

The 16 November 2020 episode was rather unusual, because of the clear ground deformation detected from the monitoring cameras and from the GBInSAR. The event started with puffing from the CC, followed by the fast propagation of a fracture from the CC to the SWC. This caused the upward tilting of the NE outer flank of the cone, forming two fractures on the NE flank and decompressing the uppermost conduit. This triggered a powerful explosion that started from the SWC (Figure 12). The explosion caused a blast spreading at first horizontally and, while the ash plume was still rising up, two pyroclastic density currents (PDCs) formed along the SdF, spreading down the slope and to the coast. The velocity of the PDC, obtained from the images of the SCT camera along the uppermost 250 m distance travelled along the SdF, was estimated at ~20 m s^{-1}, which is in the range of the values obtained for the events occurring at Stromboli in March–April 2020 [44]. The PDC reached the coast after 42 s, as detected from the SPCT camera, travelling the 1028 m of the slope at an average speed of ~25 m s^{-1}, and then expanded over the sea surface for about 250 m. The event lasted 54 s, and the ash plume observed from SPCT rose to about 690 m above the craters (Table 4).

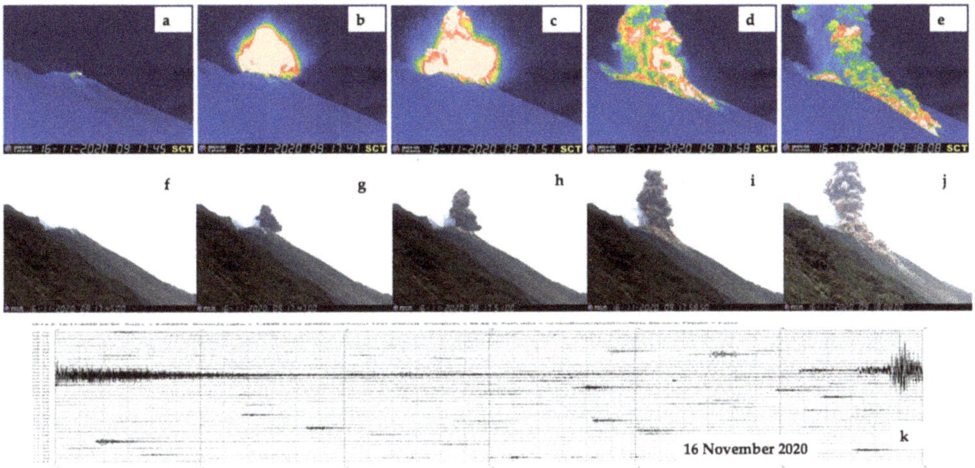

Figure 12. (**a–e**) Thermal images of the SCT camera recorded at 09:17:45 (**a**), 09:17:47 (**b**), 0917:51 (**c**), 09:17:58 (**d**), and 09:18:08 (**e**), and (**f–j**) corresponding visual images from the SQV camera of the 16 November 2020 event, view from NE. Blue is cold, white is hot. (**k**) Seismic trace of the event recorded by the IST3 station. See Figure 1 for station locations.

3.1.11. The 21 November 2020 Event

The 21 November 2020 occurred at 00:33 and was characterized by a sequence of three pulses of increasing intensity starting from the NEC and then expanding to the CC, lasting just 10 s (Figure 13). It was of very low intensity, producing an ash plume that reached ~80 m, with muzzle velocity of the ejecta of 9.48 m s^{-1} (Table 4).

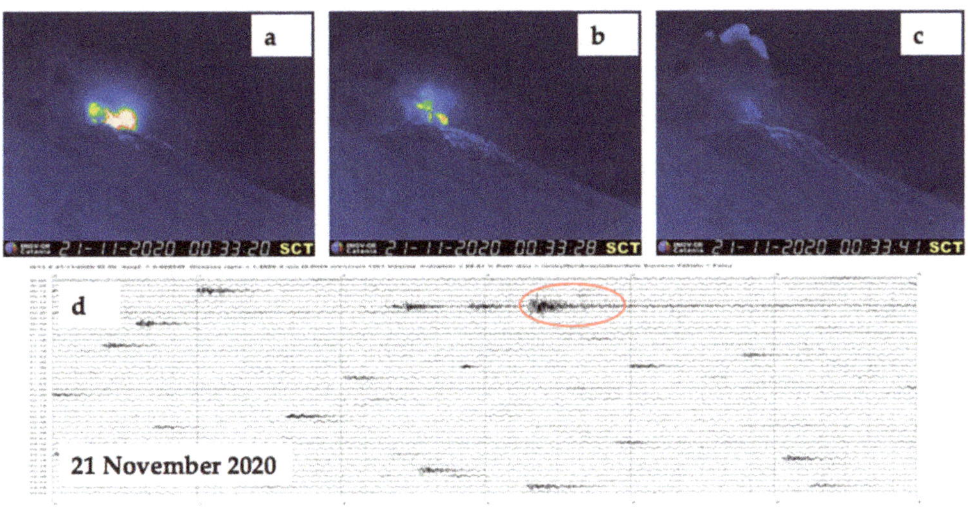

Figure 13. (**a–c**) Thermal images from the SCT camera of the 21 November 2020 event view from NE (blue is cold, white is hot), recorded at 00:33:20 (**a**), 00:33:28 (**b**), 00:33:41 (**c**) with (**d**) the corresponding seismic trace highlighted by the red oval. See Figure 1 for station locations.

3.1.12. The 6 December 2020 Event

The 6 December 2020 episode (Figure 14) comprised two pulses occurring at 05:12:44 (Figure 14a–c) and 05:13:41 (Figure 14d), both starting from the SWC crater zone but also involving the CC crater zone (Figure 14b). The eruptive plume reached a maximum height of 300 m above the crater rim, and the fallout along the SdF triggered small PDC (Figure 14c,d). The seismic trace of this episode (Figure 14e) is so continuous to be more similar to a lava fountaining than to a Strombolian-type event.

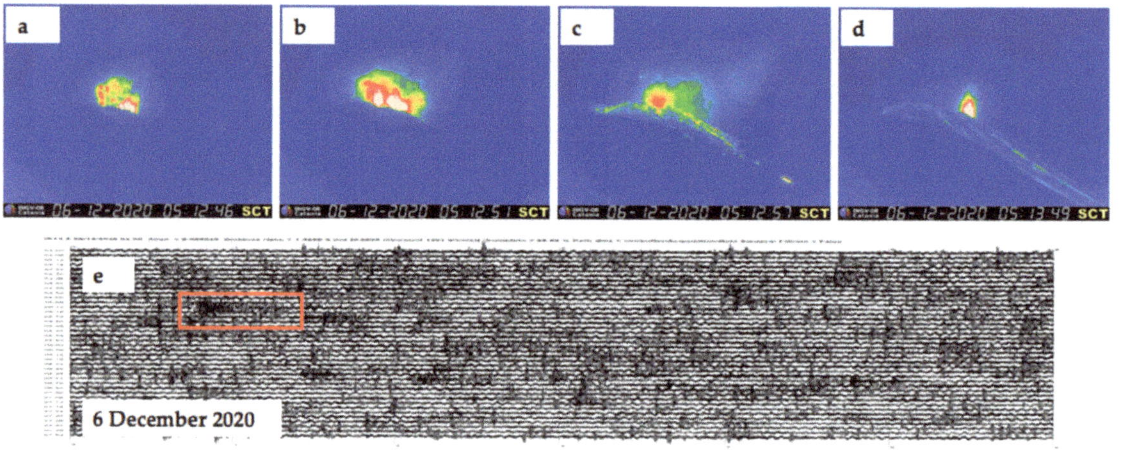

Figure 14. (**a–d**) Thermal images from the SCT camera of the 6 December 2020 event, view from NE, with (**e**) the corresponding seismic trace, indicated by the red box. In the thermal images blue is cold, white is hot.

The most important parameter to rate the magnitude and intensity of an explosion is the height of the eruptive column [82]. Other important volcanological parameters are the muzzle velocity and the duration of each explosive episode, which should be estimated using instruments that give comparable results. For this analysis, we have used the images of the SQV visual camera (see Figure 1 for camera location and Table 1 for camera features) which is the only one that recorded all the 12 eruptive events considered here. The results of muzzle velocities and explosion durations obtained from the analysis of the SQV images are reported in Table 4, together with the "V×D parameter", obtained multiplying the muzzle velocity (in m·s^{-1}) by the duration (expressed in seconds) for each event.

3.2. *VLP Size*

Seismic signals associated with the Stromboli explosions contain Very Long Period (VLP) pulses, typically in the 0.05–0.5 Hz frequency band, that are generated by the explosive mechanism [32,83–85]. These signals have a direct link with the eruptive process of both ordinary and major explosions. Components with an even longer period are recorded in the seismograms of paroxysmal explosions. Figure 15 shows the seismograms and spectrograms of paroxysmal (Figure 15a), major (Figure 15b) and ordinary (Figure 15c,d) explosion types.

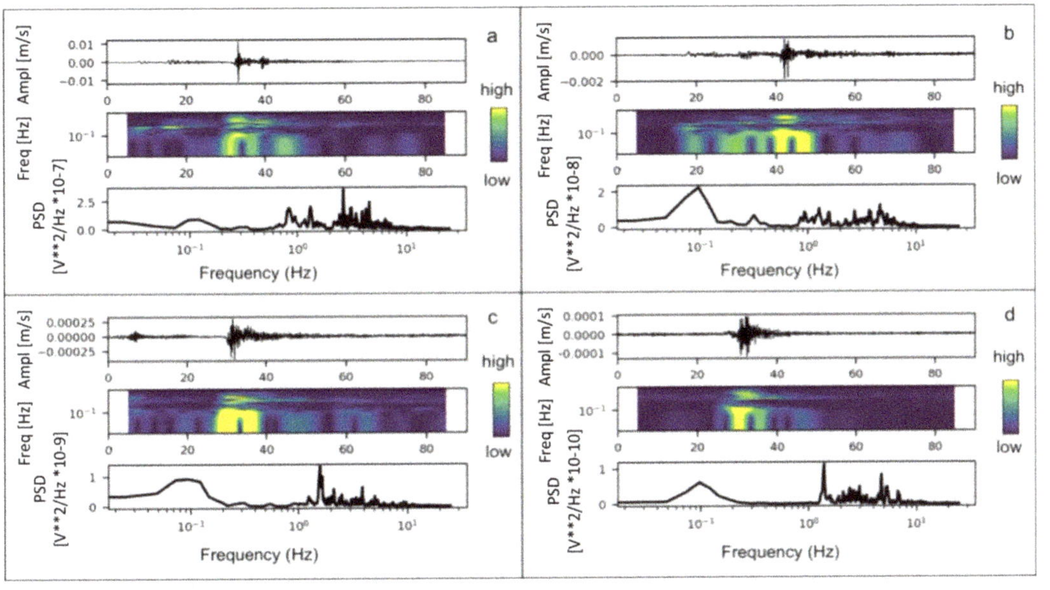

Figure 15. Seismograms (top), spectrograms (middle) and spectra (bottom) of four examples of explosions of different sizes. (**a**) The paroxysmal explosion of 19 July 2020 (03:00 UTC); (**b**) the major explosion on 10 November 2020 (20:04 UTC); (**c**) the explosion on 21 November 2020 (00:33 UTC); (**d**) an ordinary explosion on 19 July 2020 (02:54 UTC). In the horizontal axes of seismograms and spectrograms, time is expressed in seconds.

The signal amplitudes and the VLP events associated with the 12 explosions analysed in this article are very different (Figures 16 and 17).

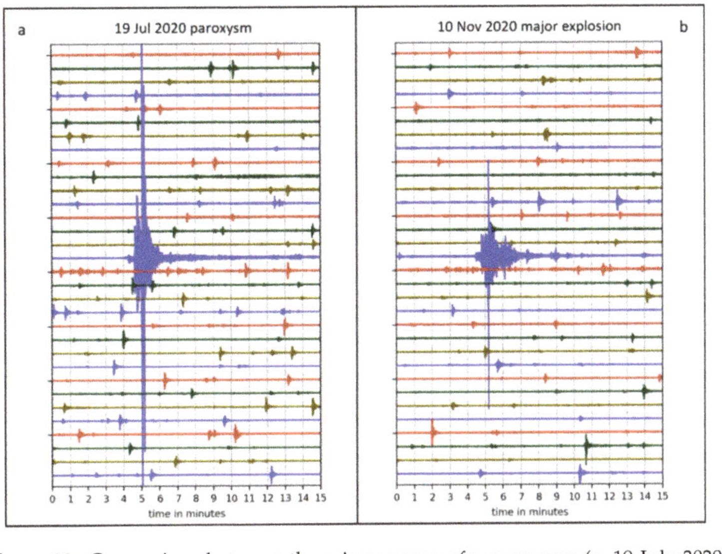

Figure 16. Comparison between the seismograms of a paroxysm (**a**, 19 July 2020) and a major explosion (**b**, 10 November 2020). Both plots (**a** and **b**) represent an 8-h signal recording of the east-west component of the STRA station (see Figure 1 for station location). The small amplitude transients that are recognizable in both plots are due to the ordinary Strombolian explosions.

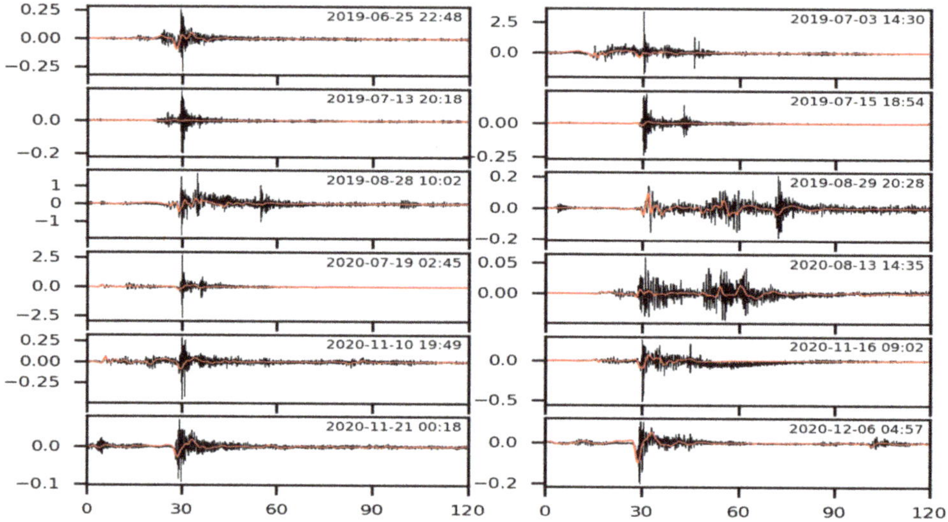

Figure 17. Seismograms of the 12 explosions, here considered as case studies, as recorded from station STRA, East component (see Figure 1 for station location). The red lines represent the filtered signal in the VLP frequency band (0.05–0.5 Hz). The vertical scale is expressed in 10^6 counts. The filtered signal is scaled by a factor of 2. The time on the horizontal axis is in seconds.

To represent the magnitude of the explosions through seismic measurements we use the VLP size introduced by Giudicepietro et al. [33]. This parameter is defined as the maximum value of the RSAM of a 30 s sliding window that moves in a 30-min time interval of signal, filtered in VLP frequency band (0.05–0.5 Hz). The 30-s window moves by 1 s steps and produces 1770 RSAM values in a 30-min time interval of signal. The maximum of these values is the VLP size of that half hour signal [33]. This analysis typically returns 48 values per day representing the size of the largest VLP event for every half hour of the day. This parameter was used in Giudicepietro et al. [32] to highlight variations in the "magnitude" of the VLPs associated with ordinary Strombolian explosions in the period preceding the paroxysm of 3 July 2019, and therefore to discover a seismic precursor of the paroxysm. However, the VLP size, which is sensitive to the amplitude and duration of the greater amplitude portion of a VLP signal, is also suitable for providing the "magnitude" of the VLP seismic transients associated with explosions of greater energy than ordinary ones such as major explosions and paroxysms. Therefore, the VLP size gives the possibility to create a relative scale of the "magnitude" of the Stromboli explosions on the basis of the seismic signal associated with them. For each explosion reported in Table 1 we computed the VLP size in 24-h long time intervals (12 h before and 12 h after the explosion). For some intervals, the signals are missing; in that case the value of the VLP size is reported as zero, for example a few hours before the explosion of 6 December 2020 (Figure 18). We normalized the VLP size values (to 100,000) with respect to the maximum, which coincides with the paroxysmal explosion of 3 July 2019. The VLP size of the 12 explosions listed in Table 3 are reported in Table 4, where they are compared with other parameters extracted from the analysis of the camera images and from the str in and tilt.

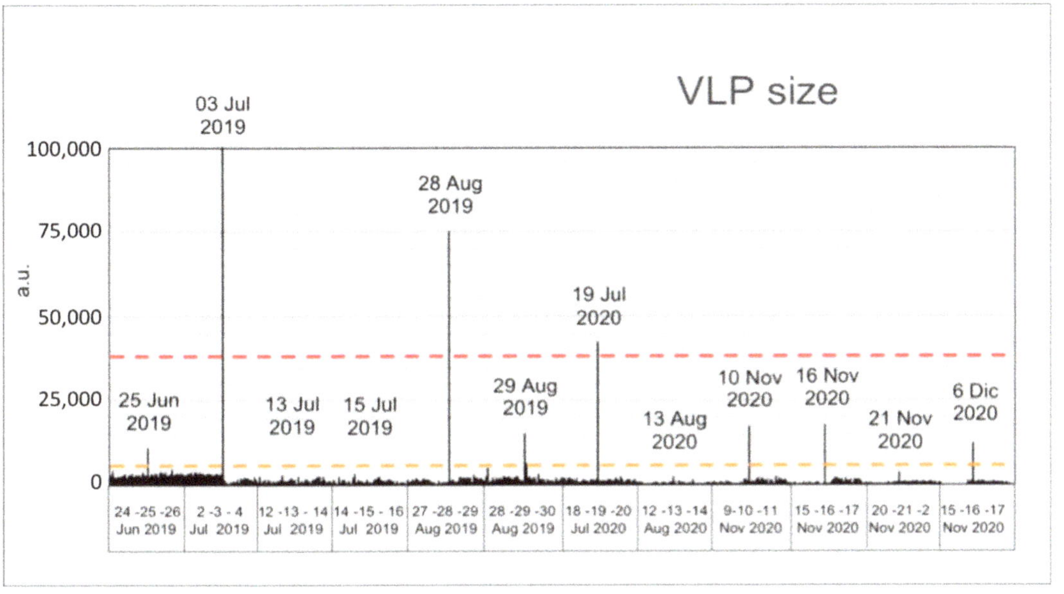

Figure 18. VLP size calculated in twelve 24-h intervals, each containing one of the twelve explosions selected as case studies. The histogram is normalized to 100,000 in arbitrary units (a.u.). The dashed orange line indicates the VLP size threshold, which separates the ordinary from the major explosions. The red dashed line indicates a possible VLP size threshold to separate major explosions from paroxysms (3 July 2019, 28 August 2019 and 19 July 2020). Note that this is not a time series in the whole considered period, but is representative only of the days reported in the abscissa axis. In this way, it is possible to observe how the VLP size of the most energetic explosive event far exceeds the VLP size of the "ordinary" explosions that occurred in the 24 h around it.

In order to approach the problem of classifying Stromboli explosions using the normalized VLP size, we chose two thresholds that separate the field of ordinary explosions from that of major ones, and the field of major explosions from paroxysms. The two thresholds are shown in Figure 18 as two dashed horizontal lines, orange and red, respectively.

The first threshold (ordinary versus major explosions) was obtained by adding 10% to the maximum value of VLP size of the ordinary explosions that fall within the analysed dataset (twelve 24-h long time intervals). Its value is 6,136 and is indicated by a horizontal dashed orange line in Figure 18. The second threshold was obtained by subtracting 10% from the VLP size of the smallest paroxysm considered among the case studies of our dataset, which is the 19 July 2020 event. The VLP size for this episode is 38,395. Applying this criterion to define the limits of the VLP size relative to ordinary, major and paroxysmal explosions, the events of 13 July 2019, 15 July 2019, 13 August 2020 and 21 November 2020 fall into the field of the ordinary explosions. This happens because their VLP size is smaller than that of the ordinary explosions recorded in other periods, for example when compared to the ordinary explosions that preceded the paroxysm of 3 July 2019 (Figure 18). The aforementioned explosions were classified as major as they are slightly larger than those on the day they occurred; therefore, they represented relative outliers.

This type of analysis, extended to longer periods, offers a fast and reliable way to define a relative quantitative scale of Stromboli explosion magnitude and can provide a criterion for the traditional distinction between ordinary, major and paroxysmal explosions which is historically linked to scenarios of possible impact on the island [12].

3.3. Borehole Geophysical Instruments (Tilt and Strain)

At Stromboli, the tiltmeters recorded signal changes during middle-term processes such as the attempt of a dike intrusion during the first months of 1995 [68] and the volcano deflation associated with the 2007 effusive eruption [86]. On the contrary, also due to the low sampling rate (1 data/minute), it is more difficult to detect clear changes associated with the impulsive and short-time events such as the explosions. However, even with few samples recorded during the explosive events, STDF can record small transient changes of short duration during the major explosions and the paroxysms (i.e., tilt variations in a few samples during the few minutes accompanying the explosion). Giudicepietro et al. [32] focused on the 1 December 2017 major explosion as a case study, and also reported the tilt data showing a very small variation of ~0.5×10^{-7} rad during the explosion. In the 2019-2020 interval considered in this study, STDF recorded tilt variations of similar magnitude during the major explosions occurring during the 10 and 16 Nov 2020 events, while no detectable tilt change was recorded for the other major explosions (Table 4). A further interesting aspect is that during the paroxysms of 3 July 2019 and 19 July 2020 (for the 28 August 2019 paroxysm, the station was out of order) the STDF tilt showed a bigger short-term transient of 0.5×10^{-6} rad (Table 4).

Table 4. Multi-parametric measurements of the 12 explosions considered as case studies useful for their classification. The duration of each event and its muzzle velocity are obtained from the SQV camera monitoring videos. Plume height (H) is considered above crater rim and measured from the fixed monitoring cameras or reported (where specified) from references. See text for further explanations.

Date Time (UT)	Muzzle Velocity (m s^{-1})	Plume H (m)	Duration (s)	VD Parameter	VLP Size (Normalized Counts)	Strain SVO (Counts)	Tilt STDF [x; y] (Micro-Strain)
25 June 2019 23:03:08	54.41	~500	8	435	11,276	~600 [1]	0, 0
3 July 2019 14:45:43	101.92	8400 [2]	160	16,307	100,000	~8000 [1]	~0.4; ~0.45
13 July 2019 20:33	29.63	110	12	356	3377	0[1]	0; 0
15 July 2019 19:09	61.63	380	18	1109	3909	0[1]	0; 0
28 August 2019 10:17:20	71.11	4000 [1] 6400 [2]	154	10,951	75,110	~8000 [1]	No data
29 August 2019 20:43:18	35.55	350	38	1351	15,437	~500 [1]	No data
19 July 2020 03:00:42	78.22	> 750	58	4537	42,661	~2000 [1]	~0.4; ~0.05
13 August 2020 14:50:27	11.86	550	64	759	2866	0[1]	0; 0
10 November 2020 20:04:21	54.50	600	20	1090	17,688	~300 [1]	~0.05; ~0.03
16 November 2020 09:17:45	54.51	1000 [1]	54	2944	18,006	~300 [1]	~0.05; ~0.04
21 November 2020 00:33:17	9.48	80	10	95	4029	0 [1]	0; 0
6 December 2020 05:12:44	11.52	300	8	92	12,778	0 [1]	0; 0

[1] The INGV monitoring reports can be found at www.ct.ingv.it. [2] Giordano & De Astis [41].

The SVO strainmeter, thanks to its high precision and to the high frequency sampling rate (50 Hz), is a powerful tool to detect transient changes associated with the explosive activity. These changes are extremely clear for the paroxysms, for which the instrument was able to detect signal variations both preceding these events by ~10 min and accompanying the explosive phase [31,52]. At SVO, the positive change (with positive change measuring compression in the rock surrounding the sensor) that preceded the explosive events by a few minutes ranged from ~8000 counts for the 3 July and 28 August 2019 events, to no change for the other smaller events. In Table 4, we report the positive strain change cumulated before the explosion as shown by the INGV weekly reports.

3.4. GBInSAR

The GBInSAR devices measured ground displacement associated with four events (Table 4): on 3 July 2019, 29 July 2019, 19 July 2020, and 10 November 2020, even though the behaviour was very different, suggesting different mechanisms and sources of deformation. The 3 July 2019 paroxysm was the only event to be preceded by a clear ground deformation detected from GBInSAR, consistent with an inflation of the summit crater terrace, which began about two and a half hours before, and which progressed until the explosion. Inflation was observed in the interferogram between 10:34 UTC and 12:13 UTC (displacement rate: 4.4 mm/h; Figure 19a), progressed in the successive interferograms (Figure 19b,c), reaching the maximum value of 44.2 mm/h 2 min before the explosion (interferogram between 14:36 UTC and 14:43 UTC (displacement rate: 42.2 mm/h; Figure 19d).

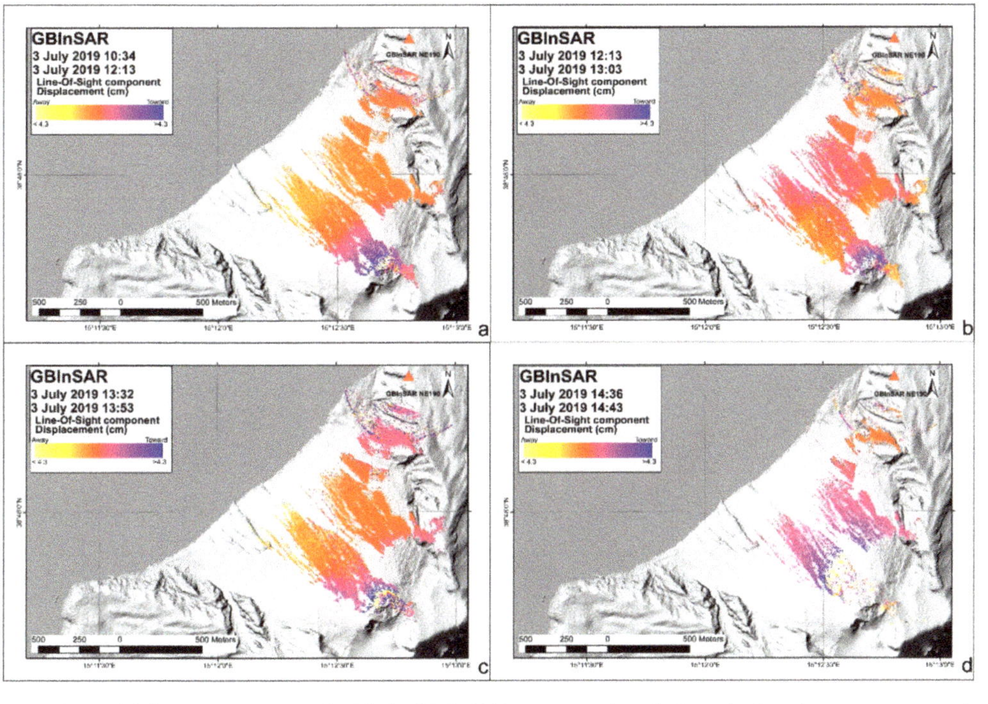

Figure 19. Ground deformation associated with the 3 July 2019 paroxysmal explosion. The interferogram, generated with GB-SAR NE190 system, revealed a progressive increase in the displacement rate, consistent with inflation of the crater terrace. (**a**) 3 July 2019 10:34-12:13 interferogram; (**b**) 3 July 2019 12:13-13:03 interferogram; (**c**) 3 July 2019 13:32-13:53 interferogram; (**d**) 3 July 2019 14:36-14:43 interferogram.

During the eruption that began on 3 July 2019 and ended on 30 August 2019, which was characterized by the outpouring of lava from the SW area of the crater terrace [33,80,81], Stromboli underwent some phases of strong explosive activity, with a number of strong Strombolian explosions (on 13 and 15 July 2019), the occurrence of another paroxysmal explosion (28 August 2019), and a major explosion on 29 August 2019. This last event was characterized by two lava fountaining explosive sequences during lava flow output (Tables 3 and 4 and Figure 8). The GBInSAR devices recorded a rapid ground movement towards the sensors (displacement rate: 17.8 mm/h; Figure 20a), followed by a long movement away from the sensors (max. displacement rate: 6.6 mm/h at 03:37 UTC; Figure 20b), compatible with an inflation-deflation cycle that began at 19:42 UTC on 29 August 2019 and ended at 06:04 UTC on 30 August 2019, coinciding with the end of the 2019 effusive eruption.

Afterwards, Stromboli was characterized by intense Strombolian activity, with some lava overflows from the crater terrace, as reported by Calvari et al. [44]. During this period, the radars recorded movements away from the sensors, compatible with the deflation of the crater terrace during some overflow events (31 March 2020; Figure 21).

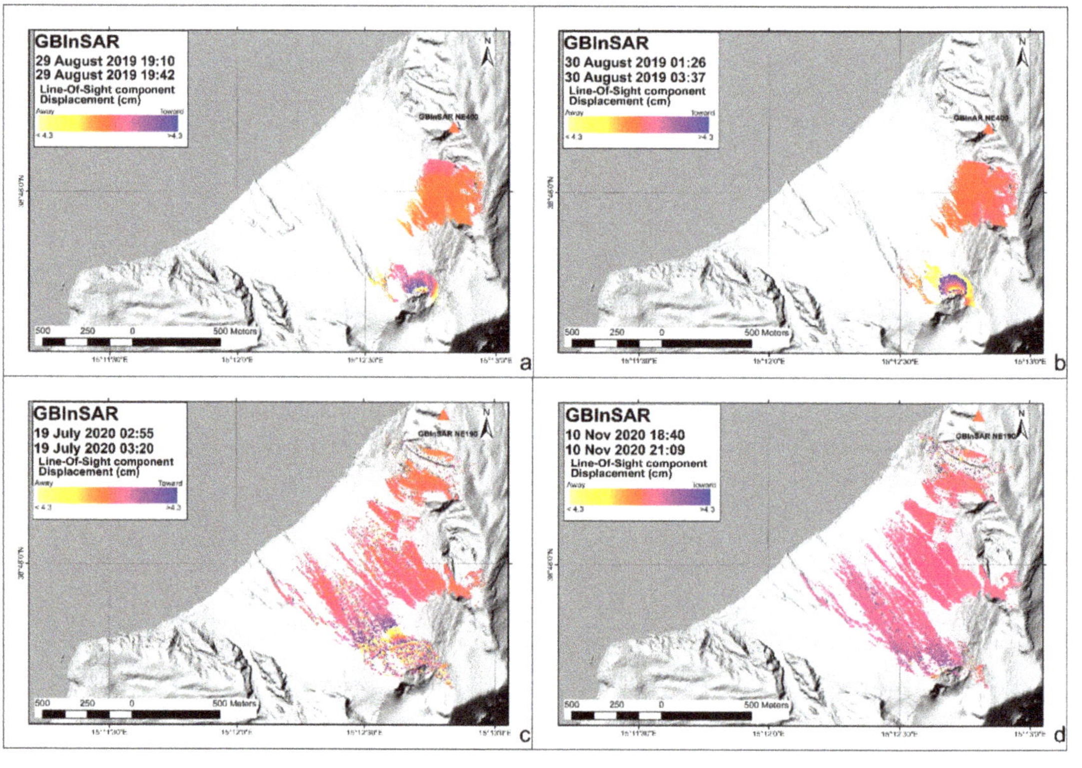

Figure 20. (**a**) 29 August 2019 (19:10-19:42) interferogram generated with GBInSAR NE400 system; (**b**) 30 August 2019 (01:26-03:37) interferogram generated with GBInSAR NE400 system; (**c**) 19 July 2020 (02:55-03:20) interferogram generated with GBInSAR NE190 system; (**d**) 10 November 2020 (18:40-21:09) interferogram generated with GBInSAR NE190 system.

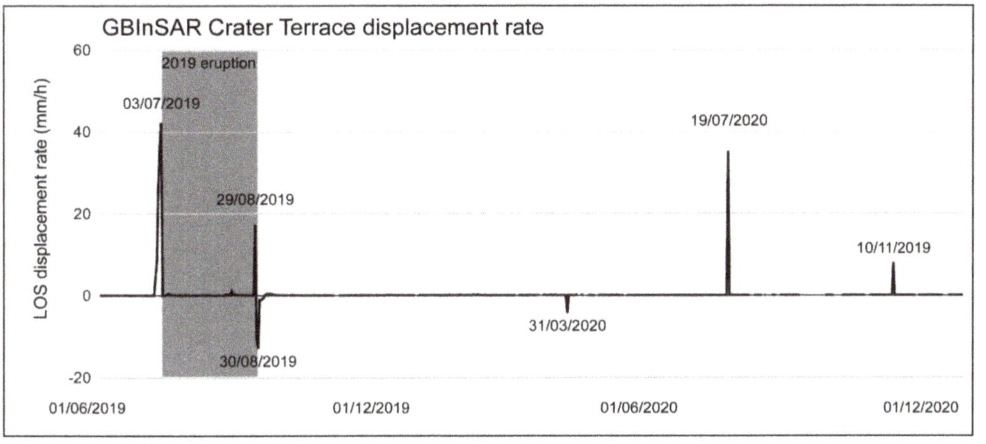

Figure 21. GBInSAR time series, with the 2019 effusive eruption in grey. It can be noted that only 3 explosive events were associated with displacement rate increases toward the sensors (3 July 2019, 19 July 2020, 10 November 2020). The increase in displacement rate, first toward then moving away from the sensors, recorded between 29 August 2019 and 30 August 2019 is associated with a more intense lava flow at the end of the 2019 eruption (inflation-deflation cycle). The signal with displacement rate increases away from the sensor recorded on 31 March 2020 was associated with overflows from the summit craters [44].

The two GBInSAR devices recorded similar deformations, albeit of different magnitudes, during the explosions of 19 July 2020 and 10 November 2020. The former event was associated with a syn-explosive displacement rate of 35.2 mm/h, whereas the latter was associated with a syn-explosive displacement rate of 8 mm/h. No deformation prior to the explosive events was recorded, suggesting that the magma volume involved by the two episodes was not large or not detectable.

4. Discussion

In the last 140 years, more than 180 high intensity explosive events have been recorded or reported at Stromboli [87], and of these only 44 events have occurred since 2003, when an integrated multiparameter monitoring system was installed on the island [88]. Of the latter, four are rated as paroxysmal explosions and 40 as major explosions [87]. In order to propose a classification scheme, the analysis of our study comprised the 12 explosive events occurring at Stromboli between 25 June 2019 and 6 December 2020. Initially, these 12 explosions were classified as two being paroxysmal (the 3 July 2019 and 28 August 2019) and 10 as major explosions.

In principle, a classification is robust if it is well related to the eruptive mechanisms and characteristics of the explosive events. To this end, we considered both the eruptive style (gas-pyroclast exit speed or muzzle velocity, event duration, height of the eruptive column, impacts, secondary effects), several associated geophysical parameters during the explosion (the magnitude of the VLP size and the tilt changes), as well as the occurrence of a precursor ground deformation signal (dilatometer, GBInSAR).

Among the physical features, an important parameter that can be used to distinguish different sizes of explosions is the height of the eruptive column [82], i.e., the height of the ash plume rising by buoyancy from the crater rim. This height is determined by the intensity of the explosion and thus by the erupted volume [56], and as such is a key discriminant between events of different magnitude and intensity. However, the INGV monitoring cameras available at the moment have a maximum field of view covering a maximum height of ~750 m above the craters, and thus cannot be used to measure the size of the eruptive columns for all paroxysms (H > 1 km) and of the strongest major explosions.

This is why, after considering the maximum height of the eruptive column, we used the muzzle velocity at the vent, considering that the most powerful explosions should also have the highest muzzle velocities, this being a function of the pressure of the expanding gases [55,82,89]. In doing so, we have used the images of the SQV monitoring camera (Table 1), which is the only one that recorded all the 12 explosive episodes analysed here. The muzzle velocity is normally attained at the very start of each explosion. An exception to this statement is the 3 July 2019 event, which started with very low speed and reached the peak velocity of ~102 m s^{-1} (Table 4) after ~6 s from the start of the main blast. This was probably caused by the degassed lava contained in the highest portion of the shallow conduit, that was pushed upwards by the gas-rich magma and erupted as lava flows spreading within the crater just before the paroxysm (Figure 4). An additional parameter related to the size of the explosive event is the duration, that can be obtained from the analysis of the videos recorded by the INGV monitoring cameras or by the seismic trace. It is worth mentioning that sometimes it could be difficult to determine this duration, such as in the case of lava fountaining (see for example Figure 8c), where the seismic signal does not show a clear end. It is less difficult from the camera images, although some problems may arise when PDC spreading along the SdF may limit sight. In our investigation the VD parameter, obtained by the multiplication of the muzzle velocities by the event durations (V × D), is well-suited to represent the power of explosive activity.

Among the geophysical information considered in this study, the VLP size, as defined by Giudicepietro et al. [33], is the chief parameter allowing us to distinguish between explosive events of different magnitude and intensity. This parameter is sufficient by itself to characterize the power of an explosive event and, moreover, has a clear volcanological correspondence with the VD parameter, as attested by the relationship shown in the graph of Figure 22. This relationship means that we can use any of them or even better a combination of them to classify explosive events at Stromboli.

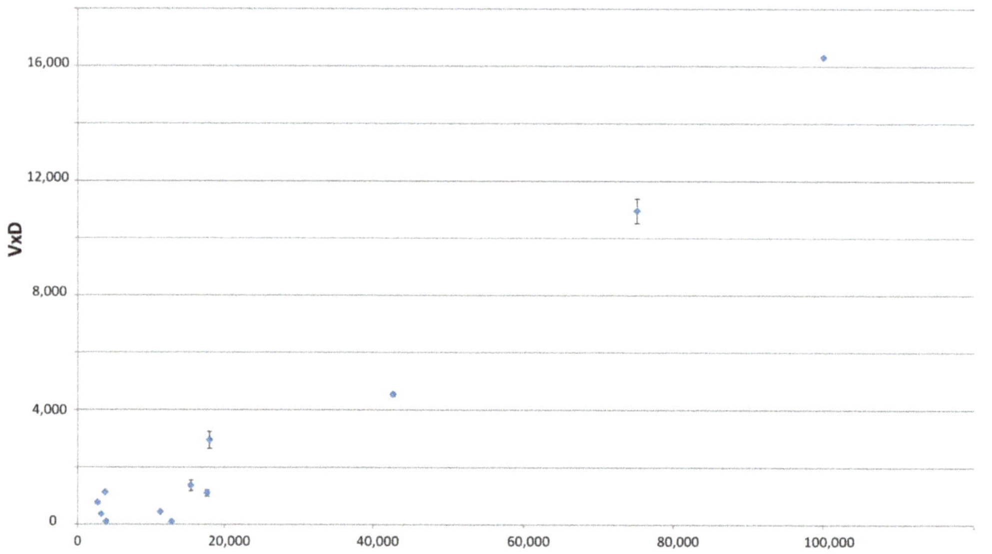

Figure 22. Graph reporting the VLP size vs. the volcanology parameters (V × D) obtained multiplying the muzzle velocity (V, in m s^{-1}) of each event by the duration (D, in s). See Table 4 for the list of values.

This interpretative tool provided by the VLP size and by the VD parameter is further supported by the geophysical measurements recorded by the borehole dilatometer and tiltmeter, which for each type of explosive event or class are characterized by a specific range of values (Table 5). A complete and useful representation of the integrate classification scheme is shown in Figure 23.

Table 5. Range of the main parameters useful to classify the explosive events of Stromboli.

Classification Scheme of Strombolian Explosions				
Explosion Class	**0**	**1**	**2**	**3**
Local explosion classification	Ordinary	Intermediate	Major	Paroxysmal
Effect/dispersal area	Crater terrace	Crater terrace/rarely top of the volcano	Top of the volcano/rarely island sectors	Island sectors/rarely more distal areas (other islands or the surrounding coasts)
Jet/plume height (m)	<100	100–300	300–1000	>1000
Duration (s)	<20	20–30	30–100	>100
Max. speed (m/s)	<10	10–30	30–70	>70
VLP size	<2000	2000–12,000	12,000–18,000	> 18,000
VD parameter	<90	90–1000	1000–4000	>4000
Tilt SVO (microstrain)	~0	~0	~0.05 × 10^{-6}	~0.5 × 10^{-6}
Strain STDF (counts)	~0	~0	5000–1000	2000–10,000

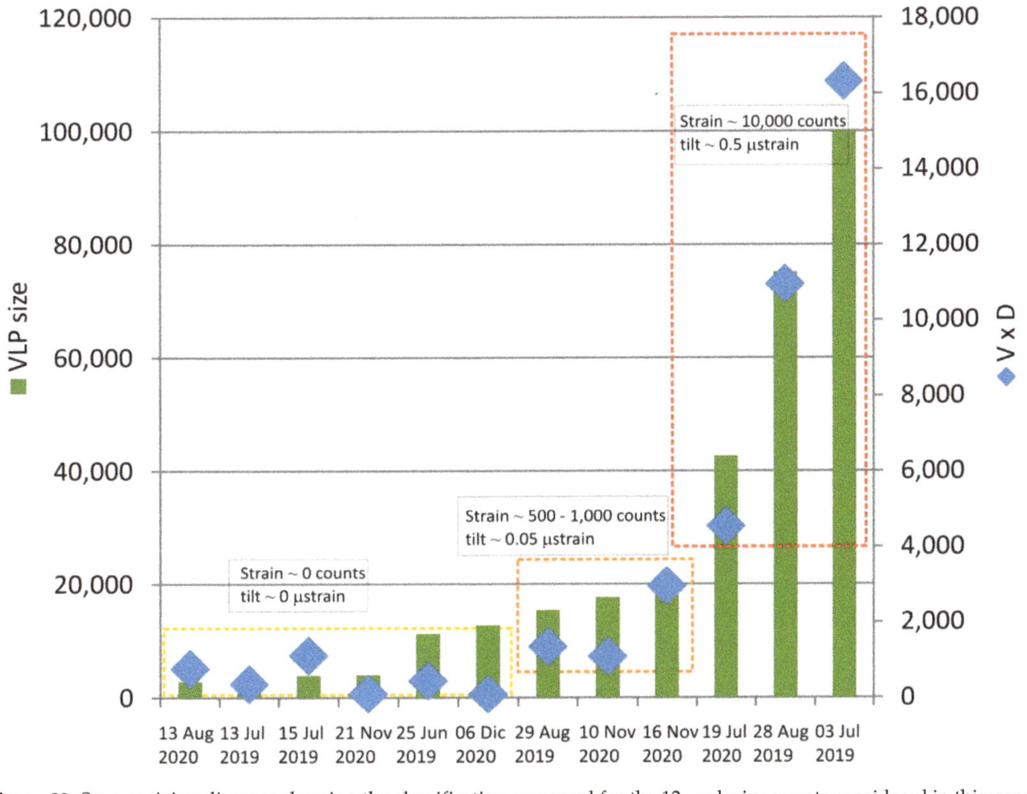

Figure 23. Summarizing diagram showing the classification proposed for the 12 explosive events considered in this paper on the basis of VLP size (left axis), and muzzle velocity by duration (right axis). See text for further explanation.

It is immediately clear that in this scheme, the explosions of 3 July 2019 and 28 August 2019, defined as paroxysmal explosions, have some characteristics in common (i.e., eruptive plume heights of several km, formation of PDCs along the SdF; high VLP size and strain), albeit with some differences. However, the classification proposed here also raises the explosive event of 19 July 2020 to the rank of paroxysmal explosion, being characterized by a high VLP size (42,661, Table 4) and a high VD parameter (4,537; Table 4, Table 5 and Figure 23). The effects of this event affected the summit of the volcano above 500 m elevation (Table 3) and influenced the NW sector of the island. Fortunately, the material erupted during the explosion fell on the SdF, thus not involving inhabited areas. At the same time, through this classification, we can exclude some events from the list of major explosions (Figure 23), which are therefore to be found in the set of "intermediate" events (13 August 2020, 13 July 2019, 15 July 2019, 25 June 2019, 21 November 2020 and 6 December 2020). These explosions had little or no effects on the summit area of the volcano (i.e., the one where tourists stop to observe the explosive activity). On the contrary, the proper major explosions (29 August 2019, 10 November 2020, and 16 November 2020; Figure 23) had important impacts in terms of ballistic blocks, spatter bombs, and tephra fall on the summit area of the volcano.

5. Conclusions

The Strombolian activity of Stromboli volcano was analysed, combining different data from monitoring cameras, seismic network, and ground deformations obtained from different remote sensing and geodetic techniques, in order to obtain a new classification scheme for different explosion intensities.

Considering the distinct sampling frequencies, the best parameters to classify these low but different intensity transient events are the VLP size and the VD parameter. The former is the maximum value of the RSAM of a 30-s sliding window that moves in a 30-min time interval of signal, filtered in VLP frequency band (0.05–0.5 Hz), whereas the latter is the product of the muzzle velocity and the explosion duration, both derived from the analysis of the monitoring camera images. These parameters are independent of each other and thus can provide the intensity of the event even in absence of the other parameter. The classification scheme identified by these two main parameters is further supported by the indication of the range of values of the dilatometer and tiltmeter recorded during different types of events (Tables 4 and 5). This work demonstrates the importance of multi-parametric monitoring systems as an objective approach towards characterizing events of varying intensity, in the context of the same eruptive style.

Author Contributions: Conceptualization, S.C., F.G. and F.D.T.; methodology, S.C., F.G., F.D.T., G.M. and A.B.; validation, S.C., F.G., F.D.T., G.M. and A.B.; formal analysis, S.C., F.G., F.D.T., G.M. and A.B.; investigation, S.C., F.G., F.D.T., G.M. and A.B.; resources, S.C. and N.C.; data curation, S.C., F.G., F.D.T., G.M. and A.B.; writing—original draft preparation, S.C., F.G., F.D.T., G.M. and A.B.; writing—review and editing, S.C., F.G., F.D.T., G.M. and A.B.; visualization, S.C., F.G., F.D.T., G.M. and A.B.; supervision, S.C., F.G., F.D.T., G.M. and A.B.; project administration, S.C. and N.C.; funding acquisition, S.C. and N.C. All authors have read and agreed to the published version of the manuscript.

Funding: This research was funded by the Project FIRST-ForecastIng eRuptive activity at Stromboli volcano: Timing, eruptive style, size, intensity, and duration, INGV-Progetto Strategico Dipartimento Vulcani 2019, (Delibera n. 144/2020; Scientific Responsibility: S.C.). This work was financially supported by the "Presidenza del Consiglio dei Ministri—Dipartimento della Protezione Civile" (Presidency of the Council of Ministers–Department of Civil Protection) (DPC-UNIFI Agreement 2019–2021; Scientific Responsibility: N.C.). The contents of this paper represent the authors' ideas and do not necessarily correspond to the official opinion and policies of the "Presidenza del Consiglio dei Ministri–Dipartimento della Protezione Civile" (Presidency of the Council of Ministers–Department of Civil Protection).

Acknowledgments: We would like to thank the Commander of the Guardia Costiera di Catania M. Privitera for making available the video recorded during the 3 July 2019 afternoon, and the tourists for their videos of the 3 July 2019 and 28 August 2019 events diffused in the internet, because these all furnished important data for the study of these episodes. We would like to acknowledge the essential work of INGV, UNIFI, and Ellegi srl LiSALab technicians for the installation and maintenance of the monitoring networks on Stromboli. We would like to thank Stephan Conway for the correction of the English style.

Conflicts of Interest: The authors declare no conflict of interest.

References

1. Parfitt, E.A. A discussion of the mechanisms of explosive basaltic eruptions. *J. Volcanol. Geotherm. Res.* **2004**, *134*, 77–107. [CrossRef]
2. Harris, A.; Ripepe, M. Temperature and dynamics of degassing at Stromboli. *J. Geophys. Res.* **2007**, *112*, B03205. [CrossRef]
3. Patrick, M.R. Dynamics of Strombolian ash plumes from thermal video: Motion, morphology, and air entrainment. *J. Geophys. Res.* **2007**, *112*, B06202. [CrossRef]
4. Patrick, M.R.; Harris, A.J.L.; Ripepe, M.; Dehn, J.; Rothery, D.A.; Calvari, S. Strombolian explosive styles and source conditions: Insights from thermal (FLIR) video. *Bull. Volcanol.* **2007**, *69*, 769–784. [CrossRef]
5. Leduc, L.; Gurioli, L.; Harris, A.; Colò, L.; Rose-Koga, E.F. Types and mechanisms of strombolian explosions: Characterization of a gas-dominated explosion at Stromboli. *Bull. Volcanol.* **2015**, *77*, 8. [CrossRef]
6. Taddeucci, J.; Edmonds, M.; Houghton, B.; James, M.R.; Vergniolle, S. Hawaiian and Strombolian eruptions. In *The Encyclopedia of Volcanoes*; Elsevier: San Diego, CA, USA, 2015; pp. 485–503.
7. Chouet, B.; Hamisevicz, N.; McGetchin, T.R. Photoballistics of volcanic jet activity at Stromboli, Italy. *J. Geophys. Res.* **1974**, *79*, 4961–4976. [CrossRef]
8. Gurioli, L.; Harris, A.J.L.; Colò, L.; Bernard, J.; Favalli, M.; Ripepe, M.; Andronico, D. Classification, landing distribution, and associated flight parameters for a bomb field emplaced during a single major explosion at Stromboli, Italy. *Geology* **2013**, *41*, 559–562. [CrossRef]
9. Taddeucci, J.; Alatorre-Ibargüengoitia, M.A.; Cruz-Vázquez, O.; Del Bello, E.; Scarlato, P.; Ricci, T. In-flight dynamics of volcanic ballistic projectiles. *Rev. Geophys.* **2017**, *55*, 675–718. [CrossRef]
10. Romagnoli, C. Characteristics and morphological evolution of the Aeolian volcanoes from the study of submarine portions. *Geol. Soc. Lond. Mem.* **2013**, *37*, 13–26. [CrossRef]
11. Taddeucci, J.; Palladino, D.M.; Sottili, G.; Bernini, D.; Andronico, D.; Cristaldi, A. Linked frequency and intensity of persistent volcanic activity at Stromboli (Italy). *Geophys. Res. Lett.* **2013**, *40*, 3384–3388. [CrossRef]
12. Barberi, F.; Rosi, M.; Sodi, A. Volcanic hazard assessment at Stromboli based on review of historical data. *Acta Vulcanol.* **1993**, *3*, 173–187.
13. Calvari, S.; Bonaccorso, A.; Madonia, P.; Neri, M.; Liuzzo, M.; Salerno, G.G.; Behncke, B.; Caltabiano, T.; Cristaldi, A.; Giuffrida, G.; et al. Major eruptive style changes induced by structural modifications of a shallow conduit system: The 2007–2012 Stromboli case. *Bull. Volcanol.* **2014**, *76*, 841. [CrossRef]
14. Valade, S.; Lacanna, G.; Coppola, D.; Laiolo, M.; Pistolesi, M.; Delle Donne, D.; Genco, R.; Marchetti, E.; Ulivieri, G.; Allocca, C.; et al. Tracking dynamics of magma migration in open-conduit systems. *Bull. Volcanol.* **2016**, *78*, 78. [CrossRef]
15. Aiuppa, A.; de Moor, J.M.; Arellano, S.; Coppola, D.; Francofonte, V.; Galle, B.; Giudice, G.; Liuzzo, M.; Mendoza, E.; Saballos, A.; et al. Tracking formation of a lava lake from ground and space: Masaya volcano (Nicaragua), 2014–2017. *Geochem. Geophys. Geosyst.* **2018**, *19*, 496–515. [CrossRef]
16. Albert, H.; Costa, F.; Di Muro, A.; Herrin, J.; Métrich, N.; Deloule, E. Magma interactions, crystal mush formation, timescales, and unrest during caldera collapse and lateral eruption at ocean island basaltic volcanoes (Piton de la Fournaise, La Réunion). *Earth Plan. Sci. Lett.* **2019**, *515*, 187–199. [CrossRef]
17. Albino, F.; Biggs, J.; Escobar-Wolf, R.; Naismith, A.; Watson, M.; Phillips, J.C.; Chigna Marroquin, G.A. Using TanDEM-X to measure pyroclastic flow source location, thickness and volume: Application to the 3rd June 2018 eruption of Fuego volcano, Guatemala. *J. Volcanol. Geotherm. Res.* **2020**, *406*, 107063. [CrossRef]
18. Battaglia, J.; Métaxian, J.-P.; Garaebiti, E. Short term precursors of Strombolian explosions at Yasur volcano (Vanuatu). *Geophys. Res. Lett.* **2016**, *43*, 1960–1965. [CrossRef]
19. Houghton, B.F.; Taddeucci, J.; Andronico, D.; Gonnermann, H.M.; Pistolesi, M.; Patrick, M.R.; Orr, T.R.; Swanson, D.A.; Edmonds, M.; Gaudin, D.; et al. Stronger or longer: Discriminating between Hawaiian and Strombolian eruption styles. *Geology* **2016**, *44*, 163–166. [CrossRef]
20. Rüdiger, J.; Tirpitz, J.-L.; Maarten de Moor, J.; Bobrowski, N.; Gutmann, A.; Liuzzo, M.; Ibarra, M.; Hoffmann, T. Implementation of electrochemical, optical and denuder-based sensors and sampling techniques on UAV for volcanic gas measurements: Examples from Masaya, Turrialba and Stromboli volcanoes. *Atmos. Meas. Tech.* **2018**, *11*, 2441–2457. [CrossRef]
21. Spampinato, L.; Ganci, G.; Hernandez, P.A.; Calvo, D.; Tedesco, D.; Perez, N.M.; Calvari, S.; Del Negro, C.; Yalire, M.M. Thermal insights into the dynamics of Nyiragongo lava lake from ground and satellite measurements. *J. Geophys. Res. Solid Earth* **2013**, *118*, 5771–5784. [CrossRef]
22. Spampinato, L.; Oppenheimer, C.; Cannata, A.; Montalto, P.; Salerno, G.G.; Calvari, S. On the time-scale of thermal cycles associated with open-vent degassing. *Bull. Volcanol.* **2012**, *74*, 1281–1292. [CrossRef]

23. Vergniolle, S.; Caplan-Auerbach, J. Basaltic thermals and Subplinian plumes: Constraints from acoustic measurements at Shishaldin volcano, Alaska. *Bull. Volcanol.* **2006**, *68*, 611–630. [CrossRef]
24. Woitischek, J.; Woods, A.W.; Edmonds, M.; Oppenheimer, C.; Aiuppa, A.; Pering, T.D.; Ilanko, T.; D'Aleo, R.; Garaebiti, E. Strombolian eruptions and dynamics of magma degassing at Yasur Volcano (Vanuatu). *J. Volcanol. Geotherm. Res.* **2020**, *398*, 106869. [CrossRef]
25. Pioli, L.; Pistolesi, M.; Rosi, M. Transient explosions at open-vent volcanoes: The case of Stromboli (Italy). *Geology* **2014**, *42*, 863–866. [CrossRef]
26. Andronico, D.; Corsaro, R.A.; Cristaldi, A.; Polacci, M. Characterizing high energy explosive eruptions at Stromboli volcano using multidisciplinary data: An example from the 9 January 2005 explosion. *J. Volcanol. Geotherm. Res.* **2008**, *176*, 541–550. [CrossRef]
27. Calvari, S.; Büttner, R.; Cristaldi, A.; Dellino, P.; Giudicepietro, F.; Orazi, M.; Peluso, R.; Spampinato, L.; Zimanowski, B.; Boschi, E. The 7 September 2008 Vulcanian explosion at Stromboli volcano: Multi-parametric characterisation of the event and quantification of the ejecta. *J. Geophys. Res. Solid Earth* **2012**, *117*, B05201. [CrossRef]
28. Rittmann, A. Der ausbruch des Stromboli am 11 September 1930. *Z. Vulkanol.* **1931**, *14*, 47–77.
29. Calvari, S.; Spampinato, L.; Lodato, L. The 5 April 2003 vulcanian paroxysmal explosion at Stromboli volcano (Italy) from field observations and thermal data. *J. Volcanol. Geotherm. Res.* **2006**, *149*, 160–175. [CrossRef]
30. Calvari, S.; Lodato, L.; Steffke, A.; Cristaldi, A.; Harris, A.J.L.; Spampinato, L.; Boschi, E. The 2007 Stromboli flank eruption: Chronology of the events, and effusion rate measurements from thermal images and satellite data. *J. Geophys. Res. Solid Earth* **2010**, *115*, B04201. [CrossRef]
31. Bonaccorso, A.; Calvari, S.; Linde, A.; Sacks, S.; Boschi, E. Dynamics of the shallow plumbing system investigated from borehole strainmeters and cameras during the 15 March 2007 Vulcanian paroxysm at Stromboli volcano. *Earth Plan. Sci. Lett.* **2012**, *357–358*, 249–256. [CrossRef]
32. Giudicepietro, F.; Calvari, S.; Alparone, S.; Bianco, F.; Bonaccorso, A.; Bruno, V.; Caputo, T.; Cristaldi, A.; D'Auria, L.; De Cesare, W.; et al. Integration of Ground-Based Remote-Sensing and In Situ Multidisciplinary Monitoring Data to Analyze the Eruptive Activity of Stromboli Volcano in 2017–2018. *Remote Sens.* **2019**, *11*, 1813. [CrossRef]
33. Giudicepietro, F.; Lopez, C.; Macedonio, G.; Alparone, S.; Bianco, F.; Calvari, S.; De Cesare, W.; Delle Donne, D.; Di Lieto, B.; Esposito, A.M.; et al. Geophysical precursors of the July-August 2019 paroxysmal eruptive phase and their implications for Stromboli volcano (Italy) monitoring. *Sci. Rep.* **2020**, *10*, 10296. [CrossRef]
34. Pistolesi, M.; Delle Donne, D.; Pioli, L.; Rosi, M.; Ripepe, M. The 15 March 2007 explosive crisis at Stromboli volcano, Italy: Assessing physical parameters through a multidisciplinary approach. *J. Geophys. Res. Solid Earth* **2011**, *116*, B12206. [CrossRef]
35. Pistolesi, M.; Rosi, M.; Pioli, L.; Renzulli, A.; Bertagnini, A.; Andronico, D. The paroxysmal event and its deposits. In *The Stromboli Volcano: An Integrated Study of the 2002–2003 Eruption*; Calvari, S., Inguaggiato, S., Puglisi, G., Ripepe, M., Rosi, M., Eds.; Geophysical Monograph Series; AGU: Washington, DC, USA, 2008; Volume 182, pp. 317–330.
36. Rosi, M.; Bertagnini, A.; Landi, P. Onset of the persistent activity at Stromboli Volcano (Italy). *Bull. Volcanol.* **2000**, *62*, 294–300. [CrossRef]
37. Lautze, N.C.; Houghton, B.F. Physical mingling of magma and complex eruption dynamics in the shallow conduit at Stromboli volcano, Italy. *Geology* **2005**, *33*, 425–428. [CrossRef]
38. Métrich, N.; Bertagnini, A.; Landi, P.; Rosi, M.; Belhadj, O. Triggering mechanism at the origin of paroxysms at Stromboli (Aeolian Archipelago, Italy): The 5 April 2003 eruption. *Geophys. Res. Lett.* **2005**, *32*, L10305. [CrossRef]
39. D'Oriano, C.; Bertagnini, A.; Pompilio, M. Ash erupted during normal activity at Stromboli (Aeolian Islands, Italy) raises questions on how the feeding system works. *Bull. Volcanol.* **2011**, *73*, 471–477. [CrossRef]
40. Bertagnini, A.; Di Roberto, A.; Pompilio, M. Paroxysmal activity at Stromboli: Lessons from the past. *Bull. Volcanol.* **2011**, *73*, 1229–1243. [CrossRef]
41. Giordano, G.; De Astis, G. The summer 2019 basaltic Vulcanian eruptions (paroxysms) of Stromboli. *Bull. Volcanol.* **2021**, *83*, 1. [CrossRef]
42. Calvari, S.; Spampinato, L.; Lodato, L.; Harris, A.J.L.; Patrick, M.R.; Dehn, J.; Burton, M.R.; Andronico, D. Chronology and complex volcanic processes during the 2002–2003 flank eruption at Stromboli volcano (Italy) reconstructed from direct observations and surveys with a handheld thermal camera. *J. Geophys. Res. Solid Earth* **2005**, *110*, B02201. [CrossRef]
43. Calvari, S.; Spampinato, L.; Bonaccorso, A.; Oppenheimer, C.; Rivalta, E.; Boschi, E. Lava effusion—A slow fuse for paroxysms at Stromboli volcano? *Earth Plan. Sci. Lett.* **2011**, *301*, 317–323. [CrossRef]
44. Calvari, S.; Di Traglia, F.; Ganci, G.; Giudicepietro, F.; Macedonio, G.; Cappello, A.; Nolesini, T.; Pecora, E.; Bilotta, G.; Centorrino, V.; et al. Overflows and Pyroclastic Density Currents in March-April 2020 at Stromboli Volcano Detected by Remote Sensing and Seismic Monitoring Data. *Remote Sens.* **2020**, *12*, 3010. [CrossRef]
45. Di Traglia, F.; Nolesini, T.; Intrieri, E.; Mugnai, F.; Leva, D.; Rosi, M.; Casagli, N. Review of ten years of volcano deformations recorded by the ground-based InSAR monitoring system at Stromboli volcano: A tool to mitigate volcano flank dynamics and intense volcanic activity. *Earth Sci. Rev.* **2014**, *139*, 317–335. [CrossRef]
46. Di Roberto, A.; Bertagnini, A.; Pompilio, M.; Bisson, M. Pyroclastic density currents at Stromboli volcano (Aeolian Islands, Italy): A case study of the 1930 eruption. *Bull. Volcanol.* **2014**, *76*, 827. [CrossRef]

47. Lodato, L.; Spampinato, L.; Harris, A.J.L.; Calvari, S.; Dehn, J.; Patrick, M. The Morphology and Evolution of the Stromboli 2002-03 Lava Flow Field: An Example of Basaltic Flow Field Emplaced on a Steep Slope. *Bull. Volcanol.* **2007**, *69*, 661–679. [CrossRef]
48. Pioli, L.; Rosi, M.; Calvari, S.; Spampinato, L.; Renzulli, A.; Di Roberto, A. The eruptive activity of 28 and 29 December 2002. In *The Stromboli Volcano: An Integrated Study of the 2002–2003 Eruption*; American Geophysical Union, Geophysical Monograph Series; Calvari, S., Inguaggiato, S., Puglisi, G., Ripepe, M., Rosi, M., Eds.; AGU: Washington, DC, USA, 2008; Volume 182, pp. 105–116. [CrossRef]
49. Di Traglia, F.; Calvari, S.; D'Auria, L.; Nolesini, T.; Bonaccorso, A.; Fornaciai, A.; Esposito, A.; Cristaldi, A.; Favalli, M.; Casagli, N. The 2014 effusive eruption at Stromboli: New insights from in-situ and remote sensing measurements. *Remote Sens.* **2018**, *10*, 2035. [CrossRef]
50. Bonaccorso, A.; Calvari, S.; Garfì, G.; Lodato, L.; Patané, D. December 2002 flank failure and tsunami at Stromboli volcano inferred by volcanological and geophysical observations. *Geophys. Res. Lett.* **2003**, *30*, 1941–1944. [CrossRef]
51. D'Auria, L.; Giudicepietro, F.; Martini, M.; Peluso, R. Seismological insight into the kinematics of the 5 April 2003 vulcanian explosion at Stromboli volcano (southern Italy). *Geophys. Res. Lett.* **2006**, *33*, L08308. [CrossRef]
52. Di Lieto, B.; Romano, P.; Scarpa, R.; Linde, A.T. Strain signals before and during paroxysmal activity at Stromboli volcano, Italy. *Geophys. Res. Lett.* **2020**, *47*, e2020GL088521. [CrossRef]
53. Landi, P.; Métrich, N.; Bertagnini, A.; Rosi, M. Recycling and "re-hydration" of degassed magma inducing transient dissolution/crystallization events at Stromboli (Italy). *J. Volcanol. Geotherm. Res.* **2008**, *174*, 325–336. [CrossRef]
54. Morrissey, M.M.; Mastin, L.G. Vulcanian Eruptions. In *Encyclopedia of Volcanoes*; Academic Press: San Diego, CA, USA, 1999; pp. 463–475.
55. Walker, G.P.L. Explosive volcanic eruptions—A new classification scheme. *Geol. Rudsch.* **1973**, *62*, 431–446. [CrossRef]
56. Sparks, R.S.J.; Bursik, M.I.; Carey, S.N.; Gilbert, J.S.; Glaze, L.S.; Sigurdsson, H.; Woods, A.V. *Volcanic Plumes*; John Wiley & Sons: Chichester, UK, 1997.
57. Calvari, S.; Cannavò, F.; Bonaccorso, A.; Spampinato, L.; Pellegrino, A.G. Paroxysmal Explosions, Lava Fountains and Ash Plumes at Etna Volcano: Eruptive Processes and Hazard Implications. *Front. Earth Sci.* **2018**, *6*, 107. [CrossRef]
58. Bonadonna, C.; Pistolesi, M.; Cioni, R.; Degruyter, W.; Elissondo, M.; Baumann, V. Dynamics of wind-affected volcanic plumes: The example of the 2011 Cordón Caulle eruption, Chile. *J. Geophys. Res.* **2015**, *120*, 1–20. [CrossRef]
59. Gaudin, D.; Taddeucci, J.; Scarlato, P.; Del Bello, E.; Ricci, T.; Orr, T.; Houghton, B.; Harris, A.; Rao, S.; Bucci, A. Integrating puffing and explosions in a general scheme for Strombolian-style activity. *J. Geophys. Res. Solid Earth* **2017**, *122*, 1860–1875. [CrossRef]
60. Newhall, C.G.; Self, S. The Volcanic Explosivity Index (VEI): An Estimate of Explosive Magnitude for Historical Volcanism. *J. Geophys. Res.* **1982**, *87*, 1231–1238. [CrossRef]
61. Houghton, B.F.; Swanson, D.A.; Rausch, J.; Carey, R.J.; Fagents, S.A.; Orr, T.R. Pushing the Volcanic Explosivity Index to its limit and beyond: Constraints from exceptionally weak explosive eruptions at Kīlauea in 2008. *Geology* **2013**, *41*, 627–630. [CrossRef]
62. Zanon, V.; Neri, M.; Pecora, E. Interpretation of data from the monitoring thermal camera of Stromboli volcano (Aeolian Islands, Italy). *Geol. Mag.* **2009**, *146*, 591–601. [CrossRef]
63. Salvatore, V.; Silleni, A.; Corneli, D.; Taddeucci, J.; Palladino, D.M.; Sottili, G.; Bernini, D.; Andronico, D.; Cristaldi, A. Parameterizing multi-vent activity at Stromboli Volcano (Aeolian Islands, Italy). *Bull. Volcanol.* **2018**, *80*, 64. [CrossRef]
64. Schmid, M.; Kueppers, U.; Cigala, V.; Sesterhenn, J.; Dingwell, D.B. Release characteristics of overpressurised gas from complex vents: Implications for volcanic hazards. *Bull. Volcanol.* **2020**, *82*, 68. [CrossRef]
65. Mills, D.; Martin, J.; Burbank, J.; Kasch, W. Network Time Protocol Version 4: Protocol and Algorithms Specification. RFC 5905. June 2010. Available online: https://www.rfc-editor.org/info/rfc5905 (accessed on 20 January 2021). [CrossRef]
66. De Cesare, W.; Orazi, M.; Peluso, R.; Scarpato, G.; Caputo, A.; D'Auria, L.; Giudicepietro, F.; Martini, M.; Buonocunto, C.; Capello, M.; et al. The Broadband Seismic Network of Stromboli Volcano, Italy. *Seismol. Res. Lett.* **2009**, *80*, 435–439. [CrossRef]
67. Orazi, M.; Martini, M.; Peluso, R. Data acquisition for volcano monitoring. *Eos Trans. Am. Geophys. Union* **2006**, *87*, 385–392. [CrossRef]
68. Bonaccorso, A. Evidence of a dyke-sheet intrusion at Stromboli volcano inferred through continuous tilt. *Geophys. Res. Lett.* **1998**, *25*, 4225–4228. [CrossRef]
69. Bonaccorso, A.; Falzone, G.; Gambino, S. An investigation into shallow borehole tiltmeters. *Geophys. Res. Lett.* **1999**, *26*, 1637–1640. [CrossRef]
70. Gambino, S.; Falzone, G.; Ferro, A.; Laudani, G. Volcanic processes detected by tiltmeters: A review of experience on Sicilian volcanoes. *J. Volcanol. Geotherm. Res.* **2014**, *271*, 43–54. [CrossRef]
71. Sacks, I.S.; Suyehiro, S.; Evertson, D.W. Sacks-Evertson strainmeter, its installation in Japan and some preliminary results concerning strain steps. *Proc. Jpn. Acad.* **1971**, *47*, 707–712. [CrossRef]
72. Martini, M.; Giudicepietro, F.; D'Auria, L.; Esposito, A.M.; Caputo, T.; Curciotti, R.; De Cesare, W.; Orazi, M.; Scarpato, G.; Caputo, A.; et al. Seismological monitoring of the February 2007 effusive eruption of the Stromboli volcano. *Ann. Geophys.* **2007**, *50*, 775–788.
73. Antonello, G.; Casagli, N.; Farina, P.; Leva, D.; Nico, G.; Sieber, A.J.; Tarchi, D. Ground-based SAR interferometry for monitoring mass movements. *Landslides* **2004**, *1*, 21–28. [CrossRef]

74. Di Traglia, F.; Battaglia, M.; Nolesini, T.; Lagomarsino, D.; Casagli, N. Shifts in the eruptive styles at Stromboli in 2010–2014 revealed by ground-based InSAR data. *Sci. Rep.* **2015**, *5*, 13569. [CrossRef]
75. Rudolf, H.; Leva, D.; Tarchi, D.; Sieber, A.J. A mobile and versatile SAR system. In Proceedings of the IEEE 1999 International Geoscience and Remote Sensing Symposium, IGARSS'99 (Cat. No.99CH36293), Hamburg, Germany, 28 June–2 July 1999; pp. 592–594.
76. Monserrat, O.; Crosetto, M.; Luzi, G. A review of ground-based SAR interferometry for deformation measurement. *ISPRS J. Photogramm. Remote Sens.* **2014**, *93*, 40–48. [CrossRef]
77. Calvari, S.; Intrieri, E.; Di Traglia, F.; Bonaccorso, A.; Casagli, N.; Cristaldi, A. Monitoring crater-wall collapse at active volcanoes: A study of the 12 January 2013 event at Stromboli. *Bull. Volcanol.* **2016**, *78*, 1–16. [CrossRef]
78. Casagli, N.; Tibaldi, A.; Merri, A.; Del Ventisette, C.; Apuani, T.; Guerri, L.; Fortuny-Guasch, J.; Tarchi, D. Deformation of Stromboli Volcano (Italy) during the 2007 eruption revealed by radar interferometry, numerical modelling and structural geological field data. *J. Volcanol. Geotherm. Res.* **2009**, *182*, 182–200. [CrossRef]
79. Schaefer, L.N.; Di Traglia, F.; Chaussard, E.; Lu, Z.; Nolesini, T.; Casagli, N. Monitoring volcano slope instability with Synthetic Aperture Radar: A review and new data from Pacaya (Guatemala) and Stromboli (Italy) volcanoes. *Earth Sci. Rev.* **2019**, *192*, 236–257. [CrossRef]
80. Marchese, F.; Genzano, N.; Neri, M.; Falconieri, A.; Mazzeo, G.; Pergola, N. A Multi-Channel Algorithm for Mapping Volcanic Thermal Anomalies by Means of Sentinel-2 MSI and Landsat-8 OLI Data. *Remote Sens.* **2019**, *11*, 2876. [CrossRef]
81. Genzano, N.; Pergola, N.; Marchese, F. A Google Earth Engine Tool to Investigate, Map and Monitor Volcanic Thermal Anomalies at Global Scale by Means of Mid-High Spatial Resolution Satellite Data. *Remote Sens.* **2020**, *12*, 3232. [CrossRef]
82. Carey, S.N.; Sparks, R.S.J. Quantitative models of the fallout and dispersal of tephra from volcanic eruption columns. *Bull. Volcanol.* **1986**, *48*, 109–125. [CrossRef]
83. Neuberg, J.; Luckett, R.; Ripepe, M.; Braun, T. Highlights from a seismic broadband array on Stromboli volcano. *Geophys. Res. Lett.* **1994**, *21*, 749–752. [CrossRef]
84. Chouet, B.; Dawson, P.; Ohminato, T.; Martini, M.; Saccorotti, G.; Giudicepietro, F.; De Luca, G.; Milana, G.; Scarpa, R. Source mechanisms of explosions at Stromboli volcano, Italy, determined from moment-tensor inversions of very- long-period data. *J. Geophys. Res.* **2003**, *108*, 2019. [CrossRef]
85. Giudicepietro, F.; Auria, L.D.; Martini, M.; Caputo, T.; Peluso, R.; de Cesare, W.; Orazi, M.; Scarpato, G. Changes in the VLP seismic source during the 2007 Stromboli eruption. *J. Volcanol. Geotherm. Res.* **2009**, *182*, 162–171. [CrossRef]
86. Bonaccorso, A.; Gambino, S.; Guglielmino, F.; Mattia, M.; Puglisi, G.; Boschi, E. Stromboli 2007 eruption: Deflation modeling to infer shallow-intermediate plumbing system. *Geophys. Res. Lett.* **2008**, *35*, L06311. [CrossRef]
87. Bevilacqua, A.; Bertagnini, A.; Pompilio, M.; Landi, P.; Del Carlo, P.; Di Roberto, A.; Aspinal, W.; Neri, A. Major explosions and paroxysms at Stromboli (Italy): A new historical catalog and temporal models of occurrence with uncertainty quantification. *Sci. Rep.* **2020**, *10*, 17357. [CrossRef]
88. Bertolaso, G.; Bonaccorso, A.; Boschi, E. Scientific Community and Civil Protection Synergy during the Stromboli 2002–2003 Eruption. In *The Stromboli Volcano: An Integrated Study of the 2002–2003 Eruption*; AGU Geophysical Monograph Series; AGU: Washington, DC, USA, 2008; pp. 387–397. [CrossRef]
89. McBirney, A.R. Factors Governing the Intensity of Explosive Andesitic Eruptions. *Bull. Volcanol.* **1973**, *37*, 443–453. [CrossRef]

remote sensing

MDPI

Article

Overflows and Pyroclastic Density Currents in March-April 2020 at Stromboli Volcano Detected by Remote Sensing and Seismic Monitoring Data

Sonia Calvari [1,*], Federico Di Traglia [2], Gaetana Ganci [1], Flora Giudicepietro [3], Giovanni Macedonio [3], Annalisa Cappello [1], Teresa Nolesini [4], Emilio Pecora [1], Giuseppe Bilotta [1], Veronica Centorrino [1], Claudia Corradino [1], Nicola Casagli [2] and Ciro Del Negro [1]

1 Istituto Nazionale di Geofisica e Vulcanologia, Osservatorio Etneo—Sezione di Catania, 95125 Catania, Italy; gaetana.ganci@ingv.it (G.G.); annalisa.cappello@ingv.it (A.C.); emilio.pecora@ingv.it (E.P.); giuseppe.bilotta@ingv.it (G.B.); veronica.centorrino@ingv.it (V.C.); claudia.corradino@ingv.it (C.C.); ciro.delnegro@ingv.it (C.D.N.)

2 Dipartimento di Scienze della Terra, Università degli Studi di Firenze, Via La Pira 4, 50121 Firenze, Italy; federico.ditraglia@unifi.it (F.D.T.); nicola.casagli@unifi.it (N.C.)

3 Istituto Nazionale di Geofisica e Vulcanologia, Osservatorio Vesuviano, 80124 Napoli, Italy; flora.giudicepietro@ingv.it (F.G.); giovanni.macedonio@ingv.it (G.M.)

4 Università degli Studi di Firenze—Centro per la Protezione Civile, Piazza San Marco 4, 50121 Firenze, Italy; teresa.nolesini@unifi.it

* Correspondence: sonia.calvari@ingv.it

Received: 3 July 2020; Accepted: 13 September 2020; Published: 16 September 2020

Abstract: Between 28 March and 1 April 2020, Stromboli volcano erupted, with overflows from the NE crater rim spreading along the barren Sciara del Fuoco slope and reaching the sea along the NW coast of the island. Poor weather conditions did not allow a detailed observation of the crater zone through the cameras monitoring network, but a clear view of the lower slope and the flows expanding in the area allowed us to characterize the flow features. This evidence was integrated with satellite, GBInSAR, and seismic data, thus enabling a reconstruction of the whole volcanic event, which involved several small collapses of the summit cone and the generation of pyroclastic density currents (PDCs) spreading along the slope and on the sea surface. Satellite monitoring allowed for the mapping of the lava flow field and the quantification of the erupted volume, and GBInSAR continuous measurements detected the crater widening and the deflation of the summit cone caused by the last overflow. The characterization of the seismicity made it possible to identify the signals that are associated with the propagation of PDCs along the volcano flank and, for the first time, to recognize the signal that is produced by the impact of the PDCs on the coast.

Keywords: Stromboli volcano; effusive activity; satellite thermal imagery; ground-based thermal imagery; cinder cone instability; pyroclastic density currents

---

1. Introduction

Rapid changes of the surface morphology often occur in open-conduit basaltic volcanoes that frequently erupt, such as Etna and Stromboli (Italy), Piton de la Fournaise (La Réunion Island), or Kilauea (Hawaii). Cinder-cones ~60 m high can form in just one or a few weeks [1,2], large lava flows can spread over roads or villages [3–6], and summit collapses following major lava withdrawal can involve large areas and result in new calderas [6–9]. However, as they are rapid in building new reliefs, the often loose and unstable mixture of breccia, ash, and spatter, which accumulates quickly and gets possibly destabilized by dike intrusion or overloading, can also suddenly collapse, affecting small portions of the summit cone [10–14] or even the volcano flank [15–18]. Magma drainage from the feeder conduit can cause significant summit collapses, such as those experienced at Stromboli in 2003 and 2007 [19–21], at Piton de la Fournaise in 2007 [8], and at Kilauea in 2018 [6,9]. Regardless of the cause, instability as well as the formation of pyroclastic density currents (PDC) at active and inhabited volcanoes can cause injuries and even loss of lives. This happened, for example, at Stromboli in 1930 and 2019 [22–25], at Unzen (Japan) in 1991 [26], at Merapi (Java) in 1994 and 2010 [27]. For this reason, PDCs are identified among the primary sources of fatalities at active volcanoes [28].

Stromboli volcano is the easternmost island of the Aeolian Archipelago (Italy; Figure 1a,b). It is known as the "Lighthouse of the Mediterranean", because of its persistent explosive activity from the summit craters, with bursts occurring every few minutes [29–31]. The summit crater of the volcano is a depression ~300 m long in a NE-SW direction (Figure 1c), ~50 m wide, and ~50 m deep, located at ~750 m elevation [14,19]. Three crater areas are located within the summit depression: the NE crater zone (NEC), the Central crater zone (CC), and the SW crater zone (SWC), each of them comprising a variable number of active vents (Figure 1c). The capacity of the uppermost feeder conduit increased after the 2002–2003 and 2007 flank eruptions [20,32], but sudden changes in the magma level may result in a greater magmastatic pressure, which could cause the destabilization and collapse of portions of the summit cone [14,33].

The 2019 eruption at Stromboli has twice shown how dangerous PDCs can be to those tourists approaching the slopes of an active volcano, even from the seaside [34]. Being less dense than water, the PDC formed by 4–5 km high explosive column collapse managed to expand on the sea surface for several hundred meters, running towards a tourist boat sailing nearby (https://www.youtube.com/watch?v=RPKgS3sPP1Y). PDCs can also trigger tsunamis that can devastate the coast, as happened at Stromboli in 1930 and 1944 [22,23]. More recently, PDCs spreading along the Sciara del Fuoco (SdF) slope formed as a consequence of high magma level within the conduit, instability, and collapse of the summit cone [14,35]. PDCs are common at Stromboli especially during the initial phases of flank eruptions, when the opening of an eruptive fissure breaches the summit craters causing the spread of a mixture of hot debris made of lava blocks, lithics and pyroclastics running down the slope and towards the sea [36–40]. More recent examples of this phenomenon occurred in 2013 and 2014 [14,40,41], caused, respectively, by the erosion of the crater rim due to overflows and the opening of an eruptive fissure.

PDCs can result from the collapse of eruptive columns during paroxysmal explosive eruptions [13,19,23,39,42] or from gravitational instability (gravity-induced PDCs); e.g., [10,12,13,23,25,43,44]. PDCs are very mobile, thus being potentially very dangerous for people living close to or on the flanks of active volcanoes. This is why the monitoring of active volcanoes is becoming more and more complex and now integrates several different disciplines, spanning from volcanology, seismicity, geochemistry, and geodesy, in an attempt to obtain a complete picture of what happens not only on the surface, but also—and more importantly—within the shallow feeding system of the volcano, allowing for the prevention and forecasting of disasters and hence avoiding the loss of lives. At Stromboli PDCs normally expand along the barren slope of the SdF [14,19,35,44]. However, during the past century, they have emplaced along the inhabited slopes at least three times; in 1906, 1930, and 1944 [22,23,25,45]. We do not have a complete stratigraphy of PDCs on Stromboli, but three such sequences have been found also during the Neostromboli period (<12.5 ka; [46–48]), these last probably representing the most widespread PDC deposits found on the island.

Recently, the volcano has been the site of two paroxysmal explosive eruptions occurred on 3 July and 28 August 2019, which produced eruptive columns 4–5 km high and PDCs. On such occasions, they triggered fires and caused a death, while spreading along the SdF as well as on the sea surface. These episodes were accompanied by a lava overflow from the SWC that started on 3 July and lasted nearly two months, ending on 30 August [24,25,34]. Two more recent overflows from the NEC occurred on 18 January and 3 February 2020, lasting just 3.0 and 1.5 h, respectively. They resulted in a short lava flow field that expanded in the uppermost SdF and did not reach the coast.

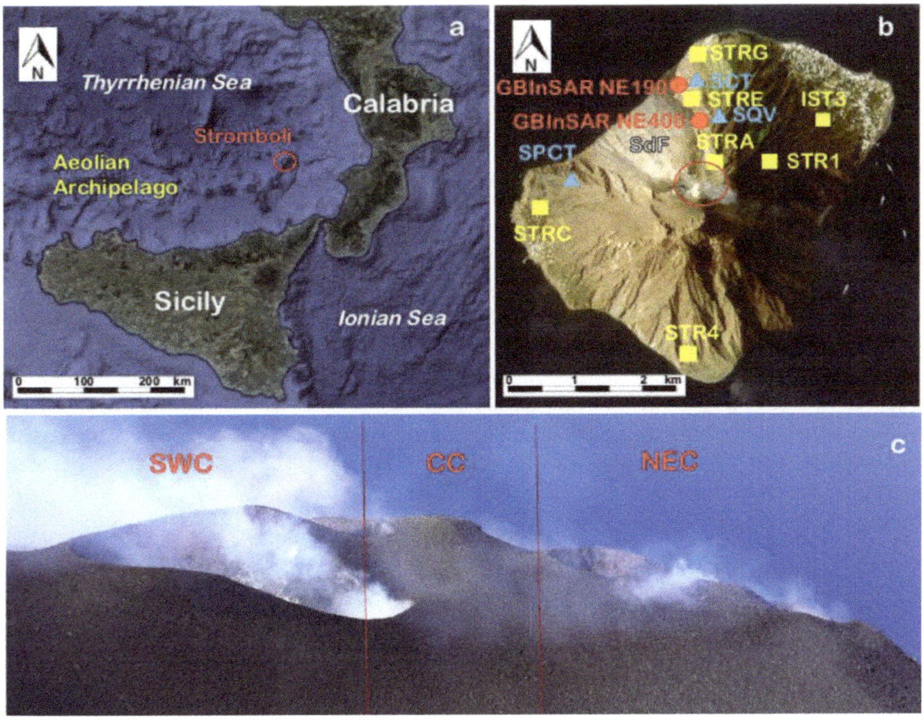

Figure 1. (**a**) Google map of southern Italy, with the red circle showing the position of Stromboli volcano, at the NE end of the Aeolian Archipelago. (**b**) Stromboli island with the position and labels of the monitoring instruments used in this study. The blue triangles are the monitoring cameras; the red circles are the Ground-Based Interferometric Synthetic Aperture Radar (GBInSAR) stations; the yellow squares are the seismic stations. The empty red circle outlines the position of the summit craters displayed in c. SdF = Sciara del Fuoco slope. (**c**) View from South of Stromboli summit craters, taken on 21 February 2020, showing the names of the active crater areas. NEC = NE Crater zone; CC = Central Crater zone; SWC = SW Crater zone. The field of view is about 300 m wide. Photo courtesy of F. Ciancitto, INGV.

The aim of this paper is to present our study of the eruptive activity occurred at Stromboli between 28 March and 1 April 2020, along with our interpretation and quantification of the eruptive processes that are based on the analysis of monitoring data, comprising time-lapse videos recorded by the camera network, satellite images, GBInSAR, and seismic stations.

2. Methods

The description of the events, the calculation of velocity for lava flows and PDCs spreading along the SdF and on the sea surface, and the count of the explosions occurring over time from the summit craters were obtained through the analysis of the videos recorded by the network of fixed monitoring cameras maintained by INGV-OE (Istituto Nazionale di Geofisica e Vulcanologia—Osservatorio Etneo). The volcano deformation was measured through two Ground-Based Interferometric Synthetic Aperture Radar (GBInSAR) devices that were installed by the University of Florence. Seismicity was analyzed using data from the broadband seismic network that was installed by INGV-OV (Istituto Nazionale di Geofisica e Vulcanologia—Osservatorio Vesuviano). Lava flow field area and volume, as well as an estimation of the effusion rate, were computed by the TechnoLab of INGV-OE while using multispectral infrared and optical satellite data.

2.1. The INGV Cameras Monitoring Network

The INGV cameras monitoring network at Stromboli volcano in March–April 2020 comprised three fixed instruments, two thermals (SCT and SPCT) and a visual (SQV). Their details are listed in Table 1 and their location is shown in Figure 1. SQV acquires at a frequency of one image every two seconds, SCT one image every second, and SPCT two images every second. The difference in acquisition frequency, as well as in the distance from the crater area and viewing angle, result in a different count of the explosions, depending on the camera taken into consideration. The number of explosions is thought to represent an expression of the magma level within the feeder conduit [19,39,49,50]. We have manually counted the total number of events occurring within the whole crater area using only the thermal cameras (SCT and SPCT), because they allowed a comparison between day and night views. However, the presence of clouds and/or dust may limit or hide visibility, as happened during most of the day on 30 March and in the early h on 1 April 2020. The calculated velocity of PDCs and lava flows along the SdF slope, obtained from the images of the monitoring cameras, are average velocities, because they were calculated dividing the whole travelled path by the time. Some of the images were affected by shadows due to fog, clouds, or ash, resulting in a non-well-defined outline of the features. The cumulative error on these spatial measurements, which was due to the poor quality of some frames, is about 2%, and it was obtained from the number and size of uncertain pixels. All of the times are expressed here as UT.

Table 1. List of the INGV monitoring cameras and of their main features.

Label	Type and Model	Location	Distance from the Craters	Optics	Field of View
SPCT	Thermal, FLIR A320	West SdF flank, 85 m a.s.l.	1698 m	90°	90° × 73° 2150 × 1613 m
SCT	Thermal, FLIR A655sc	East SdF flank, 165 m a.s.l.	1538 m	25°	25° × 19° 807 × 605 m
SQV	Visual, Sony FCB-EX480CP	East SdF flank, 390 m a.s.l.	1027 m	18×	48° (wide end) × 2.8° (tele end) 657 × 493 m

2.2. GBInSAR

Measuring surface deformation, exploiting the phase difference between two spaceborne SAR images (differential InSAR, DInSAR; [51]) makes it possible to recognize ground displacements along the satellite line of sight (LOS) direction on a centimeter-scale. Processing a long stack of images using multi temporal (MT) InSAR techniques allows for the detection of millimeter-scale displacements over long time frames through the reduction of error sources [52,53]. GBInSAR has the additional advantage of producing frequent SAR images (on the order of seconds to minutes), resulting in very high frequency slope maps and time series. Moreover, the use of GBInSAR in the Ku-band (17.0–17.1 mm radar) can penetrate dust clouds, abundant especially during collapse events, and can work with variable light and atmospheric conditions [14]. The NE portion of the summit crater terrace at Stromboli and the northern portion of the SdF are monitored by two GBInSAR devices, which are located in a stable area N of the SdF (Figure 1). The first GBInSAR (GBInSAR NE400; Model: GB-InSAR LiSALab, Ellegi srl, Rovello Porro, Italy, http://lisalab.com/home/; Revisiting time; 11 min; [54]) was installed in February 2003, during the 2002–2003 flank eruption, whereas the second device (GBInSAR NE190; Model: GB-InSAR LiSAmobile k09, Ellegi srl – LiSALaB, Rovello Porro, Italy, http://lisalab.com/home/; Revisiting time; 2 min) was installed on 14 December 2014, after the flank eruption that took place that year.

Radar images were obtained through sampling techniques; for this reason, particular frequency and spatial steps had to be selected in order to avoid ambiguity in range and cross-range [54].

The system is able to measure line-of-sight (LoS) ground displacement in the time interval between two acquisitions and the displacement is calculated from the phase difference between the back-scattered signals received at different times, through the cross-correlation between two SAR images. Range and cross-range resolution are, on average, 2 × 2 m, with a measurement precision being referred to the displacement of less than 1 mm [54]. The displacement rate is the result of the mathematical division between the displacement measured in an interferogram (referred to the difference between two SAR images) and the elapsed time between the two images, allowing for the identification of very low displacement rates (0.010–0.001 mm/h) related to the creep of the northern sector of the SdF or very fast displacement rates (up to 300 mm/h) associated with effusive vent opening [55]. The capability of InSAR to detect ground displacement depends on the persistence of phase coherence over appropriate time intervals, therefore a SAR coherence mask (threshold = 0.5) was set to mask the noisy areas of the interferogram [54]. The phase values can be affected by ambiguity (unwrapped phase), but, due to the short-elapsed time between two subsequent measurements on Stromboli volcano, the interferometric displacements were usually smaller than half wavelength, so that no unwrapping procedures were needed.

The GBInSAR phase data are useful to detect: (i) the inflation/deflation of the summit plumbing system [40,44]; (ii) small-to-large scale slope instability, in response to eruptive (over-steepening and overloading; [14]) and magmatic activity (dike intrusions; [33,56]); (iii) the gravitational re-adjustment of the talus, sometimes evolving into rock-falls [35]; (iv) thermal contraction of the lava field, mainly in areas of lower pre-effusive slope angle [57]; and, (v) persistent flank motion [58].

Averaged power (amplitude squared) images produced by the GBInSAR NE400 were used to quantitatively evaluate the changes in the NEC's morphology, as they were the closest devices to the NEC. Each analyzed image was derived-from-48-images averaged (≈ 1 every 4 h) in order to increase the signal to noise ratio and, in doing so, facilitate its interpretation. Because the NEC rim produced a shadow zone corresponding to the crater depression, it was possible to calculate the area of the crater itself as it changed over time [14]. The standard deviation was calculated as equal to 80 m^2 (see [14]). A threshold was set at ~50 dB to map the area affected by the NEC widening and narrowing collapse.

2.3. Satellite Remote Sensing Monitoring

Multispectral satellite images processing is increasingly demonstrating its potential in providing both timely event detection for volcanic effusive events and, in the case of eruption, extraction of parameters that can help tracking the lava flow [59]. Even if the volcanic features of interest are usually much smaller than the nominal pixel size of the satellite image, moderate spatial resolution sensors (~1 km) can detect emitted radiance in the mid-infrared (MIR) wavelengths, a spectral region in which high temperature events, such as active lava flows, vents, and domes, emit copious amounts of energy. Satellite infrared data represented a useful means to describe the evolution of the eruptive event occurred during 28 March–1 April at Stromboli volcano. In particular, we used the HOTSAT system [60–62] to detect the presence of thermal anomalies through the analysis of multispectral infrared images acquired by a variety of satellite sensors with a revisit time of about 12 h per satellite, such as VIIRS (Visible Infrared Imaging Radiometer Suite), providing at-nadir pixel footprint of 375 m for I-bands, SLSTR (Sea and Land Surface Temperature Radiometer) on board of SENTINEL-3, and MODIS (Moderate Resolution Imaging Spectroradiometer), both providing 1 km pixels at-nadir. The combined use of sensors that differed for spatial characteristics (from 375 m to 1 km) and different acquisition times has proved to be a robust and reliable instrument for the thermal monitoring of active volcanoes [63–65]. The HOTSAT system locates the thermal anomalies (hotspot), computes the associated radiant heat flux summing up the contribute of each hotspot pixel, and, in the case of effusive eruption, provides the Time Averaged Discharge Rate (TADR) as proportional to the radiant heat flux [66]. The conversion from radiant heat flux to TADR was performed according to Harris et al. [67] using: $TADR = Q/(\rho\ (c_p\ \Delta T + c_L\ \Delta\Phi))$, where Q is the total thermal flux obtained summing up the radiative power computed for each hotspot pixel, ρ is the lava density (2600 kg m^{-3}), c_p is the specific heat capacity (1150 J kg^{-1} K^{-1}), ΔT is the eruption temperature minus temperature at which flow stops (100–200 K), c_L is the latent heat of crystallization (3.5 × 10^5 J kg^{-1}, and ΔΦ is the volume percent of crystals that form while cooling through ΔT (30–54%).

The HOTSAT system was extended with a new module to process data acquired by Landsat 8 OLI and TIRS, Sentinel 2 MSI, and ASTER images in order to exploit higher spatial resolution multispectral images. Besides providing further information on the radiant heat flux, these data can be used to locate eruptive vents and describe the evolution of the lava flow field [68–70], based on the spatial and spectral resolution of the available bands and the phase and size of the eruption they catch. For example, in the case of an ongoing eruption, Sentinel-2 MSI, thanks to its bands in the SWIR (bands 11 and 12), can provide the position of an active vent and flow at the spatial resolution of 20 m, whereas Landsat-8 OLI, with its SWIR bands 6 and 7, can provide the same information at 30 m of spatial resolution. If a crusted lava flow is cooling, SWIR bands might not be able to detect it; on the other hand, it could still be visible in the thermal infrared TIR bands, i.e., Landsat 8 TIRS (bands 10 and 11) or ASTER (bands 10–14) at 100 and 90 m of spatial resolution, respectively. Due to the limited temporal resolution of these higher spatial resolution multispectral images, post-eruptive images occur more often than intra-eruptive ones and, in many cases, the flows cool too fast to be visible, even in the TIR bands.

Recently, the high spatial resolution and freely available information coming from the Multispectral Imager (MSI) on-board Sentinel-2 satellite has been used to facilitate the two-dimensional (2D) mapping of lava flows [71] through a new Machine Learning (ML) classifier, which discriminates the recent lava flows from pre- and post-eruptive multispectral images acquired by MSI, combined with pre-eruptive digital topography. Bands 2, 3, 4, and 8 at the spatial resolution of 10 m are used as input to the classifier. This ML approach relies on two steps: (i) a k-medoids unsupervised classifier separating input data in clusters whose pixels have the same properties; and, (ii) a Bayesian neural network mapping recent lava flows. In particular, the first step reveals pixels undergoing similar changes in time between pre- and post-eruptive images, adopting the correlation distance as a measure of similarity. Subsequently, in the second step, a small representative subset of each cluster is exploited to train the BNN, so that it provides us with the pixels belonging to the recent lava flow.

The advancement of satellite remote sensing techniques also has great potential for what concerns the three-dimensional mapping of volcanic products. Indeed, high spatial resolution data acquired in stereo, tri-stereo, or multi-view configuration (e.g., Pléiades, PlanetScope, ASTER) can be used to frequently update the topography and to estimate volcanic deposits by differencing successive topographies. Such estimates can improve the 2D mapping of lava flows while providing an independent maximum bound to lava flow volume that can be derived from the satellite infrared data [59,64,72,73]. We exploited the Pléiades constellation, which is composed of two optical satellites, Pléiades 1A and 1B, respectively, launched on December 2011 and 2012, in order to retrieve areas, volumes, and thickness distribution of the recent volcanic deposits in Stromboli. These satellites provide images at 50 cm spatial resolution in stereo and tri-stereo mode [74]. The 3D processing of the Pléiades imagery was performed using the free and open source MicMac photogrammetric library (available at http://micmac.ensg.eu), in this way 1-m Digital Elevation Models (DEMs) were obtained.

We derived two 1-m DEMs to further constrain the volume of volcanic deposits: a pre-eruptive one from the tri-stereo optical imagery acquired on 8 October 2019, and a post-eruptive one from the Pléaides-1 images of 7 April 2020. We differentiated them, so to obtain the thickness distributions of volcanic deposits emplaced between October 2019 and April 2020. The two DEMs were first co-registered using the Nuth and Kääb [75] method in order to avoid any errors that could derive from a misalignment.

2.4. The INGV Seismic Network

The seismic monitoring of Stromboli is based on a broadband seismic network [76] deployed on the island after the 2002–2003 effusive eruption. In the following years, the number of stations decreased due to the closure of some routes. Moreover, the seismic network was also damaged by the paroxysmal explosions of 2003, 2007, and 2019. The current configuration consists of seven stations (Figure 1, Table 2), managed by INGV-OV and INGV-OE. The data transmission is based on UHF radio links and the INGV wireless local area networking system. Seismic data are first transmitted to the centres in Stromboli and Lipari and then to the acquisition centres of the INGV-OV and the INGV-OE, in Napoli and Catania, respectively.

Table 2. Technical characteristics of the seismic network stations.

Station	Sensor	Sampling Rate Sps
STR1	Guralp CMG40T	50
STR4	Guralp CMG40T	50
STRA	Guralp CMG40T	50
STRC	Guralp CMG40T	50
STRE	Guralp CMG40T	50
STRG	Guralp CMG40T	50
IST3	Nanometrics Trillium120PA	100

Stromboli seismicity is typically characterized by explosion-generated signals [77–81] and by persistent volcanic tremor [82]. Seismic signals that are associated with landslides [83,84], rolling blocks, PDCs, and lava flows were also recorded in the period 28 March–1 April 2020 (Figure 2). Although the explosion signals are transient events, compared to the recordings of tectonic and volcano-tectonic earthquakes, they (Figure 2a) are characterized by an emerging onset, so that a clear arrival of the P wave cannot be recognized on the seismogram. Moreover, they show a wide frequency band (0.05–10 Hz) containing a Very Long Period (VLP) event. The volcanic tremor (Figure 2b), typical of open conduit volcanoes, has a frequency content of between 1 and 3 Hz.

The signals that are caused by the typical landslides in loose pyroclastic deposits on the SdF (Figure 2c) have a fusiform envelope and are characterized by a relatively high frequency (4–15 Hz). An increase in the occurrence of these signals was a short-term precursor of the flank effusive eruption on 2007; therefore, they have been studied in detail in various works, especially from the point of view of their automatic and early detection [83,84]. Such signals are partly due to the morphogenetic processes of the SdF slope. Moreover, they can be related to the explosive activity that throws incoherent materials on the slope, which can be easily re-mobilized, thus generating landslides. These are also favored by effusive activity, as the lava front is often a source of incoherent material, which can move on the steep slope of the volcano's flank [84]. Additionally, the signals that are caused by the rolling of large blocks (Figure 2d), in some cases, can be associated with detachments occurring at the lava flow front. They are similar to the ones associated with landslides in loose pyroclastic deposits, however some impulsive phases can be recognized in the waveform, and the spectrogram shows a less gradual onset when compared to that of the landslides in loose pyroclastic deposits. In the example of Figure 2d, the association of the seismic signal with the rolling of large blocks was based on its comparison with the images of the cameras. But an experienced seismic analyst is able to distinguish these signals, even on the basis of the visual analysis of the seismogram alone. This type of signal can also originate from rock-falls along the cliffs of the Labronzo area (North edge of the SdF). Occasionally, they were recorded on the southern side of the island, where there are steep morphologies. The signals caused by the rolling of large blocks (Figure 2d) are similar to the ones that were associated with landslides, however some impulsive phases can be recognized in the waveform, and the spectrogram shows a less gradual onset as compared to that of the landslides. These signals can also be associated with detachments occurring at the lava flow front. The signals that are linked to hot avalanches or PDCs (Figure 2e) are characterized by frequencies with a range 1–5 Hz and are due to incoherent hot materials that form massive flows. On 31 March 2020, the PDC mechanism generated seismic signals with large amplitude at the stations that are closest to the SdF slope.

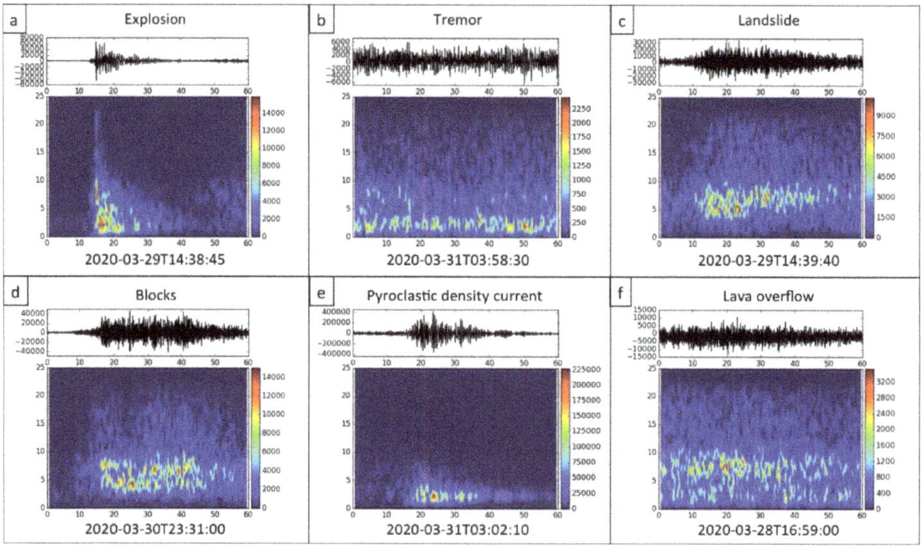

Figure 2. Examples of seismograms and spectrograms of seismic signals recorded at Stromboli in the studied period. Recording start times are reported at the bottom of each plot. The duration of the plots is one minute. (**a**) seismic signal and spectrogram produced by strombolian explosion; (**b**) seismic signal and spectrogram produced by volcanic tremor; (**c**) seismicity and spectrogram resulting from loose landslides; (**d**) seismic signal and spectrogram generated by blocks rolling down the Sciara del Fuoco slope; (**e**) seismic signal and spectrogram caused by pyroclastic density currents; (**f**) seismic trace and spectrogram recorded during the emplacement of lava flows.

3. Results

3.1. Eruptive Activity between 28 March and 1 April 2020

The eruptive activity taking place at Stromboli between 28 March and 1 April 2020 was studied using the images recorded both by satellites and by the INGV-OE monitoring cameras network. On 30 March and for the first half of the day on 1 April, poor weather conditions limited the visibility of the crater area.

The explosive activity at the summit vents was rather intense on 28 March, with 20–25 explosions/h obtained from both cameras SPCT (West flank) and SCT (East flank of the SdF, Figures 1 and 3a). The eruptive activity took place at the NEC, featuring very intense explosions generating a spherical shape of incandescent spatters, which spread more laterally than vertically and all along the crater rim, fell on the NE outer flank and rolled down the SdF slope. This relationship between magma depth and column shape has been established previously [85], thus this shape of the explosions suggested that the level of magma within the vent was very shallow—estimated in a few tens of meters. Meanwhile, the explosions produced by the SWC were essentially of hot gas and ash with collimated jets extending more vertically than laterally, thus indicating that the level of magma within this vent was rather deep—estimated in a few hundred of meters [85]. At 15:38 the first landslide of hot debris coming from the NE rim of the NEC was observed, and it was followed by a powerful explosion. A similar landslide—the second reported on that day - occurred at 15:42, and it was apparently caused by the instability of the hot debris accumulated by the explosions on the steep slope of the NEC outer crater flank. Between 16:09 and 17:00 a series of small landslides from the NEC crater rim began, whose frequency increased in time until it became almost continuous. At 17:02 a lava flow started overflowing from a vent located at the base of the NE crater rim, and both thermal cameras suddenly recorded a significant decrease of the total number of explosions/h, from the previous 20–25 to 5–15 explosions/h

(Figure 3a). At 17:44, the crater outline, as observed from SQV, showed a v-shaped cut in the crater rim, located above the effusive vent, which was caused by a collapse. The area of the missing block, estimated from the images recorded by the SQV camera, was ~110 m^2. When considering a thickness of ~10 m for the missing block, the volume of the crater rim eroded by the landslides resulted in ~1.1×10^3 m^3. The erosion caused by the spreading of the lava flows on the North flank of the cone, forming a channel that was also widening by failures and erosion of the lateral margins, was ~1000 m^2, estimated again by the images of the SQV camera. When considering a thickness of ~5 m for the collapsed area, a total volume of ~5.0×10^3 m^3 can be estimated. At 18:00, several incandescent blocks detaching from the lava flow fronts reached the sea, where formed an apron. After 19:00, the lava flow gradually decreased its output rate, and so did the number of landslides along the SdF, that had been triggered by the failure of incandescent blocks detaching from the flow fronts (Figure 3b). By midnight, the lava flow was no longer fed and, as soon as it stopped, the number of explosions per h gradually increased during the 29 March (Figure 3a). Poor visibility characterized most of 30 March between 07:00 and midnight, but from the images of SCT at 23:31 we could distinguish incandescent blocks that rolled over the NEC crater rim and down the SdF slope, signaling the start of another overflow. The number and size of the incandescent blocks rolling down the slope increased at 23:49, while two lava flows were spreading, one towards East and another towards North. The North flow was the longest and turned out to have the greater flux. The incandescent blocks detaching from the flow fronts accumulated along the coast and formed a hot talus that could be easily viewed during the early h of 31 March. At 01:46 and 01:49, two PDCs descended the SdF reaching the sea, and expanded as a cloud on the sea surface, followed by several other similar events, most of which are listed in Table 3. It is worth noting that the speed of the PDCs spreading on the sea surface was generally increasing from 6.9 m s^{-1} to 23.3 m s^{-1} between 01:50 and 02:51, and decreasing afterwards to 5.9 m s^{-1} until 03:41. The measured distance that was travelled by the PDC on the sea surface from the coast varied between 108 and 165 m, with speeds between 4.4 and 23.3 m s^{-1} (Table 3). Table 4 presents a list of several lava flows and PDCs that descended the SdF and reached the coast during 31 March. It shows the difference between the speed of the lava flows, from 2.8 to 5.2 m s^{-1}, and the PDCs, which displayed velocities that were between 12.9 and 40.3 m s^{-1}.

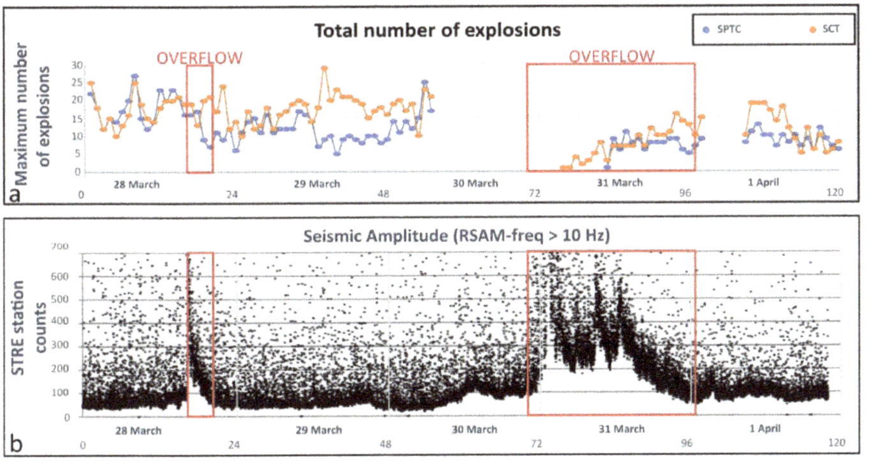

Figure 3. *Cont.*

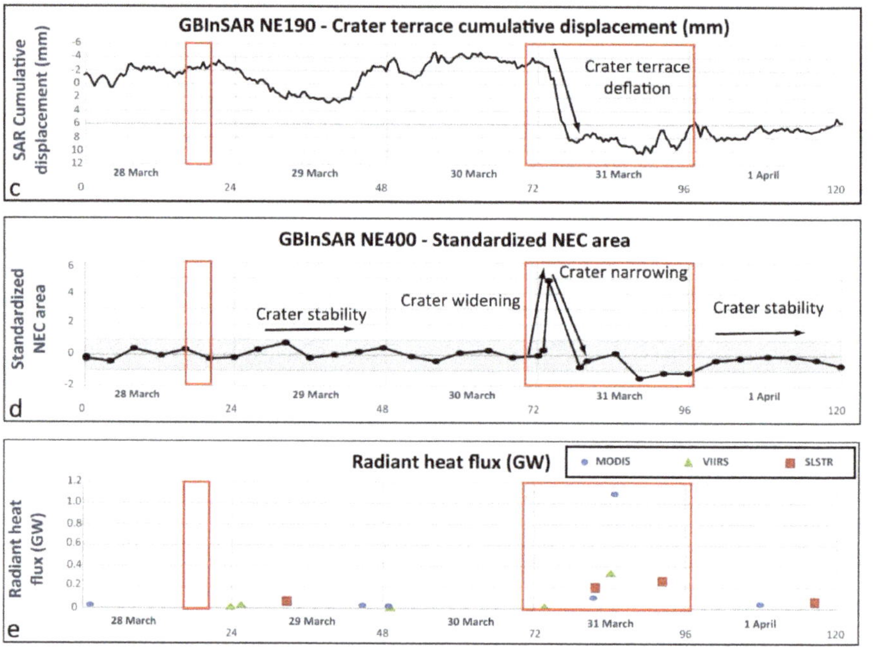

Figure 3. Comparative graphs of monitoring data collected between 28 March and 1 April 2020, with the two red rectangles displaying the duration of the two overflows from the NE crater rim occurred on 28 and 30–31 March. (**a**) Number of explosions per h detected from the thermal monitoring cameras SCT (red line) and SPCT (blue line) between 28 March and 1 April 2020. The parts with no data are caused by poor weather conditions with clouds obscuring the summit area; (**b**) Real Time Seismic Amplitude Monitoring (RSAM) of the STRE station (Figure 1) seismic signal (East–West component) filtered at a frequency > 10 Hz, calculated on 15-s windows); (**c**) GBInSAR NE190 cumulative displacement (positive values refer to the increasing distance between the sensor and the target area, representing the deflation of the summit area; (**d**) Standardized NEC area measured using GBInSAR NE400 power images (one image every four h; xn = xu μ/σ; where xn is the standardized data, xu is the original data, μ is the mean and σ is the standard deviation of the time series); and, (**e**) Radiant heat flux estimated by HOTSAT using Moderate Resolution Imaging Spectroradiometer (MODIS) (blue dots), Sea and Land Surface Temperature Radiometer (SLSTR) (red dots), and Visible Infrared Imaging Radiometer Suite (VIIRS) (green dots) data from 28 March to 1 April 2020.

Table 3. List of the pyroclastic density currents (PDCs) spreading on the sea surface on 31 March 2020, as detected from the SPCT camera. The start time indicates when the flow reached the coast, the end time when it reached the maximum distance out to sea.

Start Time (hh:mm:ss)	End Time (hh:mm:ss)	Distance on the Sea (m)	Speed (m s^{-1})
01:50:00.0	01:50:20.5	141	6.9
02:35:25.5	02:35:50.0	108	4.4
02:45:10.0	02:45:21.0	118	10.7
02:48:36.5	02:48:47.5	120	10.9
02:51:41.5	02:51:47.5	140	23.3
02:54:15.5	02:54:27.0	140	12.2
03:02:25.0	03:02:42.5	134	7.7
03:06:23.0	03:06:40.0	165	9.7
03:37:09.0	03:37:25.5	145	8.8
03:39:07.5	03:39:22.0	141	9.7
03:41:46.5	03:42:13.0	155	5.9

Table 4. List of the pyroclastic density currents (PDCs) and lava flows that descended the Sciara del Fuoco slope on 31 March 2020, as detected from the SPCT camera. The distance on the slope is intended as the path length measured along the slope from the starting point to the coast.

Start Time (hh:mm:ss)	End Time (hh:mm:ss)	Distance on the Slope (m)	Speed (m/s)	Flow Type
01:46:36.5	01:47:02.0	1028	40.3	PDC
02:34:45.5	02:35:27.0	1028	24.7	PDC
02:38:50.0	02:42:32.0	1028	4.6	Lava flow
02:40:00.0	02:45:07.0	864	2.8	Lava flow
02:50:54.5	02:51:42.0	864	18.0	PDC
02:55:13.0	02:56:03.5	740	14.8	PDC
03:00:19.0	03:01:04.0	699	15.5	PDC
03:05:43.0	03:06:23.0	740	18.5	PDC
03:30:00.0	03:31:27.0	452	5.2	Lava flow
03:41:17.0	03:41:57.0	514	12.9	PDC

Table 4 also shows how the path to the coast decreased with time with the extension of the lava flows down the slope. This happens because many PDCs were starting from the lava flow fronts by detachment of hot blocks at breaks in slope. Once the PDCs reached the coast and spread on the sea surface, most of them formed ash clouds that expanded backwards and upslope to the crater area. Not all of the lava flows and PDCs that actually occurred are reported in Table 4, but only those where the visibility was clear enough to allow for an accurate measurement of the path and speed.

The alternation between lava flows and PDCs that descended the SdF slope indicates the gradual erosion of the summit cone that is caused by the emplacement of the lava flow and by the erosion of the summit cone. After 04:44 the lava flow widened at the coastline, forming a hot apron. Lava flows and PDCs continued during the morning at a decreasing rate corresponding to the gradual decrease of the supply to the lava flows, accompanied by a gradual increase of the explosions number from the summit craters (Figure 3a). The lava flow output decreased even more after 19:00 and, by 22:30 of 31 March, the lava flow was apparently no longer fed.

Figure 3 shows a comparison between different parameters measured during the period of interest (28 March–1 April 2020). Figure 3a shows how the number of explosions per h before this second lava flow was ~25 explosions/h, it decreased to below 5 explosions/h with the start of the lava flow output, and increased again after the end of the lava flows. This observation is consistent with the erupted volume for the 30–31 March lava flow, which was much greater when compared to that of the 28 March lava flow. After the emplacement of the 30–31 March lava flow, the NE flank of the summit cone outline was significantly modified. Being concave upwards at first, it appeared convex at the end, and the eroded surface was estimated at ~730 m^2. When considering a depth of ~10 m, the eroded volume of the summit cone can be estimated to be ~7.3×10^3 m^3. This brings the total volume of the summit cone, collapsed from the uppermost NE flank between 28 and 31 March 2020, to 13.4×10^3 m^3.

Figure 3b displays the RSAM [86] of the STRE station (Figure 1) seismic signal (East–West component) filtered at a frequency >10 Hz, calculated within 15-s windows. RSAM stands for Real Time Seismic Amplitude Monitoring and it is based on the moving average of the seismic signal absolute value, optionally filtered in specific frequency bands. This parameter is sensitive to landslides, which generate frequencies >10 Hz in the seismic wave field of Stromboli that is generally dominated by frequencies <10 Hz. These signals can also be associated with lava overflows as the collapsing lava flow fronts can generate landslides. Therefore, the RSAM shown in Figure 3b clearly highlights both of the lava overflows that occurred on 28 and the 30–31 March 2020 as well as the PDCs spreading along the SdF.

Figure 3e shows the radiant heat flux registered between 28 March and 1 April. The first thermal anomalies were detected on 28 March (at 01:10) by MODIS, on 28 March (at 23:48) by VIIRS, and on 29 March (at 08:50) by SLSTR. Poor weather conditions and the short duration and magnitude of the activity prevented observation of the overflow that occurred on 28 March. The peak of activity was recorded by MODIS on 31 March at 12:45, with a radiant heat flux of about 1.1 GW, probably being associated with the maximum areal extent of the lava flow. The final thermal anomaly was detected by SLSTR on 1 April at 20:30.

3.2. GBInSAR Data

The eruptive activity reported between 28 March and 1 April 2020 was not accompanied by a long-term inflation of the summit area (Figure 3c,d and Figure 4), as was recorded for the 2012–2013 and 2014 eruptive activities [14,40,44,58]. Displacements that were recorded by the GBInSAR devices were located around the NEC and were mainly associated to the accumulation and instability of the newly emplaced material (Figure 3a). Abrupt change in deformation behavior (that is, movement away from the sensors) occurred between 01:50 on 31 March 2020 and 05:34 on 1 April 2020 (Figure 3c). Deflation was restricted to the very upper part (Figure 4b), as occurred on 27 February 2007 (lava flow), 15 February 2013 (overflow), and 7 August 2014 (lava flow). Along with summit deflation, localized movement toward the sensors was recorded in the NEC area and along the SdF, that was attributable to slope instability related to the outpouring of the overflows. In Figure 3d, it is possible to observe that the NEC area had remained roughly constant until 01:45 on 31 March 2020, when it increased almost abruptly, indicating strong crater widening. The NEC area returned quickly to its original dimensions at 06:40 on 1 April 2020.

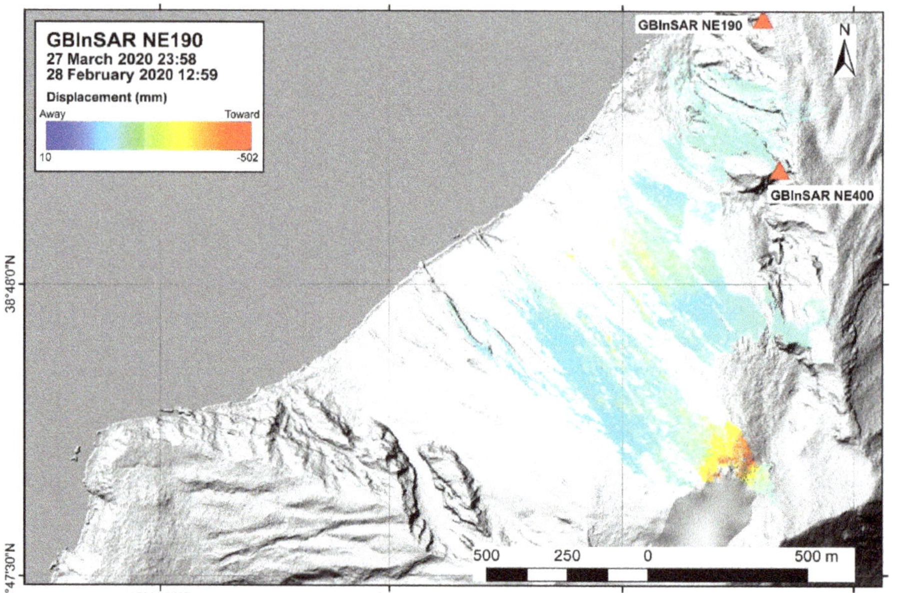

Figure 4. *Cont.*

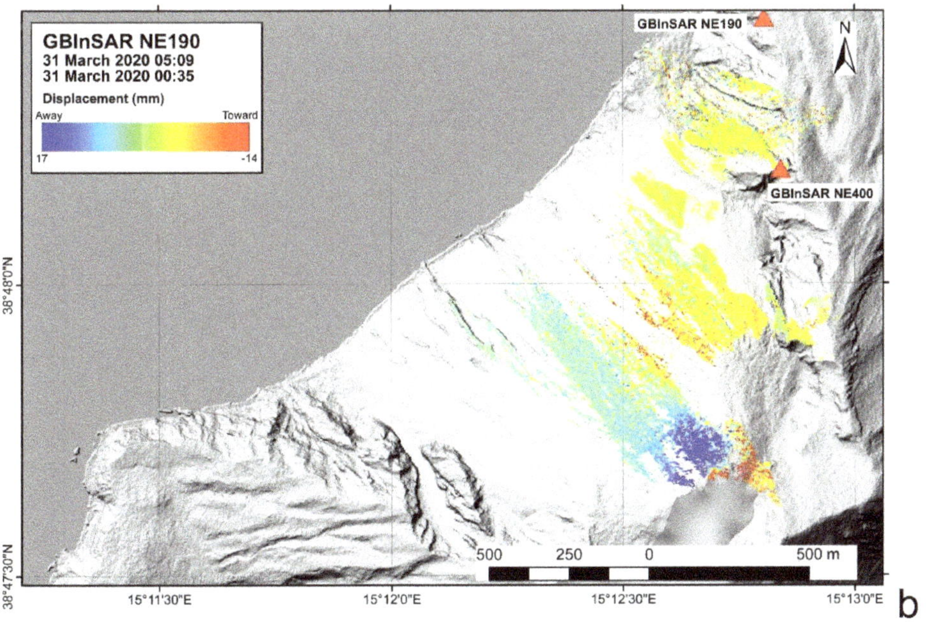

Figure 4. (**a**) 28 February 2020–27 March 2020, representing the pre-effusive period. Measured displacement was restricted to the NEC area and related to the accumulation and gravitational instability of the newly emplaced volcanic material; and, (**b**) 31 March 2020 cumulative displacement, highlighting the syn-eruptive deflation that occurred between 00:35 and 05:09.

3.3. Satellite-Derived Lava Flow Field Retrievals

During the effusive phase that occurred between 30 March and 1 April, we converted the radiant heat flux (Figure 3e) into Time-Averaged Discharge Rate (TADR), which is an estimation of the effusion rate averaged over a certain duration ([87]; Figure 5). Integrating the TADR curve, we obtained an upper and lower bound for the erupted Dense Rock Equivalent (DRE) lava volume that can be placed between 37 and 69 × 10^3 m^3. This compares to the volume of the NEC eroded during the overflows, which was estimated at 13.4 × 10^3 m^3.

The mapping of the lava overflow that occurred between 28 March and 1 April was performed through the ML classifier [72], using the Sentinel-2 MSI image acquired on 13 March as representative of the pre-eruptive, and the two images of 7 and 12 April 2020 as post-eruptive. As pre-eruptive topography, we used a 1-m Digital Elevation Model (DEM) that was generated by very high-resolution tri-stereo optical imagery acquired by the Pléiades-1 satellite constellation on 8 October 2019 [88]. Following the same steps that are discussed in [72], pixels with similar spectral properties were grouped in 50 clusters by the k-medoids unsupervised classifier. Subsequently, three pixels for each cluster were labelled to train the BNN, i.e., 150 pixels were used overall. To improve the accuracy, the lava flow field was then refined using the Pléiades image acquired on 7 April 2020, which provides a pixel resolution of 0.5 m.

Figure 6 shows the lava flow map resulting from the ML classifier. The area measures 94,500 ± 3380 m^2. The uncertainty was calculated by multiplying the satellite-derived perimeter (6760 m) by the pixel resolution of the Péiades-1 imagery (0.50 m) [69].

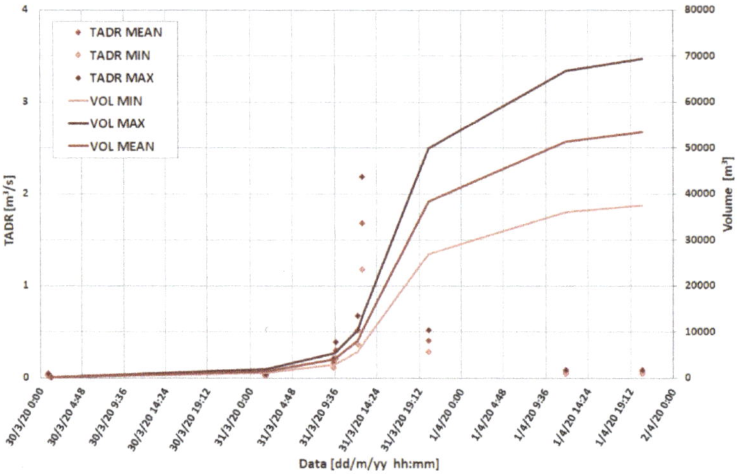

Figure 5. Minimum, mean, and maximum estimates for Time-Averaged Discharge Rate (TADR) and volume estimated during the effusive phase occurred from 30 March to 1 April 2020.

Figure 6. Google Earth view of the satellite-derived lava overflows from the NE crater rim spreading along the Sciara del Fuoco. The lava flow field (red contour) has been superimposed from the Pléiades image acquired on 7 April 2020.

In the area that was identified thanks to the ML classification (Figure 6), we found a thickness of volcanic deposits that goes from −14 m (due to the coastal erosion) to 14 m (in proximity of the NE crater) (Figure 7). The volume of the deposits accumulated near the NE crater amounts to 34,600 ± 9700 m^3. For calculating the volume of the main lava flow spreading on the SdF, we analyzed the histogram of the thicknesses (inset in Figure 7), finding a peak to 0.8 m. Being the most frequent value, we assigned it to the pixels where the DEM difference was negative, thus estimating a volume of 144,400 ± 79,000 m^3. Consequently, the total bulk volume of deposits thus amounts to 179,000 ± 89,000 m^3. The uncertainty was computed by multiplying the areas by the residual vertical accuracy outside the margins of the deposits, i.e., the standard deviation (~1.7 m) of the DEM difference in the area that was not covered by deposits.

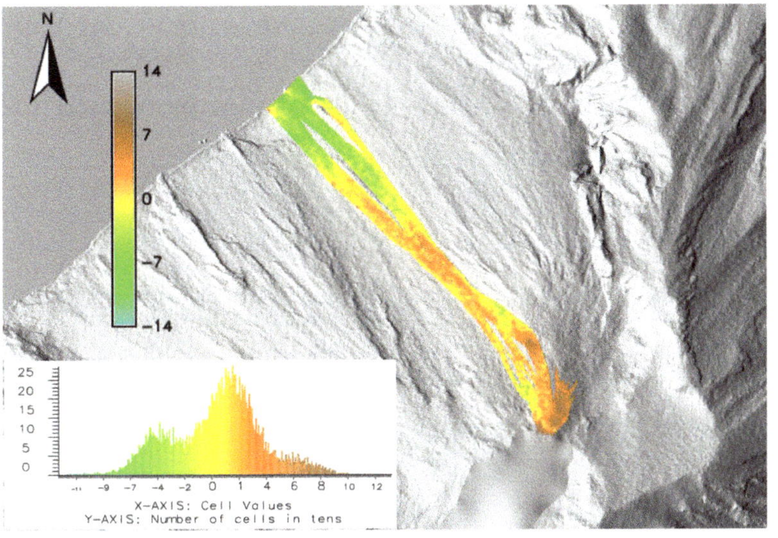

Figure 7. Three-dimensional (3D) mapping of the deposit in the area obtained using the ML classifier emplaced between 8 October 2019 and 7 April 2020.

3.4. Seismicity

The observation of large amplitude signals that were associated with PDCs, in particular those that occurred on March 31, is the main peculiarity of the seismic data recorded during the March–April 2020 eruptive crisis at Stromboli. To gain insight into the nature of these signals, we compared the seismic recordings to the images of thermal cameras (Figure 8a), timed with the same reference system as the seismic network (UTC based on GPS). The signal due to a PDC consists of a landslide-type initial part and a near monochromatic phase (peak frequency around 3 Hz) with much larger amplitude. The spectrogram (Figure 8b) clearly highlights the transition between the two phases that occurs exactly at the time of the impact of the PDC on the coast. The landslide-type signal is shown in red in Figure 8c, whereas the near monochromatic phase is drawn in blue. This observation highlights that the PDCs flowing on the ground generate signals that are similar to those typical of landslides moving on the SdF. The material accelerates on the slope and then impacts on the coastline (or on the sea surface). The impact generates the large amplitude 1–3 Hz phase.

We also focused on the seismic amplitude of STRE station, near the SdF where the PDCs flow, and STR1 station, which is relatively far from the SdF. The comparison between different seismograms produced in one h at STRE and STR1 stations (Figure 1), including the major PDCs that occurred on 31 March, is displayed in Figure 9 and indicates the explosion (E) and PDC signals. We calculated the ratio between the amplitude of the E3 explosion and PDC4 signals (top of Figure 9) that were recorded at the two stations (STRE: black; STR1: red). In order to evaluate the seismic amplitude, we calculated the average of the absolute values (RSAM [86]) of 30-s windows of both the explosion and PDC signals, starting from the onsets of the transients recorded by the two different stations (vertical component). Subsequently, we calculated the amplitude ratios of the explosion signals and the PDC signals. We obtained E3 ratio (STRE RSAM)/(STR1 RSAM) $\approx$ 2 and PDC4 ratio (STRE RSAM)/(STR1 RSAM) $\approx$ 4. This observation highlights the rapid decay of the seismic signal amplitude moving away from SdF slope and confirms that the source of the seismic phase with dominant frequency around 3 Hz is on the surface.

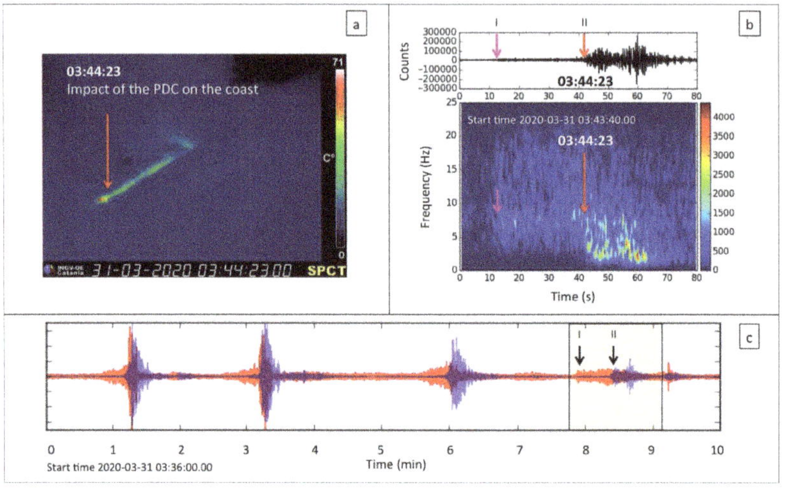

Figure 8. Comparison of a PDC seismic signal, starting at 03:43:54, 31 March 2020 (labeled "I") with the thermal camera images. (**a**) Impact of the PDC on the coast line recorded by the SPCT camera (see Figure 1 for camera location). (**b**) Seismogram and spectrogram of the PDC signal that occurred on 31 March at 03:43:54. The magenta arrows, labeled "I", indicate the onset of the landslide-type seismic signal. The red arrows, labeled "II", indicate the beginning of the 1–3 Hz frequency signal generated by the impact of the PDC on the coast line at 03:44:23 UTC. (**c**) 10 min of seismic signal highpass-filtered in frequencies >10 Hz (red) and bandpass-filtered in the 1–3 Hz band (blue) containing the recordings of some of the major PDCs that occurred on 31 March 2020. The amplitude of the highpass-filtered signal (red) is multiplied × 20. The yellow box highlights the PDCs represented in panels (**a**) and (**b**).

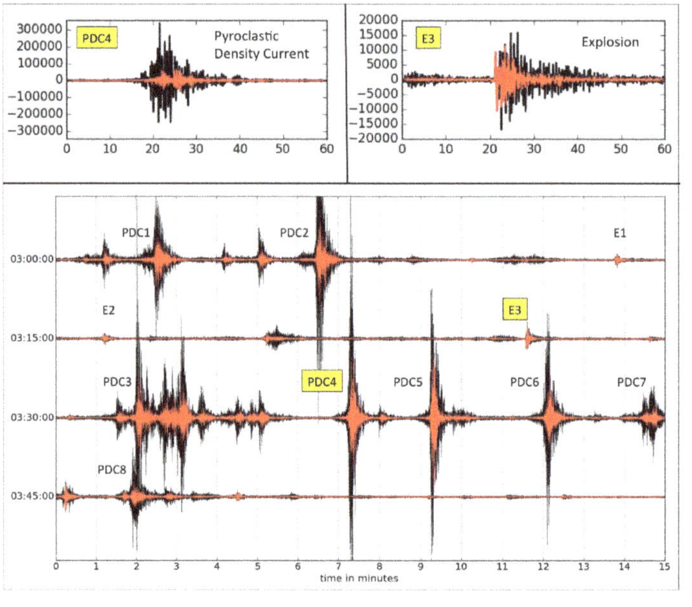

Figure 9. Comparison of one-h (2020-03-31T03: 00:00–2020-03-31T04: 00:00 UTC) of the vertical component seismic signals of the STRE (black line) and STR1 (red line) stations. The E and PDC labels

followed by an integer represent the signals of the most significant explosions and PDCs respectively, recorded in the considered time interval. At the top of the figure, the seismograms of the PDC4 and E3 explosion are shown in detail.

4. Discussion

Between 28 March and 1 April 2020, the summit craters of Stromboli volcano produced two overflows, on 28 and 30–31 March. The two episodes lasted ~7 h and ~23 h, respectively, and were accompanied by the descent of PDCs down the SdF slope and on the sea surface. Integrating several monitoring data, comprising visual and thermal images from a network of fixed cameras, ground deformation from GBInSAR, seismicity, and satellite images, we could reconstruct the sequence of events that occurred between 28 March and 1 April 2020 and understand their eruptive processes, gaining useful insights for hazard assessment.

The explosive activity, in terms of the number of explosions versus time, increased before the 28 March and 30 March lava flows (Figure 3a), which suggested that the magma level was becoming shallower within the feeder conduit [19,39,49,50] prior to the lava flow output. Lava flows were heralded and followed by rock-falls and landslides, which triggered the descent of PDCs down the SdF slope (Figure 3b). The number and duration of these events, as recorded by the seismic network, appear much greater during the 30–31 March lava flow, which lasted longer than the previous event. The first effusive episode was not preceded by significant deformation of the summit zone as detected by the GBInSAR (Figure 3c,d and Figure 4), whereas the second was accompanied by a sudden widening and narrowing of the NE crater zone (Figure 3c,d), and followed by a deflation of the summit crater terrace (max 17 mm recorded away from the sensors, Figure 4b). It should be noted that deflation of the crater terrace (Figure 3c,d) did not occur after the 28 March lava flow, confirming that its erupted volume was probably much smaller compared to the 30–31 March event. The number of explosions at the summit craters decreased after both the lava flow outputs, but this decline was greater and lasted longer on 30–31 March (Figure 3a), yet again suggesting a greater drainage of the shallow conduit, consistent with a larger erupted volume, which generated this lava flow. Hence, during the eruptive phase that took place between 28 March and 1 April at Stromboli, the shift at the summit craters from the persistent Strombolian explosions to lava flow output was heralded by an increase in the rate of explosions, and by a shallower magma level within the feeder conduit [19,39,49,50]. Conversely, the decrease of explosive activity following the lava flow output suggests that the drainage of the uppermost conduit was efficient, requiring a certain amount of time (of the order of h, Figure 3a) in order to allow the magma level to rise again after drainage in order to restore the persistent Strombolian activity at the summit craters. This is consistent with similar events, which were observed during the much longer 2002–2003 and 2007 effusive phases [19,39,89–91]. Satellite data allowed us to obtain the map of the lava flows expanding on the SdF slope between 28 March and 1 April, which was estimated at 94,500 ± 3380 m^2 (Figure 6), the radiant heat flux over time (Figure 3e), and, consequently, the TADR (Figure 5), providing an estimation of the cumulative erupted lava flow volume at 37–69 × 10^3 m^3 DRE. This compares to the volume of the NEC flank eroded during the overflows that was obtained from the camera images, resulting in 13.4 × 10^3 m^3.

The behavior of the lava flow spreading from the craters down to the steep SdF slope was consistent with an increase of the lava viscosity and of the yield strength, caused by a decrease in gas-content and bulk temperature of the flow, as well as its crystallization [92]. As a consequence, the lava flow front fragmented and hot blocks detached and descended the slope, the fragmentation generating several PDCs [19,39,44]. Again, PDCs formation was more relevant during the 30–31 March lava flow than during the previous 28 March lava flow, confirming a lower volume and/or shorter extent of the former lava flow. Poor weather conditions impeded a clear view of the summit crater area from the camera monitoring network, which only detected the mid-lower portion of the slope along which the lava flowed and PDCs were spreading. However, we could obtain the average spreading velocity of several lava flows and PDCs descending down the SdF slope (Table 4) and on the sea surface (Table 3). These data show an increase of the speed of the PDCs spreading on the sea surface from

6.9 m s^{-1} to 23.3 m s^{-1} between 01:50 and 02:51 on 31 March, which later decreased to 5.9 m s^{-1} until 03:41. This is consistent with the greater speed of PDCs and lava flows along the SdF slope that was detected during the same lapse of time (Table 4), which led us to assume that the greater PDC velocity on the sea surface was probably caused by an initial greater thermal efficiency of the PDC starting from a higher elevation and flowing along the slope above the active, well-fed and hot lava flow. This might have increased PDC mobility, allowing for it to reach the coast and spread on the sea surface at high speed. By comparison, the velocity of some PDCs emplaced in 1997 at Montserrat by dome collapses whose velocity was between 8 and 21.9 m s^{-1} [93,94], whereas much greater values of ~100 m s^{-1} were obtained for the PDC that was emplaced during the 1888 phreatic eruption at Bandai Volcano [95]. Several PDCs on 30 and 31 March at Stromboli spread on the sea surface up to a maximum distance of ~165 m from the coast, with an estimated speed up to ~23 m s^{-1} (Table 3). This velocity compares pretty well to the measurement of more than 27.8 m s^{-1} (more than 100 km h^{-1}; M. Pompilio, INGV unpublished report, 29 January 2003) obtained on a video of the PDC spreading on the sea surface for more than 100 m during the landslide and flank collapse of Stromboli on 30 December 2002. Similar PDCs spread on the sea surface at Stromboli in 2014 during the initial stages of the flank eruption, propagating for several tens of meters from the coast at speeds of 5.9 and 9.8 m s^{-1} [33].

The velocity to which the PDC expanded on the sea surface decreased along with the supply to the lava flow that was descending down the slope; at the same time, the PDC was travelling a shorter distance along the SdF slope (Table 4), thus causing a slower expansion of the PDC along the slope and reaching the coast at a lower speed. The measured distance that was travelled by the PDC on the sea surface from the coast varied between 108 and 165 m, with speeds between 4.4 and 23.3 m s^{-1} (Table 3). Hence, we understand that also small volume PDCs, like those produced during 28 March–1 April 2020 at Stromboli, can spread on the sea surface for hundred meters distance from the coast, causing a potential hazard for bathers, fishermen, or touristic boats sailing along the North coast of the island. Moreover, it must be considered that the ash cloud spreading backwards from the sea to the craters might produce ash fallout that could possibly reach tourists trekking along the SdF margins, even at lower heights.

PDCs spreading on the sea surface are not uncommon on Stromboli, and they surely represent an underestimated threat. They occurred during the subaerial and submarine landslide that caused a tsunami in December 2002 [16,17,96]; during the initial phases of flank effusive eruptions in 2002–2003, 2007, and 2014 [32,36,39]; after paroxysmal explosive eruptions triggered by column collapse [39,97,98]; and every time there is a small failure (with volumes of the order of 10^4 or 10^5 m^3) of the summit craters due to overloading or instability [33,34]. The two most impressive episodes occurred in 2019 as a result of the 4–5 km high eruptive column collapse after the paroxysmal explosions of 3 July and 28 August. On these occasions, the PDCs expanded on the sea surface for several hundred meters threatening a boat with tourists onboard (https://www.youtube.com/watch?v=RPKgS3sPP1Y). Lava flow velocities are normally much lower when compared to the mobility of PDCs, with speeds normally below 10 m s^{-1} ([99], and references therein), because of their greater viscosity (e.g., [100]). Only rarely they pose a threat to people, with the exception of few cases of lava accumulated within a crater and suddenly drained by a fissure opening, such as on Nyiragongo [101,102]. However, at Stromboli, lava flows can represent an indirect threat because they can erode the base of the summit cone and trigger summit collapses [14].

Table 4 shows the very different speed along the same SdF slope between lava flows, which had velocities between 2.8 and 5.2 m s^{-1}, and the PDCs, which displayed velocities between 12.9 and 40.3 m s^{-1}. Several authors [103–105] observed that abrupt slope changes or variations in magma supply affect the velocity of lava flows. When the slope is greater than 24°, it can result in the detachment of blocks from the flow front to form talus and, when the slope is even greater, as in the case of the SdF on Stromboli that reaches 30–35° [44,106], the tensional stresses overcome the tensional strength, so that the lava cannot flow any longer. Instead, it breaks into incandescent blocks [103] rolling down the slope and forming a PDC and a distal pile of talus.

The monitoring of PDCs along the slope of steep volcanoes, like Stromboli, is therefore of crucial importance, because they unexpectedly evolve into flank collapses possibly triggering tsunamis, as happened at Stromboli in 2002 [16,17,96], and more recently at Anak Krakatau in 2018. On that occasion, the tsunami caused over 430 fatalities, injured 14,000 people, and displaced 33,000 more along the Sunda Strait [18,107]. Interesting enough, the precursory phase of the Anak Krakatau flank collapse was characterized by an increase of the eruptive activity that lasted for 175 days, and the collapse was preceded by two seismic signals consistent with minor mass movements as well as a momentary quiescence [18,107]. Luckily, the mass movements in the eruptive crisis of 28 March–1 April 2020 at Stromboli were of modest size, not comparable to the events that were described at Anak Krakatau. However, the PDC seismic recordings of Stromboli share some characteristics with the PDCs recorded on other volcanoes, such as Merapi [108] and Unzen [109]. In general, the PDC seismic signals have frequency content between 2 and 15 Hz, similar to that of signals due to landslides in loose clastic material. When PDCs originate from dome collapse, as in the case of Unzen [109], they may contain a lower frequency component caused by the collapse of the dome. In general, the onset of PDC seismic signals emerges with a gradual increase in amplitude, which then remains constant on average for the entire time of the event. These characteristics lead to the classification of PDC seismic signals as continuous and non-transient signals [110]. In addition to these characteristics, in the Stromboli PDC seismograms we were able for the first time to identify the seismic signals caused by the impact of PDCs on the coastline. They could be due to the development of a T phase that can be generated when a landslide enters the underwater environment ([89] and references therein). However, the very well defined peak frequency around 3 Hz (Figure 2e), within the tremor frequency band (Figure 2b), can also be attributed to the resonance of the shallow conduit located in the volcano edifice at a small depth below the SdF [77], which generates the volcanic tremor of Stromboli [82].

Processing and combining multispectral infrared images that were acquired by a variety of satellite sensors with different spatial characteristics and acquisition times allowed for us to derive the radiant heat flux from 28 March at 01:10 to 1 April at 20:30, finding a peak of thermal activity of ~1.1 GW on 31 March at 12:45 (Figure 3e). Poor weather conditions, as well as the limited time duration and small magnitude of the event, did not allow for the calculation of the TADR and volume of the lava overflow erupted on 28 March, but only that erupted from 30 to 31 March, obtaining a peak of TADR of ~2.3 $m^3 s^{-1}$ on 31 March at 12:45 and a volume of ~53 × 10^3 m^3 for the 30–31 March overflow. Because it emplaced in 23 h, the Mean Output Rate (MOR) results ~0.65 $m^3 s^{-1}$. Using a ML approach, we were able to estimate a cumulated area over which the two lava overflows emplaced, which amounts to 94,500 m^2. From DEM difference, we retrieved a total average bulk volume of ~179 × 10^3 m^3 emplaced between 8 October 2019 and 7 April 2020, which reduces to a DRE volume of ~136 × 10^3 m^3 when considering an average lava vesicularity of 25% [111]. This value includes four overflows, which all occurred in the same area, on 18 January, 3 February, 28 March and 30–31 March 2020. Comparing it to the satellite-derived estimate for 30–31 March, we found a DRE cumulative volume of ~83 × 10^3 m^3 emitted during the three previous events. Due to the comparable duration (a few hours) of the overflows occurred on 18 January, 3 February, and 28 March, it is plausible to divide this volume equally, obtaining ~27.5 × 10^3 m^3 per each eruptive episode. Summing up this value with the one derived from multispectral infrared satellite images for the 30–31 March, a lava volume of ~80.5 × 10^3 m^3 could have been emitted during 28 March–1 April, eventually providing an average thickness of 0.85 m.

5. Conclusions

Even if eruption-induced mass-flows at Stromboli volcano are common, the triggering mechanisms are yet to be fully understood, also because of the great diversity of the observed phenomena. In this study, the mass-flows that were associated with overflows occurred between 28 March and 1 April 2020 have been analyzed through the use of remote sensing data, both with ground and satellite based sensors, and deriving from seismic sensors. The analysis of the videos recorded by the network of fixed monitoring cameras allowed for the description of the events, as well as the calculation of the velocity to which the lava flowed and the PDCs descended down the SdF and on the sea surface. These videos also made it possible to count the explosions occurring at the summit craters over time. Two GBInSAR devices detected both ground-deformation and morphological changes induced by the slope instability. Remote sensing data also include multispectral satellite data, used to constrain lava flow field area and volume, as well as an estimation of the effusion rate. Moreover, seismic data made it possible to characterize the various stages of the instability phenomena and, at the same time, to integrate the camera data for the description of the relationship between overflows and PDCs, and they were also useful for detecting the signal of the impact of PDCs on the sea surface.

The main results of this study can be summarized, as follows:

- before the analyzed phase, the explosive activity at the summit vents was reasonably intense (20–25 explosions/h), with a prevalence of explosions that produced coarse material in the NEC (i.e., shallow magma level in the conduit);
- the 28 March 2020 overflow was anticipated by some landslides that involved the material accumulated in the areas around the NEC (total eroded volume ~5–6 × 10^3 m^3), even if these did not generate a substantial widening of the crater itself;
- the first overflow was accompanied by a decrease of the total number of explosions/h (from the previous 20–25 to 5–15 explosions/h);
- PDCs were also generated by the crumbling of the overflow front, they reached the sea and formed an apron on the coast;
- no ground deformation was recorded before nor after the 28 March event, meaning that the lava flow volume was small;
- after the first overflow, the number of landslides detected with the seismic network decreased, while the number of explosions increased again, suggesting a new upward movement of the magma level within the conduit;
- the onset of the new overflow phase occurred on 30 March together with a new sharp reduction in the number of explosions, a new increase in the number of landslides, which produced a significant variation in the morphology of the crater and which were associated with the accumulation of incandescent material along the coast line;
- the PDCs linked to the initial phase originated from the NEC area (total eroded volume ~7.3 × 10^3 m^3), whereas, as the effusive phase progressed, the subsequent PDCs were generated directly by crumbling of lava flow front along the steep slope of the SdF;
- PDCs reached the sea with variable speed (between 12.9 and 40.3 ms^{-1}), partly flowing on the water;
- the entry into the sea of these mass-flows is associated with a strong variation in seismic signals, with the disappearance of the typical signal associated with the landslides in Stromboli (high frequency; 4–15 Hz) and the appearance of another one characterized by a large amplitude and lower frequency (1–3 Hz);
- this change in the seismic signal could be due to the PDC entrance in the underwater environment, as well as to the resonance of the Stromboli conduit, which is located in the volcano edifice, at a small depth below the SdF;
- the lava overflows that were emplaced between 28 March and 1 April covered a total area of 94,500 ± 3380 m^2;
- the volume of the deposits accumulated from October 2019 to April 2020 near the NE crater amounts to 34,600 ± 9700 m^3, whereas the volume in the overflows area was of 144,400 ± 79,000 m^3, for a total amount of 179,000 ± 89,000 m^3. Thermal satellite data also allowed for constraining the DRE lava volume between 37 and 69 × 10^3 m^3 emplaced from 30 March to 1 April 2020; integrating this result with those that were obtained from DEM difference, a lava volume of ~80.5 × 10^3 m^3 could have been emitted during 28 March–1 April.

Author Contributions: Conceptualization, S.C., F.D.T., G.G., F.G. and G.M.; methodology, S.C., F.D.T., G.G., A.C., F.G., G.M., and E.P.; validation, S.C., F.D.T., G.G., F.G. and G.M.; formal analysis, S.C., F.D.T., G.G., F.G., G.M., T.N., E.P., and A.C.; data curation, S.C., F.D.T., G.G., F.G., G.M., E.P., A.C., G.B., C.C. and V.C.; writing—original draft preparation, S.C., F.D.T., G.G., F.G., G.M., A.C., and C.D.N.; writing—review and editing, S.C., F.D.T., G.G., F.G. and G.M.; visualization, S.C., F.D.T., G.G., F.G. and G.M.; supervision, S.C., F.D.T., G.G., F.G. and G.M.; project administration, S.C., T.N., G.G., F.G. and G.M.; funding acquisition, N.C. and C.D.N. All authors have read and agreed to the published version of the manuscript.

Funding: This work has been financially supported by the "Presidenza del Consiglio dei Ministri—Dipartimento della Protezione Civile" (Presidency of the Council of Ministers–Department of Civil Protection) (DPC-UNIFI Agreement 2019–2021; Scientific Responsibility: N.C.); this publication, however, does not necessarily reflect the position and the official policies of the Department. Additional funds for paper publication have been provided by INGV-OE.

Acknowledgments: We would like to acknowledge the essential work of INGV technicians for the installation and maintenance of the monitoring networks on Stromboli. Thanks are due to National Aeronautics and Space Administration (NASA) for MODIS data (modis.gsfc.nasa.gov), to National Oceanic and Atmospheric Administration (NOAA) for VIIRS data (https://www.bou.class.noaa.gov/) and to the Sentinel Hub services (https://scihub.copernicus.eu/) for Copernicus Sentinel data. We would like to acknowledge the constructive suggestions of four reviewers who have significantly improved the quality of the paper. The English text was corrected and significantly improved by Novella Tedesco.

Conflicts of Interest: The authors declare no conflict of interest.

References

1. Calvari, S.; Pinkerton, H. Birth, growth and morphologic evolution of the "Laghetto" cinder cone during the 2001 Etna eruption. *J. Volcanol. Geotherm. Res.* **2004**, *132*, 225–239. [CrossRef]
2. Behncke, B.; Neri, M.; Pecora, E.; Zanon, V. The exceptional activity and growth of the Southeast Crater, Mount Etna (Italy), between 1996 and 2001. *Bull. Volcanol.* **2006**, *69*, 149–173. [CrossRef]
3. Del Negro, C.; Cappello, A.; Bilotta, G.; Ganci, G.; Herault, A.; Zago, V. Living at the edge of an active volcano: Risk from lava flows on Mt. Etna. *Geol. Soc. Am. Bull.* **2019**. [CrossRef]
4. Del Negro, C.; Cappello, A.; Ganci, G. Quantifying lava flow hazards in response to effusive eruption. *Geol. Soc. Am. Bull.* **2016**, *128*, 752–763. [CrossRef]
5. Harris, A.J.L.; Chevrel, M.O.; Coppola, D.; Ramsey, M.S.; Hrysiewicz, A.; Thivet, S.; Villeneuve, N.; Favalli, M.; Peltier, A.; Kowalski, P.; et al. Validation of an integrated satellite-data-driven response to an effusive crisis: The April-May 2018 eruption of Piton de la Fournaise. *Ann. Geophys.* **2019**, *62*, VO230. [CrossRef]
6. Neal, C.A.; Brantley, S.R.; Antolik, L.; Babb, J.L.; Burgess, M.; Calles, K.; Cappos, M.; Chang, J.C.; Conway, S.; Desmither, L.; et al. The 2018 rift eruption and summit collapse of Kīlauea Volcano. *Science* **2019**, *363*, 367–374. [CrossRef]
7. Acocella, V. Understanding caldera structure and development: An overview of analogue models compared to natural calderas. *Earth-Sci. Rev.* **2007**, *85*, 125–160. [CrossRef]
8. Michon, L.; Villeneuve, N.; Catry, T.; Merle, O. How summit calderas collapse on basaltic volcanoes: New insights from the April 2007 caldera collapse of Piton de la Fournaise volcano. *J. Volcanol. Geotherm Res.* **2009**, *184*, 138–151. [CrossRef]
9. Tepp, G.; Hotovec-Ellis, A.; Shiro, B.; Johanson, I.; Thelen, W.; Haney, M.M. Seismic and geodetic progression of the 2018 summit caldera collapse of Kilauea volcano. *Earth Plan. Sci. Lett.* **2020**, *540*, 116250. [CrossRef]
10. Calvari, S.; Pinkerton, H. Instabilities in the summit region of Mount Etna during the 1999 eruption. *Bull. Volcanol.* **2002**, *63*, 526–535. [CrossRef]
11. Behncke, B.; Neri, M.; Carniel, R. An exceptional case of endogenous lava dome growth spawning pyroclastic avalanches: The 1999 Bocca Nuova eruption of Mt. Etna (Italy). *J. Volcanol. Geotherm. Res.* **2003**, *124*, 115–128. [CrossRef]
12. Behncke, B.; Calvari, S.; Giammanco, S.; Neri, M.; Pinkerton, H. Pyroclastic density currents resulting from the interaction of basaltic magma with hydrothermally altered rock: An example from the 2006 summit eruptions of Mount Etna, Italy. *Bull. Volcanol.* **2008**, *70*, 1249–1268. [CrossRef]
13. Cole, P.D.; Fernandez, E.; Duarte, E.; Duncan, A.M. Explosive activity and generation mechanisms of pyroclastic flows at Arenal volcano, Costa Rica between 1987 and 2001. *Bull. Volcanol.* **2005**, *67*, 695–716. [CrossRef]

14. Calvari, S.; Intrieri, E.; Di Traglia, F.; Bonaccorso, A.; Casagli, N.; Cristaldi, A. Monitoring crater-wall collapse at active volcanoes: A study of the 12 January 2013 event at Stromboli. *Bull. Volcanol.* **2016**, *78*, 1–16. [CrossRef]
15. Christiansen, R.L.; Peterson, D.W. *The 1980 Eruptions of Mount St. Helens, Washington. Chronology of the 1980 Eruptive Activity*; US Geological Survey: Reston, VA, USA, 1981; Volume 1250, pp. 17–30.
16. Bonaccorso, A.; Calvari, S.; Garfì, G.; Lodato, L.; Patané, D. December 2002 flank failure and tsunami at Stromboli volcano inferred by volcanological and geophysical observations. *Geophys. Res. Lett.* **2003**, *30*, 1941–1944. [CrossRef]
17. Tinti, S.; Manucci, A.; Pagnoni, G.; Armigliato, A.; Zaniboni, F. The 30 December 2002 landslide-induced tsunamis in Stromboli: Sequence of the events reconstructed from the eyewitness accounts. *Nat. Hazards Earth Syst. Sci.* **2005**, *5*, 763–775. [CrossRef]
18. Walter, T.R.; Haghshenas Haghighi, M.; Schneider, F.M. Complex hazard cascade culminating in the Anak Krakatau sector collapse. *Nat. Commun.* **2019**, *10*, 4339. [CrossRef]
19. Calvari, S.; Spampinato, L.; Lodato, L.; Harris, A.J.L.; Patrick, M.R.; Dehn, J.; Burton, M.R.; Andronico, D. Chronology and complex volcanic processes during the 2002-2003 flank eruption at Stromboli volcano (Italy) reconstructed from direct observations and surveys with a handheld thermal camera. *J. Geophys. Res.-Solid Earth* **2005**, *110*, B02201. [CrossRef]
20. Neri, M.; Lanzafame, G. Structural features of the 2007 Stromboli eruption. *J. Volcanol. Geotherm. Res.* **2009**, *182*, 137–144. [CrossRef]
21. Tioukov, V.; Alexandrov, A.; Bozza, C.; Consiglio, L.; D'Ambrosio, N.; De Lellis, G.; De Sio, C.; Giudicepietro, F.; Macedonio, G.; Miyamoto, S.; et al. First muography of Stromboli volcano. *Sci. Rep.* **2019**, *9*, 6695. [CrossRef]
22. Rittmann, A. Der ausbruch des Stromboli am 11 September 1930. *Zeitschrift für vulkanologie* **1931**, *14*, 47–77.
23. Di Roberto, A.; Bertagnini, A.; Pompilio, M.; Bisson, M. Pyroclastic density currents at Stromboli volcano (Aeolian Islands, Italy): A case study of the 1930 eruption. *Bull. Volcanol.* **2014**, *76*, 827. [CrossRef]
24. Giudicepietro, F.; Lopez, C.; Macedonio, G.; Alparone, S.; Bianco, F.; Calvari, S.; De Cesare, W.; Delle Donne, D.; Di Lieto, B.; Esposito, A.M.; et al. Geophysical precursors of the July-August 2019 paroxysmal eruptive phase and their implications for Stromboli volcano (Italy) monitoring. *Sci. Rep.* **2020**, *11*, 1813. [CrossRef]
25. Salvatici, T.; Di Roberto, A.; Di Traglia, F.; Bisson, M.; Morelli, S.; Fidolini, F.; Bertagnini, A.; Pompilio, M.; Hungr, O.; Casagli, N. From hot rocks to glowing avalanches: Numerical modelling of gravity-induced pyroclastic density currents and hazard maps at the Stromboli Volcano (Italy). *Geomorphology* **2016**, *273*, 93–106. [CrossRef]
26. Yamamoto, T.; Takarada, S.; Suto, S. Pyroclastic flows from the 1991 eruption of Unzen volcano, Japan. *Bull. Volcanol.* **1993**, *55*, 166–175. [CrossRef]
27. Baxter, P.; Jenkins, S.; Seswandhana, R.; Komorowski, J.-C.; Dunn, K.; Purser, D.; Voight, B.; Shelley, I. Human survival in volcanic eruptions: Thermal injuries in pyroclastic surges, their causes, prognosis and emergency management. *Burns* **2017**, *43*, 1051–1069. [CrossRef]
28. Auker, M.R.; Sparks, R.S.J.; Siebert, L.; Crosweller, H.S.; Ewert, J. A statistical analysis of the global historical volcanic fatalities record. *J. Appl. Volcanol.* **2013**, *2*, 1–24. [CrossRef]
29. Chouet, B.; Hamisevicz, N.; McGetchin, T.R. Photoballistics of volcanic jet activity at Stromboli, Italy. *J. Geophys. Res.-Solid Earth* **1974**, *79*, 4961–4976. [CrossRef]
30. Ripepe, M.; Rossi, M.; Saccorotti, G. Image processing of explosive activity at Stromboli. *J. Volcanol. Geotherm. Res.* **1993**, *54*, 335–351. [CrossRef]
31. Patrick, M.R.; Harris, A.J.L.; Ripepe, M.; Dehn, J.; Rothery, D.A.; Calvari, S. Strombolian explosive styles and source conditions: Insights from thermal (FLIR) video. *Bull. Volcanol.* **2007**, *69*, 769–784. [CrossRef]
32. Calvari, S.; Bonaccorso, A.; Madonia, P.; Neri, M.; Liuzzo, M.; Salerno, G.G.; Behncke, B.; Caltabiano, T.; Cristaldi, A.; Giuffrida, G.; et al. Major eruptive style changes induced by structural modifications of a shallow conduit system: The 2007-2012 Stromboli case. *Bull. Volcanol.* **2014**, *76*, 841. [CrossRef]
33. Di Traglia, F.; Calvari, S.; D'Auria, L.; Nolesini, T.; Bonaccorso, A.; Fornaciai, A.; Esposito, A.; Cristaldi, A.; Favalli, M.; Casagli, N. The 2014 effusive eruption at Stromboli: New insights from in-situ and remote sensing measurements. *Remote Sens.* **2018**, *10*, 2035. [CrossRef]

34. Plank, S.; Marchese, F.; Filizzola, C.; Pergola, N.; Neri, M.; Nolde, M.; Martinis, S. The July/August 2019 Lava Flows at the Sciara del Fuoco, Stromboli–Analysis from Multi-Sensor Infrared Satellite Imagery. *Remote Sens.* **2019**, *11*, 2879. [CrossRef]
35. Di Traglia, F.; Nolesini, T.; Ciampalini, A.; Solari, L.; Frodella, W.; Bellotti, F.; Fumagalli, A.; De Rosa, G.; Casagli, N. Tracking morphological changes and slope instability using spaceborne and ground-based SAR data. *Geomorphology* **2018**, *300*, 95–112. [CrossRef]
36. Lodato, L.; Spampinato, L.; Harris, A.J.L.; Calvari, S.; Dehn, J.; Patrick, M. The Morphology and Evolution of the Stromboli 2002-03 Lava Flow Field: An Example of Basaltic Flow Field Emplaced on a Steep Slope. *Bull. Volcanol.* **2007**, *69*, 661–679. [CrossRef]
37. Martini, M.; Giudicepietro, F.; D'Auria, L.; Esposito, A.M.; Caputo, T.; Curciotti, R.; De Cesare, W.; Orazi, M.; Scarpato, G.; Caputo, A.; et al. Seismological monitoring of the February 2007 effusive eruption of the Stromboli volcano. *Ann. Geophys.* **2007**, *50*, 775–788. [CrossRef]
38. Pioli, L.; Rosi, M.; Calvari, S.; Spampinato, L.; Renzulli, A.; Di Roberto, A. The eruptive activity of 28 and 29 December 2002. In *The Stromboli Volcano: An Integrated Study of the 2002–2003 Eruption*; Calvari, S., Inguaggiato, S., Puglisi, G., Ripepe, M., Rosi, M., Eds.; American Geophysical Union Monograph Series; John Wiley & Sons: Hoboken, NJ, USA, 2008; Volume 182, pp. 105–116. ISBN 978-0-87590-447-0. [CrossRef]
39. Calvari, S.; Lodato, L.; Steffke, A.; Cristaldi, A.; Harris, A.J.L.; Spampinato, L.; Boschi, E. The 2007 Stromboli flank eruption: Chronology of the events, and effusion rate measurements from thermal images and satellite data. *J. Geophys. Res.-Solid Earth* **2010**, *115*, B04201. [CrossRef]
40. Di Traglia, F.; Battaglia, M.; Nolesini, T.; Lagomarsino, D.; Casagli, N. Shifts in the eruptive styles at Stromboli in 2010–2014 revealed by ground-based InSAR data. *Sci. Rep.* **2015**, *5*, 13569. [CrossRef]
41. Rizzo, A.L.; Federico, C.; Inguaggiato, S.; Sollami, A.; Tantillo, M.; Vita, F.; Bellomo, S.; Longo, M.; Grassa, F.; Liuzzo, M. The 2014 effusive eruption at Stromboli volcano (Italy): Inferences from soil CO2 flux and 3He/4He ratio in thermal waters. *Geophys. Res. Lett.* **2015**, *42*, 2235–2243. [CrossRef]
42. Pistolesi, M.; Delle Donne, D.; Pioli, L.; Rosi, M.; Ripepe, M. The 15 March 2007 explosive crisis at Stromboli volcano, Italy: Assessing physical parameters through a multidisciplinary approach. *J. Geophys. Res.-Solid Earth* **2011**, *116*, B12206. [CrossRef]
43. Davies, D.K.; Quearry, M.W.; Bonis, S.B. Glowing avalanches from the 1974 eruption of the volcano Fuego, Guatemala. *Geol. Soc. Am. Bull.* **1978**, *89*, 369–384. [CrossRef]
44. Di Traglia, F.; Intrieri, E.; Nolesini, T.; Bardi, F.; Del Ventisette, C.; Ferrigno, F.; Frangioni, S.; Frodella, W.; Gigli, G.; Lotti, A.; et al. The Ground-Based InSAR monitoring system at Stromboli volcano: Linking changes in displacement rate and intensity of persistent volcanic activity. *Bull. Volcanol.* **2014**, *76*, 1–18. [CrossRef]
45. Barberi, F.; Rosi, M.; Sodi, A. Volcanic hazard assessment at Stromboli based on review of historical data. *Acta Vulcanol.* **1993**, *3*, 173–187.
46. Bertagnini, A.; Landi, P. The Secche di Lazzaro pyroclastics of Stromboli volcano: A phreatomagmatic eruption related to the Sciara del Fuoco sector collapse. *Bull. Volcanol.* **1996**, *58*, 239–245. [CrossRef]
47. Giordano, G.; Porreca, M.; Musacchio, P.; Mattei, M. The Holocene Secche di Lazzaro phreatomagmatic succession (Stromboli, Italy): Evidence of pyroclastic density current origin deduced by facies analysis and AMS flow directions. *Bull. Volcanol.* **2008**, *70*, 1221–1236. [CrossRef]
48. Lucchi, F.; Francalanci, L.; De Astis, G.; Tranne, C.A.; Braschi, E.; Klaver, M. Geological evidence for recurrent collapse-driven phreatomagmatic pyroclastic density currents in the Holocene activity of Stromboli volcano, Italy. *J. Volcanol. Geotherm. Res.* **2019**, *385*, 81–102. [CrossRef]
49. Burton, M.R.; Calvari, S.; Spampinato, L.; Lodato, L.; Pino, N.A.; Marchetti, E.; Murè, F. Volcanic and seismic activity at Stromboli preceding the 2002–2003 eruption. In: Calvari S, Inguaggiato S, Puglisi G, Ripepe M, Rosi M (eds) The Stromboli volcano: An integrated study of the 2002–2003 eruption. *AGU Geophys. Monogr. Ser.* **2008**, *182*, 93–104. [CrossRef]
50. Ripepe, M.; Delle Donne, D.; Lacanna, G.; Marchetti, E.; Ulivieri, G. The onset of the 2007 Stromboli effusive eruption recorded by an integrated geophysical network. *J. Volcanol. Geotherm. Res.* **2009**, *182*, 131–136. [CrossRef]
51. Massonnet, D.; Feigl, K.L. Radar interferometry and its application to changes in the Earth's surface. *Rev. Geophys.* **1998**, *36*, 441–500. [CrossRef]
52. Ferretti, A.; Prati, C.; Rocca, F. Permanent scatterers in SAR interferometry. *IEEE Trans. Geosci. Remote Sens.* **2001**, *39*, 8–20. [CrossRef]

53. Berardino, P.; Fornaro, G.; Lanari, R.; Sansosti, E. A new algorithm for surface deformation monitoring based on small baseline differential SAR interferograms. *IEEE Trans. Geosci. Remote Sens.* **2002**, *40*, 2375–2383. [CrossRef]
54. Antonello, G.; Casagli, N.; Farina, P.; Leva, D.; Nico, G.; Sieber, A.J.; Tarchi, D. Ground-based SAR interferometry for monitoring mass movements. *Landslides* **2004**, *1*, 21–28. [CrossRef]
55. Di Traglia, F.; Nolesini, T.; Intrieri, E.; Mugnai, F.; Leva, D.; Rosi, M.; Casagli, N. Review of ten years of volcano deformations recorded by the ground-based InSAR monitoring system at Stromboli volcano: A tool to mitigate volcano flank dynamics and intense volcanic activ-ity. *Earth-Sci. Rev.* **2014**, *139*, 317–335. [CrossRef]
56. Casagli, N.; Tibaldi, A.; Merri, A.; Del Ventisette, C.; Apuani, T.; Guerri, L.; Fortuny-Guasch, J.; Tarchi, D. Deformation of Stromboli volcano (Italy) during the 2007 eruption revealed by radar interferometry, numerical modelling and structural geological field data. *J. Volcanol. Geotherm. Res.* **2009**, *182*, 182–200. [CrossRef]
57. Schaefer, L.N.; Di Traglia, F.; Chaussard, E.; Lu, Z.; Nolesini, T.; Casagli, N. Monitoring volcano slope instability with Synthetic Aperture Radar: A review and new data from Pacaya (Guatemala) and Stromboli (Italy) volcanoes. *Earth-Sci. Rev.* **2019**, *192*, 236–257. [CrossRef]
58. Di Traglia, F.; Nolesini, T.; Solari, L.; Ciampalini, A.; Frodella, W.; Steri, D.; Allotta, B.; Rindi, A.; Marini, L.; Monni, N.; et al. Lava delta deformation as a proxy for submarine slope instability. *Earth Plan. Sci. Lett.* **2018**, *488*, 46–58. [CrossRef]
59. Ganci, G.; Cappello, A.; Bilotta, G.; Del Negro, C. How the variety of satellite remote sensing data over volcanoes can assist hazard monitoring efforts: The 2011 eruption of Nabro volcano. *Remote Sens. Environ.* **2020**, *236*, 111426. [CrossRef]
60. Ganci, G.; Vicari, A.; Bonfiglio, S.; Gallo, G.; Del Negro, C. A texton-based cloud detection algorithm for MSG-SEVIRI multispectral images. *Geomat. Nat. Hazards Risk* **2011**, *2*, 279–290. [CrossRef]
61. Ganci, G.; Vicari, A.; Fortuna, L.; Del Negro, C. The HOTSAT volcano monitoring system based on a combined use of SEVIRI and MODIS multispectral data. *Ann. Geophys.* **2011**, *54*, 5. [CrossRef]
62. Ganci, G.; Bilotta, G.; Cappello, A.; Hérault, A.; Del Negro, C. HOTSAT: A multiplatform system for the satellite thermal monitoring of volcanic activity. *Geol. Soc. Lond. Spec. Publ.* **2016**, *426*, 207–221. [CrossRef]
63. Cappello, A.; Ganci, G.; Calvari, S.; Pérez, N.M.; Hernández, P.A.; Silva, S.V.; Cabral, J.; Del Negro, C. Lava flow hazard modeling during the 2014–2015 Fogo eruption, Cape Verde. *J. Geophys. Res.-Solid Earth* **2016**, *121*. [CrossRef]
64. Ganci, G.; Cappello, A.; Bilotta, G.; Hérault, A.; Zago, V.; Del Negro, C. Mapping volcanic deposits of the 2011–2015 Etna eruptive events using satellite remote sensing. *Front. Earth Sci.* **2018**, *6*, 83. [CrossRef]
65. Spampinato, L.; Oppenheimer, C.; Cannata, A.; Montalto, P.; Salerno, G.G.; Calvari, S. On the time-scale of thermal cycles associated with open-vent degassing. *Bull. Volcanol.* **2012**, *74*, 1281–1292. [CrossRef]
66. Rogic, N.; Cappello, A.; Ganci, G.; Maturilli, A.; Rymer, H.; Blake, S.; Ferrucci, F. Spaceborne EO and a Combination of Inverse and Forward Modelling for Monitoring Lava Flow Advance. *Remote Sens.* **2019**, *11*, 3032. [CrossRef]
67. Harris, A.J.L.; Blake, S.; Rothery, D.; Stevens, N. A chronology of the 1991 to 1993 Mount Etna eruption using advanced very high-resolution radiometer data: Implications for real-time thermal volcano monitoring. *J. Geophys. Res.* **1997**, *102*, 7985–8003. [CrossRef]
68. Del Negro, C.; Cappello, A.; Neri, M.; Bilotta, G.; Herault, A.; Ganci, G. Lava flow hazards at Mount Etna: Constraints imposed by eruptive history and numerical simulations. *Sci. Rep.* **2013**, *3*, 3493. [CrossRef]
69. Cappello, A.; Ganci, G.; Bilotta, G.; Hérault, A.; Zago, V.; Del Negro, C. Satellite-driven modeling approach for monitoring lava flow hazards during the 2017 Etna eruption. *Ann. Geophys.* **2019**, *61*. [CrossRef]
70. Calvari, S.; Cannavò, F.; Bonaccorso, A.; Spampinato, L.; Pellegrino, A.G. Paroxysmal Explosions, Lava Fountains and Ash Plumes at Etna Volcano: Eruptive Processes and Hazard Implications. *Front. Earth Sci.* **2018**, *6*, 107. [CrossRef]
71. Corradino, C.; Ganci, G.; Cappello, A.; Bilotta, G.; Hérault, A.; Del Negro, C. Mapping Recent Lava Flows at Mount Etna Using Multispectral Sentinel2 Images and Machine Learning Techniques. *Remote Sens.* **2019**, *11*, 1916. [CrossRef]
72. Ganci, G.; Cappello, A.; Zago, V.; Bilotta, G.; Hérault, A.; Del Negro, C. 3D Lava flow mapping of the 17–25 May 2016 Etna eruption using tristereo optical satellite data. *Ann. Geophys.* **2019**, *62*, VO220. [CrossRef]
73. Ganci, G.; Cappello, A.; Bilotta, G.; Corradino, C.; Del Negro, C. Satellite-Based Reconstruction of the Volcanic Deposits during the December 2015 Etna Eruption. *Data* **2019**, *4*, 120. [CrossRef]

74. Di Traglia, F.; Fornaciai, A.; Favalli, M.; Nolesini, T.; Casagli, N. Catching Geomorphological Response to Volcanic Activity on Steep Slope Volcanoes Using Multi-Platform Remote Sensing. *Remote Sens.* **2020**, *12*, 438. [CrossRef]
75. Nuth, C.; Kääb, A. Co-registration and bias corrections of satellite elevation data sets for quantifying glacier thickness change. *Cryosphere* **2011**, *5*, 271–290. [CrossRef]
76. De Cesare, W.; Orazi, M.; Peluso, R.; Scarpato, G.; Caputo, A.; D'Auria, L.; Giudicepietro, F.; Martini, M.; Buonocunto, C.; Capello, M.; et al. The broadband seismic network of Stromboli volcano, Italy. *Seism. Res. Lett.* **2009**, *80*, 435–439. [CrossRef]
77. Chouet, B.; Dawson, P.; Ohminato, T.; Martini, M.; Saccorotti, G.; Giudicepietro, F.; De Luca, G.; Milana, G.; Scarpa, R. Source mechanisms of explosions at Stromboli volcano, Italy, determined from moment-tensor inversions of very- long-period data. *J. Geophys. Res.-Solid Earth* **2003**, *108*, 2019. [CrossRef]
78. D'Auria, L.; Giudicepietro, F.; Martini, M.; Peluso, R. Seismological insight into the kinematics of the 5 April 2003 vulcanian explosion at Stromboli volcano (southern Italy). *Geophys. Res. Lett.* **2006**, *33*, 8. [CrossRef]
79. Esposito, A.M.; Giudicepietro, F.; D'Auria, L.; Scarpetta, S.; Martini, M.G.; Coltelli, M.; Marinaro, M. Unsupervised neural analysis of very-long-period events at Stromboli volcano using the self-organizing maps. *Bull. Seism. Soc. Am.* **2008**, *98*, 2449–2459. [CrossRef]
80. Giudicepietro, F.; D'Auria, L.; Martini, M.; Caputo, T.; Peluso, R.; De Cesare, W.; Scarpato, G. Changes in the VLP seismic source during the 2007 Stromboli eruption. *J. Volcanol. Geotherm. Res.* **2009**, *182*, 162–171. [CrossRef]
81. Calvari, S.; Büttner, R.; Cristaldi, A.; Dellino, P.; Giudicepietro, F.; Orazi, M.; Peluso, R.; Spampinato, L.; Zimanowski, B.; Boschi, E. The 7 September 2008 Vulcanian explosion at Stromboli volcano: Multi-parametric characterisation of the event and quantification of the ejecta. *J. Geophys. Res.-Solid Earth* **2012**, *117*, B05201. [CrossRef]
82. Chouet, B.; Saccorotti, G.; Martini, M.; Dawson, P.; De Luca, G.; Milana, G.; Scarpa, R. Source and path effects in the wavefields of tremor and explosions at Stromboli Volcano, Italy. *J. Geophys. Res.* **1997**, *102*, 15129–15150. [CrossRef]
83. Esposito, A.M.; Giudicepietro, F.; Scarpetta, S.; D'Auria, L.; Marinaro, M.; Martini, M. Automatic discrimination among landslide, explosion-quake, and microtremor seismic signals at Stromboli volcano using neural networks. *Bull. Seismol. Soc. Am.* **2006**, *96*, 1230–1240. [CrossRef]
84. Esposito, A.M.; D'Auria, L.; Giudicepietro, F.; Peluso, R.; Martini, M. Automatic recognition of landslides based on neural network analysis of seismic signals: An application to the monitoring of Stromboli volcano (Southern Italy). *Pure Appl. Geophys.* **2013**, *170*, 1821–1832. [CrossRef]
85. Taddeucci, J.; Valentine, G.A.; Sonder, I.; White, J.D.L.; Ross, P.-S.; Scarlato, P. The effect of pre-existing craters on the initial development of explosive volcanic eruptions: An experimental investigation. *Geophys. Res. Lett.* **2013**, *40*, 507–510. [CrossRef]
86. Endo, E.T.; Murray, T. Real-time seismic amplitude measurement (RSAM): A volcano monitoring and prediction tool. *Bull. Volcanol.* **1991**, *53*, 533–545. [CrossRef]
87. Harris, A.J.L.; Dehn, J.; Calvari, S. Lava effusion rate definition and measurement: A review. *Bull. Volcanol.* **2007**, *70*, 1–22. [CrossRef]
88. Ganci, G.; Bilotta, G.; Calvari, S.; Cappello, A.; Corradino, C.; Del Negro, C. Integrating tri-stereo Pleiades images with infrared satellite data to monitor volcanoes: The 2019 Stromboli eruption. *EGU Gen. Assem.* **2019**, 16172. [CrossRef]
89. Okal, E.A. The generation of T waves by earthquakes. *Adv. Geophys.* **2008**, *49*, 1–65.
90. Ripepe, M.; Marchetti, E.; Ulivieri, G.; Harris, A.; Dehn, J.; Burton, M.; Caltabiano, T.; Salerno, G. Effusive to explosive transition during the 2003 eruption of Stromboli volcano. *Geol. Soc. Am.* **2005**, *33*, 341–344. [CrossRef]
91. Marotta, E.; Calvari, S.; Cristaldi, A.; D'Auria, L.; Di Vito, M.A.; Moretti, R.; Peluso, R.; Spampinato, L.; Boschi, E. Reactivation of Stromboli's summit craters at the end of the 2007 effusive eruption detected by thermal surveys and seismicity. *J. Geophys. Res.* **2015**, *120*, 7376–7395. [CrossRef]
92. Lipman, P.W.; Banks, N.G. Aa Flow dynamics, Mauna Loa 1984. *US Geol. Surv. Prof. Pap.* **1987**, *1350*, 1527–1567.

93. Calder, E.S.; Luckett, R.; Sparks, R.S.J.; Voight, B. Mechanisms of lava dome instability and generation of rockfalls and pyroclastic flows at Soufrière Hills Volcano, Montserrat. *Geol. Soc. Mem. Lond.* **2002**, *21*, 173–190. [CrossRef]
94. Loughlin, S.C.; Calder, E.S.; Clarke, A.; Cole, P.D.; Luckett, R.; Mangan, M.T.; Pyle, D.M.; Sparks, R.S.J.; Voight, B.; Watts, R.B. Pyroclastic flows and surges generated by the 25 June 1997 dome collapse, Soufrière Hills Volcano, Montserrat. *Geol. Soc. Mem. Lond.* **2002**, *21*, 191–209. [CrossRef]
95. Yamamoto, T.; Nakamura, Y.; Glicken, H. Pyroclastic density current from the 1888 phreatic eruption of Bandai volcano, NE Japan. *J. Volcanol. Geotherm. Res.* **1999**, *90*, 191–207. [CrossRef]
96. Pino, N.A.; Ripepe, M.; Cimini, G.B. The Stromboli volcano landslides of December 2002: A seismological description. *Geophys. Res. Lett.* **2004**, *31*, L02605. [CrossRef]
97. Calvari, S.; Spampinato, L.; Lodato, L. The 5 April 2003 vulcanian paroxysmal explosion at Stromboli volcano (Italy) from field observations and thermal data. *Jour. Volc. Geotherm. Res.* **2006**, *149*, 160–175. [CrossRef]
98. Bonaccorso, A.; Calvari, S.; Linde, A.; Sacks, S.; Boschi, E. Dynamics of the shallow plumbing system investigated from borehole strainmeters and cameras during the 15 March 2007 Vulcanian paroxysm at Stromboli volcano. *Earth Plan. Sci. Lett.* **2012**, *357–358*, 249–256. [CrossRef]
99. Calvari, S. Understanding basaltic lava flow morphologies and structures for hazard assessment. *Ann. Geophys.* **2019**, *62*, VO217. [CrossRef]
100. Giordano, D.; Russell, J.K. Towards a structural model for the viscosity of geological melts. *Earth Plan. Sci. Lett.* **2018**, *501*, 202–212. [CrossRef]
101. Tazieff, H. An Exceptional Eruption: Mt. Niragongo, Jan. 10th, 1977. *Bull. Volc.* **1977**, *40*, 189–200. [CrossRef]
102. Komorowski, J.C. The January 2002 flank eruption of Nyiragongo volcano (Democratic Republic of Congo): Chronology, evidence for a tectonic rift trigger, and impact of lava flows on the city of Goma. *Acta Vulc.* **2003**, *15*, 27–62.
103. Borgia, A.; Linneman, S.; Spencer, D.; Morales, L.; Andre, L. Dynamics of the flow fronts, Arenal Volcano, Costa Rica. *J. Volcanol. Geotherm. Res.* **1993**, *19*, 303–329. [CrossRef]
104. Cigolini, C.; Borgia, A.; Casertano, L. Intra-crater activity, aa-block lava, viscosity and flow dynamics: Arenal Volcano, Costa Rica. *J. Volcanol. Geotherm. Res.* **1984**, *20*, 155–176. [CrossRef]
105. Kilburn, C.R.J.; Guest, J.E. Aa lavas of Mount Etna, Sicily. In *Active Lavas: Monitoring and Modelling*; Kilburn, C.R.J., Luongo, G., Eds.; UCL Press: London, UK, 1993; pp. 73–106.
106. Apuani, T.; Corazzato, C.; Cancelli, A.; Tibaldi, A. Stability of a collapsing volcano (Stromboli, Italy): Limit equilibrium analysis and numerical modelling. *J. Volcanol. Geotherm. Res.* **2005**, *144*, 191–210. [CrossRef]
107. Perttu, A.; Caudron, C.; Assink, J.D.; Metz, D.; Tailpied, D.; Perttu, B.; Hibert, C.; Nurfiani, D.; Pilger, C.; Muzli, M.; et al. Reconstruction of the 2018 tsunamigenic flank collapse and eruptive activity at Anak Krakatau based on eyewitness reports, seismo-acoustic and satellite observations. *Earth Plan. Sci. Lett.* **2020**, *541*, 116268. [CrossRef]
108. Darmawan, H.; Yuliantoro, P.; Rakhman, A.; Santoso, A.B.; Humaida, H.; Suryanto, W. Dynamic velocity and seismic characteristics of gravitational rockfalls at the Merapi lava dome. *J. Volcanol. Geotherm. Res.* **2020**, *404*, 107010. [CrossRef]
109. Uhira, K.; Yamasato, H.; Takeo, M. Source mechanism of seismic waves excited by pyroclastic flows observed at Unzen volcano, Japan. *J. Geophys. Res. Solid Earth* **1994**, *99*, 17757–17773. [CrossRef]
110. Wassermann, J. Volcano Seismology. In *IASPEI New Manual of Seismological Observatory Practice*; Bormann, P., Ed.; GFZ Potsdam: Potsdam, Germany, 2002; Volume 1, p. 42.
111. Fornaciai, A.; Landi, P.; Armienti, P. Dissolution/crystallization kinetics recorded in the 2002–2003 lavas of Stromboli (Italy). *Bull. Volcanol.* **2009**, *71*, 631–641. [CrossRef]

remote sensing

MDPI

Article

Examples of Multi-Sensor Determination of Eruptive Source Parameters of Explosive Events at Mount Etna

Valentin Freret-Lorgeril [1,*], Costanza Bonadonna [1], Stefano Corradini [2], Franck Donnadieu [3], Lorenzo Guerrieri [2], Giorgio Lacanna [4], Frank Silvio Marzano [5,6], Luigi Mereu [5,6], Luca Merucci [2], Maurizio Ripepe [4], Simona Scollo [7] and Dario Stelitano [2]

1 Department of Earth Sciences, University of Geneva, 1205 Geneva, Switzerland; costanza.bonadonna@unige.ch
2 Centro Nazionale Terremoti (CNT), Istituto Nazionale di Geofisica e Vulcanologia (INGV), 00143 Rome, Italy; stefano.corradini@ingv.it (S.C.); lorenzo.guerrieri@ingv.it (L.G.); luca.merucci@ingv.it (L.M.); dario.stelitano@ingv.it (D.S.)
3 Laboratoire Magmas et Volcans, CNRS, IRD, OPGC, Université Clermont-Auvergne, F-63000 Clermont-Ferrand, France; franck.donnadieu@uca.fr
4 Department of Earth Sciences, University of Firenze, 50124 Firenze, Italy; giorgio.lacanna@unifi.it (G.L.); maurizio.ripepe@unifi.it (M.R.)
5 Department of Information Engineering, Sapienza University of Rome, 00184 Rome, Italy; frank.marzano@uniroma1.it (F.S.M.); luigi.mereu@uniroma1.it (L.M.)
6 Centre of Excellence CETEMPS, 67100 L'Aquila, Italy
7 Istituto Nazionale di Geofisica e Vulcanologia, Osservatorio Etneo, 95125 Catania, Italy; simona.scollo@ingv.it
* Correspondence: valentin.freret-lorgeril@unige.ch

Citation: Freret-Lorgeril, V.; Bonadonna, C.; Corradini, S.; Donnadieu, F.; Guerrieri, L.; Lacanna, G.; Marzano, F.S.; Mereu, L.; Merucci, L.; Ripepe, M.; et al. Examples of Multi-Sensor Determination of Eruptive Source Parameters of Explosive Events at Mount Etna. *Remote Sens.* **2021**, *13*, 2097. https://doi.org/10.3390/rs13112097

Academic Editor: David Gomez-Ortiz

Received: 12 May 2021
Accepted: 23 May 2021
Published: 27 May 2021

Abstract: Multi-sensor strategies are key to the real-time determination of eruptive source parameters (ESPs) of explosive eruptions necessary to forecast accurately both tephra dispersal and deposition. To explore the capacity of these strategies in various eruptive conditions, we analyze data acquired by two Doppler radars, ground- and satellite-based infrared sensors, one infrasound array, visible video-monitoring cameras as well as data from tephra-fallout deposits associated with a weak and a strong paroxysmal event at Mount Etna (Italy). We find that the different sensors provide complementary observations that should be critically analyzed and combined to provide comprehensive estimates of ESPs. First, all measurements of plume height agree during the strong paroxysmal activity considered, whereas some discrepancies are found for the weak paroxysm due to rapid plume and cloud dilution. Second, the event duration, key to convert the total erupted mass (TEM) in the mass eruption rate (MER) and vice versa, varies depending on the sensor used, providing information on different phases of the paroxysm (i.e., unsteady lava fountaining, lava fountain-fed tephra plume, waning phase associated with plume and cloud expansion in the atmosphere). As a result, TEM and MER derived from different sensors also correspond to the different phases of the paroxysms. Finally, satellite retrievals for grain-size can be combined with radar data to provide a first approximation of total grain-size distribution (TGSD) in near real-time. Such a TGSD shows a promising agreement with the TGSD derived from the combination of satellite data and whole deposit grain-size distribution (WDGSD).

Keywords: tephra; remote sensing; plume height; mass eruption rate; total erupted mass; total grain-size distribution

1. Introduction

The injection of large volumes of tephra into the atmosphere during explosive eruptions has the potential to cause air traffic disruption, while the associated fallout may also impact public health, infrastructures, and various economic sectors (e.g., agriculture, tourism) [1,2]. The near real-time monitoring of active volcanoes is, thus, critical and requires strategies that are valid for a large set of eruptive conditions. At Mount

Etna (Italy), a variety of monitoring networks exists and includes a unique set of complementary remote sensing systems with ground- and satellite-based infrared instruments (e.g., [3–8]), Doppler radars [9–15], infrasound arrays [16,17], lidar [18–20] and visual cameras [20–22] (Figure 1a). The Istituto Nazionale di Geofisica e Vulcanologia, Osservatorio Etneo (INGV-OE) is also responsible for characterizing tephra-fallout deposits associated with Etna explosive events [23–27]. Originally, each strategy has been used individually to provide some estimates of key eruptive source parameters (ESPs), such as mass eruption rate (MER), plume height (H_T), total erupted mass (TEM), total grain-size distribution (TGSD), necessary to volcanic ash transportation and dispersal models (VATDMs) [3,4,12–14,16,21,28,29]. However, due to intrinsic limitations (e.g., sensor detection limits, deposit exposure), individual strategies cannot provide a comprehensive characterization of all these ESPs. Recently, multi-sensor strategies have been developed to better constrain MER, H_T, and TGSD from ground sampling, plume models, and available remote sensing systems [7,15,30–34]. Ultimately, multi-sensor strategies are also being used for the real-time determination of column height during volcano monitoring activities [22]. However, the accuracy of such combined strategies remains poorly constrained. Moreover, the sensor applicability limits have been investigated mostly during strong events and not verified for less intense plumes.

Between 2011 and 2015, Etna has produced about 50 paroxysmal events associated with the emission of fountain-fed tephra plumes. All paroxysms are characterized by highly varying ESPs with plume heights ranging between 5.2 and 17.6 km above sea level [8,33,35,36]. Such variability of eruptive processes may challenge the development of multi-sensor strategies especially because they are mostly based on strong paroxysms such as the events on 23 November 2013 [6,7,27] and 3–5 December 2015 [14,37,38]. In order to discuss the capacity of various remote sensing strategies to provide accurate ESPs regardless of eruptive conditions, here we characterize the weak paroxysm that occurred on 10 April 2011 and we compare it with the strong paroxysm of 23 November 2013. It is important to note that in addition to the interest in exploring the characterization of weak paroxysms, which are generally more frequent than strong paroxysms [35,36], such a selection was mostly driven by the availability of geophysical data as well as deposit observations. In fact, both the 10 April 2011 and the 23 November 2013 paroxysms are associated with data from two Doppler radars, visual and infrared cameras, satellite retrievals, and infrasonic signals as well as deposit sampling.

On 8 April 2011, a strombolian activity started at the New Southeast Crater (NSEC) around 06:00 UTC and lasted 2 days until the activity turned into lava fountaining at 09:12 UTC on 10 April 2011 [4,35]. This activity was weak and characterized by few sustained ballistic emissions reaching about 200 m above the crater rim [5]. A fountain-fed plume was emitted (Figure 1b) and reached altitudes of ~4 km above sea level at 10:00 UTC then about 7 km at 11:00 UTC according to the Hybrid Single-Particle Lagrangian Integrated Trajectory Model (HYSPLIT) forecast [39]. Despite this weak activity, the plume was sub-vertical due to the presence of low wind speed with an average of 7.6 m/s [4]. This eruption, that lasted 250 min [30], has been recorded by a total of six different remote sensing systems (i.e., Microwave Weather Radar—MWR, L-Band Radar—VOLDORAD-2B, Satellites—Aqua/Terra- Moderate Resolution Imaging Spectroradiometer (MODIS) and MSG-Spinning Enhanced Visible and Infrared Imager (SEVIRI), Infrasound, Visual and Infrared Cameras) (Figure 1) and the tephra-fallout deposit was sampled just after the event at 18 different locations (Figure 2).

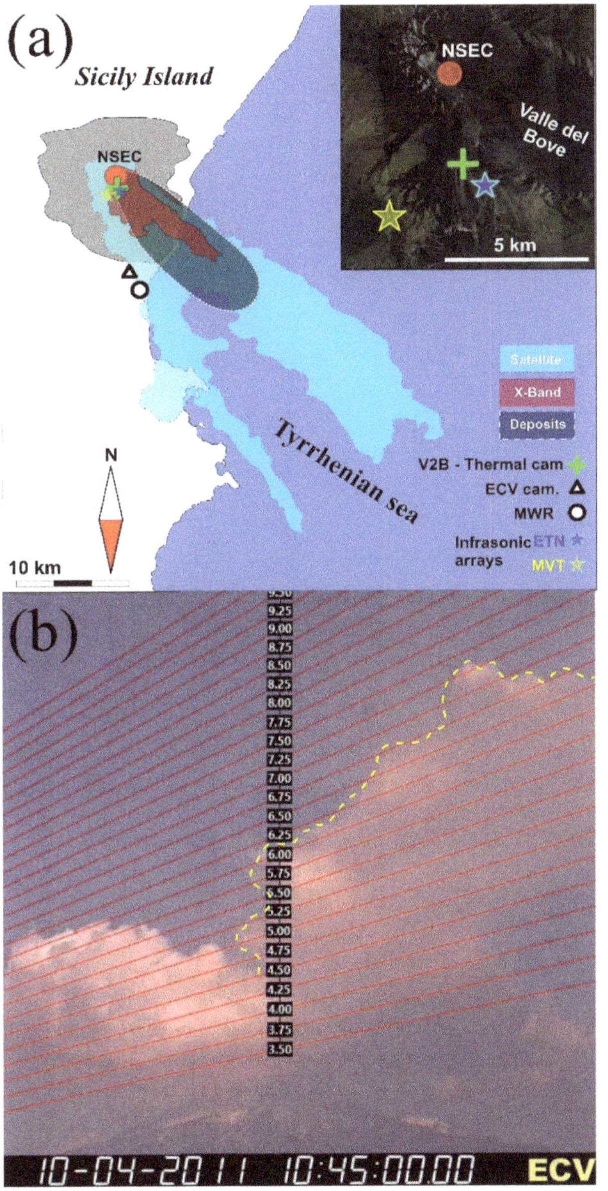

Figure 1. (**a**) Location of all remote sensing systems considered in this work (VOLDORAD 2B: V2B, Thermal camera; X-band weather radar: MWR; visible camera: ECV; Infrasound arrays on Etna ETN and Monte Vettore MVT) as well as the extension of the tephra-fallout deposit associated with the 10 April 2011 event (black area; 1 g/m^2 isomass line), and both the X-band (red area) and the satellite-derived plume/cloud margins (blue area). Note that V2B and the thermal camera have the same location. Inset: location of the remote sensors closed to the vent. (**b**) Observation from Catania visible camera (ECV) of the 10 April 2011 fountain-fed tephra plume.

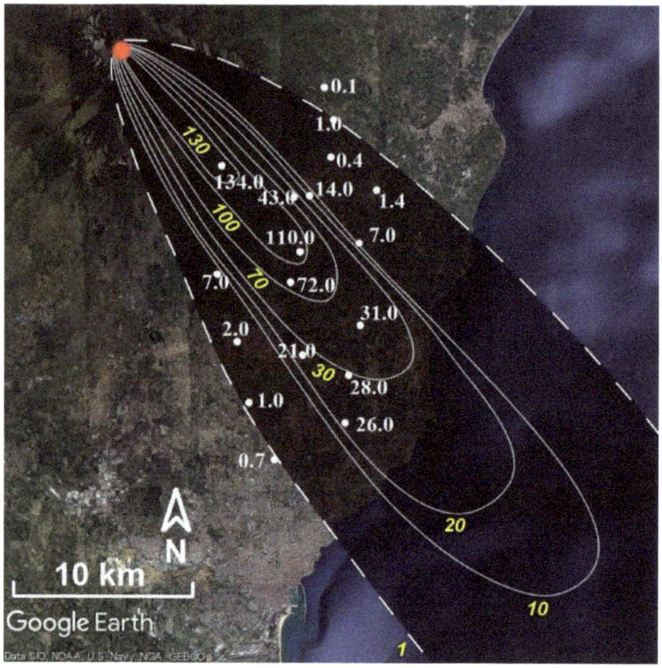

Figure 2. Isomass map of the 10 April 2011 tephra-fallout deposit. Individual sample values of ground accumulation (g/m^2) are shown in white. Isoline values are displayed in yellow.

The 23 November 2013 paroxysm is one of the largest paroxysms that occurred at Etna since 2001 [6,7,27,30–32]. Sustained lava fountains reaching heights >1 km above the vent generated a moderately weak tephra plume at altitudes of 11–12 km a.s.l. under a mean wind velocity of 17.9 m/s [4]. Bombs and blocks were carried up to 5 km from the vent and a 2-cm thick tephra deposit with cm-sized lapilli up to 20 km from the vent was observed [27]. The 23 November 2013 paroxysmal episode has been well documented and already used in the past as a case study to test multi-sensor strategies [7,15,31,33].

This paper is organized as follows. In Section 2, we present all methodologies and strategies used to derive ESPs of the 10 April 2011 and the 23 November 2013 paroxysms. We describe the results in Section 3 and discuss them in Section 4. Finally, conclusions are drafted in Section 5. See Appendix A for all acronyms and symbols used in this paper.

2. Methodology

2.1. ESPs from Tephra-Fallout Deposit

During the 10 April 2011 paroxysm, the fountain-fed plume was drifted southeastwardly by predominant winds and produced an elongated tephra-fallout deposit up to the coastline (Figures 1 and 2). In total, 18 samples were collected from 7 to 22 km from the NSEC (now simply called SEC). Unfortunately, as often happens at Etna due to the difficult access to summit areas and inside La Valle del Bove, no proximal data, i.e., <7 km from the vent, could be acquired.

2.1.1. Total Erupted Mass and Mass Eruption Rate

We compiled an isomass map from measurements of ground accumulation (g/m^2) across the entire deposit. The TEM associated with the tephra-fallout deposit (TEM_{dep}) was obtained based on the most used strategies in literature, i.e., by integrating the exponential best fit [40], the power-law best fit [41], and the Weibull best fit functions [42] of mass/area data versus the square root of isomass contours (Figure 2). The average MER_{dep} was then derived by dividing the TEM_{dep} obtained by the three integration methods by the duration of the tephra plume emission based on the two most meaningful remote sensing systems for this phase (i.e., MWR and satellite-based data; see following sections for more details).

2.1.2. Whole Deposit Grain-Size Distribution

All tephra-fallout samples were manually sieved down to 4 Φ (i.e., 63 microns). We obtained the mass of all individual sieved fractions, i.e., each half Φ class, using a 10^{-4} g resolution weighing scale (Figure A1). We determined the size distributions of all samples having a mass of ≤1 g using a BETTERSIZER morpho-grainsizer (https://www.3p-instruments.com/analyzers/bettersizer_s3_plus/; accessed on 26 May 2021). We applied the Voronoi Tessellation method of Bonadonna and Houghton [41] to compute the whole deposit grain-size distribution (WDGSD) by using a dedicated Matlab application [43].

2.2. ESPs from Doppler Radars

In this study, we consider two different ground-based pulsed Doppler radars whose respective different characteristics allow for different ESPs to be constrained. One of these radars is a scanning microwave weather radar (MWR) working at a wavelength λ of 3×10^{-2} m and located at Catania airport, 33 km south from Mount Etna (Figure 1a). The MWR is operated by the Italian Civil Protection and provides 12 different elevated scans up to a maximum distance of 80 km from the radar site and with space-time resolution of 100 m and 10 min [7,14,15,33]. In addition, we take advantage of the fixed-pointing VOLDORAD-2B (V2B), located at La Montagnola Station (2610 m a.s.l.) (Figure 1a), which has been monitoring the near-source explosive activity of Etna's summit craters since 2009 [11,12]. V2B is an L-band Doppler radar working at $\lambda = 23.5 \times 10^{-2}$ m and whose sampling rate of 5 Hz allows to record explosive ejection at high time resolution. Besides, open-access data based on Doppler radar records at Etna is made available, including about 50 eruptive episodes since 2011 [12,44].

2.2.1. Mass Parameters from MWR

Various strategies have been developed to derive the MER from Doppler radar at Mount Etna. First, assuming that MWR detects near-source eruptive jets that are vertical during paroxysm and uniform within the vent surface, we can derive the MER based on the surface flux approach (SFA; [15,33]). Similarly to the methodology of Calvari et al. [4] to derive eruptive bulk volumes from thermal infrared images at Etna (see Section 2.4.1), the SFA uses the following equation:

$$MER^{SFA}(t) = v_{exit}(t)\rho_x S \quad (1)$$

where ρ_x is the density of the detected mixture set to 14.9 ± 3 kg/m^3 for Mount Etna's lava fountains [15] and S is the eruptive vent surface (m^2). Using this approach implies that the exit velocity is linked to the whole erupted mixture, i.e., both lava fountains and tephra plumes.

An additional strategy can be used to derive the MER from the MWR. This second approach, called near surface approach (NSA; [14,33]), implies that the recorded velocity corresponds to that of tephra particles entering the beam and not the exit velocity at the vent. This method uses the following flux equation:

$$MER^{NSA}(t) = C_t(t)v_{entry}(t)A \quad (2)$$

where $MER^{NSA}(t)$ represents the MER (kg/s) based on the NSA approach, $C_t(t)$ is the tephra concentration (kg/m^3), v_{entry} is the entry velocity of particles in the radar beam (m/s) (as a first approximation $v_{entry} = v_{exit}$ of Equation (1)), and A (different from S) is the entry surface of the detected volcanic jet in the beam (m^2).

The computation of the tephra concentration (Equation (2)) detected by the MWR is based on the volcanic advanced (or Ash) radar retrieval algorithm (VARR) developed by Marzano et al. [10] and largely described in Marzano et al. [45], Mereu et al. [46], and Marzano et al. [15]. Basically, the VARR provides synthetic estimates of concentration C_t and mean diameter D_n of detected particle mixtures (see Section 2.2.3) by using a Monte Carlo approach. By entering a given tephra size class [15,47], the VARR solves equations that link the radar reflectivity factor Z (in dBZ) with both concentration and diameter [9,47]. $C_t(t)$ is obtained using the following equation:

$$C_t(t) = \beta \exp[Z(t)^{\gamma}] \quad (3)$$

with β and γ being two parameters determined by the VARR related to a given class of tephra sizes (see [46]).

In addition to the SFA and NSA, the MWR provides 3D scans of the entire eruptive columns [14,15] and first estimates of H_T. From this, we can use the top plume approach (TPA; [15]) by entering H_T in the Degruyter and Bonadonna [48] formula to derive MER^{TPA}. An additional approach that can be used with MWR data, the mass continuity approach (MCA; [15,47]), is based on the mass conservation equations and is calculated by considering the mass that enters and leaves a constrained volume above the eruptive vent (see [15,46] for more details).

2.2.2. Mass Parameters from V2B

Given that the V2B's beam is fixed and pointing right above Etna's active vents, it cannot provide direct information on the eruptive columns that are mostly fed by lava fountains. Nonetheless, as for MWR determination of mass parameters, the SFA (Equation (1)) and NSA (Equation (2)) can be used directly with V2B's data [15]. The exit velocity from V2B is calculated by taking into account the elevation angle (θ) of the radar beam with $v_{exit}=v_r/\sin(\theta)$ ([12]). In particular, $1/\sin(\theta)$ is equal to 4.45 in Equation (1) for all events that happened before December 2012 at Etna and 3.89 for the others until today [30]. Following V2B's radar beam description provided by Donnadieu et al. [12] and Freret-Lorgeril et al. [30], A in Equation (2) corresponds to the half 3-D surface of the volume sounded above the vent, which has a height of 300 m (i.e., the length of the two probed volumes above the NSEC; [30]) and a radius of 280 m. Finally, for V2B, β and γ in Equation (3) are equal to 0.8827 and 0.04625, respectively.

In addition, assuming that the tephra plume is fed by the lava fountain, the product between V2B echo power and radial velocity (v_r) measured in beam volumes above the erupting crater can also be used to calculate the MER. In fact, this product, directly proportional to the tephra mass in the bin (mass proxy), has been shown to be proportional to H_T and has been calibrated using the theoretical formula of Degruyter and Bonadonna [48] that links H_T to the MER [30]. Estimates of MER using the proxy method have been applied to 47 paroxysms that occurred at Etna between 2011 and 2015.

2.2.3. Radar Grain-Size Distribution

Dual-polarization radars, such as the MWR, provide the first estimates of grain-size distributions related to the size classes they are sensitive to. As stated above, the VARR provides an estimate of the reflectivity-weighted mean diameter D_n potentially detected by both MWR and V2B by applying the following parametric equation:

$$D_n(t) = \delta \exp[Z(t)^{\varepsilon}] \quad (4)$$

with δ and ε being two parameters, whose values depend on which class of tephra sizes is input in the VARR algorithm (see [45,46]).

Depending on the aforementioned tephra size classes, β, γ, δ, and ε parameters (Equations (3) and (4)) have different values and can provide polydisperse grain-size information from MWR measurements [46]. Starting from C_t estimates of VARR

(Equation (3)), the MWR grain-size distribution is computed as the ratio between the particle weight W_p based on the mean diameter D_n (Equation (4)) and the total particle weight W_t of the whole tephra plume. We compute W_p as the total tephra mass, for discrete increasing steps of particle size and multiply by the gravitational acceleration in kg/m^2/s. Each discrete mass value is normalized by the total particle weight W_t, computed as the integral of W_p extended to the whole particle diameter (D) range, i.e., a single class of possible sizes from 0.008 to 64 mm used in this study. In this way, we derive the GSD_{MWR} (wt%) according to the following relation:

$$W_p(D_n) = \left(\frac{4}{3}\pi\left(\frac{D_n}{2}\right)^3 \rho_p\right)g \text{ and } wt\% = \frac{W_p(D_n)}{\int_0^\infty W(D_n)\mathrm{d}D} 100 \quad (5)$$

where g is the gravitational acceleration.

2.2.4. Plume Height

Given that the VOLDORAD 2B does not capture the whole plume but only the jet region, H_T can only be determined based on the MWR located at Catania airport (Figure 1b). Plume height determination from MWR data is straightforward thanks to each radar scan of the eruptive column with a resolution of 10 min [14,15,28]. Height measurements by the MWR meet the usual detection limitations such as possible incomplete volume filling, beam scans not going high enough, small particle concentrations, and/or sizes remaining undetected on plume margins.

2.3. Mass Eruption Rate from Ground-Based Thermal Camera

The INGV-OE monitoring network includes ground-based thermal infrared cameras that have been used to indirectly determine the MER during paroxysmal events [3,4,6] (Figure 1a). By estimating the height of the thermally saturated domains in eruptive columns that are assumed to correspond to lava fountains [4] and found to be very similar to the vertical ballistic domain seen with the MWR [46], the Torricelli equation can be used to compute the source exit velocity (v_{exit} in m/s) [15,33]. Assuming that the tephra plumes at Etna are fed by lava fountains during paroxysmal activity [15,30,33], the exit velocity can be used to compute the MER using the SFA method (Equation (1); [15]).

2.4. ESPs from Satellite Retrievals

2.4.1. Plume Height

Satellite-based observations made each 15 min by the geostationary MSG-SEVIRI platform are widely used to retrieve volcanic plume heights [7,8,49,50]. At Etna, they are retrieved by applying the dark pixel procedure [51] over an area of 729 km^2 (9 $\times$ 9 pixels of 3 km resolution) centered on Etna's summit craters [6,7] (Figure 1a). Assuming that detected plumes are in thermal equilibrium with the surrounding atmosphere, the volcanic plume heights can be derived from the comparison between the darkest pixel (the pixel with the lower brightness temperature computed at 11 μm) of the selected area and the ARPA atmospheric profiles available at INGV-OE every 6 h [8,20,52]. When available, we also use data obtained by MODIS onboard the NASA Terra and Aqua polar-orbiting platforms. Despite their lower temporal resolution, i.e., three-four measurements per day over Etnean area compared with 96 or 288 SEVIRI daily images (every 15 or 5 min respectively), MODIS spatial resolution is higher with an image pixel resolution of 1 km^2. One of the main strengths of satellite-based detection is the capacity to capture volcanic plumes and clouds over large distances from their source (Figure 1a). However, as ground-based visible imagery, satellite-based ash detection suffers from the presence of meteorological water and ice clouds in the plume environment. In addition, satellite-derived plume height might present large uncertainties when plumes are too diluted to avoid ground contribution to the overall mixture temperature [52].

The discrimination between ash and ice/water vapor particles is obtained by exploiting their different absorption at 11 and 12 μm [53]. Negative brightness temperature difference (BTD), the difference between the brightness temperature at 11 and 12 μm, indicates the presence of ash and vice-versa for ice/water vapor particles. In particular, the 10 April 2011 event shows the formation of large quantities of ice particles that cover almost all the ash present in the volcanic cloud.

2.4.2. Erupted Mass and Grain-Size Distribution

Quantitative ESPs from SEVIRI data are derived using the volcanic plume retrieval algorithm to estimate the amount of fine ash and also SO_2 carried by volcanic clouds (VPR; [7,29,54–56]). This procedure allows removing the detected volcanic cloud from satellite images by a linear interpolation of the radiances at the plume edges. The comparison between the original and the interpolated images allows the estimation of the volcanic cloud transmittances at SEVIRI thermal infrared (TIR) bands centered at 8.7, 10.8, and 12 μm (Channels 7, 9, and 10). From those quantities, the particles' effective radius (Re) and the aerosol optical depth (AOD) are derived. From both Re and AOD, the ash mass per unit area (g/m^2) can be computed using the Wen and Rose [57] simplified formula. The same procedure stands for MODIS Channels 29, 31, and 32. Consequently, we can compute the GSD_{sat} by using values of particle radius in all pixels containing ash signal from the beginning to the end of the paroxysmal event. The GSD_{sat} was weighted in mass, for each SEVIRI image, using the ratio between the total mass of pixels containing particles within a certain Φ range and the total mass of the whole volcanic cloud. We computed the mean GSD_{sat} by averaging all the SEVIRI images that displayed at least 100 pixels containing ash particles. It is important to note that the ash particles retrieved in the TIR spectral range are those with effective radii (Re) between 0.5 and 10 μm (i.e., diameters comprised between 5.5 and 10 Φ).

From SEVIRI data, a mass flux (kg/s) used to determine a total erupted mass is computed from a transect perpendicular to the plume dispersal axis at 15 km from the vent and considering the wind speed derived from the ARPA profiles at the plume altitude. Finally, the MODIS-based mass fluxes are computed by applying the "traverse" approach [52,58–61] and considering the wind speed at the volcanic cloud altitude derived from the Trapani WMO atmospheric profiles (37.91 N, 12.50 E) [http://weather.uwyo.edu/upperair/sounding.html (last access on 15 April 2021)].

2.5. ESPs from Infrasound Array

At Etna, two small aperture infrasound arrays managed by the University of Florence are set up at 5500 (ETN at 2100 m a.s.l.) and 6500 m (MVT at 1800 m a.s.l.) from the summit vents for monitoring purposes (Figure 1a). Infrasonic data have been shown to be relevant at quantifying the dynamics of lava fountaining activity [62] and have been used to produce an early-warning system for paroxysmal events at Etna [16,63]. Experimental and numerical studies have been carried out to simulate the infrasound signal generated in eruptive conduits [64,65]. The acoustic waves generated in a volcanic conduit will be affected by the acoustic impedance contrast between the open-end surface of the volcanic vent and the atmosphere. A large part of the acoustic wave energy at the vent-atmosphere boundary is reflected inside the conduit as a function of the *ka* parameter defined by the acoustic wave number *k* and the effective vent radius *a* [65]. At the vent surface, acoustic pressure inside the conduit decreases drastically to equilibrate the atmospheric pressure, and, for the conservation of the flux, the acoustic velocity increases almost two times for a small value of *ka*. The propagation from inside the conduit to the atmosphere strongly influences the radiation pattern and the amplitude of the acoustic wavefield transmitted in the atmosphere. The directivity due to the vent radius and wave number for $ka < 0.43$ can be neglected and the radiation pattern is isotropic outside the vent [65].

Exit Velocities and MER

As shown by Ulivieri et al. [62] and Ripepe et al. [16], the frequency content for lava fountaining events at Mount Etna is typically below 1 Hz, as was the case of both the events described here. Considering a vent radius ranging between 5 and 20 m (i.e., 5 m, 10 m, 13.5 m, 20 m), *ka* values are ranging between 0.09 and 0.36 (see [65]), hence below 0.43. This means that the acoustic signal can be used to calculate the volumetric flux inside the conduit $q_i(t)$ considering a perfectly isotropic radiation pattern outside the vent and an insertion loss caused by topography $IL = 0$, given that the ETN array position is in the line-of-sight with the NSEC vent [65]:

$$q_i(t) = \frac{2\pi r}{(1+|R|)\rho\alpha 10^{\left(\frac{IL}{20}\right)}} \int_0^t \Delta P\left(t+\frac{r}{c}\right)\mathrm{d}t \tag{6}$$

where α is the directivity at 0° being equal to 1, ΔP is the pressure signal (Pa), t is the time (s), r is the distance between the acoustic source and the array (m), c is the speed of sound of 345 m/s, ρ is the atmosphere density (kg/m^3) and $|R|$ is the acoustic reflectance that ranges between 0.99 and 0.90 for our *ka* values (0.09–0.36). Finally, we can estimate the infrasound acoustic velocity by dividing $q_i(t)$ by the cross-section area of the eruptive vent, which needs to be constrained. We used the exit velocities retrieved from V2B signals to constrain the best-suited vent radii to compute velocities from the acoustic signal. The resulting best vent radius will also be used in the SFA (Equation (1)) for both the V2B and MWR. In addition, we used infrasound velocity to determine the independent MER using the SFA methodology [15,66].

2.6. Plume Height from Visible Camera

The monitoring network of the INGV-OE uses a set of visual cameras that record Etna's summit craters and their close environment in real-time. Images taken by these cameras have been calibrated to allow direct measurements of plume heights depending on daily weather forecast [20] and following isolines of heights above sea leveltitude derived from the Trapani WMO atmosp (Figure 1b; [21,22]). In particular, we use a visible camera located at Catania (ECV; Figure 1a) to evaluate plume height during an explosive event. Images are recorded each 1 s and provide plume height estimates with an uncertainty of ±0.5 km [21]. The main limitations of this method are the strong influence of weather and light conditions, e.g., the presence of clouds, and the incapacity to measure heights above 9 km (a.s.l.) or when plumes are drifted outside the camera's field of view during the period of 2011–2013. In order to improve the visible monitoring system and extend its use to various plume dispersal axes, a new camera was installed on the west flank of Mount Etna (i.e., Etna Bronte High Definition camera, *EBHD*). This camera, thanks to its field of view and depending on wind direction, allows a maximally visible determination of H_T up to 15 km (a.s.l.) (see Scollo et al. [22] for more details).

3. Results

3.1. Plume Height Estimates

As previously described, plume height can be independently determined at Etna based on at least three different remote sensing systems that are complementary in terms of detection limits and space-time resolution (Visible Camera, Satellite retrievals—MODIS and SEVIRI, and X-Band radar—MWR) (see Figure 3).

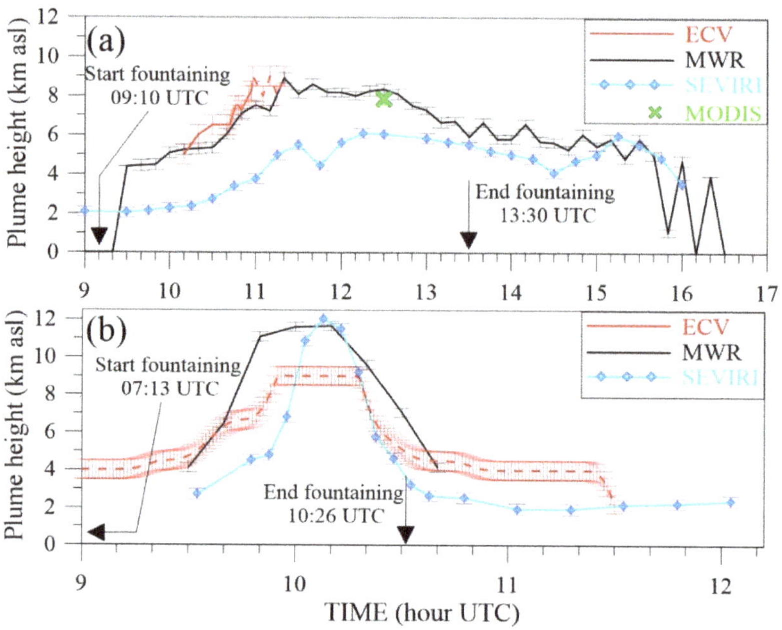

Figure 3. Plume height estimates on10 April 2011 (**a**) and 23 November 2013 (**b**) from X-band radar (MWR, black line), MODIS at 12:30 UTC (green cross), SEVIRI (blue line), and ECV (red line). Red dashed line corresponds to ECV measurements when the plume starts to leave the field of view. The start and end time of the paroxysmal activity is derived from VOLDORAD 2B signal and indicated by the black arrow. Error bars are also shown for all sensors.

Even though the determination of H_T by the visual camera (ECV) was not possible after 11:20 UTC due to the presence of meteorological clouds, continuous detection during the whole 10 April 2011 paroxysm was possible with the MWR and SEVIRI (Figure 3a). In addition, a punctual measurement obtained at 12:30 UTC using MODIS data is also available. Heights obtained from ECV images (Figure 1b) are derived from 10:10 to 11:20 UTC [22,30]. Starting from 5 km a.s.l., H_T increases rapidly and starts oscillating between 8 and 9 km a.s.l. (red dashed line in Figure 3a) from 10:58 to 11:20 UTC. After that time, the highest part of the plume leaves the camera field of view until the end of the paroxysm. As derived from MWR and SEVIRI data, H_T also increases at 09:30 UTC from 1 km above NSEC to reach maximal altitudes of 8.9 $\pm$ 0.3 km a.s.l. as seen with the MWR at 11:20 UTC, and 6.1 km a.s.l. at 12:15 UTC using SEVIRI. Despite offering data only during the increase of the paroxysmal activity, ECV's height estimates agree with MWR heights as does the 7.9 km height a.s.l. retrieved with MODIS at 12:30 UTC. On average, SEVIRI records display average H_T that are 1–2 km lower than the MWR during the paroxysmal activity. Nevertheless, both instruments present similar H_T values after the end of the paroxysm at 13:30 UTC with average values around 5.9 $\pm$ 0.4 km for MWR and 5.0 $\pm$ 0.6 km for SEVIRI until 15:15 UTC when H_T starts to decrease.

H_T on 23 November 2013 as detected by ECV started to increase from 09:15 up to 09:55 UTC to reach altitudes up to 9 $\pm$ 0.5 km a.s.l. (Figure 3b). H_T derived by the MWR and SEVIRI increased rapidly at 09:30 UTC up to similar top heights of 11.7 km a.s.l. at 10:10 UTC and 12.0 km a.s.l. at 10:07 UTC, respectively. The MWR and ECV-based H_T increase 5–10 min before satellite estimates. Over the same eruptive period between 09:30 and 10:30 UTC, mean H_T derived from the MWR, ECV and SEVIRI are close with values of 8.2 $\pm$ 3.2, 7.4 $\pm$ 1.5, and 6.9 $\pm$ 3.4 km a.s.l., respectively. The X-band detection of the tephra plume ends at 10:40 UTC, ~20 min after the end of the paroxysm while visible and satellite thermal data last up to 11:30 and 12:02 UTC, respectively.

3.2. Determination of Exit Velocity, TEM and MER

3.2.1. Mixture Exit Velocity from V2B and Infrasound

We determined the vertical velocity of the eruptive mixture above the vent as retrieved from the Doppler radar V2B and infrasound measurements. These vertical velocities can be considered as a first approximation of the source v_{exit} that is used to compute MER^{SFA} (Equation (1)) as well as the velocity v_{entry} at which the eruptive mixture enters radar beams that are used to compute MER^{NSA} (Equation (2)).

V2B and infrasound measurements result in a similar mean exit velocity when using vent radii considered in previous studies for Etna (i.e., a radius of 10 m and 13.5 m [3,15,67]). In particular, the mean exit velocity associated with V2B and infrasound is 43.7 ± 26.7 m/s and 42.6 ± 18.8 m/s for the 10 April 2011 paroxysm, and 101.1 ± 63.1 and 129.2 ± 62.7 m/s, respectively (Tables 1 and 2 and Figure 4). The values of exit velocities associated with the infrasound are averaged between the calculation for 10 m and 13.5 m vent radius. Given that neither of the two values provides a perfect match, we will use both vent radii to compute the SFA-based MER hereafter.

Table 1. 10 April 2011 paroxysm eruptive source parameters (ESPs) retrieved for all methodologies (in case of multiple strategies associated with individual sensors, mean values are indicated in bold). * duration from Calvari et al. [4]. ** Total erupted mass (TEM) derived with the power law strategy (averaged for distal integration limits of 100 and 400 km from vent). *** Mean duration from all microwave weather radar (MWR) and satellite approaches.

	Mean Exit Velocity (m/s)	Max H_T (km a.s.l.)	Method	Duration (min)	TEM (kg)	Averaged MER (kg/s)
ECV	/	>9	ECV	/	/	/
V2B	43.7 ± 26.7	/	SFA	250	5.8×10^9	$2.8 \pm 1.6 \times 10^5$
			NSA	250	2.2×10^9	$1.4 \pm 1.2 \times 10^5$
			Proxy	250	4.0×10^8	$2.7 \pm 3.5 \times 10^4$
			MEAN	**250**	$\mathbf{2.8 \pm 2.8 \times 10^9}$	$\mathbf{1.5 \pm 1.3 \times 10^5}$
MWR	/	8.9	SFA	210	1.2×10^9	$9.3 \pm 0.4 \times 10^4$
			NSA	190	2.6×10^9	$2.3 \pm 1.7 \times 10^5$
			TPA	420	4.7×10^9	$1.9 \pm 2.3 \times 10^5$
			MCA	410	6.4×10^8	$2.6 \pm 2.4 \times 10^4$
			MEAN	**308 ± 124**	$\mathbf{2.3 \pm 1.8 \times 10^9}$	$\mathbf{1.4 \pm 0.9 \times 10^5}$
Infrasound	42.6 ± 18.8	/	**SFA**	**273**	$\mathbf{2.9 \times 10^9}$	$\mathbf{2.0 \pm 0.9 \times 10^5}$
Ground-IR	/	/	**SFA**	**240 ***	$\mathbf{1.2 \times 10^9}$	$\mathbf{8.6 \pm 2.5 \times 10^4}$
SEVIRI	/	**6.1**	**TPA**	**315**	$\mathbf{6.0 \times 10^8}$	$\mathbf{2.7 \pm 2.5 \times 10^4}$
			VPR-ASH	**255**	$\mathbf{2.0 \times 10^6}$	$\mathbf{1.2 \pm 0.9 \times 10^2}$
			VPR-ICE	**370**	$\mathbf{4.9 \times 10^7}$	$\mathbf{2.1 \pm 1.3 \times 10^3}$
MODIS	/	**7.9**	**ASH**	**160**	$\mathbf{2.3 \times 10^6}$	$\mathbf{2.6 \pm 3.1 \times 10^2}$
			ICE	**170**	$\mathbf{1.4 \times 10^7}$	$\mathbf{1.7 \pm 1.7 \times 10^3}$
Deposit	/	/	**Power-law ****	**310 ± 94 *****	$4.7 \pm 2.3 \times 10^7$	$2.5 \pm 2.0 \times 10^3$
			Weibull		1.3×10^7	0.7×10^3
			Exponential		1.4×10^7	0.8×10^3
			MEAN		$\mathbf{2.5 \pm 1.9 \times 10^7}$	$\mathbf{1.4 \pm 1.1 \times 10^3}$

Table 2. 23 November 2013 paroxysm ESPs retrieved for all methodologies. * TEM derived with the power law strategy (averaged for distal integration limits of 100 and 400 km from vent). ** TEM derived by Andronico et al. [27]. *** Mean duration from all the MWR and satellite approaches.

	Mean Exit Velocity (m/s)	Max H_T (km a.s.l.)	Method	Signal Duration (min)	TEM (kg)	Averaged MER (kg/s)
ECV	/	>9	/	/	/	/
V2B	**53.8 ± 63.7 (07:00–10:30) 101.1 ± 63.1 (09:00–10:30)**	/	SFA	193	4.2×10^9	$3.4 \pm 3.8 \times 10^5$
			NSA	193	2.7×10^9	$2.3 \pm 3.5 \times 10^5$
			Proxy	193	4.3×10^9	$3.6 \pm 8.3 \times 10^5$
			MEAN	**193**	$\mathbf{3.7 \pm 0.9 \times 10^9}$	$\mathbf{3.1 \pm 0.7 \times 10^5}$
MWR	/	11.7	SFA	40	3.5×10^9	$1.5 \pm 0.4 \times 10^6$
			NSA	30	4.3×10^9	$2.6 \pm 2.2 \times 10^6$
			TPA	80	6.4×10^9	$1.3 \pm 1.5 \times 10^6$
			MCA	70	5.5×10^9	$1.3 \pm 1.5 \times 10^6$
			MEAN	**55 ± 24**	$\mathbf{4.9 \pm 1.3 \times 10^9}$	$\mathbf{1.7 \pm 0.6 \times 10^6}$
Infrasound	**129.2 ± 62.7 (09:00–10:30)**	/	**SFA**	**85**	$\mathbf{3.7 \times 10^9}$	$\mathbf{7.4 \pm 3.7 \times 10^5}$
Ground-IR	/	/	**SFA**	**130**	$\mathbf{5.8 \times 10^9}$	$\mathbf{7.5 \pm 4.7 \times 10^5}$
SEVIRI	/	12.0	**TPA**	**40**	$\mathbf{3.6 \times 10^9}$	$\mathbf{1.4 \pm 1.8 \times 10^6}$
			VPR-ASH	**120**	$\mathbf{1.3 \times 10^7}$	$\mathbf{1.6 \pm 1.8 \times 10^3}$
			VPR-ICE	**105**	$\mathbf{1.0 \times 10^7}$	$\mathbf{1.4 \pm 1.3 \times 10^3}$
Deposit	/	/	**Power–Law** *	**69 ± 35 *****	$\mathbf{1.4 \pm 0.0 \times 10^9}$	$3.4 \pm 1.7 \times 10^5$
			Weibull **		1.3×10^9	3.1×10^5
			Exponential		1.2×10^9	2.9×10^5
			MEAN		$\mathbf{1.3 \pm 0.1 \times 10^9}$	$\mathbf{3.1 \pm 0.3 \times 10^5}$

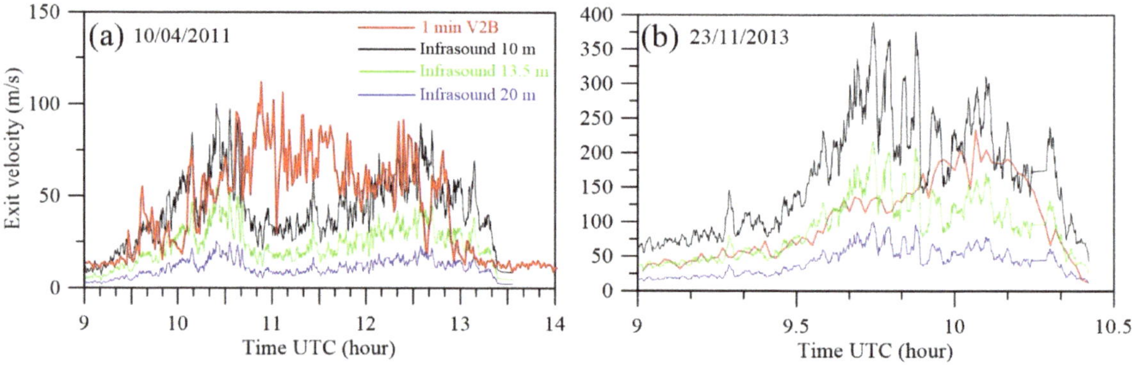

Figure 4. Mixture exit velocities determined by onfrasound (black and blue lines) and L-band Doppler radar VOLDORAD 2B (V2B, orange and red lines) during (**a**) the 10 April 2011 and (**b**) the 23 November 2013 paroxysms. In particular, the determination of the exit velocity with the infrasound is based on a vent diameter of 10 m (black line) and 13.5 m (green line) as well as 20 m (blue line).

3.2.2. TEM and MER from Tephra-Fallout Deposit

The integration of the three empirical fits of the ground mass accumulation versus square root of area contours of the 10 April 2011 event (Figure 5) results in values of TEM_{dep} of about 1.3×10^7, 1.4×10^7, and $4.7 \pm 2.3 \times 10^7$ kg using the Weibull, exponential and power-law fit, respectively (Table 1). While the exponential and the Weibull fit can be integrated between zero and infinity, two integration limits have to be selected for the power-law fit due to the associated asymptotic nature. In particular, the proximal limit is calculated as proposed by Bonadonna and Houghton [41], while, in order to characterize the associated uncertainty, the distal integration limits were set both at 100 km and 400 km where ground accumulation becomes negligible (i.e., between 10^{-3} and 10^{-4} kg/m^2 based on the thinning trend of Figure 5). In fact, given that the power-law exponent is <2 (Figure 5), the resulting volume is sensitive to the distal integration limit but not to the proximal one [41,68]. The TEM_{dep} from the power-law fit is then averaged between the values obtained with the two different distal integration limits (100 and 400 km). Given the absence of proximal data (due to difficult access) and of distal data (due to most of the deposit falling in the sea) (e.g., [26]), Weibull and exponential estimates must be considered as minimal values [68] for this tephra deposit. As an example, Spanu et al. [69] have shown for the 24 November 2006 paroxysm of Etna that a lack of sampling within the first kilometers from the crater could lead to a loss of 30% of the TEM. In this context, even though associated with the uncertainty of the integration limits, the power law fit might provide a better estimate given that it can better predict the medial and distal gradual thinning. We obtain a value of MER_{dep} by dividing each TEM_{dep} by a mean duration of 310 $\pm$ 94 min determined based on the MWR and satellite-based infrared (Table 1). Accordingly, we found MER_{dep} between 0.7 and 2.5 $\times$ 10^3 kg/s (average of 1.4 $\pm$ 1.1 $\times$ 10^3 kg/s) (Table 1; Figure 6).

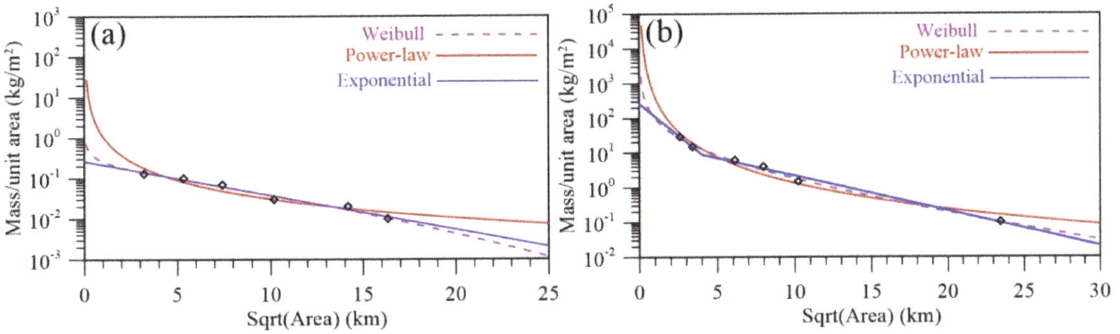

Figure 5. Variation of mass/area as a function of the square root of associated contour areas from Figure 2 for the 10 April 2011 tephra deposit (**a**) and for the 23 November 2013 tephra deposit (data from Andronico et al. [27]) (**b**). Purple dashed line, red line and blue line represent the best Weibull fit, exponential fits, and power-law fit. The power-law fit equations are y = 1.11 $x^{-1.55}$ in (**a**) and y = 408.14 $x^{-2.49}$ in (**b**).

The ground accumulation variation obtained for the 23 November 2013 tephra-fallout deposit was fitted with two exponential segments with a break in slope around 4 km as well as a Weibull and power-law function (Figure 5b). TEM_{dep} obtained by the Exponential, the Weibull, and the power-law fits (Figure 5b) are similar with values of 1.2 $\times$ 10^9, 1.3 $\times$ 10^9, and 1.4 $\pm$ 0.0 $\times$ 10^9 kg [27], respectively (Table 2). Here we fixed the distal integration of the power-law fit at 180 and 450 km, using the criteria of negligible deposit as for the 10 April event. Finally, using an average duration obtained from the SEVIRI and MWR approaches of 69 $\pm$ 35 min, we find MER_{dep} between 2.9 and 3.4 $\times$ 10^5 kg/s with an average of 3.1 $\pm$ 0.3 $\times$ 10^5 kg/s for the 23 November 2013 paroxysm. Please see the following sections for the details on the sensor selection to derive the event duration.

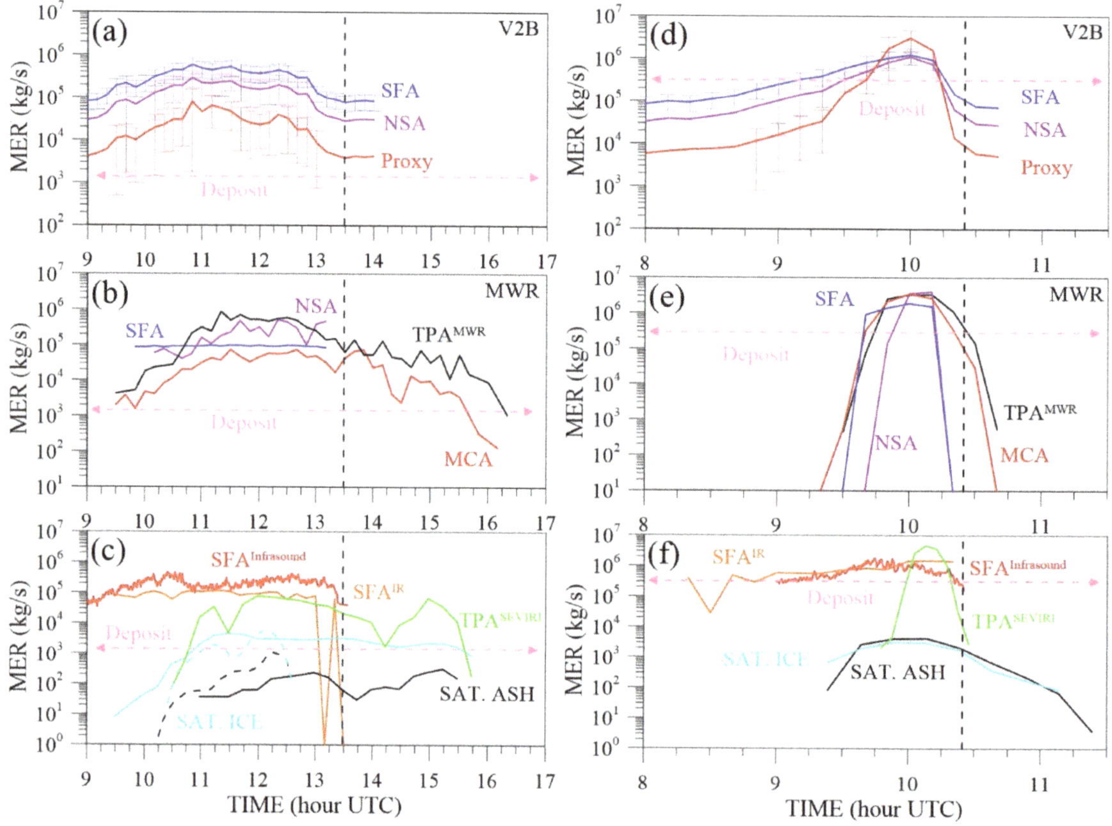

Figure 6. MER time series of 10 April 2011 (left panels) and 23 November 2013 (right panels) paroxysms. (**a–d**): SFA (blue), NSA (purple), and proxy method (red line) using V2B signal; MER values are averaged over 10 min. (**b–e**): SFA (blue), NSA (purple), MCA (red) and TPA (black) methods using the MWR signal. (**c–f**): SFA MERs based on ground-based thermal data (orange) and infrasound data (red). MER of the plume for ash and ice contents are respectively shown in black and light blue lines for SEVIRI and dashed lines for MODIS. MERs based on satellite heights (TPA) are displayed in green. Vertical dashed lines indicate the end of the lava fountain activity as derived from V2B. The deposit-derived MER_{dep} is indicated by the horizontal pink arrow (see Tables 1 and 2 for more details).

3.2.3. MER and TEM from Remote Sensing

It is important to note that sensor-derived estimates of MER and TEM presented hereafter are computed over a duration that is based on the corresponding sensor signal (see Tables 1 and 2). V2B MER methods are associated with different results: the velocity-derived NSA and SFA estimates appear relatively close, with mean values respectively of 1.4 and 2.8 $\times$ 10^5 kg/s, whereas the MER proxy derived from both power and velocity shows a significantly larger dynamics and remains one order of magnitude below for this event with an average MER of 2.7 $\times$ 10^4 kg/s (Figure 6a and Table 1). Altogether, MER values derived from all V2B strategies result in an average of 1.5 $\pm$ 1.3 $\times$ 10^5 kg/s. It is important to note that the MER associated with V2B was averaged over 10 min in order to better compare it with the MWR results.

Time series of MER from MWR start at 09:30 UTC and show a concomitant increase with V2B estimates except for MWR-SFA values that show a constant value for the whole event (Figure 6b). MWR average values of the MER range between 2.6 $\pm$ 2.4 $\times$ 10^4 and 2.3 $\pm$ 1.7 $\times$ 10^5 kg/s as based on the MCA and TPA, respectively. In between, the respective average MER from the SFA and TPA are equal to 9.3 ± 0.4 × 10^4 and 1.9 ± 2.3 × 10^5 kg/s. In total, all MWR-MER methods provide an average MER of 1.4 ± 0.9 × 10^5 kg/s. SFA-MER derived from ground-based thermal is similar to SFA-MER associated with MWR data and they are both relatively close to average values (Figure 6b,c; Table 1).

Infrasound MER values present small variations with time, with an average value of 2.0 ± 0.9 × 10^5 kg/s. Figure 6c also highlights two trends for satellite-based MER data from SEVIRI and MODIS satellites that suggest a high amount of ice in the detected tephra plume and cloud. On average, MER values corresponding to the ash content are 7 and 18 times lower than ice in MODIS and SEVIRI time series, respectively (Table 1). Interestingly, MER values based on SEVIRI observations of H_T (Figure 3) vary with time with a maximal value of 7.7 × 10^4 kg/s at 12:00 UTC (Figure 6c) and a mean value of 2.7 ± 2.5 × 10^4 kg/s, similar to mean MERs from V2B proxy and MWR NSA and MCA methods (Table 1).

All MERs from V2B are very close also for the 23 November 2013 paroxysm (Table 2). MERs from the NSA and SFA follow the same trend (Figure 6d) with mean values of 2.3 ± 3.5 × 10^5 and 3.5 ± 3.7 × 10^5 kg/s, respectively. Proxy-derived MERs remain lower than NSA and SFA MERs at the beginning of the event, following a similar trend until 09:20 UTC. It then strongly increases as a consequence of increasing echo power from the fountaining ejecta. Despite the fact that the average proxy-derived MER is similar to the SFA and NSA estimates, proxy MERs remain higher than the NSA and SFA estimates with a maximum value of 3.1 ± 1.6 × 10^6 kg/s during the paroxysm climactic phase from 09:45 to 10:15 UTC (Figure 6d).

Concerning MWR estimates, the eruptive signal started from 09:30 UTC when exit velocities and H_T estimates increased significantly (Figures 3 and 4) and lasted up to 10:40 UTC for the TPA and MCA methods, whereas the SFA and NSA estimates lasted 30–40 min up to 10:10 UTC (Figure 6e). As shown by Marzano et al. [15], all MER methods using MWR data provide very similar estimates during the climactic phase of the 23 November 2013 paroxysm with maximal MER being comprised between 1.9 × 10^6 kg/s (SFA) and 4.1 × 10^6 kg/s (NSA) (Figure 6e).

The SFA method based on infrared data provide MERs that are comprised between 2.8 × 10^4 kg/s and 1.5 × 10^6 kg/s with a signal lasting 130 min from 08:20 and 10:20 UTC (Figure 6f and Table 2). Infrasound-based MERs using the SFA are close to ground-based infrared estimates and show a maximal value of 1.8 × 10^6 kg/s. As for the 10 April 2011 event, satellite data present two trends for both plume/cloud ice and ash content whose values are, however, very close with a mean of 1.4 × 10^3 *vs* 1.6 × 10^3 kg/s (Table 2 and Figure 6f). Based on H_T derived by satellite over 40 min, i.e., when H_T becomes higher than the vent height of 3.2 km (Figure 3b), TPA-based MER are comprised between 2.1 × 10^3 kg/s at 09:47 and 4.7 × 10^6 at 10:07 UTC, respectively. The mean MER from H_T measured by satellite is also very similar to the MWR mean value (Table 2).

In terms of TEMs, all methods show very different trends of time-integrated erupted mass for the 10 April 2011 event (Figure 7a). Final values, which correspond to TEMs, converge to a value around 10^9 kg (between 6.4 × 10^8 kg for MWR MCA and 5.8 × 10^9 kg for V2B SFA, respectively), whereas deposit and satellite ash content estimates are much lower (between 2.5 ± 1.9 × 10^7 and 2.0 × 10^6 kg, respectively) (Figure 7a and Table 1). In general, SFA, NSA, and TPA-derived ESPs display higher TEM values than those retrieved with proxy and MCA for all sensors (see Table 1 and V2B-TEMs in Figure 7a). The arithmetic mean associated with all methodologies for each sensor is between 1.2 and 2.8 × 10^9 kg for V2B, MWR, infrasound, and ground-based thermal data (Table 1). For satellite mass data, two trends are still shown for both contents in ash and ice with TEMs respectively ranging between 2.0–49.0 × 10^6 kg for SEVIRI and 2.3–14.1 × 10^6 kg for MODIS data (Figure 7a; Table 1).

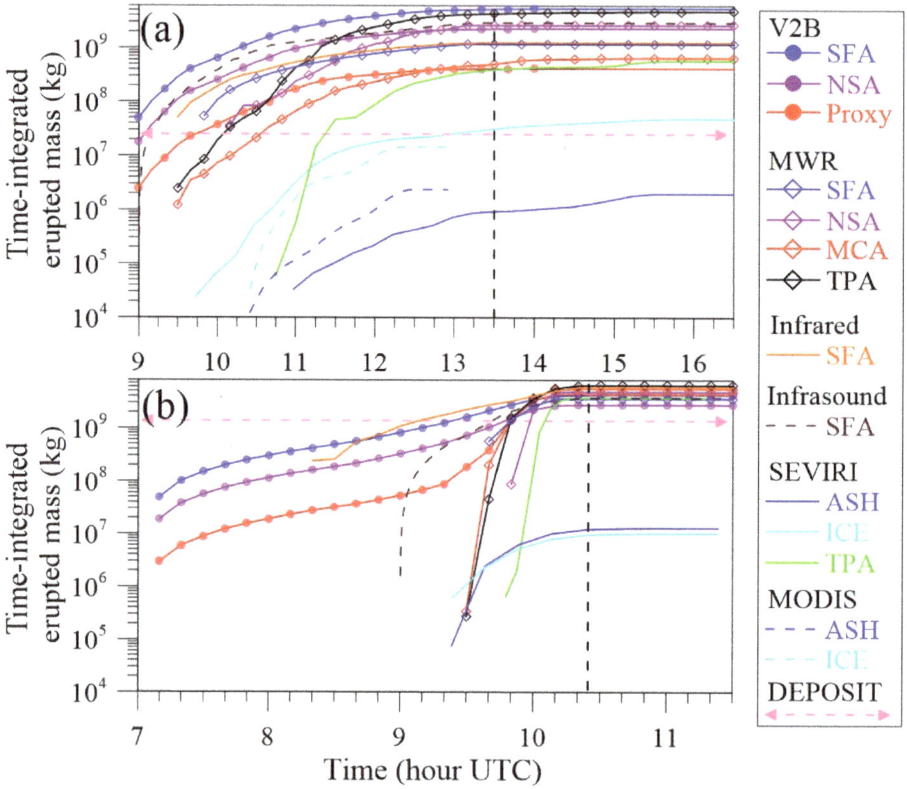

Figure 7. Time-integrated erupted mass detected by all remote sensors for the 10 April 2011 (**a**) and 23 November 2013 (**b**) paroxysms. The vertical black dashed line indicates the end of the lava fountain activity as derived from V2B. Note that each final value of time-integrated mass corresponds to each sensor TEM.

The TEMs observed for the 23 November 2013 paroxysm by V2B, MWR, infrasound and thermal camera span less than one order of magnitude between 2.7 × 10^9 kg (V2B-NSA) and 5.8 × 10^9 kg (ground-based infrared). Interestingly, all cumulative trends from infrasound, ground-based infrared and both Doppler radars are converging during the pre-climax phase toward a relatively narrow range of TEMs between 2.7 × 10^9 and 6.4 × 10^9 kg (Figure 7b and Table 2). As observed for the 10 April 2011 (Figure 7a), TEMs estimated from the satellite-derived plume/cloud, ice and ash content are lower than all other methods with very similar TEMs of 1.0–1.3 × 10^7 kg (Table 2). Nevertheless, TPA-based TEM from SEVIRI is in the same order of magnitude as all other methods with a value of 3.6 × 10^9 kg.

For both eruptive events and for all ground-based remote sensors, 0 to 10% of the TEM is emitted after the end of the fountaining activity as derived by V2B [30] and ground-based infrared [4] (Figure 7). While satellite-based TEMs are also reached by the end of the lava fountaining on 23 November 2013 (Figure 7b), 27 to 53% of the ash content and TPA-based TEMs, respectively, is detected after the end of the paroxysmal activity on 10 April 2011 (Figure 7a).

3.3. *Total Grain-Size Distributions*

3.3.1. Combination of WDGSD and GSD_{sat}

The WDGSD derived by applying the Voronoi Tessellation method on the deposit data [41] is unimodal and well sorted with a sorting coefficient of 1.41 and an Md_Φ [70] of 0.82 Φ, i.e., 0.57 mm (Figure 8a).

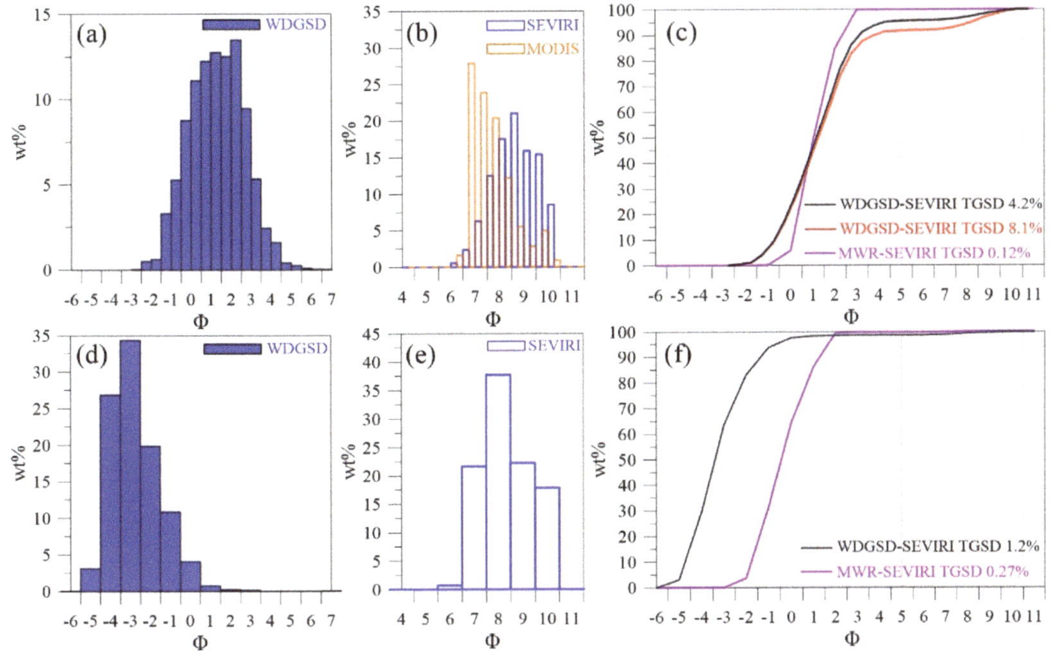

Figure 8. (**a**) Whole deposit grain-size distribution (WDGSD) of the 10 April 2011 paroxysm. (**b**) VPR-derived GSD_{sat} from SEVIRI (blue) and MODIS (orange) data. (**c**) Cumulative total grain-size distribution (TGSD) of the 10 April 2011 paroxysm; grey and red lines correspond to WDGSD-SEVIRI TGSDs obtained by considering a mass ratio of 4.2% and 8.1% between satellite and deposit TEMs, respectively (see main text for details). The purple line corresponds to the near real-time MWR-SEVIRI TGSD (see main text for details). The purple and red shaded areas correspond to the size detection limit of the MWR and SEVIRI, respectively. (**d**–**f**) Same results obtained for the 23 November 2013 paroxysm. The WDGSD in (**d**) is from Poret et al. [31].

Both SEVIRI and MODIS data can be used to provide a GSD_{sat} by using the VPR algorithm (Figure 8b). Such distributions are unimodal and very well sorted, i.e., sorting of 1.0 and 0.8, respectively, and consider that all detected material is under 20 microns of diameter; SEVIRI and MODIS GSDs display Md_Φ values of 8.3 Φ (0.003 mm) and 6.9 Φ (0.008 mm), respectively. To combine deposit and satellite data, we choose the SEVIRI GSD_{sat} that better describes the temporal variations of sizes in the detected plume/cloud instead of the MODIS punctual GSD_{sat}, hardly representative of the whole event and only available for the 10 April event. In particular, we combined all GSDs from the tephra-fallout deposit and SEVIRI based on their relative TEM proportion (see for example Bonadonna et al. [71]). In order to take into account the uncertainty associated with the determination of TEM, here we consider both the power-law estimate only and the mean of all integration approaches (power-law, Weibull, and exponential), i.e., 4.2 and 8.1%, respectively. Both resulting TGSDs are bimodal (Figure 8c) and consider both the material that felt on the ground and the very fine part of the erupted material detected by satellite, which ranges between 1 and 20 μm.

The WDGSD and GSD_{sat} of the 23 November 2013 paroxysm are shown in Figure 8d,e. The WDGSD, which was obtained by Poret et al. [31], is unimodal, coarse-grained with an Md_{Φ} of –3.4 Φ (10 mm; lapilli-sized), and well sorted. This WDGSD is particularly depleted in fine ash with no material below 63 microns. SEVIRI GSD_{sat} integrated over the whole event duration is also unimodal with an Md_{Φ} of 7.7 Φ (0.005 mm) and sorting of 1.2. Given that the estimates of TEM based on the three integration methods for the 23 November 2013 event are all very similar, WDSGD and GSD_{sat} were combined (see TGSD in Figure 8f) based on an average value of the three estimates ($1.3 \pm 0.1 \times 10^9$ kg; Table 2) resulting in a mass ratio of 1.2% with the ash TEM obtained by SEVIRI (1.3×10^7 kg) (Table 2).

3.3.2. Combination of the GSD_{MWR} and GSD_{sat}

Following the same methodology as described above, we combined the GSD_{MWR} and the SEVIRI GSD_{sat} (Figure 8c,d). Accordingly, we used the mass ratio between the arithmetic mean of MWR-based NSA and MCA TEMs (i.e., $1.6 \pm 1.4 \times 10^9$ kg and $4.9 \pm 0.8 \times 10^9$ kg, respectively for both events) and the satellite-based ash mass retrieved with the VPR (Tables 1 and 2). A very small mass ratio of 0.12% and 0.27% between SEVIRI and MWR was found for the 10 April and the 23 November paroxysms, respectively. It is important to note that, given these very small mass ratios, the GSD_{MWR} and the combined MWR-SEVIRI TGSD were very similar for both events in Figure 8c,f. Interestingly, the MWR-SEVIRI TGSD was similar to the WDGSD-SEVIRI TGSD for the 10 April 2011 event with an Md_{Φ} of 1.0 (0.5 mm) and sorting of 0.87 (Figure 8c). However, in the case of the November 2013 event, the MWR-SEVIRI TGSD with an Md_{Φ} of -0.4 (1.3 mm) was finer than the WDGSD-SEVIRI TGSD having an Md_{Φ} of -3.4 (10.6 mm) (Figure 8f).

4. Discussion

4.1. Determination of Plume Height

H_T is one of the keys and most common ESPs to be determined in real-time. In fact, active explosive volcanoes are generally monitored with visible cameras from which H_T is derived when the wind velocity is known [21,22], with satellite data following the dark pixel procedure [51] and/or with radar data [72]. For tephra forecasting and modeling purposes, H_T is an important input parameter [68,73,74] as it defines the spreading height of the volcanic cloud and strongly influences its dispersal axis and impact area [74,75] (and references therein). At Etna, H_T is typically obtained based on visible camera [21,22], satellite-based observations [7,8,39,49] (and references therein) and MWR (X-band) radar detection [13–15]. However, as shown during the 10 April 2011 and the 23 November 2013 paroxysms, the accuracy of these three techniques depends on various conditions (Table 2).

Indeed, plume height estimates at Etna based on the camera located in Catania (ECV in Figure 1) are restricted to 9 km (a.s.l.), day light, no cloudy conditions (see Figure 1a; Figure 3a), and to the camera field of view [21,22]. ECV camera cannot track H_T either when the plume dispersal axis is parallel to the camera line of sight. To overcome these limitations, a new camera was installed on the west flank of Etna volcano (i.e., Etna Bronte high definition camera, *EBHD* in Table 3 and Scollo et al. [22]) and will allow a maximally visible determination of H_T up to 15 km (a.s.l.) (Table 2), under day light and no cloudy conditions.

Thanks to its 3-D scanning capacity at different elevation angles, the MWR covers an area of 160 km wide and 20 km high [13]. Mount Etna being at ~30 km from the radar site (Figure 1a; [13–15]), the MWR is able to detect ash plumes with typical southeastward dispersion up to maximum H_T of ~12 km (a.s.l.), using the highest beam elevation angle. In addition, MWR data are exploitable for all light (day/night) and weather conditions (Table 3). Indeed, the dual-polarization capacity of the MWR allows us to discriminate ash particles from hydrometeors, which can affect the radar signature of a detected tephra plume [10,45,76].

Finally, we have shown that satellite-based H_T estimates are generally lower (Figure 3a) or delayed (Figure 3b) in comparison with those from the MWR and ECV. In the case of the 10 April 2011 weak paroxysm, the detected ash plume/cloud is not opaque and this lead to underestimating H_T when the dark pixel procedure [7,8,51,56] is applied as already observed in Scollo et al. [22]. Similarly, the delayed increase in H_T (Figure 3b) might represent the duration taken by the plume to become sufficiently opaque, i.e., allowing an accurate estimate of its height using the Dark pixel procedure (see also [7,8,22]). This observation, linked to the overall determination of MER from satellite, might suggest that the dark pixel procedure becomes accurate when the ash emission is sufficiently sustained, as observed during climactic phases of paroxysms at Etna.

4.2. Insights into Exit Velocity Measurements

The exit velocity is a critical parameter to determine MER [4,13,15,30,66] and to constrain eruptive column dynamics [77–81]. Direct evaluation of exit velocities at Etna comes from the fixed-pointing near-source Doppler radar V2B at very high time resolution [12,30]. In addition, infrasound sensors, a more common tool for monitoring active volcanoes [62,63] (and references therein), can also provide exit velocities when vent characteristics are known [65]. In fact, the determination of the exit velocity based on infrasound data requires the infrasonic type of source to be constrained (e.g., dipole, quadrupole), which is still under investigation [64,66]. As a result, the V2B values of exit velocity were used in this paper to validate the vent radius used in literature for NSEC of Etna volcano (i.e., 10 m and 13.5 m [4,15]) to be used in the calculation of MER with the SFA strategies. V2B-derived exit velocities are recorded 100–200 m above the source vent and describe the ascent of coarse lapilli and block/bombs forming the lava fountain, whereas infrasound velocities are likely to describe the gas exit velocity [65]. Even though in the jet region gas and tephra are assumed to move at the same velocity (as the tephra is carried by the expanding gas), some of the largest blocks and bombs as seen by the V2B might be slower resulting in an underestimation of the mixture velocity. It is interesting to note that these two events are separated by a long time period including many eruptions and a significant cone shape modification [35,36]. This suggests that a 10–13.5 m range is reasonable to describe the NSEC radius for paroxysms at Etna during the present cycle of activity.

4.3. ESPs of Weak and Strong Paroxysms at Etna

Paroxysms at Etna are generally composed of two main components, i.e., a lava fountain and a tephra plume which is mostly fed by the lava fountain [4,30,35,81]. The contribution of each component to the TEM, MER, and TGSD of the cumulative event can be explored using different sensors.

4.3.1. Multi-Strategy TGSD Determination

The TGSD is certainly the most challenging parameter to be retrieved in near real-time [22,82] as all remote sensors are sensitive to various tephra size ranges whose limits are difficult to constrain [83] or need to be modeled (radar-based GSDs in Figure 8c,f) [15,45,84]. Satellite thermal-infrared retrievals are sensitive to very fine ash (<20 µm) within the top ash cloud layers, whereas MWR retrievals are mostly sensitive to tephra sizes from fine ash (>25 µm, [7]) to lapilli (up to 64 mm, [45]) within the plume.

Here we show a first attempt to provide a near-real-time TGSD by combining GSD_{MWR} and GSD_{sat} from SEVIRI (Figure 8). In fact, WDGSD and SEVIRI GSD_{sat} had already been combined in the past with good results for TGSD (e.g., [31]), which, however, cannot be provided in near real-time. It is important to mention also that an MWR-SEVIRI TGSD not only can be produced in near real-time, but it can also overcome some of the limitations related to tephra-deposit sampling. In fact, it is important to bear in mind a few shortcomings of deposit sampling at Etna. First, the very proximal fraction deposited <0.5 km from the vent and contributing to building the eruptive cone, i.e., the lava fountain tephra deposit, is never sampled [35,36,81]. Second, the proximal fraction deposited

<5 km and corresponding to the coarsest part of the tephra plume GSD is also rarely sampled because of access difficulties (i.e., presence of La Valle del Bove horseshoe-shaped depression) and problems in discriminating individual deposits in periods of frequent activity [26,69]. This is also the case for the 10 April 2011 event for which the first sample was taken at 7.2 km from the vent (Figure 2). Finally, due to prevailing wind directions heading towards the East at Etna, fountain-fed tephra plumes are frequently drifted above the Tyrrhenian Sea and the distal part of tephra fallout deposits is lost (see examples in Figures 1 and 2). Accordingly, most of the paroxysm-related tephra deposits can only be sampled up to about 30 km (i.e., the coastline) (Figures 1a and 2) except for when the emitted plume is directed Southwardly (e.g., 12 January 2011 paroxysm; [26]). In contrast, GSD_{MWR} can provide information from the vent down to about 80 km from the vent depending on the size of the paroxysm.

The MWR-SEVIRI TGSD shows a promising agreement with the WDGSD-SEVIRI TGSD even though some caveats have to be considered. First, the ratio of the TEM associated with the different strategies used (i.e., tephra-fallout deposit and SEVIRI or MWR and SEVIRI) has a strong impact on the final TGSD. As an example, the large difference in TEM associated with the MWR and satellite retrievals resulted in a negligible contribution of the SEVIRI GSD_{sat} to the final TGSD for both events. However, even if the mass contribution of the very fine material below 20 microns detected by SEVIRI represents less than 0.5% of the total amount detected by MWR for both the November 2013 and the April 2011 paroxysms, the GSD_{sat} is essential for the characterization of the ash transport in the atmosphere using VATDM [31]. Second, while the GSD_{MWR} shows a good agreement with the WDGSD (whole deposit GSD) for the 10 April event, it is considerably finer with respect to the WDGSD for the 23 November event. This is mostly related to the processing of radar data. In fact, even though the MWR actually sees particles also above 8 mm, these do not represent a large portion in number, and, therefore, they disappear in the final calculation of GSD in wt%. In addition, instead of considering several size classes in the VARR as in Mereu et al. [46] and Marzano et al. [15], we used a wider single size class (from 0.008 to 64 mm) to better combine it with SEVIRI GSD data and better compare it with the WDGSD. This new procedure simplifies the data treatment, but it loses information at the tail of the distribution (i.e., particles >8 mm and <63 μm).

To conclude, both the MWR-SEVIRI TGSD and the WDGSD-SEVIRI TGSD provide important insights. The first one can be derived in near real-time and can potentially combine information on both lapilli and ash-sized particles including the very fine ash detected from SEVIRI in all weather conditions and regardless of the coastline. The second one can provide fundamental information to help better calibrate the procedure to derive the MWR-SEVIRI TGSD as well as to run VATDMs of future eruptions of similar intensity when the near real-time TGSD is not available. It is important to remember that the derivation of TEM with the different sensors/strategies is crucial to the derivation of both TGSDs in order to best combine the different contributions.

4.3.2. The Role of Signal Duration in MER and TEM Determination

As stated in Section 3.2.3., we computed all MERs based on individual sensor signal duration. In fact, each sensor has its own signal duration depending on its time resolution, on which portion of the lava fountain, plume, and/or cloud it records data for and/or on what tephra size it is the most sensitive to. The duration variability observed in Tables 1 and 2 is mostly due to the fact that the different sensors detect different phases of the eruption [5,35,85]. Three typical main phases can be detected for Etna paroxysms, as discussed below.

The first paroxysmal phase starts with lava fountaining (*Phase I* in Table 3) and is captured by a ground-based infrared camera and V2B that point directly at the area above the vent. This unstable activity produces also infrasonic waves in the conduit and at the vent that are well captured by the infrasonic array [62] and from which an early warning system has been developed [16]. The tephra emission during this phase is typically weak (mostly related to the building of the proximal cone) and is associated with low H_T. This is why V2B, infrasound, and ground-based infrared provide eruptive signal before the other systems (i.e., 20 min earlier on 10 April and 60–120 min earlier on 23 November), whose signal is based on tephra plume emission (e.g., satellite and visible camera as well as, to some extent, MWR depending on its lowest scan elevation). It is important to note that Phase I is typically preceded by mild-Strombolian activity, lasting several hours to days before the start of Etna paroxysms [16,35,62]. Such activity does not induce significant tephra emission and is mostly recorded by infrasound only.

Table 3. Summary of all ESPs that can be obtained at Etna. Blue cells: direct measurements; Green cells: derived measurements; Orange cells: measurements needing additional models; Grey cells: parameter not provided. See main text for the description of Phase I, II, and III. * with the methods of Carey and Spark [86], Rossi et al. [87], and Bonadonna and Costa [42].

Sensor/Method	Time Resolution (s)	Event Duration	Plume Height H_T	Total Erupted Mass TEM	Mass Eruption Rate MER	Total Grain-Size Distribution
L-band Doppler radar V2B	0.2	Phase I + II	Calibration based on H_T *vs* MER laws	Based on MER and duration	SFA, NSA, PROXY	Based on VARR model Single value of D_n; Detection of small lapilli to bomb/blocks (Lava fountain; Phase I and II)
X-band Doppler radar MWR	600	Phase II + III	H_T ~12 km (a.s.l.) for plumes dispersed southeastwardly	Based on MER and duration	SFA, NSA, MCA, TPA	GSD_{MWR} based on VARR model; Detection of fine ash to lapilli (Phase II and III).
Ground-based infrared	60	Phase I + II (no meteorological clouds)		Based on MER and duration	SFA	
Satellite-based infrared/visible (SEVIRI, MODIS)	900	Phase II + III (no meteorological clouds)	Based on atmospheric temperature profile	Based on MER and duration	Based on Particle Re and AOD, TPA	GSD_{sat} Resolved for particles ≤20 µm (i.e., 5.5 Φ) (Phase II and III)
Infrasound	60	Phase I + II	Based on H_T *vs* MER laws	Based on MER and duration	SFA	
Visible camera	30–60	Phase II + III (no meteorological clouds)	*ECV:* H_T <9 km (a.s.l.) *EBHD:* H_T <15 km (a.s.l.) (requires day light and no cloudy conditions)	Based on MER and duration	Based on H_T *vs* MER laws	
Tephra-fallout deposit	after the end of the eruption		Based on Md_Φ and/or largest clasts data *	Integration of best fit of mass/area vs. sqrt(area) trends (Exponential, Power-Law, Weibull)	Based on TEM and duration (Phase II + III especially if sampled >0.5 km from the vent)	WDGSD Potentially whole size range from very fine ash to block and bombs; limited to coastline (typically <30 km from vent depending on plume dispersal)

The second phase (*Phase II* in Table 3) is characterized by the emission of a sustained lava fountain-fed tephra plume. While some paroxysmal events present phases II associated with low eruptive intensities, e.g., the 10 April 2011 event, others are associated with plumes that can reach heights of 12–17 km above sea levels, such as the 23 November 2013 and the 3–5 December 2015 paroxysms [6,8,15,30,33]. This phase is well detected by all sensors including visual cameras, the MWR, and satellite-based infrared (Figure 6).

The third phase (*Phase III* in Table 3) represents the waning phase of the paroxysm [35]. While the lava fountain stops, the tephra plume and cloud emitted in Phase II are still expanding in the atmosphere. Phase III is well captured by the MWR (mostly TPA and MCA) and satellite but not by V2B, infrasound, and infrared sensors. This is the reason why, MWR and satellite signals last longer after the end of the fountaining activity, i.e., between 20 min and <2 h in both paroxysms presented herein (Figures 6 and 7).

Given that the duration of the different sensor signals is associated with different phases of the eruption, we strongly suggest calculating MER based on TEM and duration associated with the same sensor. We can also conclude that V2B, infrasound array, infrared camera, and MWR can provide information on the duration of the sustained phases of the paroxysm (i.e., Phases I and II) (Figure 7a,b). In contrast, thermal-infrared satellite and MWR signal durations are related to the presence of a tephra plume and cloud in the atmosphere, including those associated with very fine ash (i.e., Phase II and III). Moreover, most of the paroxysm TEM associated with the tephra-fallout deposit (Tables 1 and 2), is likely to be released during Phases II and III (Figure 7 and Section 3.2.3). This is why we computed deposit-based MERs using the mean signal duration as provided by MWR and satellite.

4.3.3. MER and TEM

As shown in our result section, a variety of sensors and associated strategies exist at Etna that can provide information on both the MER and TEM resulting in a large spread of values, especially for the 10 April 2011 event. The spread is mostly due to the fact that the different sensors and strategies record the 3 different phases of the paroxysm described in the previous section, and, therefore, are complementary (Table 3).

The methods that best record the tephra-plume activity (i.e., mainly during Phase II and III in Table 3) are the MWR-based MCA, the TPA (associated with all sensors), and the V2B-based proxy method (based on the Degruyter and Bonadonna [48] equation and valid for phase II; see Section 4.3.2). V2B-based proxy and MWR-based MCA methods present very similar TEMs for both paroxysms with values between 4.0×10^8 and 4.6×10^8 for the 2011 event, respectively (Table 1), against 4.6×10^9 and 5.5×10^9 kg for the 2013 event (Table 2). Over the same period of detection of the MWR signal, i.e., 09:30–10:30 UTC, V2B-based proxy MER is equal to $1.2 \pm 1.2 \times 10^6$ kg on average. This means that the MER-based MCA, V2B-based proxy, and satellite-based TPA MERs are similar for both the weak and strong paroxysms analyzed herein. In addition, these three methods present similar values to the MWR-based TPA MER for the strong 23 November paroxysm. However, they are one order of magnitude lower for the weak 10 April paroxysm. It is important to note that satellite-based H_T (from which satellite-based TPA MERs are derived) is significantly lower than H_T measured from the ECV, MWR, and MODIS for the 10 April event (Figure 3a). It is well-known that plume heights retrieved from the dark pixel procedure could be underestimated in the case of weak and non-sustained ash emission (see Section 4.1). Hence, the fact that both the MWR MCA and V2B Proxy MERs are similar to satellite-based TPA values in the case of the weak paroxysm suggests that they might also underestimate the MER during weak eruptive activity.

All sensor TEMs obtained using the SFA are relatively close, regardless of the event duration, with values between 1.2×10^9 and 5.8×10^9 kg for the 10 April 2011 against 3.2×10^9 and 5.8×10^9 kg for the 23 November 2013 paroxysm (Tables 1 and 2). In fact, SFA estimates are not very different between both weak and strong paroxysms. This is due to the fact that the exit velocity, on which the SFA strongly depends, is not the most varying parameter among all paroxysms. Indeed, over 35 paroxysms out of 48 paroxysms including a climactic phase observed by V2B between 2011 and 2015 at Etna [30], the overall mean exit velocity was equal to 125.0 m/s with a standard deviation of $\pm$ 30%. Contrastingly, the mean proxy-derived TEM was equal to 1.21×10^9 kg with a standard deviation of

±126%. This suggests that SFA estimates do not capture the real variability of intensity that exists between weak and strong paroxysms. Hence, the MER mostly based on exit velocities obtained by V2B, MWR, infrasound, and infrared might be overestimated during periods of weak activity and underestimated during intense periods, e.g., climactic phases. To better describe the variability of paroxysm intensities, approaches based on parameters related to a quantity of tephra, e.g., echo power of V2B or MWR reflectivities, should be preferred to SFA.

Regarding MWR-based and V2B-based NSA, all TEMs are similar to SFA estimates among both events (Table 1), except for the MWR-NSA TEM of the 2011 case which is up to one order of magnitude less than the other values. Indeed, NSA estimates are made by considering a given range of tephra for each radar, based on the VARR model outputs (see Section 2.2.1.; [10,15,46]). Although the dual-polarimetric capacity of the MWR allows us to model the GSD of detected tephra (Figure 8; see [46] and references therein), the size range detected by V2B inside the fountains, likely small lapilli, remains unknown. Therefore, for all paroxysms, we assume the same size range of 8×10^{-4} to 26.1 cm to determine tephra concentrations and reflectivity-weighted mean diameters from V2B (Equations (3) and (4)). Similar to exit velocities in the SFA, the upper size limit might tend to reduce the variability of the V2B signal between weak and strong paroxysms. This is the reason why, unlike V2B-based NSA estimates, MWR-based mass parameters using NSA present large differences between the weak 2011 event and the strong 2013 event, similarly to MCA values (Tables 1 and 2). Hence, without any further constraints on tephra sizes detected by V2B, SFA and Proxy methods should be preferred for V2B to NSA estimates.

Overall, it seems that all ground-based techniques capture very well the eruptive activity that occurs during the fountain-fed tephra plume activity (Phase II). The fact that most of the TEM is likely to be released during this phase (e.g., Figure 7 and [30]) induces that all masses retrieved by the ground-based sensors, hence excluding satellite and deposit data, are relatively close with mean values of $2.3 \pm 1.7 \times 10^9$ and $4.5 \pm 1.2 \times 10^9$ kg for the 10 April and the 23 November paroxysms, respectively. On the contrary, MER values are different and depend on the capacity of each sensor to monitor, in addition to phase II, either phase I (i.e., V2B, infrasound, ground-based infrared) or phase III (i.e., MWR and SEVIRI) (see Table 3).

TEM and MER based on tephra-fallout deposit analyses are one to two orders of magnitude less than all other ground-based techniques for the 10 April 2011 event but display similar values for the 23 November 2013 (Tables 1 and 2). As already mentioned, both tephra-fallout deposits were not sampled over their full extent, either in proximal nor distal areas. The power-law fits of tephra deposits associated with Etna paroxysms should be typically >2 as they are representative of small-to-moderate eruptions, as it is, in fact, the case for the 23 November event [68]. The power-law exponent of the 10 April event is <2 because of poor deposit exposure, and, therefore, the associated volume should be considered as a minimum value [68].

Satellite-based mass values represent the very fine fraction below 20 μm erupted during both paroxysms. However, if particles coarser than 20 μm are present in the detected plume/cloud, their thermal signature would be the same as that of particles of 20 μm [50,51,83]. This could lead to an underestimation of TEM in the case of coarse ash in the plume/cloud. In addition, it is important to note that when ice is present in a volcanic cloud, as was the case for both paroxysms (see ice contents in Figure 6c,f and Tables 1 and 2), the mass of ash retrieved from satellite-based infrared could also be underestimated [29,53]. Indeed, water and ice particles in the detected clouds have been shown to significantly affect the BTD and VPR procedures [7,29,53] (and references therein) and might reduce the signature of ash particles in satellite images. Taking into account the aforementioned observations, satellite-based mass estimates should be considered as minimum values for both paroxysms.

5. Conclusions

Near real-time determination of ESPs is key to the initialization of VATDMs used for near real-time forecasting of tephra dispersal and sedimentation. The comparison we made in this study between the weak 10 April 2011 and the strong 23 November 2013 paroxysms at Etna has helped to better interpret the results associated with existing approaches used to compute ESPs based on a variety of monitoring sensors (see Table 3 for a summary). In particular, this study suggests that:

(1) eruption duration, a critical parameter to convert the TEM in the MER and vice versa, is different among all sensors analyzed because it is associated with different phases of Etna's paroxysms. V2B, infrared, and infrasound signals correspond to the starting and sustained activity of the paroxysm (Phase I, i.e., lava fountaining activity, and Phase II, i.e., lava fountain-fed tephra plume activity). In contrast, the MWR and satellite signals are associated with both Phase II and the final waning phase (Phase III) related to the subsequent expansion of plume and cloud in the atmosphere with little or no tephra emission from the source vent. As a result, the MER should be derived based on the TEM and duration associated with the same sensor. In the case of TEM derived from the tephra-fallout deposit, the duration used to calculate MERs should be that associated with Phase II and III (i.e., associated with MWR and satellite signals);

(2) the three techniques currently used at Etna for the near real-time determination of H_T (visible camera, MWR, and satellite-based thermal-infrared observations) operate at various time resolutions (i.e., 1 min to 15 min). A critical application of the three techniques, including the use of visible cameras at different locations [22], allows us to assess the best value of average H_T as well as to evaluate the uncertainties associated with each remote sensor. In addition, it appears that satellite-based H_T tend to be underestimated during weak and unstable paroxysmal activity;

(3) exit velocities from V2B can be used in combination with exit velocities from infrasound to better constrain the vent radius used for MER calculations, based on the SFA. For Etna, a range of 10–13.5 m was found as the best estimate of the NSEC radius. A combination of V2B and infrared camera signal with the existing early warning system based on infrasonic data at Etna [16] has also the potential to better characterize the MER in real-time at the beginning of the paroxysmal activity, i.e., Phase I;

(4) MER approaches are based on various parameters, e.g., radar echoes, exit velocities, or H_T, and their accuracy strongly depends on the eruption intensity. Overall, approaches based on H_T (e.g., SEVIRI-TPA, MWR-TPA) or signals proportional to the quantity of detected tephra (e.g., MWR-NSA, MWR-MCA, V2B-NSA) are better suited for computing MER in a large set of eruptive intensities. As an example, MER can be constrained at various time-resolution from 0.2 s (V2B) to 10 min (MWR) for a wide range of eruptive intensities and for all weather and light conditions. Instead, SFA methods (e.g., MWR-SFA, V2B-SFA, Infrasound-SFA, Ground-IR-SFA), based on exit velocities that do not vary significantly among paroxysms, might overestimate or underestimate the MER and TEM for weak and strong paroxysms, respectively;

(5) GSD_{MWR} can be combined with GSD_{sat} to provide a TGSD in near real-time, which is strongly affected by the determination of the relative TEMs. GSD_{MWR} is representative of both the material contributing to the tephra-fallout deposit (contributing to the WDGSD) and to material that typically falls in the sea beyond the coastline (about 20 km from the vent in the case of Etna volcano). Nonetheless, a better constrain of the TEM associated with the two sensors and of the tails of the GSDs is required for operational use;

(6) the combination of the WDGSD and $GSD_{sat.}$ can be used to validate the near real-time strategy described in the previous point as well as a proxy for near real-time tephra forecasting of future eruptions of similar intensity.

Our work represents a step forward in the understanding of multi-sensor strategies applied at very active explosive volcanoes such as Mount Etna. The next step will be to better assess individual sensor sensitivities to refine ESP estimate combinations. Additional information should be taken from other paroxysms, recorded by fewer instruments, to investigate their capacity to provide ESPs in comparison with both the well-recorded weak and strong paroxysms presented herein. Such a systematic determination of remote sensor advantages and limitations should always be carried out to build multi-sensor strategies that are reliable for a large set of eruptive conditions.

Author Contributions: Conceptualization, V.F.-L., C.B., S.C., F.D., F.S.M. and S.S.; methodology, V.F.-L., C.B, S.C., F.D., G.L., F.S.M. and S.S.; validation, V.F.-L., C.B, S.C., F.D., L.G., G.L., F.S.M. and S.S.; investigation, V.F.-L., C.B., S.C., F.D., L.G., G.L., F.S.M., L.M. (Luigi Mereu) and S.S.; tephra-fall deposit data treatment, V.F.-L., C.B. and S.S.; V2B data treatment, V.F.-L. and F.D.; MWR and infrared data treatment, L.M. (Luigi Mereu) and F.S.M.; visible imagery data treatment: S.S.; satellite data treatment, S.C., L.G., L.M. (Luca Merucci) and D.S.; infrasound data treatment, G.L. and M.R.; writing—original draft preparation, V.F.-L. and C.B.; writing—review and editing, All; supervision, C.B.; project administration, C.B.; funding acquisition, C.B. All authors have read and agreed to the published version of the manuscript.

Funding: This project has received funding from the European Union's Horizon 2020 research and innovation program under grant agreement No 731070 (EUROVOLC). The work of SC, LG, DS and LM has been partially founded by the ESA project VISTA (Volcanic monItoring using SenTinel sensors by an integrated Approach), grant number 4000128399/19/I-DT, and from the INGV project Pianeta Dinamico. VOLDORAD 2B radar measurements on Etna are carried out in the frame of a collaborative research agreement between the Observatoire de Physique du Globe de Clermont-Ferrand (OPGC, Université Clermont Auvergne, Clermont-Ferrand, France), the French CNRS, and the INGV-OE. This study used the open-access Doppler radar data base of OPGC—Université Clermont Auvergne VOLDORAD (http://voldorad.opgc.fr/, accessed on 22 May 2021), also supported by EU EPOS and EUROVOLC programs and the French SNOV.

Institutional Review Board Statement: Not applicable.

Informed Consent Statement: Not applicable.

Data Availability Statement: The data presented in this study are available on request from the corresponding author.

Acknowledgments: We thank D. Lo Castro who collected tephra samples of the 10 April 2011 event together with S. Scollo. We also thank M. Prestifilippo who developed and maintains the visible calibrated camera monitoring system at INGV-OE with E. Biale, F. Ciancitto and E. Pecora. We thank M. Coltelli who was head of the volcanology team during both events studied in this paper. Samples were preserved at the sedimentology laboratory of INGV-OE thanks to the support of D. Andronico. We finally thank three anonymous reviewers who provided useful comments that helped us to greatly improved the manuscript.

Conflicts of Interest: The authors declare no conflict of interest.

Appendix A. Summary of Acronyms and Variables Used in This Study

Acronym	Term
ARPA	Agenzia Regionale per la Protezione dell'Ambiente
BTD	Brightness Temperature Difference
EBHD	Etna Bronte High Definition camera
ECV	Etna Catania Visible
ESP	Eruptive Source Parameter
ETN	Etna
GSD	Grain-Size Distribution
GSD_{sat}	Grain-Size Distribution derived from satellite retrievals
GSD_{MWR}	Grain-Size Distribution derived from X-band radar data
HYSPLIT	Hybrid Single-Particle Lagrangian Integrated Trajectory Model
INGV-OE	Istituto Nazionale di Geofisica e Vulcanologia-Osservatorio Etneo
TIR	Thermal InfraRed
IS	InfraSound
MCA	Mass Continuity Approach
MODIS	Moderate Resolution Imaging Spectroradiometer
MVT	Monte Vetore
MWR	Microwave Weather Radar
NSA	Near Surface Approach
NASA	National Aeronautics and Space Agency
NSEC	New SouthEast Crater
SEVIRI	Spinning Enhances Visible and InfraRed Imager
SFA	Surface Flux Approach
TIR	Thermal InfraRed
TGSD	Total Grain-Size Distribution
TPA	Top Plume Approach
UTC	Universal Time Coordinated
VARR	Volcanic Ash Radar Retrieval
VATDM	Volcanic Ash Transport and Dispersal Model
VPR	Volcanic Plume Retrieval
V2B	VOLDORAD 2B
WDGSD	Whole Deposit Grain-Size Distribution (i.e., GSD derived from tephra-fallout deposit)
Symbol	**Variable**
A	Entry Surface of volcanic jets in the radar beams (m^2)
c	Speed of sound (m/s) [65]
Ct	Tephra concentration (kg/m^3) [15]
D_n	Reflectivity-weighted mean radar diameter (m) [15]
H_T	Plume Top Height (km a.s.l.)
IL	Insertion loss caused by topography (dB) [65]
ka	Product between the acoustic wave number k (m^{-1}) and the vent radius a (m) [65]
MER	Mass Eruption Rate (kg/s)
q_i	Volumetric flux from infrasound (m^3/s) [65]
$\mid R \mid$	Acoustic reflectance [65]
Re	Effective radius (m)
S	Eruptive vent surface (m^2)
t	Time (s)
TEM	Total Erupted Mass (kg)
v_{entry}	Entry velocity of particles in the radar beams (m/s)
v_{exit}	Exit velocity (m/s) [12,30]
v_r	Radial velocity (m/s)
Z	Radar Reflectivity factor (dBZ)
α	Directivity of the acoustic wave at 0° [65]
$B,\gamma,\delta,\varepsilon$	VARR Parameters to be used in Equations (3) and (4) [15]
θ	Radar beam elevation angle (°) [12,30]
ΔP	Pressure signal (Pa) [65]
λ	Radar wavelength (m)
ρ	Atmosphere density (kg/m^3)
ρ_x	Mixture density (kg/m^3) [15]
Φ	Phi unit of particle sizes

Appendix B. Grain-Size Data from the 29 August 2011 Tephra-Fallout Deposit

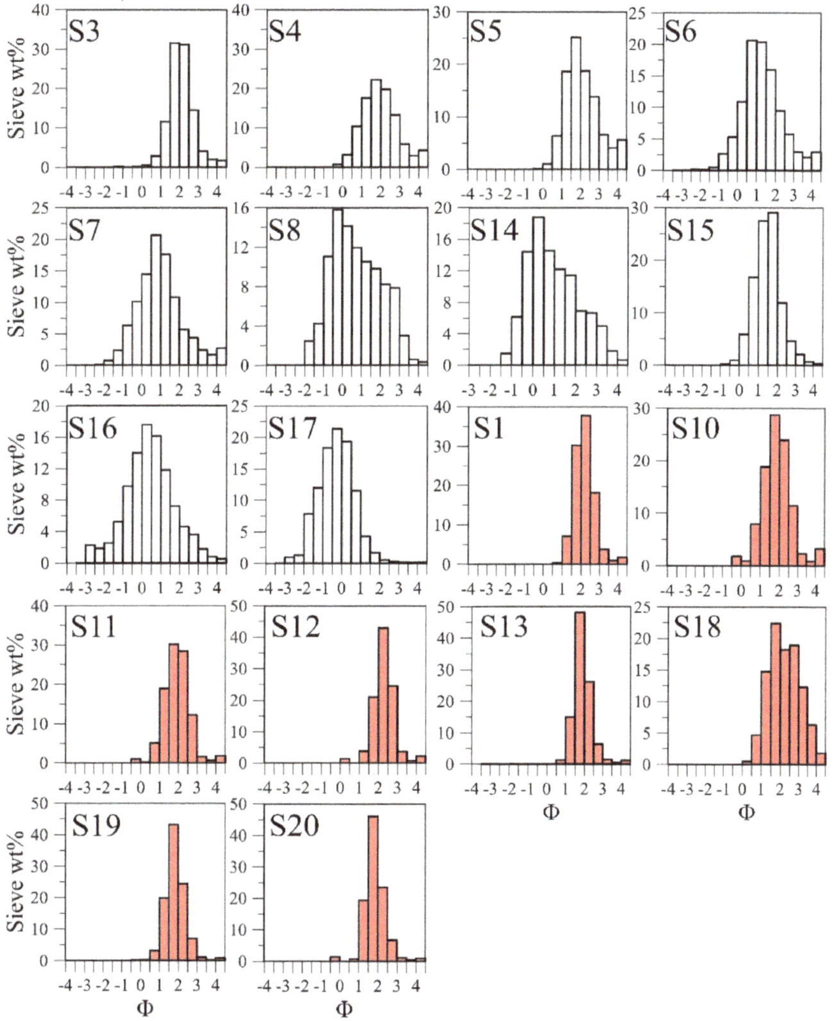

Figure A1. Individual GSDs obtained from sieving and using the BETTERSIZER morpho-grainsizer (red distributions). Sample names (i.e., from S1 to S20) are indicated on the top left of each plot.

References

1. Jenkins, S.F.; Wilson, T.; Magill, C.; Miller, V.; Stewart, C.; Blong, R.; Marzocchi, W.; Boulton, M.; Bonadonna, C.; Costa, A. Volcanic ash fall hazard and risk. In *Global Volcanic Hazard and Risk*; Loughlin, S.C., Sparks, S., Brown, S.K., Jenkins, S.F., Brown, C., Eds.; Cambridge University Press: Cambridge, UK, 2015; ISBN 978-1-107-11175-2.
2. Wilson, T.M.; Jenkins, S.; Stewart, C. Impacts from volcanic ash fall. In *Volcanic Hazards, Risk and Disasters*; Papale, P., Ed.; Elsevier: Amsterdam, The Netherlands, 2015. [CrossRef]
3. Calvari, S.; Salerno, G.G.; Spampinato, L.; Gouhier, M.; La Spina, A.; Pecora, E.; Harris, A.J.L.; Labazuy, P.; Biale, E.; Boschi, E. An unloading foam model to constrain Etna's 11–13 January 2011 lava fountaining episode. *J. Geophys. Res.* **2011**, *116*, 1–18. [CrossRef]
4. Calvari, S.; Cannavò, F.; Bonaccorso, A.; Spampinato, L.; Pellegrino, A.G. Paroxysmal Explosions, Lava Fountains and Ash plumes at Etna Volcano: Eruptive Processes and Hazard Implications. *Front. Earth Sci.* **2018**, *6*, 107. [CrossRef]

5. Bonaccorso, A.; Caltabiano, T.; Currenti, G.; Del Negro, C.; Gambino, S.; Ganci, G.; Boschi, E. Dynamics of a lava fountain revealed by geophysical, geochemical and thermal satellite measurements: The case of the 10 April 2011 Mt Etna eruption. *Geophys. Res. Lett.* **2011**, *38*, 1–7. [CrossRef]
6. Bonaccorso, A.; Calvari, S.; Linde, A.; Sacks, S. Eruptive processes leading to the most explosive lava fountain at Etna volcano: The 23 November 2013 episode. *Geophys. Res. Lett.* **2014**, *41*, 4912–4919. [CrossRef]
7. Corradini, S.; Montopoli, M.; Guerrieri, L.; Ricci, M.; Scollo, S.; Merucci, L.; Marzano, F.S.; Pugnaghi, S.; Prestifilippo, M.; Ventress, L.J.; et al. A Multi-Sensor Approach for Volcanic Ash Cloud Retrieval and Eruption Characterization: The 23 November 2013 Etna Lava Fountain. *Remote Sens.* **2016**, *8*, 58. [CrossRef]
8. Corradini, S.; Guerrieri, L.; Lombardo, V.; Merucci, L.; Musacchio, M.; Prestifilippo, M.; Scollo, S.; Silvestri, M.; Spata, G.; Stelitano, D. Proximal Monitoring of the 2011-2015 Etna Lava Fountains Using MSG-SEVIRI Data. *Geosciences* **2018**, *8*, 140. [CrossRef]
9. Marzano, F.S.; Barbieri, S.; Vulpiani, G.; Rose, W.I. Volcanic cloud retrieval by ground-based microwave weather radar. *IEEE Trans. Geosci. Rem. Sens.* **2006**, *44*, 3235–3246. [CrossRef]
10. Marzano, F.S.; Barbieri, S.; Picciotti, E.; Karlsdóttir, S. Monitoring subglacial Volcanic Eruption using Ground-Based C-Band Radar Imagery. *IEEE Trans. Geosc. Remote Sens.* **2010**, *48*, 403–414. [CrossRef]
11. Donnadieu, F. Volcanological applications of Doppler radars: A review and examples from a transportable pulse radar in L-band. In *Doppler Radar Observations—Weather Radar, Wind Profiler, Ionospheric Radar, and Other Advanced Applications*; Bech, J., Chau, J.L., Eds.; In TechOpen: London, UK, 2012; pp. 409–446. ISBN 978-957-51-0496-4.
12. Donnadieu, F.; Freville, P.; Hervier, C.; Coltelli, M.; Scollo, S.; Prestifilippo, M.; Valade, S.; Rivet, S.; Cacault, P. Near-source Doppler radar monitoring of tephra plumes at Etna. *J. Volcanol. Geotherm. Res.* **2016**, *312*, 26–39. [CrossRef]
13. Montopoli, M. Velocity profiles inside volcanic clouds from three-dimensional scanning microwave dual polarization Doppler radars. *J. Geophys. Res. Atmos.* **2016**, *121*, 7881–7900. [CrossRef]
14. Vulpiani, G.; Ripepe, M.; Valade, S. Mass discharge rate retrieval combining weather radar and thermal camera observations. *J. Geophys. Res. Solid Earth* **2016**, *121*, 5679–5695. [CrossRef]
15. Marzano, F.S.; Mereu, L.; Scollo, S.; Donnadieu, F.; Bonadonna, C. Tephra Mass Eruption Rate from Ground-based X-Band and L-Band Microwave Radars during the 23 November 2013 Etna Paroxysm. *IEEE Trans. Geosc. Remote Sens.* **2020**, *58*, 3314–3327. [CrossRef]
16. Ripepe, M.; Marchetti, E.; Delle Donne, D.; Genco, R.; Innocenti, L.; Lacanna, G.; Valade, S. Infrasonic Early Warning System for Explosive Eruption. *J. Geophys. Res. Solid Earth* **2018**, *123*, 9570–9585. [CrossRef]
17. Sciotto, M.; Cannata, A.; Prestifilippo, M.; Scollo, S.; Fee, D.; Privitera, E. Unravelling the links between seismo-acoustic signals and eruptive parameters: Etna lava fountain case study. *Sci. Rep.* **2019**, *9*, 16417. [CrossRef] [PubMed]
18. Scollo, S.; Boselli, A.; Coltelli, M.; Leto, G.; Pisani, G.; Spinelli, N.; Wang, X. Monitoring Etna volcanic plumes using a scanning lidar. *Bull. Volcanol.* **2012**, *74*, 2382–2395. [CrossRef]
19. Scollo, S.; Boselli, A.; Coltelli, M.; Leto, G.; Pisani, G.; Prestifilippo, M.; Spinelli, N.; Wang, X. Volcanic ash concentration during the 12 August 2011 Etna eruption. *Geophys. Res. Lett.* **2015**, *42*, 2634–2641. [CrossRef]
20. Scollo, S.; Prestifilippo, M.; Spata, G.; D'Agostino, M.; Coltelli, M. Monitoring and forecasting Etna volcanic plumes. *Nat. Hazard Earth Syst. Sci.* **2009**, *9*, 1573–1585. [CrossRef]
21. Scollo, S.; Prestifilippo, M.; Pecora, E.; Corradini, S.; Merucci, L.; Spata, G.; Coltelli, M. Height estimation of the 2011–2013 Etna lava fountains. *Ann. Geophys.* **2014**, *57*, 0214. [CrossRef]
22. Scollo, S.; Prestifilippo, M.; Bonadonna, C.; Cioni, R.; Corradini, S.; Degruyter, W.; Rossi, E.; Silvestri, M.; Biale, E.; Carparelli, G.; et al. Near-Real-Time Tephra Fallout Assessment at Mt. Etna, Italy. *Remote Sens.* **2019**, *11*, 2987. [CrossRef]
23. Scollo, S.; Del Carlo, P.; Coltelli, M. Tephra fallout of 2001 Etna flank eruption: Analysis of the deposit and plume dispersion. *J. Volcanol. Geotherm. Res.* **2007**, *160*, 147–164. [CrossRef]
24. Andronico, D.; Scollo, S.; Caruso, S.; Cristaldi, A. The 2002-03 Etna explosive activity: Tephra dispersal and features of the deposits. *J. Geophys. Res.* **2008**, *113*, B04209. [CrossRef]
25. Andronico, D.; Scollo, S.; Cristaldi, A.; Ferrari, F. Monitoring ash emission episodes at Mt. Etna: The 16 november 2006 case study. *J. Volcanol. Geotherm. Res.* **2009**, *180*, 123–134. [CrossRef]
26. Andronico, D.; Scollo, S.; Cristaldi, A.; Lo Castro, M.D. Representivity of incompletely sampled fall deposits in estimating eruption source parameters: A test using the 12–13 January 2011 lava fountain deposit from Mt. Etna volcano, Italy. *Bull. Volcanol.* **2014**, *76*, 861. [CrossRef]
27. Andronico, D.; Scollo, S.; Cristaldi, A. Unexpected hazards from tephra fallouts at Mt Etna: The 23 November 2013 lava fountain. *J. Volcanol. Geotherm. Res.* **2015**, *304*, 118–125. [CrossRef]
28. Marzano, F.S.; Lamantea, M.; Montopoli, M.; Herzog, M.; Graf, H.; Cimini, D. Microwave remote sensing of Plinian eruption due to the Grímsvötn Icelandic volcano on May 2011. *Rem. Sens. Env.* **2013**, *129*, 168–184. [CrossRef]
29. Guerrieri, L.; Merucci, L.; Corradini, S.; Pugnaghi, S. Evolution of the 2011 Mt. Etna ash and SO_2 lava fountain episodes using SEVIRI data and VPR retrieval approach. *J. Volcanol. Geotherm. Res.* **2015**, *291*, 63–71. [CrossRef]
30. Freret-Lorgeril, V.; Donnadieu, F.; Scollo, S.; Provost, A.; Fréville, F.; Ghéhenneux, Y.; Hervier, C.; Prestifilippo, M.; Coltelli, M. Mass Eruption Rates of Tephra Plumes During the 2011-2015 Lava Fountain Paroxysms at Mt. Etna From Doppler Radar Retrievals. *Front. Earth Sci.* **2018**, *6*, 73. [CrossRef]

31. Poret, M.; Corradini, S.; Merucci, L.; Costa, A.; Andronico, D.; Montopoli, M.; Vulpiani, G.; Freret-Lorgeril, V. Reconstructing volcanic plume evolution integrating satellite and ground-based data: Application to the 23 November 2013 Etna eruption. *Atmos. Chem. Phys.* **2018**, *18*, 4695–4714. [CrossRef]
32. Poret, M.; Costa, A.; Andronico, D.; Scollo, S.; Gouhier, M.; Cristaldi, A. Modelling eruption source parameters by integrating field, ground-based and satellite-based data: The case of the 23rd February 2013 Etna paroxysm. *J. Geophys. Res. Solid Earth* **2018**, *123*, 5427–5450. [CrossRef]
33. Mereu, L.; Scollo, S.; Bonadonna, C.; Freret-Lorgeril, V.; Marzano, F.S. Multisensor Characterization of the Incandescent Jet Region of Lava Fountain-Fed Tephra Plumes. *Remote Sens.* **2020**, *12*, 3629. [CrossRef]
34. Scollo, S.; Boselli, A.; Corradini, S.; Leto, G.; Guerrieri, L.; Merucci, L.; Prestifilippo, M.; Sanchez, R.Z.; Sannino, A.; Stelitano, D. Multi-Sensor Analysis of a Weak and Long-Lasting Volcanic Plume Emission. *Remote Sens.* **2020**, *12*, 3866. [CrossRef]
35. Behncke, B.; Branca, S.; Corsaro, R.A.; De Beni, E.; Miraglia, L.; Proietti, C. The 2011–2012 summit activity of Mount Etna: Birth, growth and products of the new SE crater. *J. Volcanol. Geotherm. Res.* **2014**, *270*, 10–21. [CrossRef]
36. De Beni, E.; Behncke, B.; Branca, S.; Nicolosi, I.; Carluccio, R.; D'Ajello Caracciolo, F.; Chiappini, M. The continuing story of Etna's New Southeast Crater (2012–2014): Evolution and volume calculations based on field surveys and aerophotogrammetry. *J. Volcanol. Geotherm. Res.* **2015**, *303*, 175–186. [CrossRef]
37. Bonaccorso, A.; Calvari, S. A new approach to investigate an eruptive paroxysmal sequence using camera and strainmeter networks: Lessons from the 3–5 December 2015 activity at Etna volcano. *Earth Planet. Sci. Lett.* **2017**, *475*, 231–241. [CrossRef]
38. Corsaro, R.A.; Andronico, D.; Behncke, B.; Branca, S.; Caltabiano, T.; Ciancitto, F.; Cristaldi, A.; De Beni, E.; La Spina, A.; Lodato, L.; et al. Monitoring the December 2015 summit eruption of Mt. Etna (Italy): Implications on eruptive dynamics. *J. Volcanol. Geotherm. Res.* **2017**, *341*, 53–69. [CrossRef]
39. Boichu, M.; Clarisse, L.; Péré, J.C.; Herbin, H.; Goloub, P.; Thieuleux, F.; Ducos, F.; Clerbaux, C.; Tanré, C. Temporal variations of flux and altitude of sulfur dioxide emissions during volcanic eruptions: Implications for long-range dispersal of volcanic clouds. *Atmos. Chem. Phys.* **2015**, *15*, 8381–8400. [CrossRef]
40. Pyle, D.M. The thickness, volume, and grainsize of tephra fall deposits. *Bull. Volcanol.* **1989**, *51*, 1–15. [CrossRef]
41. Bonadonna, C.; Houghton, B.F. Total grain-size distribution and volume of tephra-fall deposits. *Bull. Volcanol.* **2005**, *67*, 441–456. [CrossRef]
42. Bonadonna, C.; Costa, A. Estimating the volume of tephra deposits: A new simple strategy. *Geology* **2012**, *40*, 415–418. [CrossRef]
43. Biass, S.; Bonadonna, C. A quantitative uncertainty assessment of eruptive parameters derived from tephra deposits: The example of two large eruptions of Cotopaxi volcano, Ecuador. *Bull. Volcanol.* **2011**, *73*, 73–90. [CrossRef]
44. Donnadieu, F.; Freville, P.; Rivet, S.; Hervier, C.; Cacault, P. The Volcano Doppler radar data base of Etna (VOLDORAD 2B). Université Clermont Auvergne—CNRS. 2015. Available online: http://www.obs.univ-bpclermont.fr/SO/televolc/voldorad/bddtr.php (accessed on 22 May 2021).
45. Marzano, F.S.; Picciotti, E.; Vulpiani, G.; Montopoli, M. Synthetic signatures of volcanic ash cloud particles from X-Band dual-polarization radar. *IEEE Trans. Geosc. Remote Sens.* **2012**, *50*, 193–211. [CrossRef]
46. Mereu, L.; Marzano, F.S.; Montopoli, M.; Bonadonna, C. Retrieval of Tephra Size Spectra and Mass Flow Rate From C-Band Radar During the 2010 Eyjafjallajökull Eruption, Icelande. *IEEE Trans. Geosc. Remote Sens.* **2015**, *53*, 5644–5660. [CrossRef]
47. Sauvageot, H. *Radar Meteorology*; Artech House: Boston, MA, USA, 1992; ISBN 0890063184.
48. Degruyter, W.; Bonadonna, C. Improving on mass flow rate estimates of volcanic eruptions. *Geophys. Res. Lett.* **2012**, *39*, L16308. [CrossRef]
49. Corradini, S.; Spinetti, C.; Carboni, E.; Tirelli, C.; Buongiorno, M.F.; Pugnaghi, S.; Gangale, G. Etna tropospheric ash retrieval and sensitivity analysis using Moderate Resolution Imaging Spectroradiometer measurements. *J. Appl. Remote Sens.* **2008**, *2*, 023550. [CrossRef]
50. Gouhier, M.; Eychenne, J.; Azzaoui, N.; Guillin, A.; Deslandes, M.; Poret, M.; Costa, A.; Husson, P. Low efficiency of large volcanic eruptions in transporting very fine ash into the atmosphere. *Sci. Rep.* **2019**, *9*, 1449. [CrossRef] [PubMed]
51. Prata, A.J.; Grant, I.F. Retrieval of microphysical and morphological properties of volcanic ash plumes from satellite data: Application to Mt. Ruapehu, New Zealand. *Q. J. Royal Meteorol. Soc.* **2001**, *127*, 2153–2179. [CrossRef]
52. Corradini, S.; Guerrieri, L.; Stelitano, D.; Salerno, G.; Scollo, S.; Merucci, L.; Prestifilippo, M.; Musacchio, M.; Silvestri, M.; Lombardo, V.; et al. Near Real-Time Monitoring of the Christmas 2018 Etna Eruption Using SEVIRI and Products Validation. *Remote Sens.* **2020**, *12*, 1336. [CrossRef]
53. Prata, A.J. Infrared radiative transfer calculations for volcanic ash clouds. *Geophys. Res. Lett.* **1989**, *16*, 1293–1296. [CrossRef]
54. Pugnaghi, S.; Guerrieri, L.; Corradini, S.; Merucci, L.; Arvani, B. A new simplified procedure for the simultaneous SO2 and ash retrieval in a tropospheric volcanic cloud. *Atmos. Meas. Tech.* **2013**, *6*, 1315–1327. [CrossRef]
55. Pugnaghi, S.; Guerrieri, L.; Corradini, S.; Merucci, L. Real time retrieval of volcanic cloud particles and SO2 by satellite using an improved simplified approach. *Atmos. Meas. Tech.* **2016**, *9*, 1–10. [CrossRef]
56. Corradini, S.; Pugnaghi, S.; Pscini, A.; Guerrieri, L.; Merucci, L.; Picchiani, M.; Chini, M. Volcanic Ash and SO_2 retrievals using synthetic MODIS TIR data: Comparison between inversion procedures and sensitivity analysis. *Ann. Geophys.* **2014**, *57*, 2.
57. Wen, S.; Rose, W.I. Retrieval of sizes and total masses of particles in volcanic clouds using AVHRR bands 4 and 5. *J. Geophys. Res.* **1994**, *99*, 5421–5431. [CrossRef]

58. Pugnaghi, S.; Gangale, G.; Corradini, S.; Buongiorno, M.F. Etna sulfur dioxide flux monitoring using ASTER-TIR data and atmospheric observations. *J. Volcanol. Geotherm. Res.* **2006**, *152*, 74–90. [CrossRef]
59. Corradini, S.; Merucci, L.; Prata, A.J. Retrieval of SO2 from thermal infrared satellite measurements: Correction procedures for the effects of volcanic ash. *Atmos. Meas. Tech.* **2009**, *2*, 177–191. [CrossRef]
60. Merucci, L.; Burton, M.; Corradini, S.; Salerno, G. Reconstruction of SO2 flux emission chronology from space-based measurements. *J. Volcanol. Geotherm. Res.* **2011**, *206*, 80–87. [CrossRef]
61. Theys, N.; Campion, R.; Clarisse, L.; Brenot, H.; van Gent, J.; Dils, B.; Corradini, S.; Merucci, L.; Coheur, P.-F.; Van Roozendael, M.; et al. Volcanic SO2 Fluxes Derived from Satellite Data: A Survey Using OMI, GOME-2, IASI and MODIS. *Atmos. Chem. Phys.* **2013**, *13*, 5945–5968. [CrossRef]
62. Ulivieri, G.; Ripepe, M.; Marchetti, E. Infrasound reveals transition to oscillatory discharge regime during lava fountaining: Implication for early warning. *Geophys. Res. Lett.* **2013**, *40*, 3008–3013. [CrossRef]
63. Marchetti, E.; Ripepe, M.; Campus, P.; Le Pichon, A.; Vergoz, J.; Lacanna, G.; Mialle, P.; Héreil, P.; Husson, P. Long range infrasound monitoring of Etna volcano. *Sci. Rep.* **2019**, *9*, 18015. [CrossRef] [PubMed]
64. Sanchez, C.; Ripepe, M.; Lacanna, G.; Poggi, P. Experimental modeling of mass eruption rates from acoustics wave. *J. Acoust. Soc. Am.* **2019**, *146*, 3076. [CrossRef]
65. Lacanna, G.; Ripepe, M. Modeling the Acoustic Flux inside the Magmatic Conduit by 3D-FDTD Simulation. *J. Geophys. Res. Solid Earth* **2020**, *125*, e2019JB018849. [CrossRef]
66. Ripepe, M.; Bonadonna, C.; Folch, A.; Delle Donne, D.; Lacanna, G.; Marchetti, E.; Hoskuldsson, A. Ash-plume dynamics and eruption source parameters by infrasound and thermal imagery. *Earth Planet. Sci. Lett.* **2013**, *366*, 112–121. [CrossRef]
67. Carbone, D.; Zuccarello, L.; Messina, A.; Scollo, S.; Rymer, H. Balancing bulk gas accumulation and gas output before and during lava fountaining episodes at Mt. Etna. *Sci Rep.* **2015**, *5*, 18049. [CrossRef]
68. Bonadonna, C.; Biass, S.; Costa, A. Physical characterization of explosive volcanic eruptions based on tephra deposits: Propagation of uncertainties and sensitivity analysis. *J. Volcanol. Geotherm. Res.* **2015**, *296*, 80–100. [CrossRef]
69. Spanu, A.; de' Michieli Vitturi, M.; Barsotti, S. Reconstructing eruptive source parameters from tephra deposit: A numerical study of medium-sized explosive eruptions at Etna volcano. *Bull. Volcanol.* **2016**, *78*, 1–19. [CrossRef]
70. Inman, D.L. Measures for describing the size distribution of sediments. *J. Sed. Petrol.* **1952**, *22*, 125–145.
71. Bonadonna, C.; Genco, R.; Gouhier, M.; Pistolesi, M.; Cioni, R.; Alfano, F.; Hoskuldsson, A.; Ripepe, M. Tephra sedimentation during the 2010 Eyjafjallajökull eruption (Iceland) from deposit, radar, and satellite observations. *J. Geophys. Res. Lett.* **2011**, *116*, B12202. [CrossRef]
72. Petersen, G.N.; Bjornsson, H.; Arason, P.; von Löwis, S. Two weather radar time series of the altitude of the volcanic plume during the May 2011 eruption of Grímsvötn, Iceland. *Earth Syst. Sci. Data* **2012**, *4*, 121–127. [CrossRef]
73. Folch, A. A review of tephra transport and dispersal models: Evolution, current status, and future perspectives. *J. Volcanol. Geotherm. Res.* **2012**, *235–236*, 96–115. [CrossRef]
74. Witham, C. Assessment of the impact of radar height data on model forecasts for Grímsvötn 2011. In *Statistical Assessment of Dispersion Model Sensitivity*; Deliverable report D8.5 of the EU FUTUREVOLC project; Beckett, F., Witham, C., Devenish, B., Eds.; 2015; Available online: http://futurevolc.hi.is/sites/futurevolc.hi.is/files/Pdf/Deliverables/fv_d8_5_to_submit_low.pdf (accessed on 1 January 2020).
75. Beckett, F.M.; Witham, C.S.; Leadbetter, S.J.; Crocker, R.; Webster, H.N.; Hort, M.C.; Jones, A.R.; Devenish, B.J.; Thomson, D.J. Atmospheric Dipersion Modelling at the London VACC: A review of Developments since the 2010 Eyjafjallajökull Volcano Ash Cloud. *Atmosphere* **2020**, *11*, 352. [CrossRef]
76. Lacasse, C.; Karlsdóttir, S.; Larsen, G.; Soosalu, H.; Rose, W.I.; Ernst, G.G.J. Weather radar observations of the Hekla 2000 eruption cloud, Iceland. *Bull. Volcanol.* **2004**, *66*, 457–473. [CrossRef]
77. Morton, B.; Taylor, G.; Turner, J. Gravitational turbulent convection from maintained and instantaneous sources. *Proc. Math. Phys. Eng. Sci.* **1956**, *234*, 1–23. [CrossRef]
78. Sparks, R.S.J. The dimensions and dynamics of volcanic eruption columns. *Bull. Volcanol.* **1986**, *48*, 3–15. [CrossRef]
79. Mastin, L.G. A user-friendly one-dimensional model for wet volcanic plumes. *Geochem. Geophys. Geosys.* **2007**, *8*, Q03014. [CrossRef]
80. De' Michieli Vitturi, M.; Neri, A.; Barsotti, S. PLUME-MoM 1.0: A new integral model of volcanic plumes based on the method of moments. *Geosci. Model. Dev.* **2015**, *8*, 2447–2463. [CrossRef]
81. Snee, E.; Degruyter, W.; Bonadonna, C.; Scollo, S.; Rossi, E.; Freret-Lorgeril, V. A model for buoyant tephra plumes couples to lava fountains with an application to paroxysmal eruptions at Mount Etna, Italy. *J. Geophys. Res. Solid Earth* **2021**, e2020JB021360. [CrossRef]
82. Bonadonna, C.; Folch, A.; Loughlin, S.; Puempel, H. Future developments in modelling and monitoring of volcanic ash clouds: Outcomes from the first IAVCEI-WMO workshop on Ash Dispersal Forecast and Civil Aviation. *Bull. Volcanol.* **2012**, *74*, 1–10. [CrossRef]
83. Stevenson, J.; Millington, S.; Beckett, F.M.; Swindles, G.; Thordarson, T. Understanding the discrepancy between tephrochronology and satellite infrared measurements of volcanic ash. *Atmos. Meas. Tech.* **2015**, *8*, 2069–2091. [CrossRef]
84. Gouhier, M.; Donnadieu, F. Mass estimations of ejecta from Strombolian explosions by inversion of Doppler radar measurements. *J. Geophys. Res.* **2008**, *113*, B10202. [CrossRef]

85. Alparone, S.; Andronico, D.; Lodato, L.; Sgroi, T. Relationship between tremor and volcanic activity during the Southeast Crater eruption of Mount Etna in early 2000. *J. Geophys. Res.* **2003**, *108*, 2241. [CrossRef]
86. Carey, S.; Sparks, R.S.J. Quantitative models of the fallout and dispersal of tephra from volcanic eruption columns. *Bull. Volcanol.* **1986**, *48*, 109–125. [CrossRef]
87. Rossi, E.; Bonadonna, C.; Degruyter, W. A new strategy for the estimation of plume height from clast dispersal in various atmospheric and eruptive conditions. *Earth. Planet. Sci. Lett.* **2019**, *505*, 1–12. [CrossRef]

remote sensing

MDPI

Article

Anatomy of a Paroxysmal Lava Fountain at Etna Volcano: The Case of the 12 March 2021, Episode

Sonia Calvari *, Alessandro Bonaccorso and Gaetana Ganci

Istituto Nazionale di Geofisica e Vulcanologia, Osservatorio Etneo-Sezione di Catania, 95125 Catania, Italy; alessandro.bonaccorso@ingv.it (A.B.); gaetana.ganci@ingv.it (G.G.)
* Correspondence: sonia.calvari@ingv.it

Abstract: On 13 December 2020, Etna volcano entered a new eruptive phase, giving rise to a number of paroxysmal episodes involving increased Strombolian activity from the summit craters, lava fountains feeding several-km high eruptive columns and ash plumes, as well as lava flows. As of 2 August 2021, 57 such episodes have occurred in 2021, all of them from the New Southeast Crater (NSEC). Each paroxysmal episode lasted a few hours and was sometimes preceded (but more often followed) by lava flow output from the crater rim lasting a few hours. In this paper, we use remote sensing data from the ground and satellite, integrated with ground deformation data recorded by a high precision borehole strainmeter to characterize the 12 March 2021 eruptive episode, which was one of the most powerful (and best recorded) among that occurred since 13 December 2020. We describe the formation and growth of the lava fountains, and the way they feed the eruptive column and the ash plume, using data gathered from the INGV visible and thermal camera monitoring network, compared with satellite images. We show the growth of the lava flow field associated with the explosive phase obtained from a fixed thermal monitoring camera. We estimate the erupted volume of pyroclasts from the heights of the lava fountains measured by the cameras, and the erupted lava flow volume from the satellite-derived radiant heat flux. We compare all erupted volumes (pyroclasts plus lava flows) with the total erupted volume inferred from the volcano deflation recorded by the borehole strainmeter, obtaining a total erupted volume of ~3×10^6 m^3 of magma constrained by the strainmeter. This volume comprises ~1.6×10^6 m^3 of pyroclasts erupted during the lava fountain and 2.4×10^6 m^3 of lava flow, with ~30% of the erupted pyroclasts being remobilized as rootless lava to feed the lava flows. The episode lasted 130 min and resulted in an eruption rate of ~385 m^3 s^{-1} and caused the formation of an ash plume rising from the margins of the lava fountain that rose up to 12.6 km a.s.l. in ~1 h. The maximum elevation of the ash plume was well constrained by an empirical formula that can be used for prompt hazard assessment.

Keywords: Etna volcano; paroxysmal explosive and effusive episodes; ash plume; remote sensing; volcano monitoring; volcanic hazard

Citation: Calvari, S.; Bonaccorso, A.; Ganci, G. Anatomy of a Paroxysmal Lava Fountain at Etna Volcano: The Case of the 12 March 2021, Episode. *Remote Sens.* **2021**, *13*, 3052. https://doi.org/10.3390/rs13153052

Academic Editor: Andrew McGonigle

Received: 14 June 2021
Accepted: 30 July 2021
Published: 3 August 2021

1. Introduction

Explosive eruptions of mafic magmas produce lava fountains whose heights depend on the exsolved volatile content of the magma, its erupted mass flux, and the geometry of the vent, either an elongated eruptive fissure or a near circular conduit [1]. Lava fountains were typical at Kilauea volcano during the 1959–1960, 1969–1970, and 1983–2008 eruptions [2–5], being characterized by vertical jets of gas and incandescent pyroclasts rising several hundred meters above the vent. This activity is also common at Etna volcano, with several such explosive phases occurring in 2000 [6–8], 2001, 2002–2003 [9–11], and in 2011–2015 [12–16]. The last paroxysmal lava fountain sequence started on 13 December 2020, and is still going on as of 2 August 2021. A recent study, based on a catalogue of the explosive paroxysmal episodes that occurred at Etna since 1986 (and updated to 1 April 2021), showed a general marked increase in the release of seismic energy over time [17]. This is in agreement with a general trend of increasing heights and volume of the lava fountains and ash plumes from 2000 until now [6,8,10–12,14,15].

The summit of Etna volcano comprises four main craters: Voragine, Bocca Nuova, NE Crater, and SE Crater, with the addition of the newly formed New SE Crater (NSEC) built up on top of the SE Crater since 2011 [14,18]. Lava fountains from the summit craters or from eruptive fissures have often preceded major flank effusive activity, such as at Etna in 2001 and 2002–2003; thus, they were considered as possible precursors [10,11]. However, the several lava fountain events between 2011 and 2015 [12,14–17,19] were not followed by flank eruptions. This observation, combined with the estimation of erupted volumes from the monitoring camera analyses [14,15], allowed recognizing that Etna displayed a steady-state behavior for at least four decades [20–23]. Therefore, in cases such as the 2011–2015 lava fountains, this explosive activity can represent a modality of magma discharge able to maintain the steady-state. At Etna volcano, the average output rate was estimated at 0.8 $m^3 s^{-1}$, or 25 × 10^6 m^3 per year [21,22]. On this basis, Bonaccorso and Calvari [22] found that the magma stored within the plumbing system can be released either through a high number of lava fountains, erupting volumes of ~2–3 × 10^6 m^3 each [12,14,15,22], or with a small number of flank effusive eruptions, each normally releasing ~30–60 × 10^6 m^3 of lava [21,22]. Considering this conceptual model, it is not surprising that on 13 December 2020, after about 18 months of eruptive pause [24–27], the volcano entered a new eruptive phase characterized by 57 lava fountain episodes (as of 2 August 2021) accompanied, preceded, or followed by short-lasting effusive phases. Lava fountains, being characterized at Etna by heights of a few km, and always accompanied by several km high ash plumes generated from the same vent, raise serious concern among the local population. In fact, an ash fallout up to ~38 kg/m^2 [10] has a strong impact on the viability, on the stability of roofs, on the air traffic (the Catania airport is a major international hub), on agriculture, on water contamination, and on the health of the local population [28–32].

The release of ash plumes during mild basaltic explosive activity was described and analyzed only recently, when two distinct eruption styles were identified for Strombolian explosions [33]. Type 1 Strombolian eruptions consist of coarse ballistic scoria (cm/dm-scale) and a relatively ash-free gas plume. Type 2 Strombolian eruptions consist of an ash-rich plume, with or without additional ballistic scoria. What determines type 2 ash-rich Strombolian behavior is the sliding of loose clastic material into the vent [33–35], or rheological changes in the uppermost magma column [36,37]. The formation of ash plume during lava fountaining is even more questioned and often not considered in lava fountain models [38,39]. This is because lava fountaining is considered as mostly characterized by coarse ballistics falling around the vent and building spatter cones or spatter ramparts, with small amounts of ash being released in the atmosphere [40–42].

The formation of ash plumes is a common feature at Etna volcano, always accompanying lava fountain eruptions [6–15,17]. Conversely, ash plume is lacking during Strombolian and intermediate explosive activity [43], but is released by the summit craters during rare phreatomagmatic activity [44–46]. In addition, weak and dilute ash plume may form during summit collapses, occurring within the craters or pit crater formations [44,47,48].

Several authors described a transitional explosive stage in between the Strombolian and lava fountain [6,14,39,43], occurring when the discrete countable Strombolian bursts increase in number and frequency, shifting to the continuous lava fountaining regime [14,43]. This transition can be either abrupt or gradual [6]. At Etna, lava fountains are always associated to the formation of ash plumes [6,7,10,11,14,15], and as soon as the lava fountain regime starts, we observe the release of ash from the upper part of the lava fountain that rises for kilometers above the crater, feeding a sustained ash plume [42]. However, the way a lava fountain expands vertically into a sustained ash plume, as well as the timing of this process, is still little constrained. Nonetheless, sustained eruptive columns and ash plumes are of great concern to local authorities in Sicily and elsewhere [17,49–51] because the ash plumes may expand well beyond the national country. As an example, Etna's ash plumes also affected Malta and Greece [25,45,52,53] for several days after the end of the eruption [54], and its gas and ash plumes even travelled the entire globe [55]. Even worse, the eruption of Mt St Helens in 1982 [49], Pinatubo in 1991 [56], and the Icelandic volcano Eyjafjallajökull in 2010 caused air traffic disruption for about a month across Europe [50,51].

In this paper, we have analyzed, in detail, one of the most powerful (and best recorded) lava fountains taking place between 13 December 2020, and 20 July 2021, namely the episode of 12 March 2021, which occurred during good weather conditions and in the daytime (Figure 1), and enabled the collection of excellent data from the ground and satellite. We present details on the lava fountains and their connections with ash plume and lava flow field formation, as well as persistence and decline gathered from a network of ground-based monitoring cameras and from high temporal resolution satellite sensors (e.g., SEVIRI, MODIS, and VIIRS). We compare the timing and volumes obtained from these devices with results from the reference borehole strainmeter [57,58], in order to highlight key processes that characterize the phenomenon and are useful for hazard assessment. In particular, we analyze the formation and growth of the lava fountain and of the associated lava flow field, the timing, and how the fountain feeds the ash column and eruptive plume. Our focus was to acquire parameters that could be useful for prompt hazard assessment.

Figure 1. Photo of Etna paroxysmal episode on 12 March 2021. The shot was taken at 08:50 UTC from the main road (Etnea Street) crossing N–S the city of Catania. The view is from south (photo by A.B.).

2. Methods

2.1. Camera Networks

We used the monitoring camera network installed and maintained by the Istituto Nazionale di Geofisica e Vulcanologia (INGV) Osservatorio Etneo-Sezione di Catania, comprising thermal and visible cameras, in order to detect and quantify the phases of eruptive activity. Our aim was to define the timing of its changes, as well as the height of the lava fountains and ash plume, the erupted volume of pyroclastics and the expansion of the lava flow field, and of their timing—parameters that are essential for hazard assessment at a frequently erupting volcano. The labels of the cameras used in this paper, as well as their main features, viewing direction, and average distance from the craters, are listed in Table 1, and their positions are shown in Figure 2. The height of the lava fountains was obtained from the thermal cameras ENT and EBT located on the S and NW flanks of the volcano, respectively. The error in the height measurement is ±50 m [14,15]. These heights were used to calculate the erupted volume of pyroclasts, following the method developed by Calvari et al. [14,15]. This method consists in measuring the lava fountain height on thermal images with a 1-minute time lapse, and applying the Equation (1):

$$v = (2gh)^{0.5} \tag{1}$$

for the calculation of the flux of gas and pyroclasts through the vent section. In Equation (1), v is the velocity of the mixture comprising gas plus pyroclasts, g is the acceleration of gravity, and h is the lava fountain height, expressed in meters above the crater rim. The NSEC vent section, following Calvari et al. [14,15], was considered circular, with a vent diameter of 30 m. By integrating the velocity of the gas plus pyroclasts mixture over the entire duration of the lava fountain, multiplied by the vent section area, and extracting from the final value the 0.18%, which represents the average amount of pyroclasts within the fluid mixture [14,15], we obtain the volume of pyroclasts erupted during the lava fountain episode. It is worth noting that the growth of the NSEC cinder cone during one single lava fountain episode is not enough to affect our measurements of lava fountain or ash plumes [18].

Table 1. List of the INGV monitoring cameras used in this paper and their main features. The field of view is considered at the crater rim.

Label	Type	Location	Distance from the Craters (km)	Frame Rate	Field of View
ENT	Thermal FLIR A40M	Nicolosi, South flank 730 m a.s.l.	15.0	2 frames/s	24° (horizontal) 18° (vertical)
EBT	Thermal FLIR A320	Bronte, NW flank 85 m a.s.l.	13.5	2 frames/s	25° (horizontal) 18.8° (vertical)
EMCT	Thermal FLIR A320	Mt. Cagliato, East flank 1160 m a.s.l.	8.3	2 frames/s	25° (horizontal) 18.8° (vertical)
EMOT	Thermal FLIR A320	Montagnola, South flank 2600 m a.s.l.	3.0	1 frame/s	25° (horizontal) 18.8° (vertical)
EMCH	Visible Vivotec IP8172	Mt. Cagliato, East flank 390 m a.s.l.	8.3	2 frames/min	33°~93° (horizontal), 24°~68° (vertical)
ECV	Visible Canon VC-C4R	Catania Nesima, South flank 35 m a.s.l.	26.7	1 frame/2 s	3–47.5° (horizontal and vertical)
ECVH	Visible Vivotec IP8172	Catania Nesima, South flank 35 m a.s.l.	27.0	1 frame/min	33°~93° (horizontal), 24°~68° (vertical)
EBVH	Visible Vivotec IP8172	Bronte, NW flank 163 m a.s.l.	13.5	1 frame/min	33°~93° (horizontal), 24°~68° (vertical)

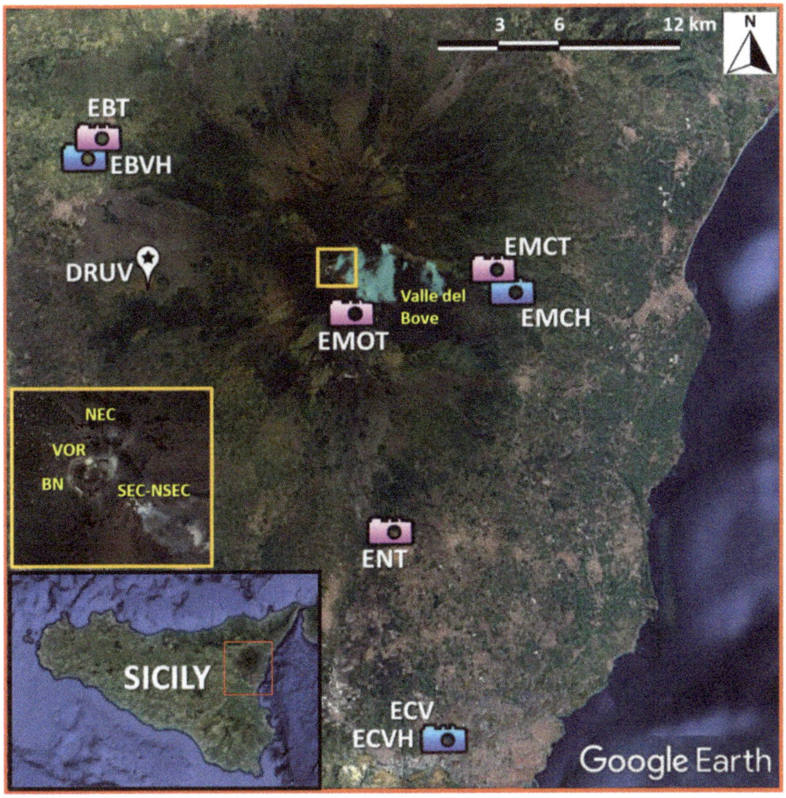

Figure 2. Map of INGV monitoring cameras and borehole strainmeter (DRUV) used in this paper. The pale blue area shows the portion of topography that can be imaged by the EMCT and EMCH cameras, used for the emplacement of the 12 March 2021, lava flow field. The Northeast Crater (NEC), Voragine (VOR), Bocca Nuova (BN), Southeast Crater (SEC), and New Southeast Crater (NSEC) summit craters are indicated in the yellow frame inset. The blue color is for visible cameras, the purple color is for thermal cameras.

The height of the ash plume was measured from the calibrated [59–61] visible cameras ECV and EBVH located on the S and NW flank of the volcano, respectively. ECV has a maximum vertical field of view of ~9.0–9.5 km above sea level (a.s.l.), whereas EBVH allows the detection of ash plumes up to ~12.5 km a.s.l., depending on the wind speed and direction [59–61].

The EMCT thermal camera (Table 1 and Figure 2) was used to follow the lava flow emplacement associated with the lava fountaining activity [62,63]. This camera is located ~8.3 km east of the summit craters (Figure 2, Table 1). Thermal images acquired from EMCT are currently received in real time and stored as RGB files. A routine was implemented to automatically process the images. The images are reprojected on the topography, considering the position and orientation of the camera. To detect the active portion of the lava flow, a threshold is set at 245 for the red channel. This threshold was found by considering the histogram images of recorded values in the presence—or not—of lava flow. Concerning the topography, a digital elevation model derived from Pleiades images and updated in 2020 is taken into account.

2.2. Satellite Thermal Monitoring

Low spatial-high temporal resolution satellite images (1–3 km pixel at nadir, 6 h—up to 5 min frequency), such as those acquired by SEVIRI, MODIS, and VIIRS, are currently used to follow the eruptive activity at Mount Etna. Due to the short-lived nature of the lava fountains that occurred to date in 2021, SEVIRI aboard the geostationary Meteosat Second Generation, providing information at 15 to 5 min sample times, is the best sensor to describe the evolution of the eruptive phenomena [62,64,65]. The thermal anomalies related to the volcanic activity can be located in the satellite images by processing the middle infrared (MIR) channel that is particularly sensitive to high temperature events. The automatic system HOTSAT [66] was used to process these data. Besides locating the thermal anomalies, HOTSAT also computes the radiant heat flux by quantifying the thermal anomaly in each image. From a temporal sequence of images, a radiant heat flux curve can be retrieved, and the timing of an eruptive event can be determined. In the case of effusive events, this curve can provide an estimation of the eruption rate, i.e., it can be converted into a time averaged discharge rate (TADR; [64,67–69]), TADR being an essential parameter for defining the size and magnitude of a volcanic eruption [40,68]. This conversion entails some assumptions [70], among which the thermal steady state needs to be reached [71]. Lava fountains are very fast and transient events, so converting the radiant heat flux into TADR is not possible. Indeed, during the climax phase, saturation and plume obscuration occur, increasing the uncertainties on the peak values of radiant heat flux. To overcome these limitations, the method developed by Ganci et al. [65] was applied here. This method considers the surface temperature for a stagnant, stable, cooling lava surface as a function of time following the solution of the Stefan cooling problem [68,72]. The satellite-derived radiant heat flux depends on the radiative heat flux density due to the surface temperature and the area of cooling lava. The erupted volume of lava is hence computed by modeling the cooling curve apparent in the satellite-derived radiant heat flux curve. A minimum and a maximum range of thickness are assumed for the lava flow field, and the actual curve is constrained between two modeled curves by using the Nelder–Mead algorithm.

SEVIRI, MODIS, and VIIRS data were also used to compute the volcanic Ash Cloud Top Height (ACTH). In this work, this value was derived by comparing the brightness temperature for the pixels contaminated by the volcanic plume with atmospheric temperature profiles. Data for the atmospheric profiles were downloaded with hourly frequency, regridded to a regular lat–long grid of 0.25 degrees, from ERA5, the fifth generation ECMWF reanalysis for the global climate and weather (available at https://cds.climate.copernicus.eu/, accessed on 2 August 2021). The radiances acquired in the thermal infrared were corrected for atmospheric effect by using the MODTRAN (MODerate resolution atmospheric TRANsmission) model and converted to brightness temperatures by using the Planck law. We computed the Brightness Temperature Difference (BTD) between channels IR12.0 and IR10.8 to highlight the presence of ash/SO_2 plume, so we also computed the area of the volcanic cloud as seen from space for each image. In order to compare the temperature at the top of the volcanic cloud with the temperature of the atmospheric profile, we developed a MATLAB routine that interpolated the atmospheric profile at the measured value of temperature and provided the correspondent height. The interpolation is made through the MATLAB function *spline*. The method assumes that the top layer volcanic ash cloud behaves as a blackbody, and it is opaque; the assumption can cause significant overestimation of the cloud top temperature and, therefore, underestimation of the volcanic ash height if there are multilayer clouds under the top volcanic ash layer. Moreover, for high clouds near the tropopause and at high latitudes, the method can lead to errors because the rate of temperature change with height is small [73]. However, reanalysis of regional atmospheric products was used for ACTH estimations at the Etna volcano during recent eruptions [25,52]; the results of these models for lava fountains at Mount Etna were also validated with other ground-based approaches [59]. Finally, higher spatial resolution images, such as those acquired by Sentinel-2 MSI, Landsat 8, and Aster, were used to locate and map the active or recently emplaced lava flow field [74,75].

2.3. High Precision Strain from the Borehole Dilatometer

A network composed of deep borehole dilatometers was installed on Etna in 2011 (two stations) and 2014 (further two stations). The dilatometers measure the volumetric strain of the rocks where they are installed, reaching nominal resolution of 10^{-10} to 10^{-11}, and guaranteeing a frequency range from 10^{-7} to >20 Hz. The instruments are usually installed into deep drilled holes (depth > 100 m) to reduce environmental noise, mainly the thermoelastic strain effects, to better exploit their high sensitivity. The instruments are coupled to the rock by using expansive cement and they require final calibration after installation. The calibrations are usually performed by comparing the recorded strain with the estimated reference signals, such as those produced by lunar tides, mainly the diurnal O1 (25.82 h) and the semidiurnal M2 (12.42 h) [58]. Other approaches are also implemented by comparing the recorded dynamic strain amplitude of long-period surface waves from strong distant earthquakes [76] or by direct comparison of the strain recorded by the borehole dilatometer with the seismic strain of teleseismic waves, recorded by a nearby broadband seismic array [77]. A detailed description of the installations, instrumental in situ calibrations, and main results are fully described by Bonaccorso et al. [76,78]. In this study, we used the signal from the most precise station, namely DRUV. This dilatometer was installed at a depth of 172.5 m within a very massive basalt layer in the mid-western flank of the volcano at about 10 km away from the summit craters (Figure 2). All previously cited calibration approaches were successfully applied to the strain recorded at this station obtaining the same calibration coefficient [58,76–78]. This is considered the reference station since, as testified by the in situ calibrations, it has a >20 times more precise sensitivity than the other stations.

3. Eruptive Activity before the 12 March 2021, Paroxysm

In the recent years, Etna volcano often displayed sequences of lava fountain events, mostly occurring from the SE Crater (SEC), and more recently from the NSEC [6,8,9,11,17]. These are characterized by the development of associated ash plumes and short-lived lava overflows from the crater rim [6–11,17]. Several periods of lava fountain activity characterized the growth of the SEC: in 1989 (16 lava fountains), in 1998–1999 (22), in 2000 (64), in 2001 (15), and in 2013–2015 (49) [8,14,15,17]. The first, most relevant of these sequences occurred in 2000, when during the six months (spanning between January and June) the SEC produced 64 such episodes [8]. This episodic activity was triggered by more primitive and gas-rich magma entering the SEC reservoir, where it mixed with the resident and more evolved magma, giving rise to a gas bubble foam layer accumulated at about 1.5 km depth below the erupting crater [7,8,57]. In general, paroxysmal episodes taking place close in time are generally impulsive and characterized by rapid waxing and waning phases compared to the episodes more distant in time that show a slower pattern [17].

Following the short flank eruption on Etna in December 2018 [25,27], the volcano had another effusive phase from the summit craters between 30 May and 6 June 2019, when some short fissures opened at the base of the NSEC, feeding a lava flow field that spread eastwards [26,79]. Once this eruptive activity ended, the summit craters of the volcano displayed a mild Strombolian explosive activity with occasional ash emission. On 18 July 2019, an effusive vent opened at the base of the NSEC, producing a small lava flow that spread NE for a few kilometers. This lava flow stopped on the evening of 20 July 2019. Another effusive vent opened on 27 July 2019, at the S base of the NSEC, producing a lava flow that spread towards SW and S for several hundred meters and stopped the next day. The Strombolian explosive activity at the summit craters continued during the year, accompanied by occasional and pulsating ash emissions, and producing an intra-crater cinder cone and a several hundred-meter-long lava flow within the Voragine crater in September 2019. From October 2019, the summit craters of the volcano displayed a mild

Strombolian explosive activity with occasional dilute ash emission [80]. In December 2019, the explosive activity increased in intensity, with bomb spatter and ballistics falling on the outer flanks of the NSEC, and a lava flow erupted from the Voragine crater, spreading within the nearby Bocca Nuova crater. This eruptive activity continued in 2020, intensifying during February–early March 2020 [80], when up to three scoria cones built up within the Voragine crater by March 2020. A complex lava flow field fed by the Voragine vents was spreading within the nearby Bocca Nuova crater, lasting until the end of April 2020. The Strombolian explosive activity continued at all summit craters with occasional dilute ash emissions, increasing in May 2020, and forming ash clouds rising several hundred meters above the craters. Several such ash clouds were also observed from June and August 2020 until 13 December 2020, when the first episode of lava fountaining occurred at the NSEC, accompanied by lava flow output from the crater rim spreading S. The collapse of a portion of the crater rim caused three pyroclastic density currents (PDC) spreading S, SW and SE from the base of the cone for several hundred meters. The explosive activity climaxed into an additional lava fountaining episode on 14 December 2020, and a lava flow from the NSEC spreading S on 15 December 2020. Additional lava fountaining episodes and lava flow outputs occurred on: 21 and 22 December 2020, 18 January 2021; on 16, 17, 19, 20, 22, 23, 24, and 28 February 2021; on 2, 4 (two episodes), 7, 9, 12, 14, 17, 19, 23, and 31 March 2021, on 19, 21, 22, 24, 25, 26 (three episodes), 27, 28 (four episodes), and 30 May 2021, on 2, 4, 12, 16, 17, 19, 20, 22, 23 (two episodes), 24, 25 (two episodes), 26, 27, and 28 June 2021; and on 1, 4, 6, 8, 14, 20 and 31 July 2021 (see activity reports of INGV-OE at https://www.ct.ingv.it/index.php/monitoraggio-e-sorveglianza/prodotti-del-monitoraggio/bollettini-settimanali-multidisciplinari, accessed on 2 August 2021).

4. Results

The chronology of the eruptive events was gathered from the analysis of the images acquired by the INGV monitoring network, comprising visible and thermal cameras (Table 1 and Figure 2) and allowing a view of the volcano from various distances and directions, and by satellite. All times indicated in this paper are UTC.

4.1. Eruptive Events and Characterization of the Lava Fountain and Ash Plume

The Strombolian activity at the NSEC, observed from the INGV network of monitoring cameras (Figure 2 and Table 1), started on 12 March 2021, at 02:35, gradually increasing in frequency, height, and intensity of the jets. Table 2 summarizes the timing of the events as observed by the monitoring cameras and satellite, also complemented by strain information. At 06:40 the activity became transitional between Strombolian and lava fountaining (Figure 3a), and at 07:00 a lava flow breached the lower eastern rim of the crater spreading E within the Valle del Bove depression (VdB; Figures 2 and 3b). At 07:22 a second overflow occurred from the NE margin of the crater rim, expanding on the N flank of the crater (Figure 3b). At 07:35, the explosive activity became a clear lava fountaining, accompanied by significant ash emission and fast formation and growth of an eruptive column (Figure 3b). The third lava flow started spreading SE from 09:20 (Figure 3c).

Table 2. Timing of the 12 March 2021, eruptive episode retrieved from the INGV monitoring cameras, from the satellite and strain signal. IER = instantaneous effusion rate [68].

Time UTC	Eruptive Activity	Notes
02:35	Strombolian activity started at the NSEC	Figure 7
02:56	First satellite thermal anomaly detected and Strombolian activity intensified	Figure 7
05:00	Plume already reached 4000 m a.s.l.	Figures 5 and 6a
06:40	Transitional activity between Strombolian and lava fountaining started at NSEC	Figures 3a and 6b
07:00	Lava flow output started from the E rim of NSEC, expanding E	Figure 3a
07:22	A second lava flow started from the NE rim of NSEC, expanding N, strain suddenly declining	Figure 3b
07:35	Lava fountaining phase with significant ash emission and formation of an eruptive column, 20–30 m s−1 of muzzle velocity	Figures 3b and 6d
07:35–08:56	Radiant heat flux increased from satellite, muzzle velocity of ~70 m s−1, strain rate increased	Figure 9
08:14	Maximum muzzle velocity of 133 m s−1 reached by the lava fountain	
08:30	Maximum plume height of 11.3 km a.s.l. detected from EBVH	Figures 4 and 5
08:47–08:49	Maximum elevation reached by the lava fountain of 3000 m above the crater rim (from EBT), and of 2400 m above the crater rim (from ENT) and peak IER of 276 m3 s−1	Figures 4b and 5
08:57	Peak value of 35 GW measured from satellite and peak value from strain	Figures 5 and 9
09:05–09:30	The ash plume started declining in height	Figure 5
09:20	A third lava flow started from the SE rim of NSEC, expanding SE	Figure 3c
09:45	Lava fountain ended, strain declining stopped	Figure 3c
10:10–10:15	The ash plume dropped below 6.0 km a.s.l.	Figure 5
10:54	Explosive paroxysm ended	Figure 3c
12:00	Lava flow output ended and lava flow field cooling	

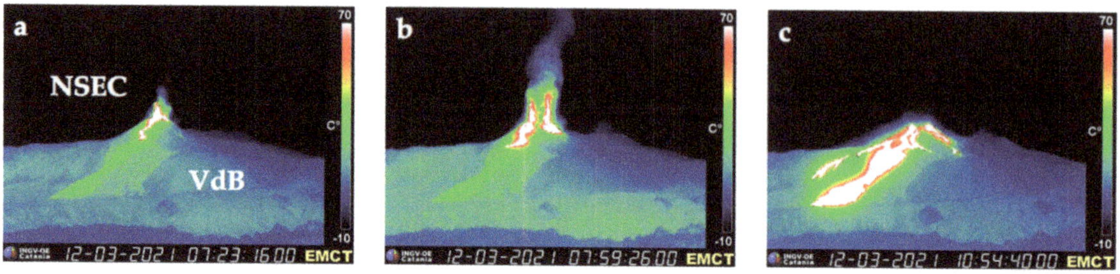

Figure 3. The output of three lava flows from the crater rim of the New Southeast Crater (NSEC), observed by the EMCT thermal camera located on the E flank of the volcano, on 12 March 2021. North is on the right, south is on the left. See Figure 2 for camera location and Table 1 for its characteristics. (**a**) Thermal image recorded at 07:23:16 showing the first lava flow (white) spreading E along the western wall of the Valle del Bove (VdB). (**b**) Thermal image recorded at 07:59:26 showing the lava fountain with the ash plume (blue), and the second lava flow spreading N (white, right of the image). (c) Thermal image recorded at 10:54:40 showing the crater inactive, and in white the three lava branches slowly expanding.

The height of the lava fountain, detected from the ENT (S flank) and EBT (NW flank, Figure 2 and Table 1) thermal cameras, gradually increased up to 08:49, when it reached the maximum elevation of 2400 m above the crater rim and the peak instantaneous effusion rate (IER, averaged over a shorter lapse of time than the effusion rate [68]) of 276 $m^3\ s^{-1}$ (Table 2). The muzzle velocity, obtained from the EMOT camera which is closer to the summit vents, was only 20–30 $m\ s^{-1}$ during the initial phase of fountaining, increased to ~70 $m\ s^{-1}$ after 08:05, and reached the peak of 133 $m\ s^{-1}$ at 08:14 (Table 2). Then the fountain height decreased, stopping at 09:45 (Table 2), after a duration of 130 min (Figure 3). The average fountain height, calculated from the values measured every minute with the ENT camera, was 1149 m.

The height of the ash plume was measured on the frames overlapped on the visible images recorded by the ECV and EBVH visible cameras, located on the S and NW flank of the volcano, respectively (Figures 2 and 4, Table 1), using the calibrated images automatically provided by the procedure developed by Corradini et al. [59] and Scollo et al. [60,61]. The images of the ash plume are displayed in Figure 4, and the results of the ash plume heights against time are shown in Figure 5, where they are compared with the heights of the lava fountains retrieved from the ENT and EBT thermal cameras.

Figure 4. Heights and shapes of the 12 March 2021 ash plume at 08:55 imaged from (**a**) the EBVH calibrated camera located on the NW flank of the volcano where it reached 11.0 km above sea level (red dotted line) and with the lava fountain measuring 2.6 km above the crater rim (yellow dotted line), from (**b**) the uncalibrated ECVH camera located on the S flank of the volcano, with the red box displaying the area imaged in (**c**) and related to the ECV calibrated camera field of view, located in the same position as the ECVH, and having a maximum vertical field of view up to 9.5 km elevation above sea level, and displaying the elevation of the lava fountain (yellow dotted line) estimated at 2.3 km above the crater.

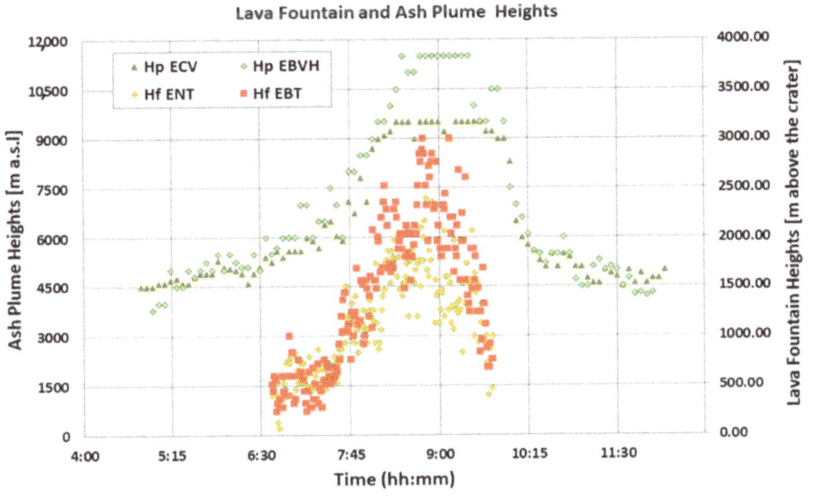

Figure 5. Heights of the lava fountain (Hf, in m above the crater rim) obtained from the ENT and the EBT thermal cameras, compared to the heights of the ash plume (Hp, in m above sea level) obtained from the ECV and EBVH visible cameras against time (hh:mm, UTC).

It is worth noting that although the lava fountain phase started at 07:35, the elevation of the plume was more than 4000 m above sea level (a.s.l.) much earlier, and at least from dawn at 05:00 when the ash plume became visible (Figures 5 and 6, Table 2). This was probably owing to the heat released by the Strombolian activity that started at the NSEC at 02:35 (Figure 3a and Table 2). However, at that time, it was a weak plume bent eastward and comprising mostly diluted reddish ash (Figure 6a). At 06:40, as soon as the explosive activity became transitional [4,14,43] between Strombolian and lava fountaining, the ash plume rose to 5300 m a.s.l. and changed inclination (Figure 6b), suddenly becoming more vertical as a result of an increased IER~120 m^3 s^{-1} and displaying increasing water vapor condensation at the top (Figure 6c–e).

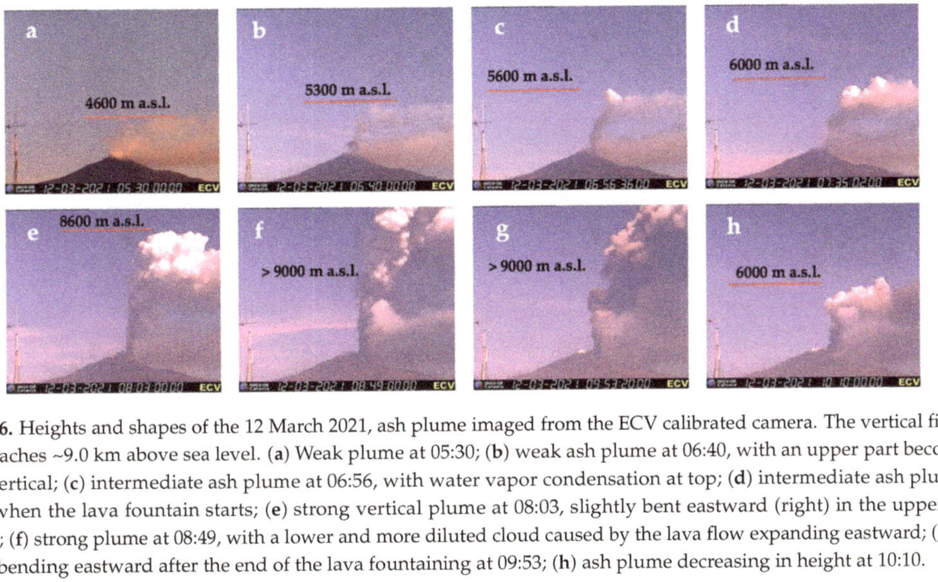

Figure 6. Heights and shapes of the 12 March 2021, ash plume imaged from the ECV calibrated camera. The vertical field of view reaches ~9.0 km above sea level. (**a**) Weak plume at 05:30; (**b**) weak ash plume at 06:40, with an upper part becoming more vertical; (**c**) intermediate ash plume at 06:56, with water vapor condensation at top; (**d**) intermediate ash plume at 07:35, when the lava fountain starts; (**e**) strong vertical plume at 08:03, slightly bent eastward (right) in the uppermost portion; (**f**) strong plume at 08:49, with a lower and more diluted cloud caused by the lava flow expanding eastward; (**g**) ash plume bending eastward after the end of the lava fountaining at 09:53; (**h**) ash plume decreasing in height at 10:10.

The amount of water vapor condensation at the top of the eruptive column increased even more after 07:35, when the eruptive activity became lava fountaining (Figure 6d–f). The increased IER~153 $m^3 s^{-1}$ resulted in the formation of a strong plume extending vertically above the vent, with only the uppermost portion being bent eastward by the wind (Figure 6e). The ash plume went beyond the ECV camera field of view (i.e., more than 9.0–9.5 km a.s.l.) as soon as the lava fountain attained its peak IER of 252–276 $m^3 s^{-1}$ at 08:47–08:49 (Table 2 and Figure 6f). At this stage, the lava flow field spreading eastward increased its speed significantly, as observed by satellite (Table 2) and also by the appearance of a lower steam cloud produced by the heat released by the lava flow (Figure 6f). The ash plume started to drop several minutes after the lava fountaining stopped (Figure 6g), but its disappearance was evident only after 10:10 (Figure 6h), about 30 min after the end of the lava fountaining phase.

The maximum plume elevation was not recorded by the ECV camera because its maximum field of view reaches ~9.0–9.5 km a.s.l. [60], but probably also the EBVH camera gave a slightly underestimated maximum ash plume elevation, given that the maximum elevation of 11.5 km a.s.l. was observed at 08:30 (Table 2 and Figure 5), whereas the maximum elevation of the lava fountains was attained at 08:49 from ENT (2.4 km above the crater rim, Table 2; average value 1.15 km).

The lava fountain heights decreased soon after having reached the peak values (i.e., at 08:50 from ENT; Figure 5), whereas the ash plume started decreasing in height at 09:05 from ECV and at 09:30 from EBVH, with a delay of 18–43 min (Figure 5).

Given that ash plume can be a serious threat to airport and airplane viability due to the proximity of Mount Etna with the Catania international airport (~32 km), we need to estimate the maximum elevation that the ash plume can attain, as well as its direction, as soon as possible, in order to provide prompt advice to the Civil Protection and Air Traffic Authorities. In this regard, the average lava fountain height is a key parameter because it allows us to estimate the maximum plume elevation as soon as the peak height of the lava fountain is reached. From the data recorded during the 2011–2013 lava fountains from NSEC, Calvari et al. [15] proposed the following empirical equation:

$$H_P = 5.26\, H_F + 6.83 \quad (2)$$

where H_P is the maximum height reached by the ash plume, and H_F is the average height of the lava fountain. Considering an average lava fountain height of 1.15 km above the crater rim, and applying the formula (2) by Calvari et al. [15], the estimated maximum plume height is 12.9 km, close to the real value of 11.5 km a.s.l. estimated from the EBVH monitoring camera (Figure 4a).

The difference in elevation for the lava fountain and ash plume obtained from the different cameras can be due to the irregular shapes of the lava fountain and ash plume, and/or to the ash partially obscuring the sight at the thermal image. Although the lava fountain suddenly stopped at 09:45, the ash plume was above 9.0 km a.s.l. at least for 10 additional minutes (until 09:55), and started to decrease below 6.0 km a.s.l. only after 10:10–10:15, with a delay of about 30 min (Figure 5).

On the basis of the lava fountain heights measured from the ENT camera, we estimated the volume of pyroclasts erupted during the lava fountain episode, following the method proposed by Calvari et al. [14,15]. The resulting volume is ~1.6×10^6 m^3, which, averaged over the 130 min of duration of the event, gave a time-averaged discharge rate (TADR; [68]) of ~209 $m^3 s^{-1}$ and a peak instantaneous effusion rate (IER, [68]) of 276 $m^3 s^{-1}$ recorded at 08:49.

4.2. Satellite Thermal Data

Processing satellite images enables us to derive: (i) the timing of eruptive activity as seen from space; an estimation of the (ii) area; and (iii) volume for the lava flow field and of (iv) the top height for the ash plume. The first thermal anomaly, related to the 12 March eruptive episode, was detected by SEVIRI at 02:56 when the Strombolian activity intensified

(Table 2). This anomaly was followed by a constant increase in the satellite-derived radiant heat flux mainly related to the lava flow spreading, as also visible from the lava flow area increase imaged by the EMCT thermal camera (Figure 7b). A first increase in the radiant heat flux signal was recorded at 6:42 and a second sharp increase occurred at 08:12, with a peak value of ~25 GW at 08:57 (Table 2). After this a constant decrease, related to the cooling of the lava flow, was observed. Due to the low spatial resolution of SEVIRI images at Etna volcano, we are not able to distinguish between the radiant heat flux curve coming from the lava fountain and the contribution related to the lava flow field. Most of the thermal signal is due to the lava flow emplacement as shown by the comparison with the lava flow field growth recorded by the EMCT thermal camera (Figure 7a,b). This happens for two reasons: the first is that the SEVIRI pixel over the summit crater saturates, and the second one is that the eruptive column above the lava fountain covers the view from the satellite. Applying the method by Ganci et al., [64], the satellite-derived cooling curve was modeled and a lava flow volume of ~2.4 × 10^6 m^3 was estimated.

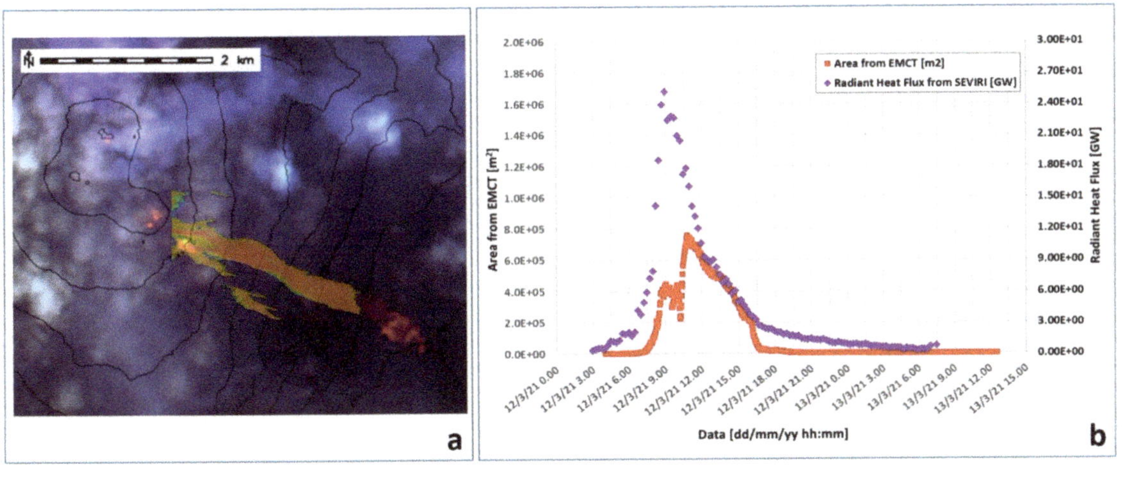

Figure 7. (**a**) Map of the lava flow derived from the thermal camera EMCT superimposed on the RGB composite (Band 12, Band 11, Band 5 at 20 m spatial resolution) of the Sentinel-2 image acquired on 13 March at 09:50. The projected map of the lava flow is cut in order to remove the jet portion of the lava fountain. (**b**) Thermally anomalous area detected from the EMCT thermal camera against time (red squares), compared with the radiant heat flux curve retrieved from SEVIRI (blue diamonds).

Figure 7b shows the SEVIRI-derived radiant heat flux versus the active lava flow area as imaged by the EMCT thermal camera. We found a slight difference between the two signals at the beginning and at the end because the oblique view from the camera missed the thermal activity at the crater area and part of the lava flow emplaced below ~1900 m a.s.l. that was instead visible by satellite. During the fountaining, the ash plume partially prevented the view of the lava flow field from EMCT (see Figure 6g,h), until 10:45 when the maximum value of 0.75 × 10^6 m^2 was reached. Figure 7a shows the lava flow area as imaged from EMCT superimposed on the RGB composite obtained from Band 12, Band 11, Band 5 (20 m spatial resolution) of the Sentinel-2 image acquired on 13 March at 09:50. The EMCT-derived map of the lava flow was retrieved considering all the images acquired by the camera with a portion of active lava flow from 12 March at 04:26 to 13 March at 05:40. We found an overlap of 97% between the projected thermal camera lava flow surface and the one visible from the Sentinel-2 image for the portion visible from the EMCT camera (Figure 7a). From the Sentinel-2 image, we derived a whole lava flow field area that resulted of about 1.17 × 10^6 m^2. Combining the SEVIRI-derived volume and the Sentinel-2 derived area, we found an average thickness of ~2.1 m for the lava flow field.

Figure 8 shows the ash cloud top height obtained from a VIIRS image acquired on 12 March 2021, at 10:46. The result is a maximum height of 12.6 km a.s.l. (Figure 8).

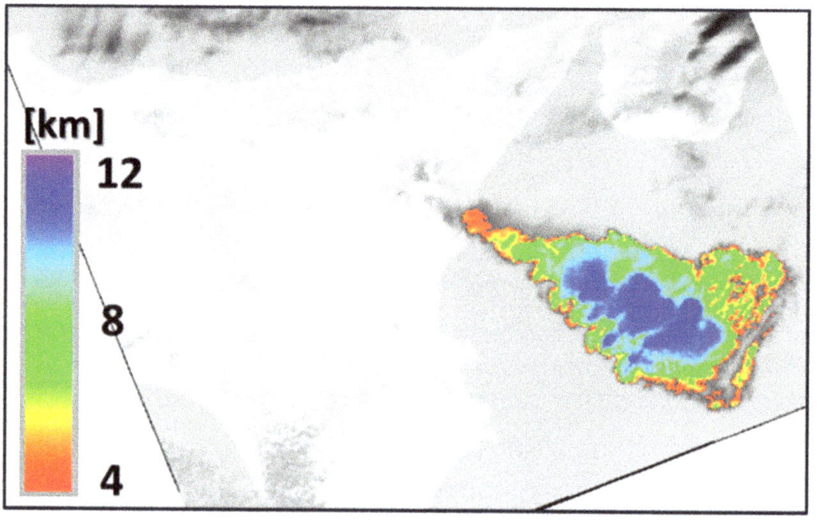

Figure 8. Ash Cloud Top Height (ACTH) computed from a VIIRS image of 12 March 2021, at 10:46.

By processing SEVIRI data, we were also able to follow the ash cloud dispersion during the eruptive episode. The plume top area was visible by SEVIRI at 8:15 and the area increased until 10:45 with an almost constant velocity between 0.1 and 0.2 $km^2\ s^{-1}$ (Figure 9). At 11:00, more than one hour after the end of the lava fountaining (Table 2), the plume top area started decreasing and separating from the volcano.

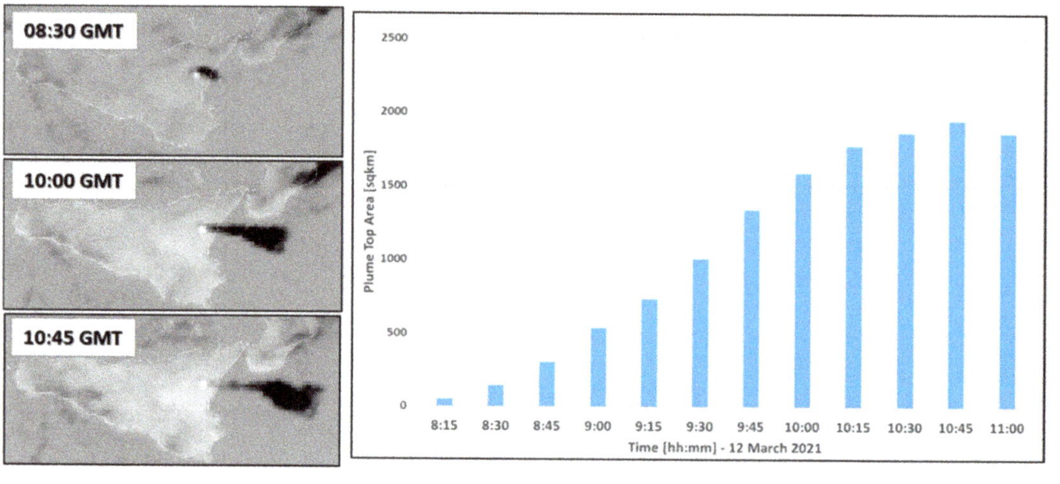

Figure 9. On the left: three SEVIRI scenes (channel IR 10.8) acquired on 12 March 2021 at 8:30, 10:00, and 10:45. On the right: histogram of the plume top area during 12 March 2021 from 8:15 to 11:00.

4.3. Strain

The DRUV reference station is located 10 km away from the summit craters (Figure 2), but has a very high sensitivity allowing to clearly detect the small strain variations (~0.2 microstrain) caused by the activity of the lava fountains at that distance. During the paroxysmal phase, the strain signal showed a negative variation which corresponded to a decompression of the medium surrounding the instrument. A weak variation started at 06:40, during the transitional phase (from Strombolian to lava fountaining activity) and the small lava flows occurrence. Then the strain increased its rate during the most intense lava fountain phase (07:35–08:57; Table 2). The strain signal continued to decrease until 09:46, cumulating a change of 0.18 microstrain. This variation is of the same order of magnitude as those recorded during the 2011–2013 paroxysmal episodes, with a value a little greater than the average of these events, which was 0.15 microstrain [58]. The strain rate, calculated as the strain change per 1 min sampling rate unit, reached a maximum value at 08:57. The strain recorded is shown in Figure 10a, where it is compared to the heat flux measured by satellite, and the strain rate against heat flux is shown in Figure 10b. In Figure 10, four phases can clearly be identified: (1) Strombolian activity and an initial weak effusive phase producing the beginning of the thermal release, but without strain change; (2) increase of the Strombolian activity (i.e., transitional activity between Strombolian and lava fountaining) in which the strain starts to decrease (i.e., decompression begins) and the strain rate increases; (3) start of the lava fountain phase ejecting at a high mass rate. This phase is characterized by a strong increment in the thermal contribution and by a marked decompression recorded by the strain. The maximum of the strain rate at 08:57 is coincident with the maximum of the radiant heat flux and lava fountain height (Table 2 and Figure 5); (4) after 08:56, the strain rate began to decrease, indicating that the lava fountain intensity was going to decrease and, therefore, the turning point of the strain rate represented the exact moment at which the eruptive activity started to decline; (5) at 09:45, the strain change reached the minimum, indicating that the lava fountain finished, the magma was no longer emitted, and the strain no longer recorded decompression (only the regular lunar tides), while the slow cooling of the hot material of the effused portion caused a slow exponential decrease in the thermal contribution detected by satellite.

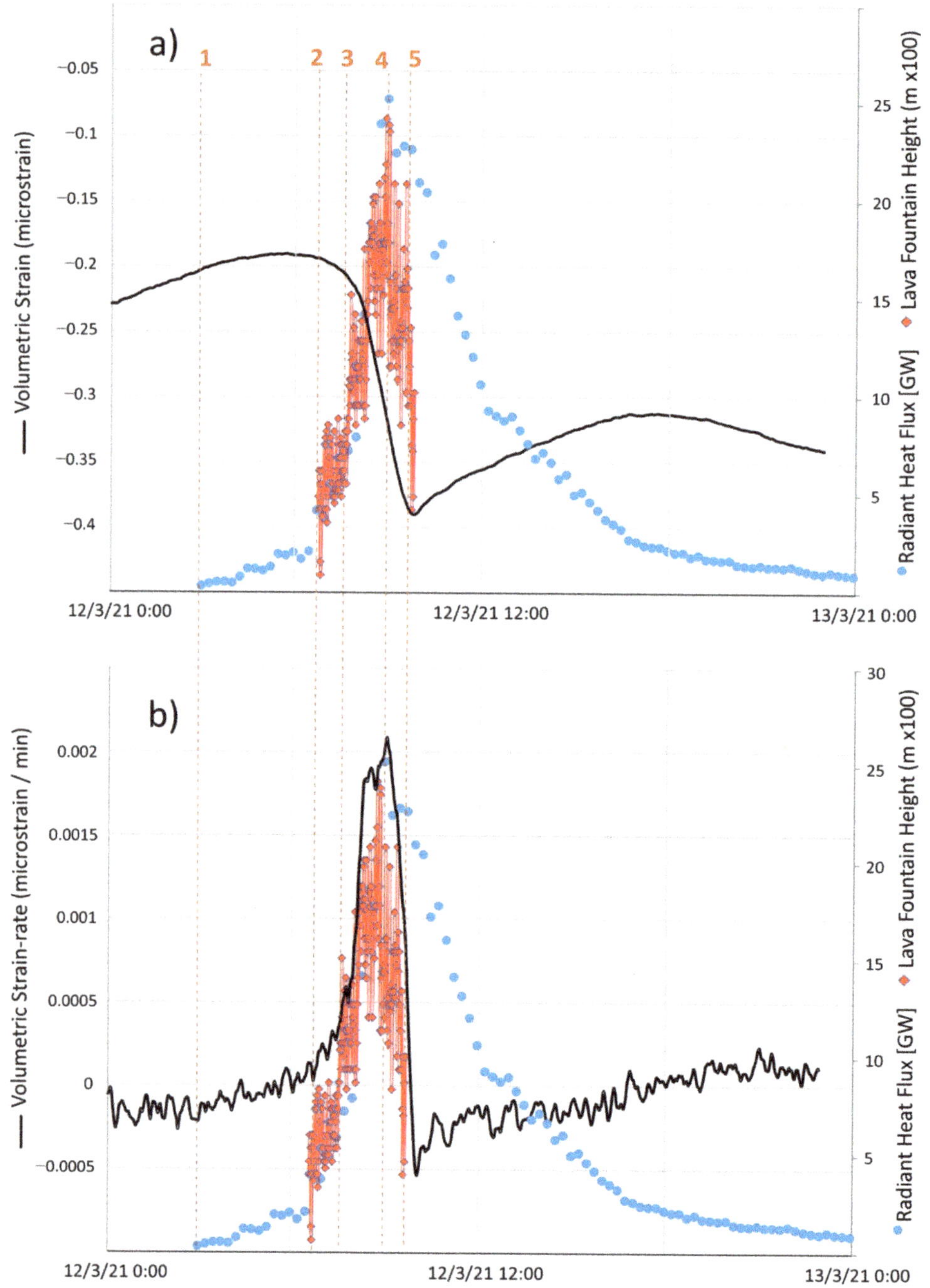

Figure 10. (**a**) Comparison of the radiant heat flux (GW, blue diamonds) detected by SEVIRI satellite with the strain signal (black line, in microstrain) and (**b**) the strain signal recorded at the DRUV station (black line, in microstrain per minute). (1) Strombolian activity and initial weak effusive phase; (2) transitional activity between Strombolian and lava fountaining; the strain starts to decrease and the strain rate increases; (3) start of the lava fountain phase; (4) the strain rate began to decrease indicating that the lava fountain intensity was going to decrease; (5) the strain change reached the minimum indicating that the lava fountain finished.

It is interesting to observe that the strain signal provided correct timings of the start and end of the lava fountain phase (Figure 10a), with times in agreement with those obtained from the camera frames. The strain rate marked the intensity regime of the explosive phase, and when the sign inversion occurred (at the beginning of phase 4, Figure 10a) there was a precise indication that the lava fountain began to decline (5, Figure 10b).

5. Discussion

The explosive mechanism of the lava fountains at Etna is generally understood as the "foam model" [81,82], which takes account of a rapid and violent ascent of a bubble foam layer previously accumulated at a shallow depth [7,8,14,83]. In this paper we have analyzed ground, satellite data and high precision strain signals collected during the 12 March lava fountain episode at Etna volcano in order to characterize the formation and growth of the lava fountain and of the associated lava flow field, and the way the fountain feeds the ash column and eruptive plume. The aim was to acquire parameters that could be useful for hazard assessment.

The cameras allowed us to observe the phenomenon from the ground and provided precise information on the characteristics of the lava fountain, on the subsequent eruptive column, ash plume development, and height. In particular, they constrained the total amount of the erupted fluid (gas plus pyroclasts) during the lava fountaining phase, and from this value we extracted the volume of pyroclasts as 0.18% of the total [14,15]. Thermal satellite analyses enable estimating the thermal energy and lava flows erupted during and after the lava fountains. In addition, satellite images reveal the size and elevation of the ash plume and their changes in time. The strain measures the response of the volcanic edifice to the decompression caused by the eruptive activity and provides constraints on the timing and total erupted volume. In general, the approach of integrating these various observations allowed us to obtain robust constraints to characterize the phenomenon.

In particular, in this study we have described the Strombolian activity at the vent, which began on 12 March 2021 at 02:35 (Table 2), gradually increasing with time in intensity and frequency of the bursts. Only after 06:40, i.e., after about 4 h of growing explosive activity, did the Strombolian activity pass to transitional explosions [14,39,43], and at 07:35, about one hour later, became lava fountaining. This transition corresponds to an increase in coalescence between gas bubbles [39] that drives the change from countable discrete explosions (Strombolian activity), revealing a bubbly flow regime within the conduit, to the transitional activity [14,39,43], indicative of a slug flow regime within the conduit, to the uncountable oscillations of a lava fountaining typical of a sustained annular flow regime [84]. It is at this stage—namely when lava fountaining is fully developed—that abundant ash is released from the fountain margins to feed the ash plume, suggesting further passage from an annular flow regime to a dispersed flow regime [84]. The lava fountaining phase showed a growing muzzle velocity that started from 20–30 m s^{-1}, rapidly grew to 70 m s^{-1}, and peaked at 133 m s^{-1} (Table 2). These values are in the range of Etna's previous paroxysmal events [15,31,85].

Considering that the wind speed during the lava fountaining episode ranged between ~5 and 10 m s^{-1}, at an altitude between 3 and 10 km (ERA5 Reanalysis available at https://cds.climate.copernicus.eu/, accessed on 2 August 2021), the results by Calvari et al. [15] is confirmed, namely that wind speed up to 10 m s^{-1} leads to a strong to intermediate plume rising vertically above the crater or slightly bending in the wind direction. This shape has a lower impact on the local population because ash fallout is mainly concentrated around the vent, but has a greater impact on aviation because the plume reaches greater elevation [15]. In the case of the 12 March 2021, paroxysmal episode, the ash plume rose to the maximum elevation of 11.3 km a.s.l. at 08:30, as detected from the monitoring cameras (Table 2), but grew even further away from the volcano and reached 12.6 km at 10:46 (Figure 8), as detected by the satellite. Thus, even more than one hour after the end of the lava fountaining, the ash plume was still threatening the airplanes path (Figures 8 and 9). Given that the maximum elevation of the lava fountaining was detected 10–15 min after the top plume height detected from the EBVH camera (Table 2), it is possible that the maximum elevation of the ash plume was slightly higher than that detected by the EBVH ground camera even close to the volcano. This is confirmed by the 12.6 km detected by satellite at 10:46 (Figure 8) and is in agreement with what was predicted by the Equation (2) [15] that estimated a maximum ash plume elevation of 12.9 km.

The lava fountaining paroxysm ejected 1.6×10^6 m^3 pyroclasts at maximum IER of 276 m^3 s^{-1}. Most of the pyroclastic material erupted by the fountaining fell around the vent, further increasing the size of the NSEC cinder cone, as happened in the recent past [14,15,18,86]. Together with the lava fountaining, a lava flow also erupted from the crater rim, spreading a volume of 2.4×10^6 m^3 over a surface of 1.2×10^6 m^2 and travelling for 3.7 km eastward for a few hours.

The strain, based on the change cumulated during the lava fountain, was particularly useful for giving an estimate of the total erupted volumes, comprising both the pyroclasts erupted during the lava fountaining phase and the lava flows. In fact, considering the strain changes recorded during the lava fountains occurring at Etna on 2011–2013, Bonaccorso et al. [57] inferred a near spherical source of radius 0.5 km located below the crater area at a depth close to the sea level. This source represents the shallow storage where gas-rich magma is trapped and then violently ejected through the lava fountains. During the lava fountain, this source deflated and its volume changed by 2×10^6 m^3; due to the compressibility of the magma that accommodates a further amount of magma, the total volume of magma expelled was ~2.5×10^6 m^3 [57]. This was considered the representative volume for a lava fountain producing the mean strain change recorded for the 2011–2013 lava fountains, which was a 0.15 microstrain. Since the expected strain caused by a depressurizing spherical source is linearly related to the volume change of this source [87], we can use the results obtained by Bonaccorso et al. [57] to estimate the volume emitted by the NSEC lava fountains by the amplitude of the strain change at DRUV. For the 12 March 2021, episode, the 0.18 microstrain corresponds to a total emitted volume of ~3 $\times 10^6$ m^3, comprising both lava flows and pyroclasts. This value results in an average eruption rate of 385 m^3 s^{-1}. However, by summing up the satellite-derived lava flow volume and the thermal camera derived pyroclastic volumes, a value of 4×10^6 m^3 is obtained. We argue that the lava fountain heights analysis measured a quantity of magma that partly flowed as lava flow and partly fell into the cone as pyroclasts. Indeed, by comparing the satellite-derived and the thermal camera derived volumes, the quantity of magma ejected into the fountain and falling into the lava flow field is about 1×10^6 m^3, while about 0.6×10^6 m^3 is related to pyroclasts. This value is comparable with the growth of the NSEC cone already measured during 2011–2013 [14], but also with a recent DEM difference computation by using Pleiades data (6.4×10^6 m^3 during 12 eruptive episodes; Ganci et al. unpublished data).

It is noteworthy that sustained ash plumes at the Etna volcano always accompany lava fountains [10,11,14,15,17,25,52–54,59]. Ash plumes cause the greatest concern on the civil protection authorities because they can attain up to 12–14 km in elevation [52,88] causing severe threat to the air traffic, while still expanding in the atmosphere, but also great damage to infrastructures, viability, and public health upon falling on the ground [30]. The ash plume forms as soon as the explosive activity shifts from transitional to lava fountain (stage 3 in Figure 10), most often within 30 min from the paroxysmal start [14,15,25]. It corresponds to an acceleration of the jet that is responsible for the peak IER (Table 2) and the greater cooling and fragmentation of pyroclasts [25,33,44]. The heat released during the lava fountain phase is sufficient to rise a large volume of fine-grained pyroclasts (ash) up into the atmosphere, causing it to spread for several hours around the volcano, and to travel distances of several tens of kilometers [89].

A numerical study involving explosive eruptions, carried out at the Etna volcano [90], estimated that the mass deposited over a distance of 1 and 100 km from the vent represents 30% of the emitted pyroclastic mass. Thus, it is noteworthy that this amount was sufficient to feed an ash plume 11–13 km high (Table 2, Figures 4 and 5) expanding for hundreds of kilometers from the vent (Figures 8 and 9), and covering a sky surface of up to 1900 km^2 and a distance of ~140 km from the vent in two and half hours. This matches a plume-cloud expanding for 60 km from the vent in just 50 min during a previous paroxysmal episode [89], and fully displays the hazard posed by the ash cloud to aviation safety and circulation also because of growing airline traffic [91–93]. It is, thus, of paramount importance to use monitoring data to develop simple equations, such as those used during recent effusive eruptions at Etna [94,95], which might allow for fast and reliable estimates useful for hazard assessment during the earlier phases of an explosive paroxysm. In this respect, application of the formula proposed by Calvari et al. [15], to estimate the maximum vertical extent of the ash plume once the lava fountaining phase stabilized, proved to be effective when applied to the 12 March 2021, paroxysm. In fact, this formula estimated an ash plume maximum elevation of 12.9 km a.s.l., which is a value very close to the 12.6 km a.s.l. estimated from the satellite images.

6. Conclusions

Our results have essential implications in regard to hazard assessments at Etna during paroxysmal explosive phases. They confirm the role of wind speed [15] in determining if a strong, intermediate, or weak ash plume forms, with wind speeds below 10 m s^{-1} favoring the formation of strong to intermediate, taller vertical plumes, which cause most of the pyroclastic fallout around the vent. Our results on the 12 March 2021, episode confirm the possibility of estimating the maximum ash plume elevation using the formula proposed by Calvari et al. [15], given that the maximum plume elevation obtained by satellite (12.6 km) was very close to the 12.9 km estimated by the empirical formula. Integrating results from the ground monitoring cameras, satellite, and strainmeters, we obtained an estimation of the total erupted volume of 3×10^6 m^3, of which 1.6×10^6 m^3 erupted as pyroclasts, with $\sim 1 \times 10^6$ m^3 of the volume of pyroclasts flowing together with the lava flows to comprise a lava flow field extending over a surface of $\sim 1.17 \times 10^6$ m^2, with a volume of 2.4×10^6 m^3. Considering the duration of 130 min for the episode, there was an average eruption rate of 385 m^3 s^{-1} for this event, comprising both pyroclasts and lava flows. Our results show that extreme caution must be applied when calculating the volume erupted during paroxysmal episodes, combining data obtained from monitoring cameras and satellites. In fact, by comparing these results with the strain changes at the shallow magma source, we have shown how a significant portion of pyroclasts ($\sim 1 \times 10^6$ m^3) flowed along the flanks of the NSEC cone to feed the lava flows. This corresponds to ~33% of the total volume erupted by the paroxysmal episode.

Although the duration of the eruptive event was rather short (130 min), the expansion of the ash cloud continued for the following hours, reaching the maximum elevation detected by satellite about 1 h after the end of the paroxysm (Figures 8 and 9). The ash cloud expanded in the atmosphere and eventually detached from the volcano 1.5 h after the end of the paroxysm (Figure 9). This result must be taken into account when organizing air traffic immediately after the end of an explosive paroxysm.

Author Contributions: Conceptualization, S.C.; methodology, S.C., G.G., A.B.; validation, S.C., G.G., A.B.; formal analysis, S.C., G.G., A.B.; investigation, S.C., G.G., A.B.; resources, S.C.; data curation, S.C., G.G., A.B.; visible and thermal ground imagery data treatment, S.C., G.G.; satellite data treatment, G.G.; ground deformation data treatment, A.B.; writing—original draft preparation, S.C., G.G., A.B.; writing—review and editing, S.C., G.G., A.B.; visualization, S.C., G.G., A.B.; supervision, S.C., G.G., A.B.; project administration, S.C.; funding acquisition, S.C. All authors have read and agreed to the published version of the manuscript.

Funding: This research was funded by the Project FIRST-ForecastIng eRuptive activity at Stromboli volcano: timing, eruptive style, size, intensity, and duration, INGV-Progetto Strategico Dipartimento Vulcani 2019, (Delibera n. 144/2020; Scientific Responsibility: S.C.). The research has moreover benefited from funding provided by the Italian Presidenza del Consiglio dei Ministri—Dipartimento della Protezione Civile (DPC), All. B2-Task 11 "Real-time quantification of Etna's eruptive activity from fixed thermal cameras and satellite data". (Scientific Responsibility: G.G.) and Task 9 "Ottimizzazione dell'acquisizione dei segnali ad alta precisione degli strainmeter installati in pozzo sull'Etna" (Scientific Responsibility: A.B.). A.B. also benefited from the EC 298H2020-FET OPEN project grant agreement n. 863220 "SiC optical nano-strain-meters for pico-detection in Geosciences" (SiC nano for picoGeo). This paper does not necessarily represent DPC's official opinions and policies.

Data Availability Statement: The cameras and strainmeter data used in this study are property of INGV-OE (Istituto Nazionale di Geofisica e Vulcanologia—Osservatorio Etneo, Sezione di Catania). They can be made available, upon reasonable request, asking to the corresponding author. The satellite data processed and presented in this study are openly available.

Acknowledgments: We would like to thank the scientists and technicians from INGV-OE for monitoring the network maintenance, especially E. Pecora and M. Prestifilippo for providing information essential for this work. We are grateful to EUMETSAT for SEVIRI data (https://data.eumetsat.int), to the National Aeronautics and Space Administration (NASA) for MODIS data (modis.gsfc.nasa.gov), to the National Oceanic and Atmospheric Administration (NOAA) for VIIRS data (https://www.bou.class.noaa.gov/), and to the Sentinel Hub services (https://scihub.copernicus.eu/) (accessed on 2 August 2021) for Copernicus Sentinel data. The English style was corrected by Stephan Conway.

Conflicts of Interest: The authors declare no conflict of interest.

References

1. Wilson, L.; Parfitt, E.A.; Head, J.W., III. Explosive volcanic eruptions-VIII. The role of magma recycling in controlling the behaviour of Hawaiian-style lava fountains. *Geophys. J. Int.* **1995**, *121*, 215–225. [CrossRef]
2. Swanson, D.A.; Duffield, W.A.; Jackson, D.B.; Peterson, D.W. *Chronological Narrative of the 1969–71 Mauna Ulu Eruption of Kilauea Volcano, Hawai'i*; U.S. Geological Survey Professional Paper 1056; U.S. Government Printing Office: Washington, DC, USA, 1979.
3. Wolfe, E.W.; Neal, C.A.; Banks, N.G.; Duggan, T.J. Geological observations and chronology of eruptive events. In *The Puu Oo eruption of Kilauea Volcano, Hawaii*; Episodes 1 through 20, January 3, 1983, through June 8, 1984; U.S. Geological Survey Professional Paper 1463; U.S. Government Printing Office: Washington, DC, USA, 1988.
4. Parfitt, E.A.; Wilson, L. The 1983–86 Pu'u'O'o eruption at Kilauea Volcano, Hawaii: A study of dike geometry and eruption mechanisms for a long-lived eruption. *J. Volcanol. Geotherm. Res.* **1994**, *59*, 179–205. [CrossRef]
5. Parcheta, C.E.; Houghton, B.F.; Swanson, D.A. Contrasting patterns of vesiculation in low, intermediate, and high Hawaiian fountains: A case study of the 1969 Mauna Ulu eruption. *J. Volcanol. Geotherm. Res.* **2013**, *255*, 79–89. [CrossRef]
6. Alparone, S.; Andronico, D.; Lodato, L.; Sgroi, T. Relationship between tremor and volcanic activity during the Southeast Crater eruption on Mount Etna in early 2000. *J. Geophys. Res.* **2003**, *108*, B52241. [CrossRef]
7. Allard, P.; Burton, M.; Murè, F. Spectroscopic evidence for a lava fountain driven by previously accumulated magmatic gas. *Nature* **2005**, *433*, 407–410. [CrossRef] [PubMed]
8. Andronico, D.; Corsaro, R.A. Lava fountains during the episodic eruption of South–East Crater (Mt. Etna), 2000: Insights into magma-gas dynamics within the shallow volcano plumbing system. *Bull. Volc.* **2011**, *73*, 1165–1178. [CrossRef]
9. Behncke, B.; Neri, M. The July-August 2001 eruption of Mt. Etna (Sicily). *Bull. Volcanol.* **2003**, *65*, 461–476. [CrossRef]
10. Andronico, D.; Branca, S.; Calvari, S.; Burton, M.R.; Caltabiano, T.; Corsaro, R.A.; Del Carlo, P.; Garfì, G.; Lodato, L.; Miraglia, L.; et al. A multi-disciplinary study of the 2002–03 Etna eruption: Insights for into a complex plumbing system. *Bull. Volcanol.* **2005**, *67*, 314–330. [CrossRef]
11. Andronico, D.; Scollo, S.; Caruso, S.; Cristaldi, A. The 2002–03 Etna explosive activity: Tephra dispersal and features of the deposits. *J. Geophys. Res.* **2008**, *113*, B04209. [CrossRef]
12. Bonaccorso, A.; Calvari, S. A new approach to investigate an eruptive paroxysmal sequence using camera and strainmeter networks: Lessons from the 3–5 December 2015 activity at Etna volcano. *Earth Plan. Sci. Lett.* **2017**, *475*, 231–241. [CrossRef]
13. Bonaccorso, A.; Calvari, S.; Linde, A.; Sacks, S. Eruptive processes leading to the most explosive lava fountain at Etna volcano: The 23 November 2013 episode. *Geophys. Res. Lett.* **2014**, *41*, 4912–4919. [CrossRef]
14. Calvari, S.; Salerno, G.G.; Spampinato, L.; Gouhier, M.; La Spina, A.; Pecora, E.; Harris, A.J.L.; Labazuy, P.; Biale, E.; Boschi, E. An unloading foam model to constrain Etna's 11–13 January 2011 lava fountaining episode. *J. Geophys. Res.* **2011**, *116*, B11207. [CrossRef]
15. Calvari, S.; Cannavò, F.; Bonaccorso, A.; Spampinato, L.; Pellegrino, A.G. Paroxysmal Explosions, Lava Fountains and Ash Plumes at Etna Volcano: Eruptive Processes and Hazard Implications. *Front. Earth Sci.* **2018**, *6*, 107. [CrossRef]

16. Ganci, G.; Cappello, A.; Bilotta, G.; Herault, A.; Zago, V.; Del Negro, C. Mapping volcanic deposits of the 2011–2015 Etna Eruptive events using satellite remote sensing. *Front. Earth Sci.* **2018**, *6*, 83. [CrossRef]
17. Andronico, D.; Cannata, A.; Di Grazia, G.; Ferrari, F. The 1986–2021 paroxysmal episodes at the summit craters of Mt. Etna: Insights into volcano dynamics and hazard. *Earth Sci. Rev.* **2021**, *220*, 103686. [CrossRef]
18. Behncke, B.; Stefano Branca, S.; Corsaro, R.A.; De Beni, E.; Miraglia, L.; Proietti, C. The 2011–2012 summit activity of Mount Etna: Birth, growth and products of the new SE crater. *J. Volcanol. Geoth. Res.* **2014**, *270*, 10–21. [CrossRef]
19. Andronico, D.; Scollo, S.; Cristaldi, A.; Lo Castro, M.D. Representativity of incompletely sampled fall deposits in estimating eruption source parameters: A test using the 12–13 January 2011 lava fountain deposit from Mt. Etna volcano, Italy. *Bull. Volcanol.* **2014**, *76*, 861. [CrossRef]
20. Wadge, G.; Guest, J.E. Steady-state magma discharge at Etna 1971–81. *Nature* **1981**, *294*, 548–550. [CrossRef]
21. Harris, A.J.L.; Steffke, A.; Calvari, S.; Spampinato, L. Thirty years of satellite-derived lava discharge rates at Etna: Implications for steady volumetric output. *J. Geophys. Res.* **2011**, *116*, B08204. [CrossRef]
22. Bonaccorso, A.; Calvari, S. Major effusive eruptions and recent lava fountains: Balance between erupted and expected magma volumes at Etna volcano. *Geophys. Res. Lett.* **2013**, *40*, 6069–6073. [CrossRef]
23. Bonaccorso, A.; Aloisi, M. Tracking magma storage: New perspectives from 40 years (1980–2020) of ground deformation source modeling on etna volcano. *Front. Earth Sci.* **2021**, *9*, 638742. [CrossRef]
24. Aloisi, M.; Bonaccorso, A.; Cannavò, F.; Currenti, G.; Gambino, S. The 24 December 2018 eruptive intrusion at Etna volcano as revealed by multidisciplinary continuous deformation networks (CGPS, borehole strainmeters and tiltmeters). *J. Geophys. Res.* **2020**, *125*, e2019JB019117. [CrossRef]
25. Calvari, S.; Bilotta, G.; Bonaccorso, A.; Caltabiano, T.; Cappello, A.; Corradino, C.; Del Negro, C.; Ganci, G.; Neri, M.; Pecora, E.; et al. The VEI 2 Christmas 2018 Etna eruption: A small but intense eruptive event or the starting phase of a larger one? *Remote Sens.* **2020**, *12*, 905. [CrossRef]
26. De Beni, E.; Cantarero, M.; Neri, M.; Messina, A. Lava flows of Mt Etna, Italy: The 2019 eruption within the context of the last two decades (1999–2019). *J. Maps.* **2020**, *17*, 65–76. [CrossRef]
27. Mattia, M.; Bruno, V.; Montgomery-Brown, E.; Patanè, D.; Barberi, G.; Coltelli, M. Combined seismic and geodetic analysis before, during, and after the 2018 Mount Etna eruption. *Geochem. Geophys. Geosyst.* **2020**, *21*, e2020GC009218. [CrossRef]
28. Kampouri, A.; Amiridis, V.; Solomos, S.; Gialitaki, A.; Marinou, E.; Spyrou, C.; Georgoulias, A.K.; Akritidis, D.; Papagiannopoulos, N.; Mona, L. Investigation of Volcanic Emissions in the Mediterranean: "The Etna–Antikythera Connection". *Atmosphere* **2021**, *2*, 40. [CrossRef]
29. Wilson, T.M.; Cole, J.W.; Stewart, C.; Cronin, S.J.; Johnston, D.M. Ash storms: Impacts of wind-remobilised volcanic ash on rural communities and agriculture following the 1991 Hudson eruption, southern Patagonia, Chile. *Bull. Volc.* **2011**, *73*, 223–239. [CrossRef]
30. Andronico, D.; Del Carlo, P. PM10 measurements in urban settlements after lava fountain episodes at Mt. Etna, Italy: Pilot test to assess volcanic ash hazard to human health. *Nat. Hazards Earth Syst. Sci. Discuss.* **2016**, *16*, 29–40. [CrossRef]
31. Donnadieu, F.; Freville, P.; Hervier, C.; Coltelli, M.; Scollo, S.; Prestifilippo, M.; Valade, S.; Rivet, S.; Cacault, P. Near-source Doppler radar monitoring of tephra plumes at Etna. *J. Volcanol. Geotherm. Res.* **2016**, *312*, 26–39. [CrossRef]
32. Marzano, F.S.; Mereu, L.; Scollo, S.; Donnadieu, F.; Bonadonna, C. Tephra Mass Eruption Rate From Ground-Based X-Band and L-Band Microwave Radars During the November 23, 2013, Etna Paroxysm. *IEEE Trans. Geosci. Rem. Sens.* **2020**, *58*, 3314–3327. [CrossRef]
33. Patrick, M.R.; Harris, A.J.L.; Ripepe, M.; Dehn, J.; Rothery, D.; Calvari, S. Strombolian explosive styles and source conditions: Insights from thermal (FLIR) video. *Bull. Volcanol.* **2007**, *69*, 769–784. [CrossRef]
34. Murata, K.J.; Dondoli, C.; Saenz, R. The 1963–65 eruption of Irazú Volcano, Costa Rica (the period of March 1963 to October 1964). *Bull. Volcanol.* **1966**, *29*, 765–796. [CrossRef]
35. Booth, B.; Walker, G.P.L. Ash deposits from the new explosion crater, Etna 1971. *Trans. R. Soc. Lond.* **1973**, *A274*, 147–161.
36. Taddeucci, J.; Pompilio, M.; Scarlato, P. Conduit processes during the July–August 2001 explosive activity of Mt Etna (Italy): Inferences from glass chemistry and crystal size distribution of ash particles. *J. Volcanol. Geotherm. Res.* **2004**, *137*, 33–54. [CrossRef]
37. Lautze, N.; Houghton, B.F. Physical mingling of magma and complex eruption dynamics in the shallow conduit at Stromboli Volcano, Italy. *Geology* **2005**, *33*, 425–428. [CrossRef]
38. Head, J.W.; Wilson, L. Lava fountain heights at Pu'u O'o, Kilauea, Hawaii: Indicators of amount and variations of ex- solved magma volatiles. *J. Geophys. Res.* **1987**, *92*, 13715–13719. [CrossRef]
39. Parfitt, E.A.; Wilson, L.; Neal, C.A. Factors influencing the height of Hawaiian lava fountains: Implications for the use of fountain height as an indicator of magma gas content. *Bull. Volcanol.* **1995**, *57*, 440–450. [CrossRef]
40. Walker, G.P.L. Explosive volcanic eruptions—a new classification scheme. *Geol. Rundsch.* **1973**, *62*, 431–446. [CrossRef]
41. Parfitt, E.A. A study of clast size distribution, ash deposition and fragmentation in a Hawaiian-style volcanic eruption. *J. Volcanol. Geotherm. Res.* **1998**, *84*, 197–208. [CrossRef]
42. Sparks, R.S.J.; Bursik, M.I.; Carey, S.N.; Gilbert, J.S.; Glaze, L.S.; Sigurdsson, H.; Woods, A.W. *Volcanic Plumes*; John Wiley Sons: Hoboken, NJ, USA, 1997; p. 574. ISBN 0-471-93901-3.

43. Spampinato, L.; Calvari, S.; Oppenheimer, C.; Lodato, L. Shallow magma transport for the 2002–03 Mt. Etna eruption inferred from thermal infrared surveys. *J. Volcanol. Geotherm. Res.* **2008**, *177*, 301–312. [CrossRef]
44. Taddeucci, J.; Pompilio, M.; Scarlato, P. Monitoring the explosive activity of the July–August 2001 eruption of Mt. Etna (Italy) by ash characterization. *Geophys. Res. Lett.* **2002**, *29*, 71-1–71-4. [CrossRef]
45. Dellino, P.; Kyriakopoulos, K. Phreatomagmatic ash from the ongoing eruption of Etna reaching the Greek island of Cefalonia. *J. Volcanol. Geotherm. Res.* **2003**, *126*, 341–345. [CrossRef]
46. Calvari, S.; Pinkerton, H. Birth, growth and morphologic evolution of the "Laghetto" cinder cone during the 2001 Etna eruption. *J. Volcanol. Geotherm. Res.* **2004**, *132*, 225–239. [CrossRef]
47. Calvari, S.; Pinkerton, H. Instabilities in the summit region of Mount Etna during the 1999 eruption. *Bull. Volcanol.* **2002**, *63*, 526–535. [CrossRef]
48. Andronico, D.; Spinetti, C.; Cristaldi, A.; Buongiorno, M.F. Observations of Mt. Etna volcanic ash plumes in 2006: An integrated approach from ground-based and polar satellite NOAA–AVHRR monitoring system. *J. Volcanol. Geotherm. Res.* **2009**, *180*, 135–147. [CrossRef]
49. Harris, D.M.; Rose, W.I. Estimating particle size, concentrations, and total mass of ash in volcanic clouds using weather radar. *J. Geoph. Res.* **1983**, *88*, 10969–10983. [CrossRef]
50. Spinetti, C.; Barsotti, S.; Neri, A.; Buongiorno, M.F.; Doumaz, F.; Nannipieri, L. Investigation of the complex dynamics and structure of the 2010 Eyjafjallajokull volcanic ash cloud using multispectral images and numerical simulations. *J. Geophys. Res. Atmos.* **2013**, *118*, 4729–4747. [CrossRef]
51. Durig, T.; Gudmundsson, M.T.; Karmann, S.; Zimanowski, B.; Dellino, P.; Rietze, M.; Buttner, R. Mass eruption rates in pulsating eruptions estimated from video analysis of the gas thrust-buoyancy transition—A case study of the 2010 eruption of Eyjafjallajökull, Iceland. *Earth Planets Space* **2015**, *67*, 180. [CrossRef]
52. Corradini, S.; Montopoli, M.; Guerrieri, L.; Ricci, M.; Scollo, S.; Merucci, L.; Marzano, F.S.; Pugnaghi, S.; Prestifilippo, M.; Ventress, L.J.; et al. Multi-Sensor Approach for volcanic ash cloud retrieval and eruption characterization: The 23 November 2013 Etna Lava Fountain. *Remote Sens.* **2016**, *8*, 58. [CrossRef]
53. Corradini, S.; Guerrieri, L.; Stelitano, D.; Salerno, G.; Scollo, S.; Merucci, L.; Prestifilippo, M.; Musacchio, M.; Silvestri, M.; Lombardo, V.; et al. Near real-time monitoring of the Christmas 2018 Etna eruption using SEVIRI and products validation. *Remote Sens.* **2020**, *12*, 1336. [CrossRef]
54. Corradini, S.; Guerrieri, L.; Brenot, H.; Clarisse, L.; Merucci, L.; Pardini, F.; Prata, A.J.; Realmuto, V.J.; Stelitano, D.; Theys, N. Tropospheric volcanic SO_2 mass and flux retrievals from satellite. The Etna December 2018 eruption. *Remote Sens.* **2021**, *13*, 2225. [CrossRef]
55. Athanassiadou, M. The Mt Etna SO_2 eruption in December 2015—The view from space. *Weather* **2016**, *71*, 273–278. [CrossRef]
56. Bluth, G.J.S.; Doiron, S.D.; Schnetzler, C.C.; Krueger, A.J.; Walter, L.S. Global tracking of the SO2 cloud from the June, 1991 Mount Pinatubo eruptions. *Geophys. Res. Lett.* **1992**, *19*, 151–154. [CrossRef]
57. Bonaccorso, A.; Calvari, S.; Currenti, G.; Del Negro, C.; Ganci, G.; Linde, A.; Napoli, R.; Sacks, S.; Sicali, A. From source to surface: Dynamics of Etna's lava fountains investigated by continuous strain, magnetic, ground and satellite thermal data. *Bull. Volcanol.* **2013**, *75*, 690. [CrossRef]
58. Bonaccorso, A.; Currenti, G.; Linde, A.; Sacks, S. New data from borehole strainmeters to infer lava fountain sources (Etna 2011–2012). *Geophys. Res. Lett.* **2013**, *40*, 3579–3584. [CrossRef]
59. Corradini, S.; Guerrieri, L.; Lombardo, V.; Merucci, L.; Musacchio, M.; Prestifilippo, M.; Scollo, S.; Silvestri, M.; Spata, G.; Stelitano, D. Proximal monitoring of the 2011–2015 Etna lava fountains using MSG-SEVIRI data. *Geosciences* **2018**, *8*, 140. [CrossRef]
60. Scollo, S.; Prestifilippo, M.; Pecora, E.; Corradini, S.; Merucci, L.; Spata, G.; Coltelli, M. Eruption column height estimation of the 2011–2013 Etna lava fountains. *Ann. Geophys.* **2014**, *57*, S0214.
61. Scollo, S.; Prestifilippo, M.; Bonadonna, C.; Cioni, R.; Corradini, S.; Degruyter, W.; Rossi, E.; Silvestri, M.; Biale, E.; Carparelli, G.; et al. Near-real-time tephra fallout assessment at Mt. Etna, Italy. *Remote Sens.* **2019**, *11*, 2987. [CrossRef]
62. Ganci, G.; James, M.R.; Calvari, S.; Del Negro, C. Separating the thermal fingerprints of lava flows and simultaneous lava fountaining using ground-based thermal camera and SEVIRI measurements. *Geophys. Res. Lett.* **2013**, *40*, 5058–5063. [CrossRef]
63. Corradino, C.; Ganci, G.; Cappello, A.; Bilotta, G.; Calvari, S.; Del Negro, C. Recognizing eruptions of mount etna through machine learning using multiperspective infrared images. *Remote Sens.* **2020**, *12*, 970. [CrossRef]
64. Ganci, G.; Vicari, A.; Cappello, A.; Del Negro, C. An emergent strategy for volcano hazard assessment: From thermal satellite monitoring to lava flow modelling. *Remote Sens. Environ.* **2012**, *119*, 197–207. [CrossRef]
65. Ganci, G.; Harris, A.J.L.; Del Negro, C.; Guéhenneux, Y.; Cappello, A.; Labazuy, P.; Calvari, S.; Gouhier, M. A year of lava fountaining at Etna: Volumes from SEVIRI. *Geophys. Res. Lett.* **2012**, *39*, L06305. [CrossRef]
66. Ganci, G.; Vicari, A.; Fortuna, L.; Del Negro, C. The HOTSAT volcano monitoring system based on combined use of SEVIRI and MODIS multispectral data. *Ann. Geophys.* **2011**, *54*, 5. [CrossRef]
67. Ganci, G.; Cappello, A.; Bilotta, G.; Corradino, C.; Mangiameli, M.; Mussumeci, G.; Del Negro, C. 3D Lava flow mapping in volcanic areas using multispectral and stereo optical satellite data. *AIP Conf. Proc.* **2020**, *2293*, 300003. [CrossRef]
68. Harris, A.J.L.; Dehn, J.; Calvari, S. Lava effusion rate definition and measurement: A review. *Bull. Volcanol.* **2007**, *70*, 1–22. [CrossRef]

69. Harris, A.J.L.; Flynn, L.P.; Keszthelyi, L.; Mouginis-Mark, P.J.; Rowland, S.K.; Resing, J.A. Calculation of lava effusion rates from Landsat TM data. *Bull. Volcanol.* **1998**, *60*, 52–71. [CrossRef]
70. Wright, R.; Blake, S.; Harris, A.; Rothery, D. A simple explanation for the space-based calculation of lava eruption rates. *Earth Plan. Sci. Lett.* **2001**, *192*, 223–233. [CrossRef]
71. Garel, F.; Kaminski, E.; Tait, S.; Limare, A. An experimental study of the surface thermal signature of hot subaerial isoviscous gravity currents: Implications for thermal monitoring of lava flows and domes. *J. Geophys. Res.* **2012**, *117*, B02205. [CrossRef]
72. Hon, K.; Kauahikaua, J.; Denlinger, R.; Mackay, K. Emplacement and inflation of pahoehoe sheet flows: Observations and measurements of active lava flows on Kilauea Volcano, Hawaii. *Geol. Soc. Am. Bull.* **1994**, *106*, 351–370. [CrossRef]
73. Prata, A.J.; Bluth, G.J.S.; Rose, W.I.; Schneider, D.J.; Tupper, A.C. Comments on failures in detecting volcanic ash from a satellite-based technique. *Remote Sens. Environ.* **2001**, *78*, 341–346. [CrossRef]
74. Del Negro, C.; Cappello, A.; Ganci, G. Quantifying lava flow hazards in response to effusive eruption. *Geol. Soc. Am.* **2015**, *128*, 752–763. [CrossRef]
75. Corradino, C.; Ganci, G.; Cappello, A.; Bilotta, G.; Herault, A.; Del Negro, C. Mapping recent lava flows at Mount Etna using multispectral sentinel-2 images and machine learning techniques. *Remote Sens.* **2019**, *11*, 1916. [CrossRef]
76. Bonaccorso, A.; Linde, A.; Currenti, G.; Sacks, S.; Sicali, A. Theborehole dilatometer network of Mount Etna: A powerful tool to detect and infer volcano dynamics. *J. Geophys. Res. Solid Earth* **2016**, *121*, 4655–4669. [CrossRef]
77. Currenti, G.; Zuccarello, L.; Bonaccorso, A.; Sicali, A. Borehole volumetric strainmeter calibration from a nearby seismic broadband array at Etna volcano. *J. Geophys. Res.* **2017**, *122*, 7729–7738. [CrossRef]
78. Bonaccorso, A.; Currenti, G.; Linde, A.; Sacks, S.; Sicali, A. Advances in understanding intrusive, explosive and effusive processes as revealed by the borehole dilatometer network on Mt. Etna volcano. *Front. Earth Sci.* **2020**, *7*, 357. [CrossRef]
79. Marchese, F.; Neri, M.; Behncke, B. Main morpho-structural changes and eruptions of Etna in 2016–2019 captured by satellite observations. In Proceedings of the 22nd EGU General Assembly, Online Event, 4–8 May 2020; p. EGU2020-13909. [CrossRef]
80. Genzano, N.; Marchese, F.; Neri, M.; Pergola, N.; Tramutoli, V. Implementation of robust satellite techniques for volcanoes on ASTER data under the google earth engine platform. *Appl. Sci.* **2021**, *11*, 4201. [CrossRef]
81. Jaupart, C.; Vergniolle, S. Laboratory models of Hawaiian and Strombolian eruptions. *Nature* **1988**, *331*, 58–60. [CrossRef]
82. Jaupart, C.; Veirgniolle, S. The generation and collapse of foam layer at the roof of a basaltic magma chamber. *J. Fluid Mech.* **1989**, *203*, 347–380. [CrossRef]
83. Bonaccorso, A.; Cannata, A.; Corsaro, R.A.; Di Grazia, G.; Gambino, S.; Greco, F.; Miraglia, L.; Pistorio, A. Multi-disciplinary investigation on a lava fountain preceding a flank eruption: The 10 May 2008 Etna case. *Geochem. Geophys. Geosyst.* **2011**, *12*, 7. [CrossRef]
84. Jaupart, C. Magma ascent at shallow levels. *Encycl. Volcanoes* **1999**, *1*, 237–245.
85. Dubosclard, G.; Cordesses, R.; Allard, P.; Hervier, C.; Coltelli, M.; Kornprobst, J. First testing of a volcano Doppler radar (Voldorad) at Mt. Etna. *Geophys. Res. Lett.* **1999**, *26*, 3389–3392. [CrossRef]
86. De Beni, E.; Behncke, S.; Branca, I.; Nicolosi, R.; Carluccio, F.; D'Ajello, C.; Chiappini, M. The continuing story of Etna's new southeast crater (2012–2014): Evolution and volume calculations based on field surveys and aerophotogrammetry. *J. Volcanol. Geotherm. Res.* **2015**, *303*, 175–186. [CrossRef]
87. Roeloffs, E.A.; Linde, A.T. Borehole observations and continuous strain and fluid pressure. In *Volcano Deformation Geodetic Measurments Techniques*; Dzurisin, D., Ed.; Springer: Berlin/Heidelberg, Germany, 2007; pp. 305–322. [CrossRef]
88. Vulpiani, G.; Ripepe, M.; Valade, S. Mass discharge rate retrieval combining weather radar and thermal camera observations. *J. Geophys. Res. Solid Earth* **2016**, *121*, 5679–5695. [CrossRef]
89. Pailot-Bonnétat, S.; Harris, A.J.L.; Calvari, S.; De Michele, M.; Gurioli, L. Plume height time-series retrieval using shadow in single spatial resolution satellite images. *Remote Sens.* **2020**, *12*, 3951. [CrossRef]
90. Spanu, A.; de'Michieli Vitturi, M.; Barsotti, S. Reconstructing eruptive source parameters from tephra deposit: A numerical study of medium-sized explosive eruptions at Etna volcano. *Bull. Volcanol.* **2016**, *78*, 59. [CrossRef]
91. Casadevall, T.J. Volcanic ash and aviation safety: Proceedings of the First International Symposium on Volcanic Ash and Aviation Safety. *USGS Bull.* **1994**, *2047*, 450–469.
92. Rose, W.I.; Delene, D.J.; Schneider, D.J.; Bluth, G.J.S.; Krueger, A.J.; Sprod, I.; McKee, C.; Davies, H.L.; Erns, G.G.J. Ice in the 1994 Rabaul eruption cloud: Implications for volcano hazard and atmospheric effects. *Nature* **1995**, *375*, 477–479. [CrossRef]
93. Marzano, F.S.; Picciotti, E.; Montopoli, M.; Vulpiani, G. Inside Volcanic Clouds. Remote Sensing of Ash Plumes Using Microwave Weather Radars. *Am. Meteorol. Soc.* **2013**, *94*, 1567–1586. [CrossRef]
94. Bonaccorso, A.; Calvari, S.; Boschi, E. Hazard mitigation and crisis management during major flank eruptions at Etna volcano: Reporting on real experience. In *Detecting, Modelling and Responding to Effusive Eruptions*; Harris, A.J.L., De Groeve, T., Garel, F., Carn, S.A., Eds.; Special Publications (IAVCEI) Series; Geological Society of London: London, UK, 2015; Volume 426, pp. 447–461. ISBN 978-1-86239-736-1. [CrossRef]
95. Solana, M.C.; Calvari, S.; Kilburn, C.R.J.; Gutierrez, H.; Chester, D.; Duncan, A. Supporting the development of procedures for communications during volcanic emergencies: Lessons learnt from the Canary Islands (Spain) and Etna and Stromboli (Italy). In *Advances in Volcanology, Observing the Volcano World, Volcano Crisis Communication*; Fearnley, C.J., Bird, D.K., Haynes, K., McGuire, W.J., Jolly, G., Eds.; Springer: Cham, Switzerland, 2017; pp. 289–305. ISBN 978-3-319-44095-8. [CrossRef]

remote sensing

MDPI

Article

Comparison between Automated and Manual Detection of Lava Fountains from Fixed Monitoring Thermal Cameras at Etna Volcano, Italy

Sonia Calvari [1,*] and Giuseppe Nunnari [2]

1 Istituto Nazionale di Geofisica e Vulcanologia, Osservatorio Etneo—Sezione di Catania, Piazza Roma 2, 95125 Catania, Italy
2 Dipartimento di Ingegneria Elettrica, Elettronica e Informatica, Università degli Studi di Catania, Viale A. Doria 6, 95122 Catania, Italy; giuseppe.nunnari@dieei.unict.it
* Correspondence: sonia.calvari@ingv.it

Abstract: The Etna volcano is renowned worldwide for its extraordinary lava fountains that rise several kilometers above the vent and feed eruptive columns, then drift hundreds of kilometers away from the source. The Italian Istituto Nazionale di Geofisica e Vulcanologia-Osservatorio Etneo (INGV-OE) is responsible for the monitoring of Mt. Etna, and for this reason, has deployed a network of visible and thermal cameras around the volcano. From these cameras, INGV-OE keeps a keen eye, and is able to observe the eruptive activity, promptly advising the civil protection and aviation authorities of any changes, as well as quantifying the spread of lava flows and the extent of pyroclastic and ash plumes by using a careful analysis of the videos recorded by the monitoring cameras. However, most of the work involves analysis carried out by hand, which is necessarily approximate and time-consuming, thus limiting the usefulness of these results for a prompt hazard assessment. In addition, the start of lava fountains is often a gradual process, increasing in strength from Strombolian activity, to intermediate explosive activity, and eventually leading to sustained lava fountains. The thresholds between these different fields (Strombolian, Intermediate, and lava fountains) are not clear cut, and are often very difficult to distinguish by a manual analysis of the images. In this paper, we presented an automated routine that, when applied to thermal images and with good weather conditions, allowed us to detect (1) the starting and ending time of each lava fountain, (2) the area occupied by hot pyroclasts, (3) the elevation reached by the lava fountains over time, and (4) eventually, to calculate in real-time the erupted volume of pyroclasts, giving results close to the manual analysis but more focused on the sustained portion of the lava fountain, which is also the most dangerous. This routine can also be applied to other active volcanoes, allowing a prompt and uniform definition of the timing of the lava fountain eruptive activity, as well as the magnitude and intensity of the event.

Keywords: automated detection; remote sensing; lava fountains; Etna volcano

Citation: Calvari, S.; Nunnari, G. Comparison between Automated and Manual Detection of Lava Fountains from Fixed Monitoring Thermal Cameras at Etna Volcano, Italy. *Remote Sens.* **2022**, *14*, 2392. https://doi.org/10.3390/rs14102392

Academic Editor: Fumio Yamazaki

Received: 23 March 2022
Accepted: 13 May 2022
Published: 16 May 2022

1. Introduction

New data and interpretations have emerged of the geodynamics of the eastern Sicily point to Etna as a volcano, undergoing an evolutionary phase where a future increase in highly energetic explosive activity is possible [1]. As a matter of fact, the last three decades of Etna's activity were characterized by frequent highly explosive eruptions, here called paroxysms [2,3]. Paroxysms at Etna are characterized by lava fountaining lasting 1–2 h, reaching the height of 1–3 km above the crater, and generating conspicuous and lengthy ash plumes that can drift hundreds of kilometers from the vents [4,5], often accompanied by short-lasting lava overflows from the crater rim [6,7]. During the last few decades, Etna volcano underwent several eruptions characterized by lava flows mainly from the summit vents, alternating with short-lasting but powerful explosive episodes [2,3,8–10]. In particular, between 2011 and 2015, Etna produced more than 50 such eruptions [3,10–12], releasing a cumulative erupted volume of a similar order to a major flank eruption [2], which was normally ten times greater than summit activity [13]. Given that explosive paroxysms can have a major impact on aviation [14], on road and traffic conditions, and also on the villages on the slope of the volcano [15–17], it is of paramount importance for a volcano observatory such as the Istituto Nazionale di Geofisica e Vulcanologia-Osservatorio Etneo (INGV-OE) to be able to raise an early warning as soon as possible, and then to advise the civil protection authorities of its possible impact on human activities [18–21]. The first and most important parameter to be detected as soon as possible is the timing of start and end of any impending activity, and this information needs to be completed with the extent of the ash plume, lava flows and lava fountains, and with the volume erupted [3,6,8,12,22,23]. The volume erupted during a lava fountain (LF) episode quantifies the magnitude of the event, whereas the eruption rate determines its intensity [24]. The information gathered from the monitoring system is then used to inform the civil protection of the magnitude and intensity of the event, and also in models routinely used for the prediction of the extent and distribution of the eruption products, which, at Etna, often comprise both lava flows [25,26] and pyroclastics [27–29]. An automated procedure to map the lava flows from the images of the thermal monitoring cameras was recently developed [6,7], whereas the LF detection and the estimation of the erupted volumes were normally carried out by manual analysis of the images [3,8,30]. The pyroclastic volume estimated by thermal images was compared to the total erupted volume estimated by strain and with the lava flow volume erupted during each episode as estimated by satellite [30], and an error of ~20% was estimated for the calculated fluid volume, comprising gas plus pyroclastics [31].

A sustained LF normally gives rise to an eruptive column comprising three main portions: a lower and innermost zone called gas-thrust region with the highest velocity at the exit of the erupting vent; an intermediate zone, where convective movements of the hot mixture of gas and tephra allow ingestion of the cold surrounding air, thereby slowing down the spreading hot jet, and an uppermost zone reaching the buoyancy zone and consisting of the laterally spreading umbrella region [32,33]. In addition, the eruptive columns at Etna can be distinguished into weak plumes if bent in the wind direction due to wind speeds greater than 10 m s^{-1}, or strong to intermediate plumes when rising vertically above the vent or slightly bent in the wind direction for wind speeds lower than 10 m s^{-1} [3].

LF heights were determined following several different methods. One of the earliest, applied at Kilauea, was carried out by digitizing film from 8 mm time-lapse movie cameras deployed on the ground, and using a few theodolite measurements as calibration points for the film [34]. At Etna, a C++ code was developed in order to explore the INGV-OE thermal image library for image processing [35], applying appropriate thresholds, and converting the color images into a binary black and white image over which the maximum vertical extent of the LF can be easily retrieved. A less automated but sometimes more precise system involves the visual and manual analysis of each image [3,8], given that the thermal images can be affected by low gas, weather or ash clouds [36–40] that may reduce the automated measurement of the LF height, or the LF jet may be inclined [3,8,30,35]. Calibrated images of visible cameras can also be used to estimate the vertical extent of proximal ash plumes associated with the lava fountains [12,22].

One of the most difficult challenges in volcanology is to determine when an eruption is over, especially when it includes multiple episodes and long pauses [41], although sometimes, a gradual decline of the mass eruption rate may anticipate the end of the eruption [42]. During the lava fountain activity at Etna, the start and ending time, as well as any early warning alarm, is given on the basis of the volcanic tremor and infrasound [18,43–46]. However, a volcanic tremor does not allow us to calculate the volume erupted [3,8], and does not provide information about the extension of the lava fountains and ash plume [5,22]. Another useful device is the borehole strainmeter, which allows calculating the total erupted volume [47,48], comprising both lava flows and pyroclastics [30]. However, in order to assess the impact on the population, the amount of solely the pyroclastics component erupted during a lava fountain event needs to be established, because this affects the stability of roofs, the cleaning up of roads and motorways, the impact on the nearby Catania international airport, and the health effects on the local population [49,50]. Automated routines for volcanic activity detection and characterization have been recently developed [23,45,51,52], and they will probably resolve most of the issues related with early warning alarms. In this paper, we presented a new automated routine that, when applied to the images recorded by the thermal monitoring cameras, allowed us to calculate (1) lava fountain height, (2) area of the lava fountain jet, and (3) volume of the erupted tephra, using the formula applied by Calvari et al. [3,8]. We compared these results with those obtained by a manual analysis, discussing limits and advantages, and future possible improvements.

2. Methods

The INGV-OE thermal camera network for monitoring Etna volcano includes four fixed, continuously operating thermal cameras located on the flanks of the volcano at different distances from and looking towards the summit (Figure 1, Table 1). The images recorded by these cameras are transmitted to the INGV-OE Operative Room and displayed in real-time to allow continuous monitoring of the volcano. From this perspective, operators have the task of recognizing the type of event as early as possible. Thermal cameras are remote-sensing ground-based fixed devices that have significantly improved INGV-OE's observational capabilities. They allowed us to monitor the summit area continuously, and to identify and locate eruptive events. Only poor weather conditions, and especially thick clouds of water vapor, gas, and/or volcanic ash [37–40] may affect the visibility of the cameras, and consequently, the reliability of the acquired images, by partially or totally hiding what is happening in the monitored area. When, by manual examination of images, we recognized that visibility was limited to a few frames or interfering clouds were low and only partially obscured the lava fountain (LF), a linear interpolation was carried out on the data [3]. The manual analysis of the thermal images followed what was described by Calvari et al. [3,8].

Table 1. List of the INGV monitoring thermal cameras used in this paper and their main features. The field of view is considered at the crater rim.

Label	Type	Location	Distance from the Craters (km)	Frame Rate	Field of View
ENT	Thermal FLIR A40M	Nicolosi, South flank 730 m a.s.l.	15.0	2 frames/s	320 × 240 pixels
EBT	Thermal FLIR A320	Bronte, NW flank 85 m a.s.l.	13.5	2 frames/s	25° × 18.8°
EMCT	Thermal FLIR A320	Mt. Cagliato, East flank 1160 m a.s.l.	8.3	2 frames/s	320 × 240 pixels
EMOT	Thermal FLIR A320	Montagnola, South flank 2600 m a.s.l.	3.0	1 frame/s	320 × 240 pixels

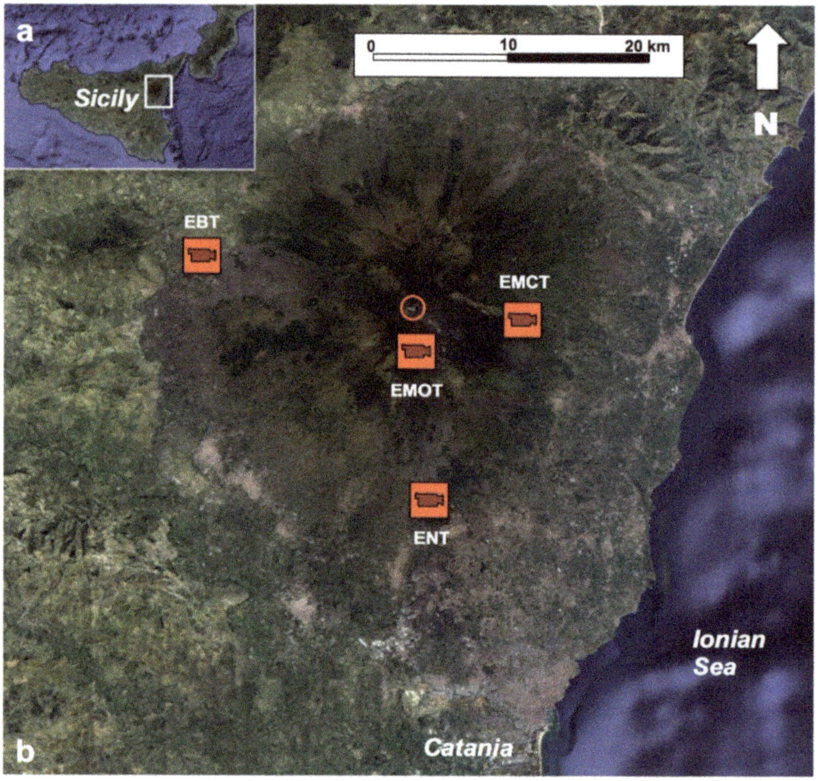

Figure 1. (**a**) Google map of Sicily, with the white rectangle showing the area magnified in (**b**), which is the Google map of Etna volcano showing the position of the INGV thermal monitoring cameras used in this study. The red circle indicates the position of the summit craters producing the lava fountain activity.

A sequence of LF episodes occurred at Mt. Etna between 2020 and 2022, and the list of these events is given in Table 2. We have analyzed the images of the thermal cameras that recorded the event (Figure 1, Table 2), and chose for each episode the one offering the best view and the entire vertical extension of the LF, as a function of the LF size, wind direction, and consequent ash plume fallout. In cases of rotating ash plumes, we used the integration of images from more than one camera. The manual analysis of the camera images allowed retrieving the starting and ending time of each episode, the duration expressed in minutes and seconds, the maximum height of the lava fountain and its average value, as well as the erupted volume of pyroclastics and the time-averaged discharge rate (TADR, [53]). The volume of pyroclastics was calculated by following the method by Calvari et al. [3,8], based on the measurements of the LF heights at time lapses of 1 min, and considering a constant vent radius of 15 m and a pyroclastic ratio of 0.18% of the total erupted fluids comprising gas plus pyroclastics. All these results, obtained from the manual analysis of the thermal camera images, are reported in Table 2.

Table 2. Paroxysmal explosive episodes occurring at Etna volcano between 2020 and 2022 (updated to 23 February 2022). The first column reports the episode's progressive number. All times are expressed in UTC. LF = lava fountain. TADR = time averaged discharge rate [53]. Data from episode # 32 are lacking because there was no visibility from any of the thermal monitoring cameras.

Ep. #	Date	Starting Time (hh:mm)	Ending Time (hh:mm)	Duration (in Minutes and in Seconds)	Label of Cam-era Used	Max LF Height (m above the Crater)	Mean LF Height (m above the Crater)	LF Volume ($\times 10^6$ m^3)	TADR (m^3 s^{-1})
1	13 December 2020	22:00	22:48	48 min. 2880 s	ENT	514	231	0.24	84
2	13–14 December 2020	23:58	00:11	13 min. 780 s	ENT	400	235	0.07	90
3	14 December 2020	01:02	01:40	38 min. 2280 s	ENT	286	121	0.13	59
4	21 December 2020	09:11	09:59	48 min. 2880 s	EBT	3080	1296	0.58	201
5	22 December 2020	03:05	05:13	128 min. 7680 s	ENT	800	295	0.72	93
6	18 January 2021	19:38	21:03	85 min. 5100 s	ENT	1067	343	0.52	101
7	16 February 2021	16:11	17:02	51 min. 3060 s	EMCT	1560	757	0.46	150
8	17–18 February 2021	22:32	00:51	139 min. 8340 s	EMCT	1230	358	0.82	98
9	19 February 2021	08:16	10:06	110 min. 6600 s	EMCT	1365	492	0.78	118
10	20–21 February 2021	21:32	01:15	223 min. 13,380 s	EMCT	1500	386	1.43	107
11	22–23 February 2021	21:17	00:03	166 min. 9960 s	ENT EMOT EMCT	3667	686	1.19	120
12	23 February 2021	03:52	04:50	58 min. 3480 s	ENT EMOT EMCT	900	337	0.32	92
13	24 February 2021	18:56	21:41	165 min. 9900 s	ENT	1800	649	1.37	139
14	28 February 2021	07:31	08:34	63 min. 3780 s	ENT	3600	1376	0.70	185
15	2 March 2021	11:23	14:50	207 min. 12,420 s	EMOT	606	278	1.03	83
16	4 March 2021	01:30	04:10	160 min. 9600 s	ENT	600	204	0.75	78

Table 2. *Cont.*

Ep. #	Date	Starting Time (hh:mm)	Ending Time (hh:mm)	Duration (in Minutes and in Seconds)	Label of Cam-era Used	Max LF Height (m above the Crater)	Mean LF Height (m above the Crater)	LF Volume ($\times 10^6$ m^3)	TADR (m^3 s^{-1})
17	4 March 2021	07:11	09:32	141 min. 8460 s	ENT	3233	1275	1.58	186
18	7 March 2021	04:10	07:01	171 min. 10,260 s	EMOT EBT	4000	638	1.07	104
19	9–10 March 2021	23:55	02:46	171 min. 10,260 s	ENT EMCT	1860	655	1.44	140
20	12 March 2021	07:35	09:45	130 min. 7800 s	ENT EBT	2400	1149	1.63	209
21	14–15 March 2021	23:12	01:42	150 min. 9000 s	ENT	1333	670	0.53	59
22	17 March 2021	01:30	04:57	207 min. 12,420 s	ENT	1533	538	1.58	128
23	19 March 2021	08:18	10:13	115 min. 6900 s	EMOT	629	171	0.75	108
24	23–24 March 2021	21:33	08:19	646 min. 38,760 s	EMCT	1333	456	4.56	118
25	31 March– 1 April 2021	19:30	08:53	803 min. 48,180 s	EMCT	630	241	4.10	85
26	19 May 2021	00:50	04:25	215 min. 12,900 s	ENT	667	482	1.59	124
27	21 May 2021	00:50	02:44	114 min. 6840 s	EMCT	1533	683	0.99	145
28	22 May 2021	20:27	22:08	101 min. 6060 s	ENT	1200	649	0.87	143
29	24 May 2021	20:25	21:49	84 min. 5040 s	ENT	1467	831	0.82	162
30	25 May 2021	18:20	18:53	33 min. 1980 s	ENT	533	317	0.20	102
31	26 May 2021	10:20	11:10	50 min. 3000 s	ENT	1267	627	0.41	137
32	27 May 2021	12:00	13:00	60 min. 3600 s	EMCT	Poor visibility	Poor visibility	Poor visibility	Poor visibility
33	28 May 2021	06:30	07:27	57 min. 3420 s	ENT	800	433	0.40	116
34	28 May 2021	16:05	16:11	6 min. 360 s	ENT	333	295	0.04	112

Table 2. *Cont.*

Ep. #	Date	Starting Time (hh:mm)	Ending Time (hh:mm)	Duration (in Minutes and in Seconds)	Label of Cam-era Used	Max LF Height (m above the Crater)	Mean LF Height (m above the Crater)	LF Volume ($\times 10^6$ m^3)	TADR (m^3 s^{-1})
35	28 May 2021	19:48	20:50	62 min. 3720 s	ENT	1000	601	0.51	138
36	30 May 2021	04:20	05:44	84 min. 5040 s	ENT	1000	589	0.69	137
37	2 June 2021	08:30	10:46	136 min. 8160 s	EBT	1640	924	1.38	170
38	4 June 2021	16:12	17:40	88 min. 5280 s	EMCT	1170	665	0.76	143
39	12 June 2021	20:00	21:46	106 min. 6360 s	EMCT	810	438	0.73	115
40	14 June 2021	21:15	22:21	66 min. 3960 s	EMCT	870	419	0.45	112
41	16 June 2021	11:37	12:38	61 min. 3660 s	ENT	1733	673	0.52	142
42	17 June 2021	22:40	23:55	75 min. 4500 s	EMCT	1140	404	0.50	111
43	19 June 2021	18:47	19:35	48 min. 2880 s	EMCT	1140	572	0.38	131
44	20 June 2021	22:40	23:44	64 min. 3840 s	ENT	2467	892	0.63	165
45	22 June 2021	02:30	03:45	75 min. 4500 s	ENT	2000	848	0.71	159
46	23 June 2021	02:44	03:19	35 min. 2100 s	ENT	1867	1035	0.38	183
47	23 June 2021	18:00	19:12	72 min. 4320 s	ENT	2933	1301	0.60	153
48	24 June 2021	09:45	10:26	41 min. 2460 s	ENT EMOT	1733	825	0.39	157
49	25 June 2021	00:38	01:48	70 min. 4200 s	ENT	2333	876	0.64	153
50	25 June 2021	18:40	19:20	40 min. 2400 s	ENT	1133	691	0.36	149
51	26 June 2021	16:00	16:38	38 min. 2280 s	ENT	1600	772	0.35	155
52	27 June 2021	08:53	09:43	50 min. 3000 s	ENT	1600	674	0.43	143

Table 2. *Cont.*

Ep. #	Date	Starting Time (hh:mm)	Ending Time (hh:mm)	Duration (in Minutes and in Seconds)	Label of Cam-era Used	Max LF Height (m above the Crater)	Mean LF Height (m above the Crater)	LF Volume ($\times 10^6$ m^3)	TADR ($m^3\ s^{-1}$)
53	28 June 2021	14:25	15:30	65 min. 3900 s	EBT	2390	1211	0.75	193
54	1–2 July 2021	22:50	00:27	97 min. 5820 s	ENT	1800	804	0.91	156
55	4 July 2021	15:15	16:50	95 min. 5700 s	ENT	1467	873	0.94	164
56	6 July 2021	22:16	23:44	88 min. 5280 s	EBT	3270	1673	1.20	227
57	8 July 2021	20:35	22:12	97 min. 5820 s	EBT	2710	1242	1.12	192
58	14 July 2021	10:45	12:30	105 min. 6300 s	EBT	2230	1097	1.15	183
59	20 July 2021	05:10	08:11	181 min. 10,860 s	EBT	3510	1533	2.26	208
60	31 July 2021	19:44	23:37	233 min. 13,980 s	EBT	3830	1473	2.83	202
61	9 August 2021	02:07	04:11	124 min. 7440 s	EBT	2390	1280	1.48	199
62	29 August 2021	16:24	17:55	91 min. 5460 s	EBT	2390	1310	1.10	202
63	21 September 2021	07:21	08:35	74 min. 4440 s	ENT	2333	1234	0.88	199
64	23 October 2021	08:20	10:17	117 min. 7020 s	ENT	4000	1844	1.63	232
65	10 February 2022	18:40	21:56	196 min. 11,760 s	ENT	5714	2160	2.88	245
66	21 February 2022	11:11	12:50	99 min. 5940 s	ENT	4057	1865	1.39	234
				Average Duration (min./s)		**Average Max LF height (m)**	**Average Mean LF height (m)**	**Average LF volume ($\times 10^6$ m^3)**	**Average TADR ($m^3\ s^{-1}$)**
				120/7171		1815	784	0.993	144.75

The dataset consisted of 66 episodes of LF recorded between 13 December 2020 and 21 February 2022 at Mt Etna by using the four thermal cameras of the INGV-OE monitoring network listed in Table 1, and whose position is shown in Figure 1. Additional technical details on the thermal sensors can be found in Calvari et al. [3,29]. Original data were provided as *.avi* format files, each containing about 5 min of recorded volcanic activity. Information concerning the name of the camera and the starting time of each file were embedded in its name. For instance, the filename *EMOT 20210319–082500.avi*, refers to a file recorded by the camera named EMOT, starting on 19 March 2021 at time 08:25:00. All times are UTC. Other information, such as the duration, the frame-rate, and other useful video properties, such as width, height, bits per pixel, and the video format were embedded in the file object. Files were pre-processed in order to crop the color bar and the information about the acquisition time and camera name, which are normally embedded in the frame.

In order to detect the presence of LF, each frame was converted from the original RGB format to grayscale and further binarized by adopting a threshold luminance value, specified as a value in the range [0, 1]. In this way, the hottest objects, such as newly erupted or cooling down products, will be represented in the binarized image as white areas, while all the others will be represented in black (Figure 2). Thus, in the absence of hot objects, such as new ejected volcanic matter, hot spots or cooling lava, the binarized images will result completely black. However, as mentioned above, it should be noted that hot objects may not be detected due to the presence of a thick cloud cover. Clearly, the choice of the threshold parameter played a crucial role. In fact, although on one hand, it allowed filtering unwanted information, due, for example, to warming of the monitored area by the Sun, on the other hand, it could also remove useful information, especially in the phase of emergence of an LF episode. Unfortunately, there were no optimal criteria for the choice of this parameter which was therefore made for each camera, adopting the traditional trial and error method. Each binarized image was processed, and detected objects, represented in white color in Figure 2c, were measured in order to obtain:

- the areas A_i, $i = 1, N$, A_i being the area in pixels of the i-th object and N the number of recognized objects;
- the coordinates (x_i, y_i) of the centroid, of the i_th object, x_i and y_i being the horizontal and vertical coordinates, respectively.

The overall process, starting from the original image (Figure 2a), through the grayscale (Figure 2b), the binarized (Figure 2c), and finally, to the labeled image (Figure 2d) is reported, as an example, in Figure 2. In particular, the centroids of detected hot objects are shown by the red asterisks in Figure 2d.

The presence of multiple objects, even from an individual LF episode, was due to the fact that the volume occupied by an LF does not have a uniform temperature, as can be seen from Figure 2a where the original colors of the thermal image indicate different temperatures in a scale starting from blue (0 °C) to white (>100 °C). However, for practical reasons, as different hot objects belong to the same individual LF, it is a good practice to consider all detected objects as a single one. In our case, this was obtained by summing up the areas of all detected objects, and calculating the coordinates of a single centroid by a weighted average of the coordinates of the individual centroids, as expressed in Equation (1).

$$x = \sum_{i=1}^{N} w_i x_i; y = \sum_{i=1}^{N} w_i y_i,\ w_i = \frac{A_i}{\sum_{i=1}^{N} A_i} \quad (1)$$

where A_i is the area of the i-th object, N is the number of detected objects, and w_i is the normalized area of the i-th object. It is straightforward to say that, with considerable approximation, due to the fact that the LF occurs in 3-D volume, while images refer to a 2-D area, the estimated area A is, in some way, related to the volume of hot matter, while the y-coordinate of its centroid may be related with the mean elevation. Of course, A and its centroids' coordinates (x,y), originally expressed in pixel units, can be converted into geographical units by appropriate conversion constants, depending on the position of the considered camera with respect to the monitored area. The graphical representation of the area *A* (Figure 3a) and its mean altitude (Figure 3b) for an LF episode occurring on Mt. Etna on 24 February 2021, is shown in Figure 3.

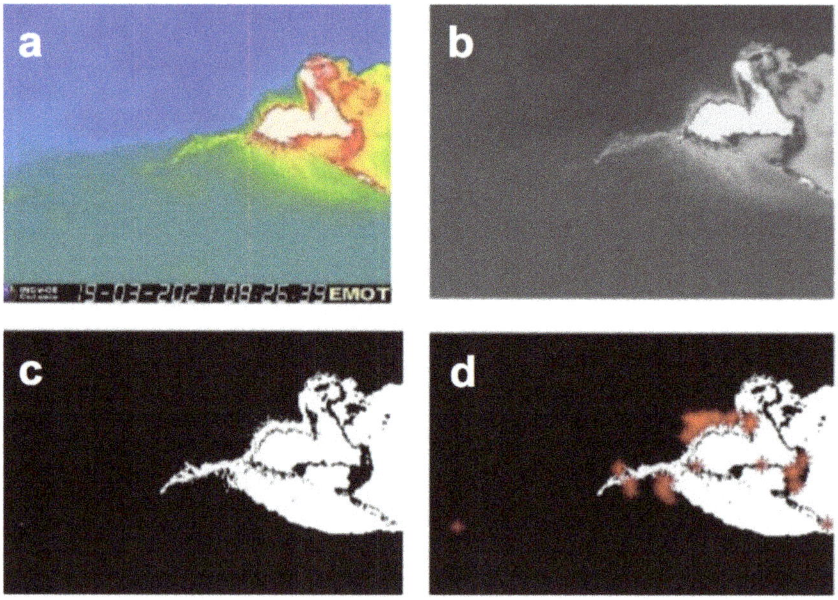

Figure 2. (**a**) The original RGB thermal image from the EMOT camera, with temperatures in °C comprised between blue (0 °C) and white (>100 °C); (**b**) the cropped grayscale; (**c**) the binarized image, and (**d**) the binarized image with computed centroids (red asterisks), representing the center of mass of the hot objects detected by the threshold process.

Figure 3 can be interpreted as follows: during 24 February 2021, a mass of previously erupted material was cooling down around the crater area. This can be deduced from the fact that the *y*-coordinate of the centroids (Figure 3b) fluctuates around zero during the time interval from 00:00 to about 19:25. In Figure 3b, for convenience, the zero value of the mean altitude has been arbitrarily set to the average value of the *y*-coordinate, measured throughout the recording period. The lack of signal in the mean altitude (in the reported example, for instance, around 14:20 and 14:40) is due to the thick cloud cover which prevented viewing of the cooling mass. After 19:15, an LF appeared, as can be seen from both the area signal (Figure 3a) and the increasing mean altitude of the center of mass (Figure 3b). The LF continued until about 20:00, after which, the erupted material began a cooling process. It should be observed that while the value of the area slowly decreased, the value of the mean altitude of the hot objects almost instantaneously decreased when the LF ended. As can be seen from Figure 3, an LF can therefore be recognized by the characteristic bell-shaped distribution of both the area and mean altitude time series. This suggests a criterion for identifying the time mark to be associated with the beginning and the end of the paroxysmal phenomenon, by using a Change Point Detection (CPD) algorithm, which will be explained in Section 3.2.

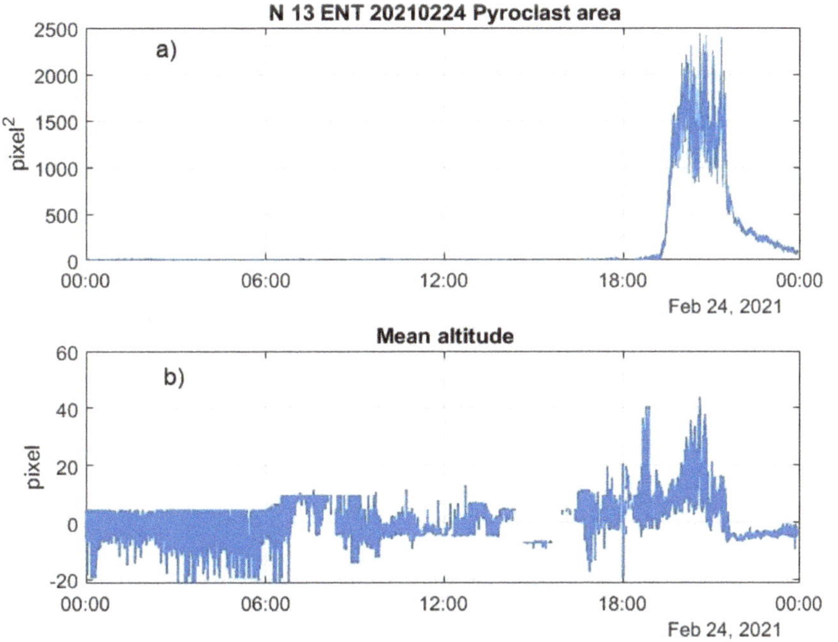

Figure 3. Area (**a**) and mean altitude (**b**) of the lava fountain episode at Mt Etna on 24 February 2021.

3. Results

3.1. Manual Estimation of the Eruptive Activity

Of the 66 LF episodes that occurred between 13 December 2020 and 21 February 2022, only one could not be detected because of poor weather conditions: the episode of 27 May 2021, when there was poor visibility from all the monitoring cameras. In all the other cases, visibility was more than 80% of the duration of the episode. Thus, for the short lapses of time when there was no visibility, the LF heights were linearly interpolated. The total volume of erupted pyroclastics during the 65 episodes was ~65 × 10^6 m^3. The average duration of the 65 episodes listed in Table 2 was of 120 min (minimum 6 min, maximum 803 min), or 7171 s (minimum 360 s, maximum 48,180 s); the average erupted volume of pyroclastics was 0.99 × 10^6 m^3, with 41 × 10^3 m^3 as minimum value and 4.6 × 10^6 m^3 as maximum value; the TADR, calculated for the only pyroclastic portion of the episodes and for the whole duration of each paroxysmal event, was 145 m^3 s^{-1} on average, spanning from a minimum of 59 m^3 s^{-1} and a maximum of 245 m^3 s^{-1}; maximum LF heights above the vents were 1815 m on average, spanning from a peak value of 5714 m to a minimum of 333 m, and the average heights of the LFs were between 115 m and 778 m, with a peak value of 2160. All values are listed in Table 2.

From a volcanological point of view, it is worth noting that if we exclude the two outliers of 23 and 31 March 2021, which emitted more than 4 × 10^6 m^3 of tephra, the volume of pyroclastics which erupted during the LF episodes occurring between mid-December 2020 and February 2022 increased with time. This can clearly be seen from Figure 4, where we report the distribution of the erupted volumes with time and its linear trend. In addition, the time-averaged discharge rate (TADR) and its trend increased with time (Figure 5). These observations, i.e., of the increase of erupted volume of pyroclastics and rate of eruption with time, suggested that the sequence of paroxysmal events was not yet over [31].

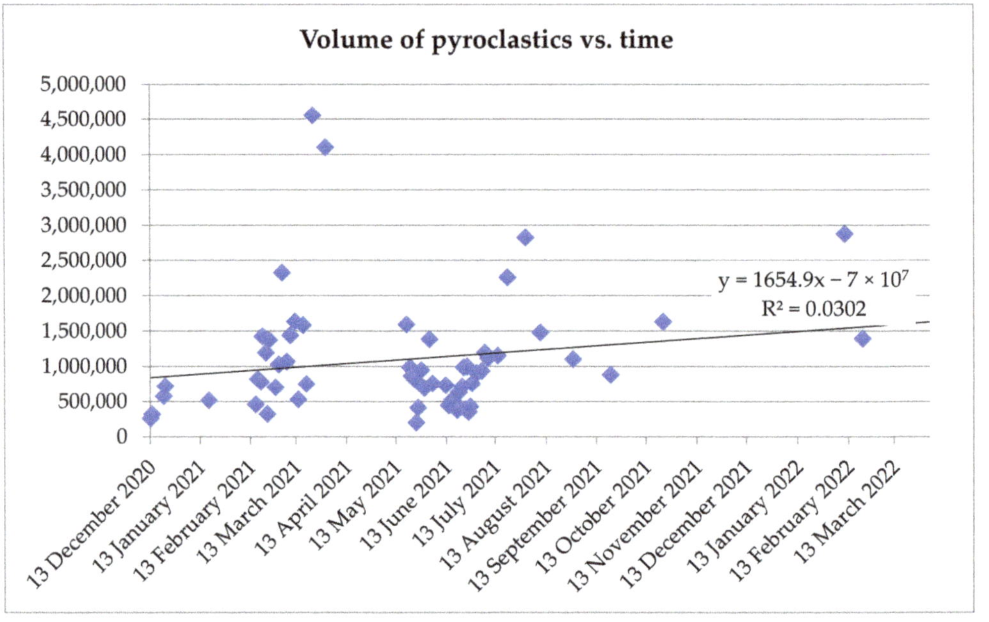

Figure 4. Volume of pyroclastics (*y*-axis, expressed in m^3) against time (*x*-axis, date) for the 65 explosive episodes listed in Table 2, together with a linear trend and its formula.

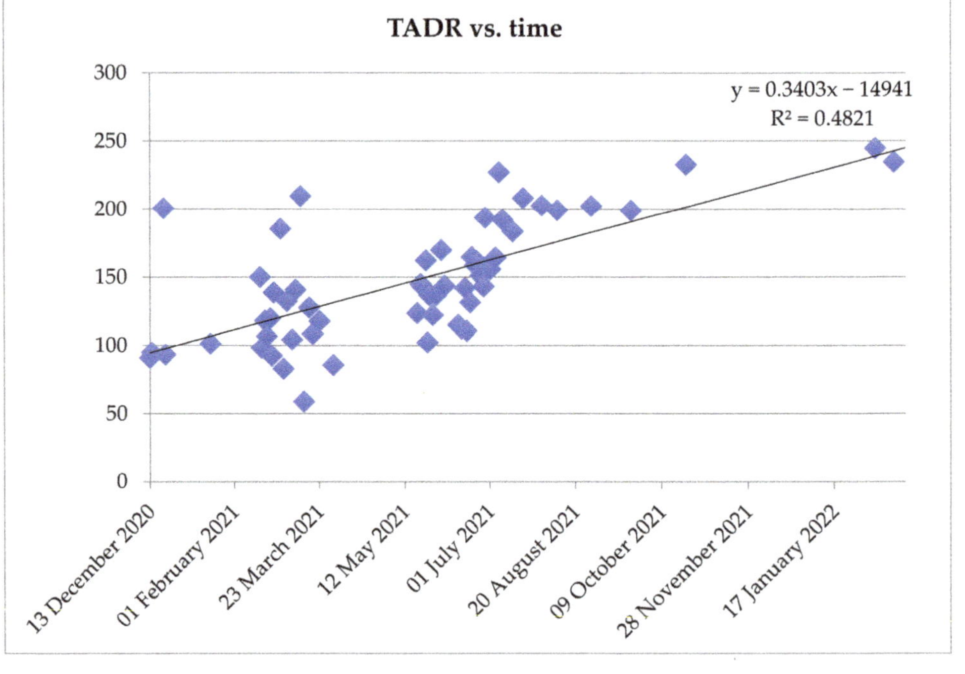

Figure 5. Time-averaged discharge rate of pyroclastics (TADR, *y*-axis, expressed in $m^3\ s^{-1}$) erupted during the lava fountain episodes against time (*x*-axis, date). The graph displays the data for the 65 explosive episodes listed in Table 2, together with a linear trend and its formula.

3.2. Change Point Detection

A change point represents a transition between different states in a process that generates the time series. Change point detection (CPD) can be defined as the problem of choosing between two alternatives: no change or indeed, a change occurred. CPD algorithms are traditionally classified as online or offline [54]. Offline algorithms consider the whole data set at once and try to recognize where the change occurred. Thus, the aim in this case, is to identify all the sequence change points in batch mode. In contrast, online, or in real-time, algorithms run concurrently with the process they are monitoring, processing each data point as it becomes available, with the goal of detecting a change point as soon as possible after it occurs, ideally before the next data point arrives. In practice, no CPD algorithm operates in perfect real-time because it must wait for new data before determining if a change point occurred. However, different online algorithms require different amounts of new data before a change point can be detected. Based on this observation, an online algorithm, which needs at least ε samples in the new batch of data to be able to find a change, is usually denoted as ε-real time. Therefore, offline algorithms can be viewed as ∞-real time whereas the best online algorithm is 1-real time, because for every data point, it can predict whether or not a change point occurs before the new data point. Smaller ε values may lead to stronger, prompter CPD algorithms. To find a change point in a time series, a global optimization approach can be used with the following basic algorithm:

1. Choose a point and divide the signal into two sections.
2. Compute an empirical estimate of the desired statistical property for each section.
3. At each point within a section, measure how much the property deviates from the empirical estimate, and at the end, add the deviation for all points.
4. Add the deviations section-to-section to find the total residual error.
5. Vary the location of the division point until the total residual error attains a minimum.

As noted above, the search for a change point k can be formulated as an optimization problem where the cost function $J(k)$ to minimize it can be written, in the general case as:

$$J(k) = \sum_{i=1}^{k-1} \Delta(x_i; \chi([x_1, \ldots, x_{k-1}])) + \sum_{i=k}^{N} \Delta(x_i; \chi([x_k, \ldots, x_N])) \quad (2)$$

where $\{x_1, x_2, \ldots x_N\}$ is the time series, χ is the chosen statistic, and Δ is the deviation measurement. In particular, when χ is the mean, the cost function assumes the following form:

$$J(k) = \sum_{i=1}^{k-1} \left(x_i - \langle x \rangle_1^{k-1}\right)^2 + \sum_{i=k}^{N} \left(x_i - \langle x \rangle_k^{N}\right)^2 \quad (3)$$

where the symbol $\langle \cdot \rangle$ indicates the mean operator.

Another aspect to be considered, when formulating the optimization problem, is that signals of practical interest have more than one change point. Furthermore, the number of change points K is often not known a priori. To handle these features, the cost function can be generalized as:

$$J(k) = \sum_{r}^{K-1} \sum_{i=k_r}^{k_{r+1}-1} \Delta(x_i; \chi([x_k, \ldots, x_{r+1}])) + \beta K \quad (4)$$

where k_0 and k_K are, respectively, the indexes of the first and the last sample of the signal. In the expression (4), the term βK is a penalty term, linearly increasing with the number of change points K, which avoids the problem of overfitting [55]. Here, β represents a positive coefficient that weights the number of searched change points. Indeed, in an extreme case (i.e., $\beta = 0$), $J(K)$ reaches the minimum value (i.e., 0) when every point becomes a change point (i.e., $K = N$).

The algorithm described above for a univariate time series, can easily be extended to the case of a multivariate time series, which was the case, for instance, of a data set recorded by a GPS network [54]. In this case, the cost function was evaluated, of course, over the whole set of available time series. The software considered in this work was implemented in MATLAB based on the CPD algorithms described in [55,56]. The package can help the user at various levels. The lowest level is to consider the software to obtain the time series of area A and mean altitude MA of detected hot objects from images of volcanic activity, following the algorithm described in the previous section. Subsequently, the user, based on a visual inspection of these time series, can indicate presumed times for the starting and ending times of the LF, and request the software to calculate other quantities of interest such as the volumes emitted. Another possibility is to leave the software to search for the transition times from Strombolian to paroxysmal activity using the CPD algorithm, possibly selecting the statistic to be used to perform the detection (i.e., abrupt changes in the mean, in the variance or in the slope). While the latter possibility is preferable when the time series shows clear LF episodes, the former is more suitable when the automated detection of the transition from Strombolian to paroxysmal activity and vice versa can be more problematic due to external noise sources (e.g., poor visibility or interference by other kinds of hot objects).

3.3. *Timing the Lava Fountains Occurring at Etna during 2020–2022*

The main advantages of performing a computer process analysis of LF images are the following:

- the user can quickly analyze the content of the image files recorded over days, an operation which, carried out manually, requires a considerable amount of time;
- the user can speed up the computation of key quantities such as height and duration of the LF, which are necessary for the calculation of the volumes of erupted material);
- it is possible to implement algorithms for automatically timing the transition from Strombolian to paroxysmal activity, which is otherwise left to human judgment, gaining in uniformity and repeatability;
- in case of lack of visibility, since it is necessary to proceed with interpolation of the data, a rather difficult operation to perform manually, the user can resort to automated interpolation techniques (e.g., linear interpolation, nearest, etc.).

However, it should not be overlooked that quantitative measurements of the LF parameters, starting from the images, require overcoming several non-negligible difficulties. First of all, the aforementioned lack of visibility can make the reliability of the measurements poor. In fact, interpolation techniques can help in cases of limited amounts of data, but obviously cannot replace them when significant amounts are missing.

Furthermore, since the LFs are recognized as hot objects, they can be confused with hot objects of other kinds. Hot areas are very often formed due to the sunlight reflection of both the ground (Figure 6a) and the clouds (Figure 6c). At other times, the Sun itself was included in the images as it travels its natural orbit (Figure 6d,e). Other kinds of hot objects, which could be confused with LFs, were cooling lava flows (Figure 6b). Moreover, different hot objects can combine their effects with those of the LFs (Figure 6e,f). Some of these effects could be avoided by using special cameras, but this is not always possible. Since the current state of development of the software does not allow a reliable distinction between the noise and the LF signal, the user, for the purpose of determining the start and end times, can limit the search space, so as to exclude particularly noisy periods.

Normally, the area signal is smoother than the mean altitude signal and therefore, usually when determining the start and end time of an LF using the CPD algorithm, it is preferable to consider the time series of the areas. However, in some cases, it may be useful to consider the series of mean altitude, as described in the following example. For instance, consider the area signal shown in Figure 7a, which refers to an LF occurring on Etna on 20 February 2021. As can be seen, due to the presence of a cooling lava flow, the area signal slowly decreases (thermal hysteresis), making it difficult to accurately perform the automated detection of the ending time of this LF episode. However, this shortcoming can be overcome by performing the CPD timing on the mean altitude signal, which is not affected by the thermal hysteresis (see Figure 7b).

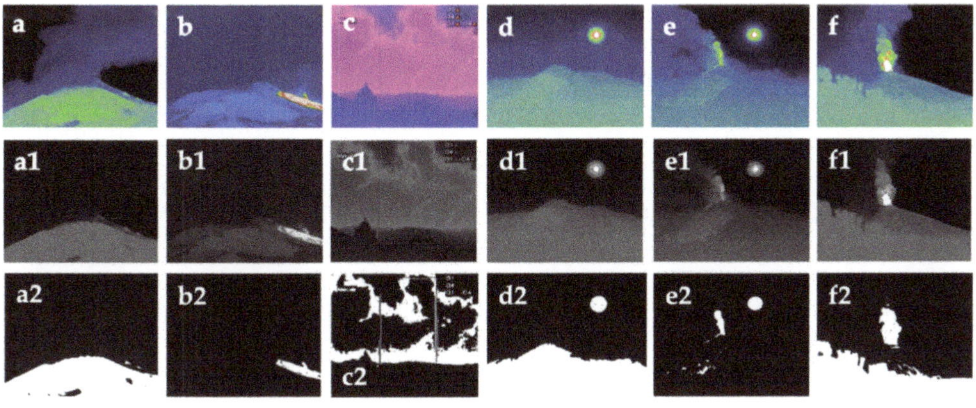

Figure 6. (**a**) Effect of the Sun reflection on the ground slope (green area), EMOT camera; (**b**) cooling lava flow (white area), EMOT camera; (**c**) Effect of the Sun on the clouds (pink area), ENT camera; (**d**) the Sun in the camera field of view (white circle), EMCT camera; (**e**) Combined effects: lava fountain in presence of the Sun in the field of view, EMCT camera; (**f**) lava fountain and Sun reflection on the vegetation in the foreground, EBT camera. (**a1**–**f1**) are the corresponding gray images, and (**a2**–**f2**) are the corresponding black and white binarized images.

Here, the term 'timing' will be used to indicate the estimation of the starting and ending time of an LF episode. In particular, for this LF, while performing the CPD timing from the area signal (Figure 8a), the end of the LF is estimated to be t_{end} = 06:38, whereas considering the mean altitude signal, the end of the LF is estimated t_{end} = 00:56, which is closer to that manually estimated (t_{end} = 01:15) and reported in Table 2. This difference was caused by the greater curvature that can be seen in Figure 8a, which was determined by the cooling lava flow area.

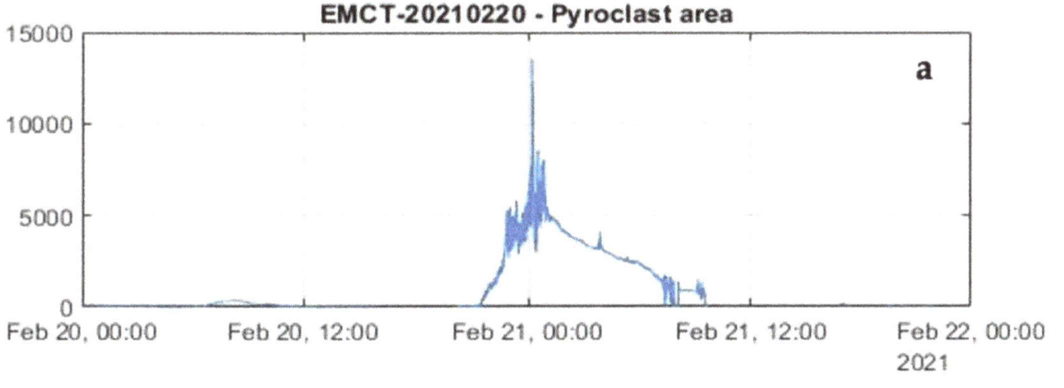

Figure 7. *Cont.*

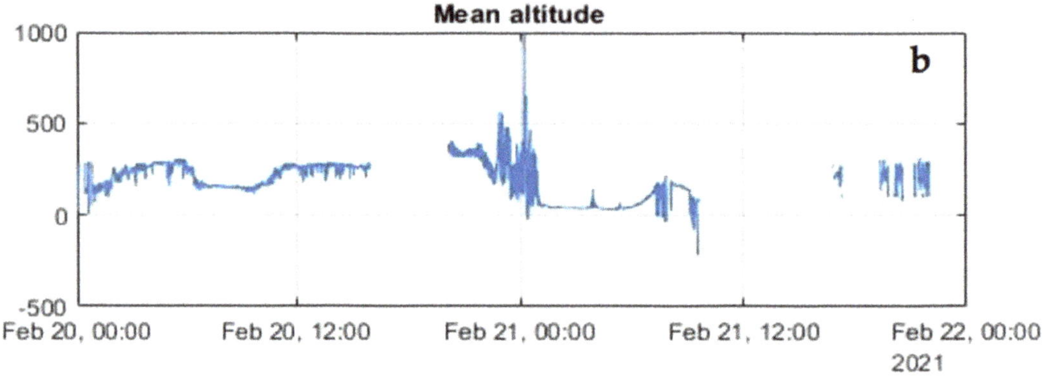

Figure 7. (**a**) Area (in pixel^2) and (**b**) mean altitude (in pixels above the crater rim) against time of the lava fountain on 20–21 February 2021 at Mt. Etna, retrieved from the EMCT camera.

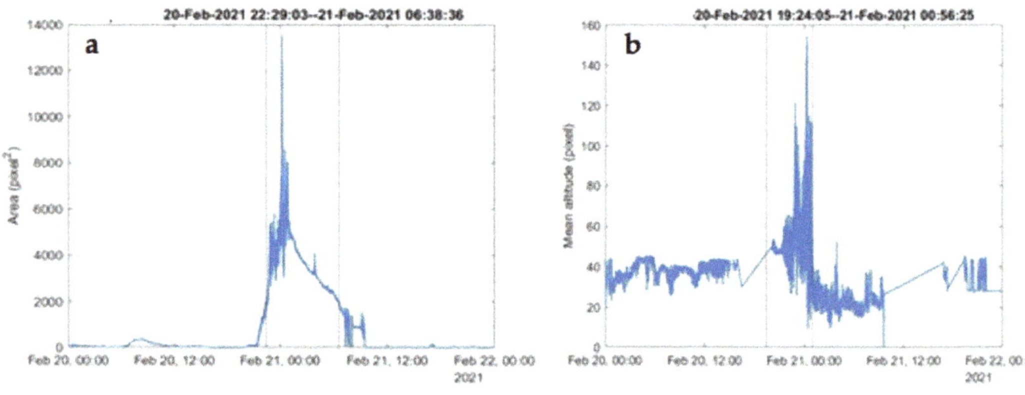

Figure 8. (**a**) Area (in pixel^2) and (**b**) mean altitude above the crater rim (in pixels, measured above the crater rim) with the timing (gray vertical lines) of start and end of the lava fountain episode. See text for explanation.

3.4. Timing the Lava Fountains by a Gaussian Function-Based Approach

In some cases, assuming that a typical LF has a time distribution of area and mean altitude, which roughly has a bell shape, it might be useful to approximate the measured data by using a Gaussian function. This can be useful, for example, when the images are affected by thick clouds passing through the field of view of the camera, generating a trend such as those shown in Figure 9a, which refers to the LF episode on 13 December 2020 and was observed from the ENT camera.

For this LF episode, the manually estimated starting and ending times were 22:00 and 22:48, respectively. However, from Figure 9a, it can be seen that precisely between these two times, the recorded signal is discontinuous due to poor visibility, but nevertheless, it can be seen that the area signal shows a well-detectable peak. Fitting the area samples, it is possible to obtain the results shown in Figure 9b and thus estimate the start and end dates of the LF to be 22:02 and 22:20, respectively, which are closer to the manually assessed corresponding times.

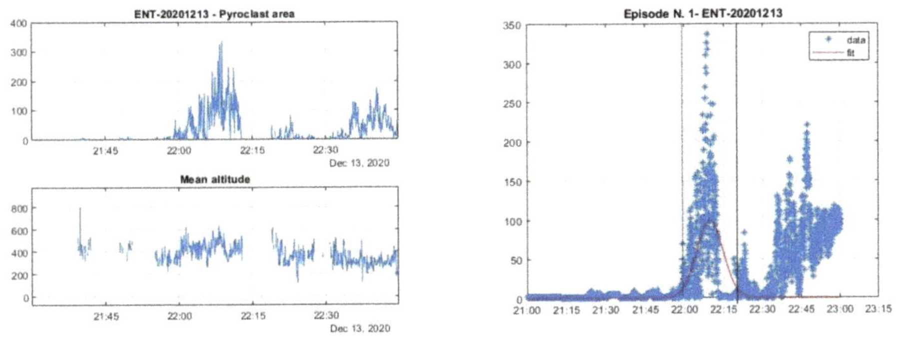

(**a**) (**b**)

Figure 9. (**a**) Area (in pixel^2) and mean altitude (in pixels, measured above the crater rim) against time of the 13 December 2020 lava fountain episode as retrieved from the ENT camera, and (**b**) Gaussian interpolation (red line) of the pyroclastic area against time of the 13 December 2020 lava fountain episode retrieved from the ENT camera.

Timing of an LF episode, after having carried out the approximation of the curve by means of a Gaussian function, is simply established by using a threshold approach: the starting time is set as the one in which the recorded data exceed, for the first time, a threshold of the normalized function height. Similarly, the end time is established as the one in which the recorded signal falls, for the first time, below the threshold. In this paper, the threshold value has been set, after a trial and error approach, to be 25% of the maximum value.

Of course, the Gaussian approach can be used for timing LF as an alternative to the CPD one, even when the visibility problems described above do not exist, as shown in the example of Figure 10.

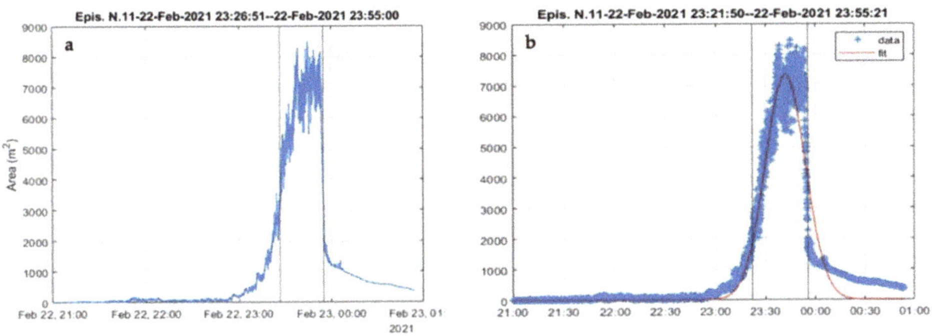

Figure 10. Graphs showing the timing of the LF episode N. 11, which occurred on 22–23 February 2021 (Table 2) obtained from (**a**) the CPD method and from (**b**) the Gaussian method. The black vertical lines indicate the start and end time, and the red line in (**b**) is the Gaussian interpolation. Table 2 shows that the manual method indicated the LF episode took place between 21:17 on 22 February and 00:03 of 23 February 2021. In accordance with the manual method, (**a**) shows that the CPD automated timing indicated that the sustained phase of the LF developed between 23:26 and 23:55 on 22 February, and similarly, the Gaussian method (**b**) indicated a timing comprised between 23:21 and 23:55 on 22 February.

The software package presented in the previous sections was considered to perform the timing of the 65 LF episodes in the data set reported in Table 2. Moreover, for each episode, the heights of the LF at 1 min intervals were used to calculate the total fluid erupted volume from Equations (5) and (6), which included both gas and pyroclastics [8,31]:

$$U = (2gH)^{0.5} \tag{5}$$

$$V = U \cdot A_V \cdot D \tag{6}$$

In expression (5), U is the mean fluid exit velocity at the vent, H is the LF height, and g is the gravity acceleration, while in expression (6), V is the fluid volume (gas + pyroclastic) erupted by the LF, A_V is the vent section area, and D is the duration of the LF in seconds. The vent surface area was calculated by assuming a circular vent with a diameter of 30 m [8] and supposed to be constant. Moreover, we have computed the volume V_2 of pyroclastics from the total erupted volume V (gas + pyroclastics), considering 0.18% as the ratio between the volumes of magma and fluids within the eruptive column as typical for Etna's fountains [3].

According to expression (6), the estimated volumes depend on the mean fluid exit velocity U and LF duration D, for assigned values of the vent surface area. Thus, the performances of the automated approach will depend on its readability to estimate height H and duration D of the LF episodes. Concerning the reliability of the automated estimating H, the comparison with the corresponding manual reading, for a few episodes of the data set, is shown in Figure 11.

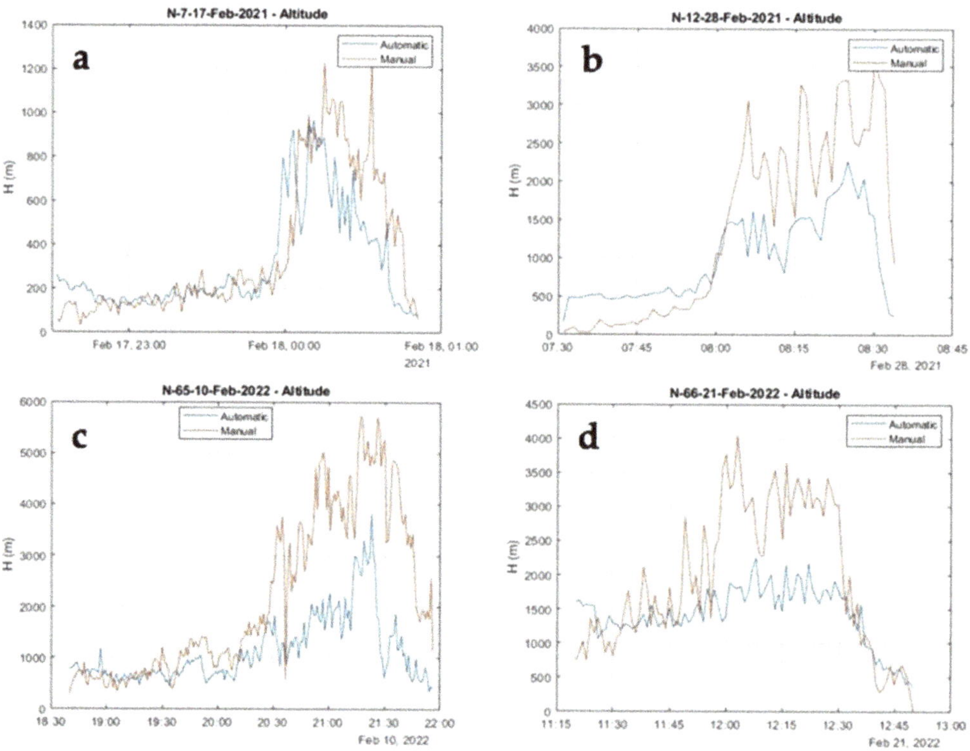

Figure 11. Time series of heights estimated by using the automated and the manual approaches for a few episodes of the data set against time, with the blue line for the automated, and the orange line for the manual detection: (**a**) Episode 7, 17–18 February 2021; (**b**) Episode 12, 28 February 2021; (**c**) Episode 65, 10 February 2022, and (**d**) Episode 66, 21 February 2022.

It can be seen in Figure 11 that manual height readings normally have a greater range than those automatically estimated. Here, it should be borne in mind that, as expressed in Equation (1), the automatically measured heights are a weighted average of the centroids, while those measured manually are normally taken as the maximum height of the lava fountain jet taken along the spreading direction. Considering that the heights represent the only geometric element on which the volumes of erupted material depend, it follows that with the automated estimation, these will normally be slightly less when compared to the manual ones, but with the advantage of immediacy. It is also necessary to bear in mind that for the purposes of estimating the volume of erupted matters, it is not so much the precise values of the instantaneous heights that are relevant, but their average value, which therefore also depends on the estimated duration for each individual episode. The comparison among the mean heights for the whole LF episodes after estimating the duration D by using both the CPD and Gaussian approaches are shown in Figure 12. In this figure, the abscissa is the integer N ranging from 1 to 66, i.e., number of LF episodes in the considered dataset (Table 2). The episode 32 is lacking because of poor visibility from all the monitoring cameras (Table 2).

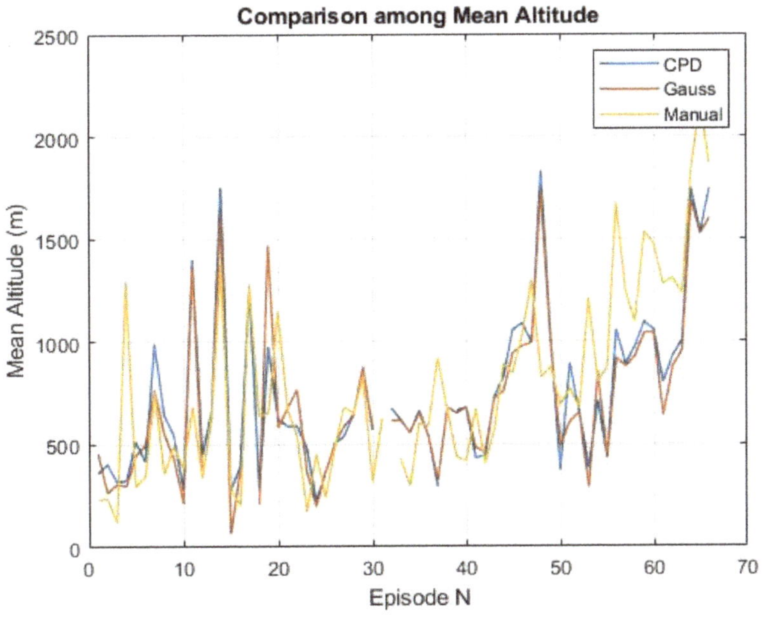

Figure 12. Mean altitude for each of the 65 lava fountain episodes estimated after having established the timing of start and end for each recorded time series by using both the CPD (blue line) and the Gaussian (red line) approach. Their values were compared with the corresponding manually estimated values (yellow line). Episode 32 was lacking because of poor visibility from the cameras (see Table 2).

In more detail, Figure 12 shows that for most of the episodes, there is a good agreement between the average mean heights estimated for each episode, not only between those obtained by using the CPD and Gaussian approaches, but also between these and those manually estimated. Altitudes obtained with the manual approach are, on average, 12% lower than those obtained by using the CPD approach, and 7% lower than the Gaussian one.

As regards to the duration D for each episode, the comparison between the automated and the manual estimation is shown in Figure 13.

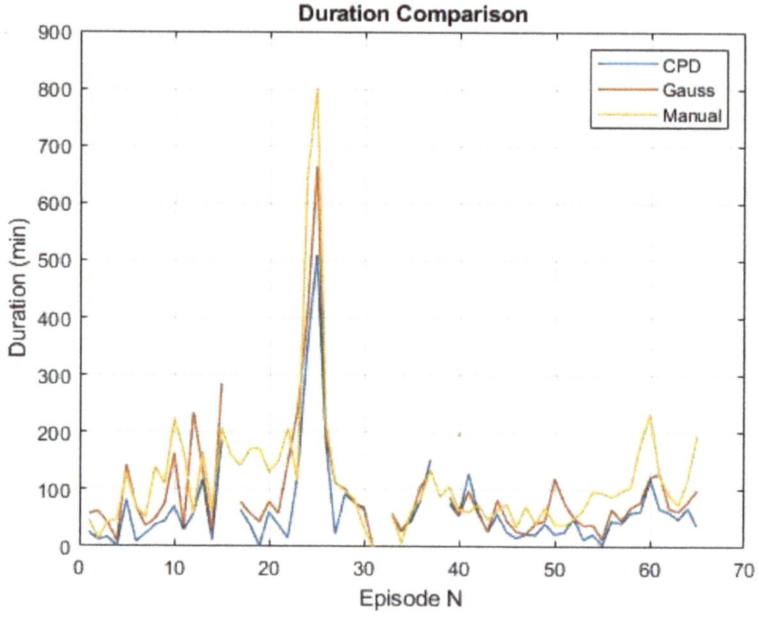

Figure 13. Estimated duration performed after timing the LF episodes by the CPD (blue line) and Gaussian (red line) approaches, in comparison with the manual (black line) ones.

Figure 13 shows that the durations obtained by using both the methods for automated timing are generally in good agreement with each other, as well as lower than those obtained manually. In more detail, duration manually estimated was on average about 34% higher, with a standard deviation of 77%, than those estimated by using the CPD, and about 3% higher, with a standard deviation of 93%, than the Gaussian one. To justify the discrepancy, it is worth noting that the automated approaches generally identify the sustained part of each LF, while the manual approach is not able to clearly distinguish the threshold of the intermediate phase preceding the sustained portion of the LF [3,8].

The comparison between estimated volumes and TADR by the three methods are reported in Figure 14a,b. A good agreement between the automated and manual estimation is apparent, bearing in mind that the automated volumes are usually smaller than the manual ones, because the durations refer to the sustained phase of the LF, while the manual and automated TADR are in good agreement because this feature is computed as the ratio between volumes and duration of each LF episode. In more detail, manual estimated volumes are, on average, about 26% higher, with a standard deviation of about 77% than those estimated by using the CPD approach, and 13% lower than the Gaussian, with a standard deviation of about 107%. The TADR manually estimated is about 15% lower that the CPD, with a standard deviation of 77% and 8% lower than the Gaussian, with a standard deviation of 33%.

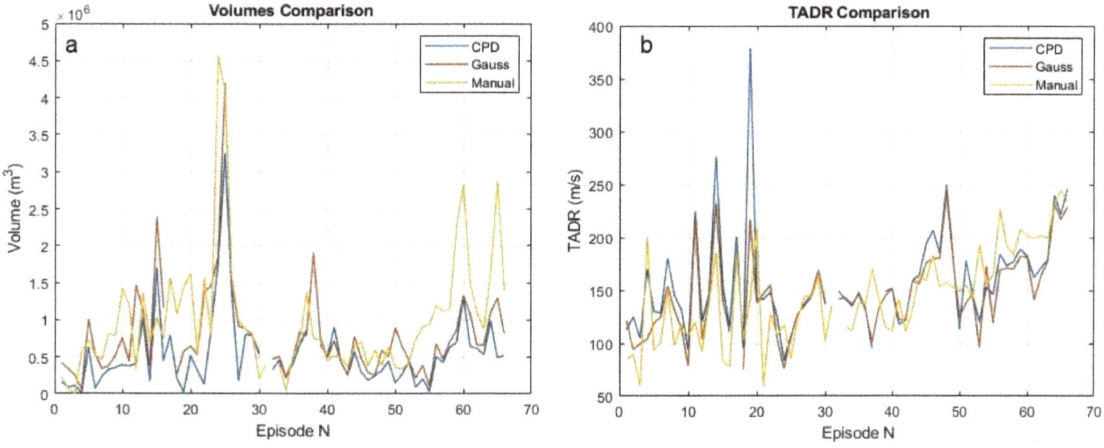

Figure 14. (**a**) Comparison among volumes, and (**b**) comparison among TADR obtained from the manual and automated approaches.

Figure 15 shows the differences between the values obtained by the manual and automated routines, and indicates the good agreement between heights of the lava fountain (Figure 15a) and TADR (Figure 15d), and the discrepancies between duration (Figure 15b) and calculated erupted volume of pyroclastics (Figure 15c).

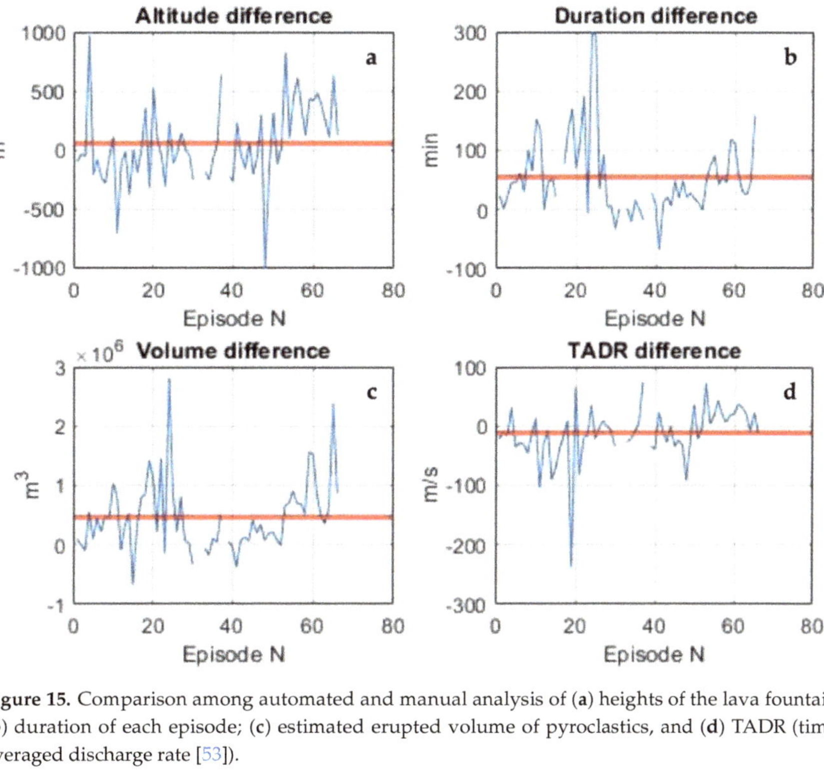

Figure 15. Comparison among automated and manual analysis of (**a**) heights of the lava fountain; (**b**) duration of each episode; (**c**) estimated erupted volume of pyroclastics, and (**d**) TADR (time-averaged discharge rate [53]).

Considering the results obtained from the automated routines, the total volume of erupted pyroclastics during the 65 episodes computed by using the automated CPD approach was ~34.5 × 10^6 m^3, with a minimum of ~0.012 × 10^6 m^3 and a maximum of ~3.25 × 10^6 m^3. The average duration of the sustained part of LF was of ~65.7 min (minimum 1 min, maximum 509 min). The TADR, calculated only for the pyroclastic portion of the episodes and for the whole duration of each paroxysmal event, was 159 m^3 s^{-1} on average, spanning from a minimum of 84 m^3 s^{-1} and a maximum of 379 m^3 s^{-1}; LF average height above the vents was 745 m, spanning from a peak value of 1834 m to a minimum of 223 m.

4. Discussion

In this paper, we have presented an automated routine that might help volcanic observatories such as INGV-OE to detect (1) the starting and ending time of an LF episode, (2) the LF heights, (3) the erupted volumes, and (4) the TADR, saving time and especially providing consistency and uniform data extraction from thermal monitoring videos. This would allow a prompt understanding of the state of the volcano, and of the magnitude and intensity of each explosive paroxysm as soon as it ended, allowing a timely volcanic hazard assessment. In addition, both our automated routines, based on the CPD and Gaussian interpolation, proved to be reliable in constraining the climax phase of the paroxysm leading to a sustained eruptive column, which is the phase posing the greater hazard for its impact on aviation and the population. Conversely, the manual analysis had clear difficulties in distinguishing the threshold between the intermediate phase and the LF sustained phase [3,8,30].

However, in order to routinely use the algorithms proposed here, it is necessary to overcome some limits that we described earlier and illustrated in Figure 6. The first shortcoming arose from the detection of unwanted objects falling in the field of view of the eruption, such as the Sun or the surfaces it irradiated. In distinguishing this anomalous pattern, the Gaussian interpolation might help, which would reveal and remove any deviation from the normal trend. A more common problem, and one that is hard to handle, is the cloud interference, with clouds resulting from water droplets, ash or gases filtering or obscuring the thermal images [36–40]. In the cases of clouds partially obscuring the field of view, it was still possible to interpolate the missing data, provided that they represent a small percentage of the total duration of the episode, which was the procedure also carried out with manual analysis. However, when the cloud cover was too continuous and extended in many directions, such as the episode of 27 May 2021 (Table 2), there was no way to retrieve any useful data, and an estimation of the erupted volume can only be performed by considering the timing obtained from the seismicity or from borehole strainmeters [17,43–45,47,48], and multiplying this for the average TADR estimated for each single episode occurring during the whole period lasting from 13 December 2020 to 21 February 2022. Thus, considering for the episode # 32 of 27 May 2021, the duration of 60 min (=3600 s) estimated from the seismicity, and multiplying this time for the TADR averaged over the 65 episodes (Table 2; ~146 m^3 s^{-1}), we obtained an estimated volume of ~0.53 × 10^6 m^3, which was in line with, and slightly below, the average of the other LF events here considered (Table 2). This brought the total erupted volume of pyroclastics or tephra, erupted between 13 December 2020 and 21 February 2022, to ~65.2 × 10^6 m^3.

Considering the manually obtained results, from a volcanological point of view, it is worth noting that, if we excluded the two outliers of 23 and 31 March 2021, which emitted more than 4 × 10^6 m^3 of tephra (Table 2), the volume of pyroclastics erupted during the LF episodes, which occurred between mid-December 2020 and February 2022, increased with time (Figure 4), and also, the time-averaged discharge rate (TADR) increased with time (Figure 5). Figures 4 and 5 display a wide variability of values, and although this variability might hide shorter eruptive cycles, it is however clear that the trend of TADR and erupted volume increased with time. These observations, i.e., of the increase of erupted volume of pyroclastics and rate of eruption with time, would suggest that the sequence of paroxysmal events was not yet over [31], and urges reliable and automated routines to be promptly developed, tested, and applied to the analysis of the LF episodes.

5. Conclusions

In this paper, we have presented the timing of start and end for 65 of the 66 LF episodes which took place at Etna volcano between 13 December 2020 and 21 February 2022, together with their duration, maximum and average LF heights, erupted volume of pyroclastics, and TADR (Table 2), obtained by manual analysis of the monitoring thermal images recorded by the INGV-OE network. We have then presented two automated routines, based on the CPD and Gaussian interpolation, that analyzed the thermal images and provided a fast and reliable way to obtain the same parameters acquired manually, in a timely way. The results obtained with the automated and manual routines are comparable (Figure 15), thus suggesting that a complete automation of the process is feasible. However, our analysis also highlighted important shortcomings arising from the presence of unwanted hot objects comprised in the field of view of the explosive episode that may lead to false results. Moreover, the presence of ash, weather, and gas clouds caused important interference with the data analysis, and might have reached the point of a complete obscuration of the field of view, as in the case of the episode #32 of 27 May 2021 (Table 2). We have shown that the Gaussian interpolation may limit the errors caused by a partial view, but more studies are necessary before this analysis can be routinely used for monitoring purposes. The results of our study showed an increasing magnitude (erupted volume) and intensity (TADR) of the explosive events in the period here considered (see Figures 4 and 5), and this issue would urge a faster and reliable analysis to be obtained as soon as possible, thus motivating the work presented here.

Author Contributions: Conceptualization, S.C. and G.N.; methodology, S.C. and G.N.; software, G.N.; validation, S.C. and G.N.; formal analysis, S.C. and G.N.; investigation, S.C. and G.N.; resources, S.C. and G.N.; data curation, S.C. and G.N.; writing—original draft preparation, S.C. and G.N.; writing—review and editing, S.C. and G.N.; visualization, S.C. and G.N.; supervision, S.C. and G.N.; project administration, S.C.; funding acquisition, S.C. All authors have read and agreed to the published version of the manuscript.

Funding: This research was funded by the Project FIRST—ForecastIng eRuptive activity at Stromboli volcano: timing, eruptive style, size, intensity, and duration, INGV-Progetto Strategico Dipartimento Vulcani 2019 (Delibera n. 144/2020; Scientific Responsibility: S.C.).

Data Availability Statement: The videos of eruptive activity used in this paper belong to the Istituto Nazionale di Geofisica e Vulcanologia, Osservatorio Etneo – Sezione di Catania and are used for monitoring purposes. Selected videos can be made available upon request to the first author of this paper.

Acknowledgments: We would like to thank the INGV-OE scientists and technicians for the monitoring network maintenance, and especially Michele Prestifilippo for providing information essential for this work. We thank Stephan Conway for revising the English style.

Conflicts of Interest: The authors declare no conflict of interest.

References

1. Barreca, G.; Branca, S.; Corsaro, R.A.; Scarfì, L.; Cannavò, F.; Aloisi, M.; Monaco, C.; Faccenna, C. Slab detachment, mantle flow, and crustal collision in eastern Sicily (southern Italy): Implications on Mount Etna volcanism. *Tectonics* **2020**, *39*, e2020TC006188. [CrossRef]
2. Bonaccorso, A.; Calvari, S. Major effusive eruptions and recent lava fountains: Balance between expected and erupted magma volumes at Etna volcano. *Geophys. Res. Lett.* **2013**, *40*, 6069–6073. [CrossRef]
3. Calvari, S.; Cannavò, F.; Bonaccorso, A.; Spampinato, L.; Pellegrino, A.G. Paroxysmal Explosions, Lava Fountains and Ash Plumes at Etna Volcano: Eruptive Processes and Hazard Implications. *Front. Earth Sci.* **2018**, *6*, 107. [CrossRef]
4. Calvari, S.; Bilotta, G.; Bonaccorso, A.; Caltabiano, T.; Cappello, A.; Corradino, C.; Del Negro, C.; Ganci, G.; Neri, M.; Pecora, E.; et al. The VEI 2 Christmas 2018 Etna Eruption: A Small but Intense Eruptive Event or the Starting Phase of a Larger One? *Remote Sens.* **2020**, *12*, 905. [CrossRef]

5. Pailot-Bonnétat, S.; Harris, A.J.L.; Calvari, S.; De Michele, M.; Gurioli, L. Plume Height Time-Series Retrieval Using Shadow in Single Spatial Resolution Satellite Images. *Remote Sens.* **2020**, *12*, 3951. [CrossRef]
6. Ganci, G.; Harris, A.J.L.; Del Negro, C.; Guéhenneux, Y.; Cappello, A.; Labazuy, P.; Calvari, S.; Gouhier, M. A year of lava fountaining at Etna: Volumes from SEVIRI. *Geophys. Res. Lett.* **2012**, *39*, L06305. [CrossRef]
7. Ganci, G.; James, M.R.; Calvari, S.; Del Negro, C. Separating the thermal fingerprints of lava flows and simultaneous lava fountaining using ground-based thermal camera and SEVIRI measurements. *Geophys. Res. Lett.* **2013**, *40*, 5058–5063. [CrossRef]
8. Calvari, S.; Salerno, G.G.; Spampinato, L.; Gouhier, M.; La Spina, A.; Pecora, E.; Harris, A.J.L.; Labazuy, P.; Biale, E.; Boschi, E. An unloading foam model to constrain Etna's 11-13 January 2011 lava fountaining episode. *J. Geophys. Res.* **2011**, *116*, B11207. [CrossRef]
9. De Beni, E.; Behncke, B.; Branca, S.; Nicolosi, I.; Carluccio, R.; Caracciolo D'Ajello, F.; Chiappini, M. The continuing story of Etna's New Southeast Crater (2012–2014): Evolution and volume calculations based on field surveys and aerophotogrammetry. *J. Volcanol. Geotherm. Res.* **2015**, *303*, 175–186. [CrossRef]
10. Andronico, D.; Cannata, A.; Di Grazia, G.; Ferrari, F. The 1986–2021 paroxysmal episodes at the summit craters of Mt. Etna: Insights into volcano dynamics and hazard. *Earth-Sci. Rev.* **2021**, *220*, 103686. [CrossRef]
11. Freret-Lorgeril, V.; Donnadieu, F.; Scollo, S.; Provost, A.; Fréville, P.; Guéhenneux, Y.; Hervier, C.; Prestifilippo, M.; Coltelli, M. Mass eruption rates of tephra plumes during the 2011–2015 lava fountain paroxysms at Mt. Etna from doppler radar retrievals. *Front. Earth Sci.* **2018**, *6*, 73. [CrossRef]
12. Corradini, S.; Guerrieri, L.; Lombardo, V.; Merucci, L.; Musacchio, M.; Prestifilippo, M.; Scollo, S.; Silvestri, M.; Spata, G.; Stelitano, D. Proximal Monitoring of the 2011–2015 Etna Lava Fountains Using MSG-SEVIRI Data. *Geosciences* **2018**, *8*, 140. [CrossRef]
13. Neri, M.; Casu, F.; Acocella, V.; Solaro, G.; Pepe, S.; Berardino, P.; Sansosti, E.; Caltabiano, T.; Lundgren, P.; Lanari, R. Deformation and eruptions at Mt. Etna (Italy): A lesson from 15 years of observations. *Geophys. Res. Lett.* **2009**, *36*, L02309. [CrossRef]
14. Guffanti, M.; Casadevall, T.J.; Budding, K. *Encounters of Aircraft with Volcanic Ash Clouds: A Compilation of Known Incidents, 1953–2009*; Version 1.0, Series 545, U.S. Geological Survey Data; U.S. Geological Survey: Reston, VA, USA, 2010; 12p., plus 4 appendixes including the compilation Database. Available online: http://pubs.usgs.gov/ds/545 (accessed on 22 March 2022).
15. Small, C.; Naumann, T. The global distribution of human population and recent volcanism. *Glob. Environ. Chang. Part B Environ. Hazards* **2001**, *3*, 93–109. [CrossRef]
16. Horwell, C.J.; Baxter, P.J. The respiratory health hazards of volcanic ash: A review for volcanic risk mitigation. *Bull. Volcanol.* **2006**, *69*, 1–24. [CrossRef]
17. Pardini, F.; Corradini, S.; Costa, A.; Esposti Ongaro, T.; Merucci, L.; Neri, A.; Stelitano, D.; de' Michieli Vitturi, M. Ensemble-Based Data Assimilation of Volcanic Ash Clouds from Satellite Observations: Application to the 24 December 2018 Mt. Etna Explosive Eruption. *Atmosphere* **2020**, *11*, 359. [CrossRef]
18. Alparone, S.; Andronico, D.; Lodato, L.; Sgroi, T. Relationship between tremor and volcanic activity during the Southeast Crater eruption on Mount Etna in early 2000. *J. Geophys. Res.* **2003**, *108*, 2241. [CrossRef]
19. Alparone, A.; Andronico, D.; Sgroi, T.; Ferrari, F.; Lodato, L.; Reitano, D. Alert system to mitigate tephra fallout hazards at Mt. Etna Volcano, Italy. *Nat. Hazards* **2007**, *43*, 333–350. [CrossRef]
20. Cannavò, F.; Aranzulla, M.; Scollo, S.; Puglisi, G. A New GNSS-Based Approach for Volcanic Crater Location During Lava Fountains. *IEEE Geosci. Remote Sens. Lett.* **2019**, *16*, 697–701. [CrossRef]
21. Cannavò, F.; Cannata, A.; Cassisi, C.; Di Grazia, G.; Montalto, P.; Prestifilippo, M.; Privitera, E.; Coltelli, M.; Gambino, S. A multivariate probabilistic graphical model for real-time volcano monitoring on Mount Etna. *J. Geophys. Res.* **2017**, *122*, 3480–3496. [CrossRef]
22. Scollo, S.; Prestifilippo, M.; Pecora, E.; Corradini, S.; Merucci, L.; Spata, G.; Coltelli, M. Eruption column height estimation of the 2011–2013 Etna lava fountains. *Ann. Geophys.* **2014**, *57*, S0214. [CrossRef]
23. Corradino, C.; Ganci, G.; Cappello, A.; Bilotta, G.; Calvari, S.; Del Negro, C. Recognizing Eruptions of Mount Etna through Machine Learning using Multiperspective Infrared Images. *Remote Sens.* **2020**, *12*, 970. [CrossRef]
24. Walker, G.P.L. The Taupo pumice: Product of the most powerful known (Ultraplinian) eruption? *J. Volcanol. Geother. Res.* **1980**, *8*, 69–94. [CrossRef]
25. Del Negro, C.; Cappello, A.; Ganci, G. Quantifying lava flow hazards in response to effusive eruption. *Geol. Soc. Am. Bull.* **2016**, *128*, 752–763. [CrossRef]
26. Del Negro, C.; Cappello, A.; Bilotta, G.; Ganci, G.; Hérault, A.; Zago, V. Living at the edge of an active volcano: Risk from lava flows on Mt. Etna. *Geol. Soc. Am. Bull.* **2020**, *132*, 1615–1625. [CrossRef]
27. Scollo, S.; Prestifilippo, M.; Spata, G.; D'Agostino, M.; Coltelli, M. Monitoring and forecasting Etna volcanic plumes. *Nat. Hazards Earth Syst. Sci.* **2009**, *9*, 1573–1585. Available online: http://www.nat-hazards-earth-syst-sci.net/9/1573/2009/ (accessed on 22 March 2022). [CrossRef]
28. Scollo, S.; Prestifilippo, M.; Bonadonna, C.; Cioni, R.; Corradini, S.; Degruyter, W.; Rossi, E.; Silvestri, M.; Biale, E.; Carparelli, G.; et al. Near-Real-Time Tephra Fallout Assessment at Mt. Etna, Italy. *Remote Sens.* **2019**, *11*, 2987. [CrossRef]
29. Scollo, S.; Boselli, A.; Corradini, S.; Leto, G.; Guerrieri, L.; Merucci, L.; Prestifilippo, M.; Zanmar Sanchez, R.; Sannino, A.; Stelitano, D. Multi-Sensor Analysis of a Weak and Long-Lasting Volcanic Plume Emission. *Remote Sens.* **2020**, *12*, 3866. [CrossRef]
30. Calvari, S.; Bonaccorso, A.; Ganci, G. Anatomy of a Paroxysmal Lava Fountain at Etna Volcano: The Case of the 12 March, 2021, Episode. *Remote Sens.* **2021**, *13*, 3052. [CrossRef]

31. Bonaccorso, A.; Calvari, S. A new approach to investigate an eruptive paroxysmal sequence using camera and strainmeter networks: Lessons from the 3–5 December2015 activity at Etna volcano. *Earth Planet. Sci. Lett.* **2017**, *475*, 231–241. [CrossRef]
32. Sparks, R.S.J.; Bursik, M.I.; Carey, S.N.; Gilbert, J.S.; Glaze, L.S.; Sigurdsson, H.; Woods, A.W. *Volcanic Plumes*; John Wiley: New York, NY, USA, 1997; 574p.
33. Bonadonna, C.; Phillips, J.C. Sedimentation from strong volcanic plumes. *J. Geophys. Res.* **2003**, *108*, 2340. [CrossRef]
34. Orr, T.R.; Ulrich, G.E.; Heliker, C.; De Smither, L.G.; Hoffmann, J.P. *The Pu'u 'Ō'ō Eruption of Kīlauea Volcano, Hawai'i—Episode 21 Through Early Episode 48, June 1984–April 1987*; U.S. Geological Survey Scientific Investigations Report; United States Geological Survey: Reston, VA, USA, 2018; Volume 5109, 107p. [CrossRef]
35. Carbone, D.; Zuccarello, L.; Messina, A.; Scollo, S.; Rymer, H. Balancing bulk gas accumulation and gas output before and during lava fountaining episodes at Mt. Etna. *Sci. Rep.* **2015**, *5*, 18049. [CrossRef] [PubMed]
36. Andò, B.; Pecora, E. An advanced video-based system for monitoring active volcanoes. *Comput. Geosci.* **2006**, *32*, 8591. [CrossRef]
37. Ball, M.; Pinkerton, H. Factors affecting the accuracy of thermal imaging cameras in volcanology. *J. Geophys. Res.* **2006**, *111*, B11203. [CrossRef]
38. Sawyer, G.M.; Burton, M.R. Effects of a volcanic plume on thermal imaging data. *Geophys. Res. Lett.* **2006**, 33, L14311. [CrossRef]
39. Spampinato, L.; Calvari, S.; Oppenheimer, C.; Boschi, E. Volcano surveillance using infrared cameras. *Earth Sci. Rev.* **2011**, *106*, 63–91. [CrossRef]
40. Harris, A.J.L. *Thermal Remote Sensing of Active Volcanoes*; Cambridge University Press: Cambridge, UK, 2013; 717p.
41. Sheldrake, T.E.; Sparks, R.S.J.; Cashman, K.V.; Wadge, G.; Aspinall, W.P. Similarities and differences in the historical records of lava dome-building volcanoes: Implications for understanding magmatic processes and eruption forecasting. *Earth-Sci. Rev.* **2016**, *160*, 240–263. [CrossRef]
42. Kauahikaua, J.; Mangan, M.; Heliker, C.; Mattox, T. A quantitative look at the demise of a basaltic vent: The death of Kupaianaha, Kilauea Volcano, Hawai'i. *Bull. Volcanol.* **1996**, *57*, 641–648. [CrossRef]
43. Cannata, A.; Montalto, P.; Patané, D. Joint analysis of infrasound and seismic signals by cross wavelet Atmospheric transform: Detection of Mt. Etna explosive activity. *Nat. Hazards Earth Syst. Sci.* **2013**, *13*, 1669–1677. [CrossRef]
44. Cannata, A.; Montalto, P.; Privitera, E.; Russo, G.; Gresta, S. Tracking eruptive phenomena by infrasound: May 13, 2008 eruption at Mt. Etna. *Geophys. Res. Lett.* **2009**, *36*, L05304. [CrossRef]
45. Nunnari, G. Clustering activity at Mt Etna based on volcanic tremor: A case study. *Earth Sci. Inform.* **2021**, *14*, 1121–1143. [CrossRef]
46. Moschella, S.; Cannata, A.; Di Grazia, G.; Gresta, S. Insights into lava fountain eruptions at Mt. Etna by improved source location of the volcanic tremor. *Ann. Geophys.* **2018**, *61*, VO446. [CrossRef]
47. Bonaccorso, A.; Currenti, G.; Linde, A.; Sacks, S.; Sicali, A. Advances in Understanding Intrusive, Explosive and Effusive Processes as Revealed by the Borehole Dilatometer Network on Mt. Etna Volcano. *Front. Earth Sci.* **2020**, *7*, 357. [CrossRef]
48. Bonaccorso, A.; Carleo, L.; Currenti, G.; Sicali, A. Magma Migration at Shallower Levels and Lava Fountains Sequence as Revealed by Borehole Dilatometers on Etna Volcano. *Front. Earth Sci.* **2021**, *9*, 740505. [CrossRef]
49. Barsotti, S.; Andronico, D.; Neri, A.; Del Carlo, P.; Baxter, P.J.; Aspinall, W.P.; Hincks, T. Quantitative assessment of volcanic ash hazards for health and infrastructure at Mt. Etna (Italy) by numerical simulation. *J. Volcanol. Geotherm. Res.* **2010**, *192*, 85–96. [CrossRef]
50. Horwell, C.J.; Sargent, P.; Andronico, D.; Lo Castro, M.D.; Tomatis, M.; Hillman, S.E.; Michnowicz, S.A.K.; Fubini, B. The iron-catalysed surface reactivity and health-pertinent physical characteristics of explosive volcanic ash from Mt. Etna, Italy. *J. Appl. Volcanol.* **2017**, *6*, 12. [CrossRef]
51. Corradino, C.; Amato, E.; Torrisi, F.; Calvari, S.; Del Negro, C. Classifying Major Explosions and Paroxysms at Stromboli Volcano (Italy) From Space. *Remote Sens.* **2021**, *13*, 4080. [CrossRef]
52. Fenner, D.; Rümpker, G.; Li, W.; Chakraborty, M.; Faber, J.; Köhler, J.; Stöcker, H.; Srivastava, N. Automated Seismo-Volcanic Event Detection Applied to Stromboli (Italy). *Front. Earth Sci.* **2022**, *10*, 809037. [CrossRef]
53. Harris, A.J.L.; Dehn, J.; Calvari, S. Lava effusion rate definition and measurement: A review. *Bull. Volcanol.* **2007**, *70*, 1–22. [CrossRef]
54. Nunnari, G.; Cannavó, F. Online detection of offsets in GPS time series. *Earth Sci. Inform.* **2021**, *14*, 267–276. [CrossRef]
55. Killick, R.; Fearnkead, P.; Eckley, I. Optimal detection of changepoints with linear computational cost. *J. Am. Stat. Assoc.* **2012**, *107*, 1590–1598. [CrossRef]
56. Lavielle, M. Using penalized contrasts for the change-point problem. *Signal. Process.* **2015**, *85*, 1501–1510. [CrossRef]

remote sensing

MDPI

Article

A New Analysis of Caldera Unrest through the Integration of Geophysical Data and FEM Modeling: The Long Valley Caldera Case Study

Fabio Pulvirenti [1,*], Francesca Silverii [2] and Maurizio Battaglia [3,4]

1 School of Remote Sensing and Geomatics Engineering, Nanjing University of Information Science and Technology, 219 Ningliu Road, Pukou District, Nanjing 210044, China
2 Department of Physics of Earthquakes and Volcanoes, German Research Centre for Geosciences (GFZ), Helmholtzstraße 6/7, 14467 Potsdam, Germany; silverii@gfz-potsdam.de
3 U.S. Geological Survey, Volcano Disaster Assistance Program, NASA Ames Research Center, Moffett Field, CA 94035, USA; maurizio.battaglia@uniroma1.it
4 Department of Earth Sciences, Sapienza—University of Rome, 00185 Rome, Italy
* Correspondence: fabiopulvirenti@yahoo.it

Abstract: The Long Valley Caldera, located at the eastern edge of the Sierra Nevada range in California, has been in a state of unrest since the late 1970s. Seismic, gravity and geodetic data strongly suggest that the source of unrest is an intrusion beneath the caldera resurgent dome. However, it is not clear yet if the main contribution to the deformation comes from pulses of ascending high-pressure hydrothermal fluids or low viscosity magmatic melts. To characterize the nature of the intrusion, we developed a 3D finite element model which includes topography and crust heterogeneities. We first performed joint numerical inversions of uplift and Electronic Distance Measurement baseline length change data, collected during the period 1985–1999, to infer the deformation-source size, position, and overpressure. Successively, we used this information to refine the source overpressure estimation, compute the gravity potential and infer the intrusion density from the inversion of deformation and gravity data collected in 1982–1998. The deformation source is located beneath the resurgent dome, at a depth of 7.5 ± 0.5 km and a volume change of 0.21 ± 0.04 km^3. We assumed a rhyolite compressibility of 0.026 ± 0.0011 GPa^{-1} (volume fraction of water between 0% and 30%) and estimated a reservoir compressibility of 0.147 ± 0.037 GPa^{-1}. We obtained a density of 1856 ± 72 kg/m^3. This density is consistent with a rhyolite melt, with 20% to 30% of dissolved hydrothermal fluids.

Keywords: numerical modeling; Long Valley Caldera; deformation and gravity joint inversion; topography correction; heterogenous crust; FEM; source parameters; intrusion density

Citation: Pulvirenti, F.; Silverii, F.; Battaglia, M. A New Analysis of Caldera Unrest through the Integration of Geophysical Data and FEM Modeling: The Long Valley Caldera Case Study. *Remote Sens.* **2021**, *13*, 4054. https://doi.org/10.3390/rs13204054

Academic Editors: Sonia Calvari, Eugenio Sansosti, Alessandro Bonaccorso, Annalisa Cappello and Flora Giudicepietro

Received: 12 August 2021
Accepted: 25 September 2021
Published: 11 October 2021

1. Introduction

The Long Valley Caldera (LVC), located in east-central California on the western edge of the Basin and Range Province and at the base of the Sierra Nevada frontal fault escarpment, is an east-west elongated oval depression formed by the eruption of the Bishop Tuff, 767,100 ± 900 years ago (Figure 1). Beginning in the late 1970s, the caldera entered a period of unrest, without any eruptions, that continues to the present time (e.g., Figure 3 in [1]). The unrest episodes include recurring earthquake swarms beneath the South Moat Seismic Zone (SMSZ) and the Sierra Nevada (SN) block, accelerated inflation of the central Resurgent Dome (RD), variations in the geothermal system and gas emissions around the flanks of Mammoth Mountain (MM) on the southwest margin of the caldera ([1] and references therein).

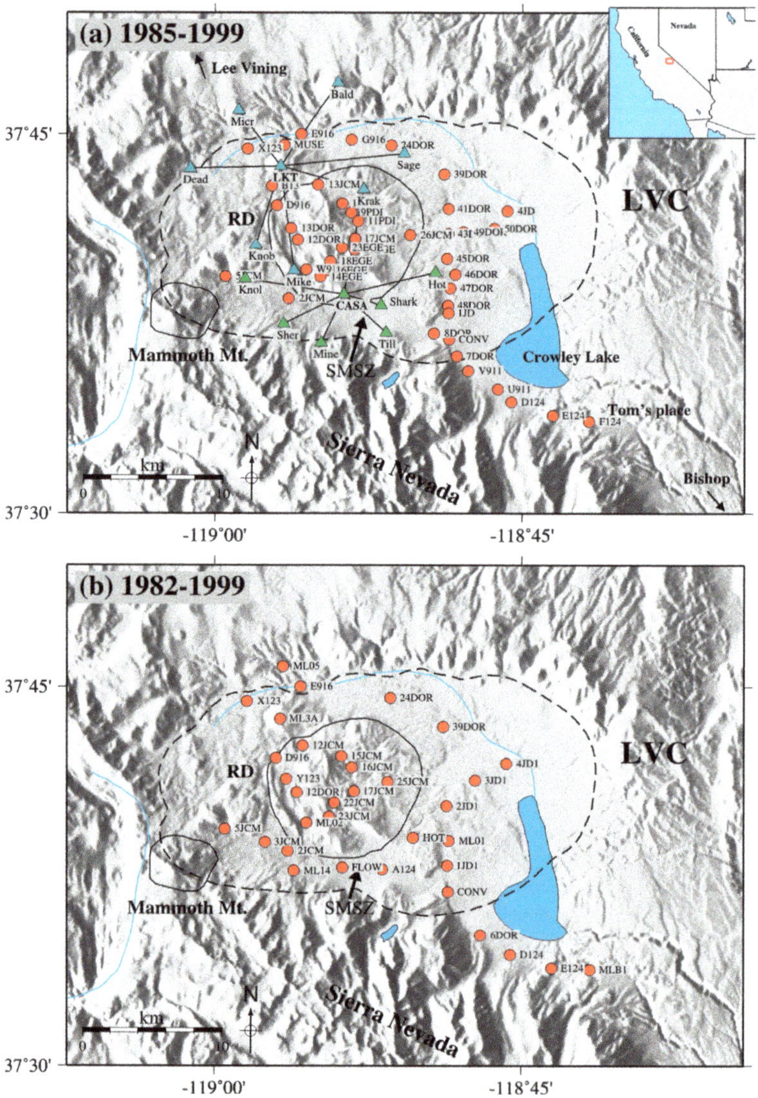

Figure 1. Map of Long Valley Caldera (LVC) and geodetic monitoring networks. Solid black lines represent the area of Resurgent Dome (RD) and Mammoth Mountain. Black dashed line outlines the LVC area. (**a**) Sites occupied in 1985–1999. Red circles are leveling stations. Green and cyan triangles refer to EDM baselines referred to common end-points CASA and LKT, respectively. (**b**) Sites occupied in 1982–1999. Red circles are leveling stations. No EDM data are available in this period.

The U.S. Geological Survey (USGS) began an intensive effort to monitor the unrest in LVC between 1975 and 1983 with the setup of new leveling lines in 1982, a two-color Electronic Distance Measurement (EDM) network in 1983, trilateration arrays in 1979, a dense seismic network in 1982, and a high-precision gravity network in 1982. Continuous Global Positioning System (GPS) measurements have been made since 1993. Both ground-based and space geodesy (including satellite interferometry) observations reveal a consistent radial and upward deformation pattern, centered at the RD and decreasing with radial distance (e.g., [2–8]).

Inferences about the cause of inflation from deformation data indicate that the main inflation source has been relatively stable since the 1980s and consists of a quasi-vertical prolate ellipsoid centered beneath the Resurgent Dome at a depth between 5 and 14 km [4,7–17]. Secondary sources of deformation include a deeper (9–15 km) source beneath the south moat [2,4,11], a small, north-northeast trending dike beneath MM [11,17], and a right-lateral strike-slip motion on west-northwest striking faults in the SMSZ [4].

The processes driving the unrest at LVC remain unclear, with the main likely source of unrest being either a magmatic intrusion into the upper crust [1], or pulses of high-pressure hydrous fluid intrusion into the upper crust [18]. Geologic and petrologic evidence support the hypothesis that the LVC rhyolitic magmatic system is moribund and that the magma body that fed the caldera-forming eruption may now be in the final stages of crystallization. The most recent eruptions along the Inyo Craters/Mono Domes chain and Mammoth Mountain have been fed by a different magmatic system. All of the eruptions inside the LVC have been rhyolitic, with the most recent eruption ~100 ka in the west moat. There has been no eruption on the resurgent dome over the last 500 ka. No significant seismicity and no emission of CO_2 or other magmatic gases has been recorded beneath the resurgent dome. Finally, the drilling of the resurgent dome found temperatures of only 100° at a depth of 3 km [18]. On the other hand, several pieces of geophysical evidence point to a possible magma intrusion as the cause of the present unrest. Multiple seismic imaging studies using different techniques (e.g., teleseismic tomography and full-waveform ambient noise tomography) highlighted large low velocity zones in the middle and lower crust, which have been interpreted as evidence of the presence of a partial melt. Different geodetic data (both ground- and satellite-based) measuring deformation at LVC since 1979, have not recorded any substantial deflation episodes yet. This might instead be expected if the inflation involved the injection of hydrothermal fluids with poroelastic swelling followed by diffusion, as observed at other calderas, such as Yellowstone and Campi Flegrei [1].

While ground deformation can provide insights about volume changes in the underground reservoir, it cannot constrain the mass of the intrusions and therefore discriminate between magma and hydrous fluid intrusion. Combined deformation and gravity measurements can be used to infer the density of the intrusive fluids and better define the source of unrest [19–27]. Given the density difference between silicate melts (~2300 kg/m^3) and hydrothermal fluids (~800 kg/m^3, [28]), density estimates can, in principle, be used to distinguish between these two possible sources of caldera unrest.

Gravity measurements at LVC have been conducted yearly between 1980 and 1985 and repeated in 1998 and 1999 [28]. In this period, the RD experienced a quasi-steady uplift, with accelerated phases in 1989–1990 and 1997–1998, when the most rapid deformation occurred (e.g., [4,13]). These data have been analyzed, together with different kinds of deformation records (EDM, leveling, GPS, InSAR) in different studies using analytical models and considering increasing complexities, from point source to tilted finite ellipsoidal source, from homogeneous to vertically layered elastic half-space [12,28–31]. The results of these studies suggest that gravity data are more compatible with the addition of a magma intrusion than pulses high-pressure hydrous fluids.

In this paper, we consider the 1982–1999 unrest period. This time has the best and most complete gravity dataset. We perform numerical computations based on the finite element method (FEM), exploring the effect of topography and realistic medium heterogeneities on the parameters (e.g., location, depth, density) of the source of unrest. We first invert EDM and leveling data from 1985 to 1999 in order to constrain the location, depth, and geometry of the unrest source. We then use the inferred source to model the deformation and gravity changes between 1982 and 1999, and to compute the source volume change, and density.

In Section 2, we present the methods including data, model setup and model computations; in Section 3, we show the results; in Section 4, we discuss our findings and conclusions.

2. Methods

2.1. Data

We used the data from the Long Valley Caldera GIS Database ([13]; https://doi.org/10.3133/ds81, accessed on 15 February 2021). The database includes extensive geologic, monitoring, and topographic datasets from studies conducted in Long Valley caldera between 1975 and 2001. The unrest is investigated using three sets of data: baseline length changes (an approximation of horizontal deformation) from two-color EDM, vertical deformation from a combination of GPS and leveling, and gravity changes.

The two-color EDM network consists of two sets of seven baselines. The first set is formed by the sites Hot, Knol, Krak, Mine, Shark, Sher and Till, observed from the central monument CASA (green triangles in Figure 1a). The second set is formed by the sites Bald, Dead, Knob, Krak, Micr, Mike and Sage, observed from the LKT monument (cyan triangles in Figure 1a). Measurements at these baselines span the 1985–1999 inflation period, which is included in the targeted time in this work (Supplementary Materials, Table S1). The methods used to extract the displacement and its error for each of the baselines are described in [11,17]. The EDM deformation data that were used are from [13].

Vertical deformation (uplift) measurements were taken during different leveling surveys along the 65-km-long line along Hwy 365 from Tom's Place to Lee Vining, and along several other routes within the caldera, and are obtained by combining leveling and GPS data (Figure 1a). Complete leveling of the caldera occurred each summer from 1982 to 1986, and in 1988 and 1992. In 1999, reference [13] occupied 44 leveling benchmarks with GPS to bring up to date the direct measurement of vertical deformation. The data sets employed in this work consist of the 44 benchmarks with leveling and GPS for the period 1985–1999 [13], and 34 benchmarks with leveling and GPS for the period 1982–1999 [28] (red circles respectively in Figure 1a,b; Supplementary Materials, Tables S2 and S3). The benchmark C916, located near Lee Vining (Mono Lake), is the elevation datum for the vertical deformation. The standard error for each elevation difference is calculated according to [13].

The Long Valley caldera gravity monitoring network is centered near Tom's Place (the primary reference station) and extends from the Sierra Nevada west of Lee Vining, CA, southeastward to a station in the White Mountains east of Bishop, CA [32]. Gravity data (gravity changes, noise from the water table and gravity corrected for the water table and free-air effect) are from [28], Supplementary Materials, Table S4. In Section 2.4, we employ the gravity data corrected for water table and free air contribution to estimate the density of the intrusion.

2.2. Model Setup

We develop a three-dimensional (3D) numerical model using the finite element method (FEM) and the software COMSOL Multiphysics (www.comsol.com, accessed on 15 February 2021). The geometry refers to a Cartesian reference system and is composed of a domain of 120 km $\times$ 130 km. The model is 40 km deep (up to the Moho depth in the area [33,34]), with zero depth corresponding to the sea level. The chosen size represents a crust portion which includes the LVC and a significant part of its surroundings (Figure 2a).

Inside the domain, we assume the existence of an internally pressurized ellipsoidal prolate cavity that we invert for its location, dimensions, and overpressure (Figure 2b; see Section 2.3).

Pressurized cavities of simple geometry can mimic/approximate the crustal stress field produced by the actual source. None of these geometries reproduced an actual source. The actual deformation source, beneath the resurgent dome, is probably a network of fractures filled with fluids (or magma) ascending from the crystallizing Pleistocene pluton below [18].

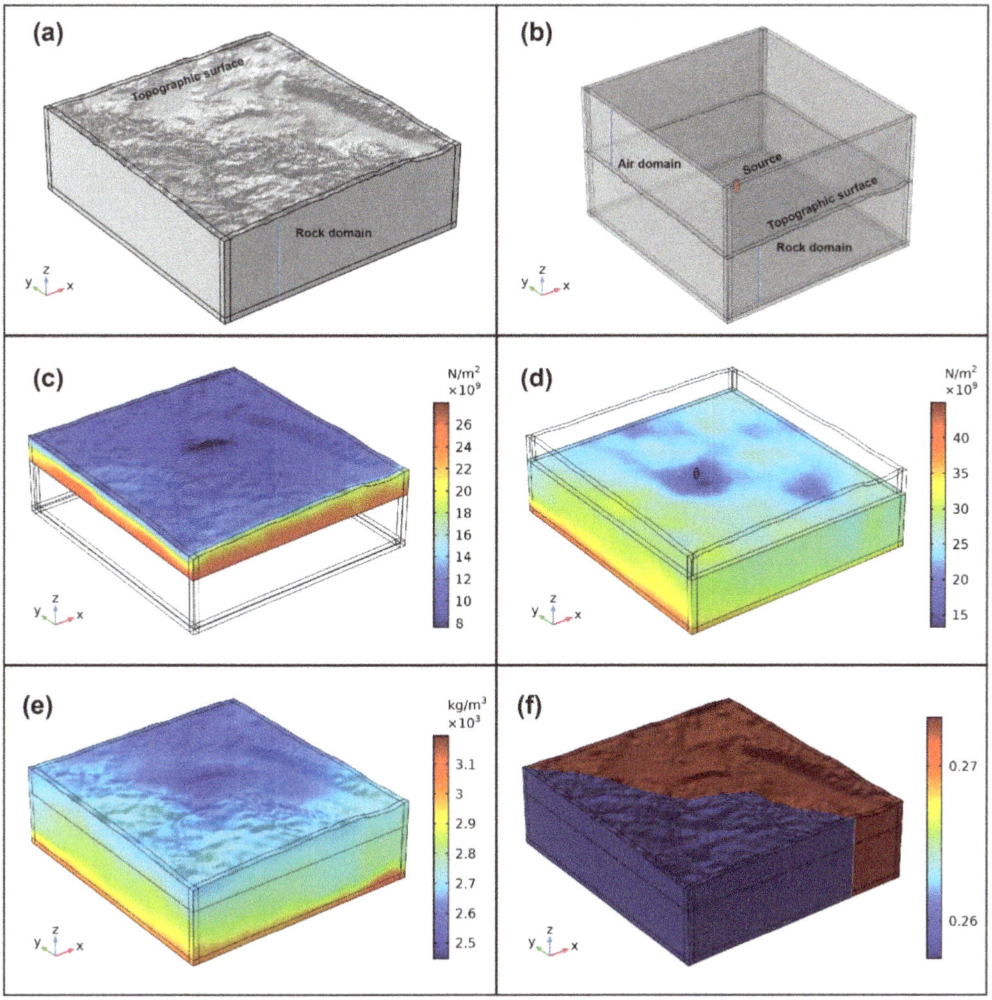

Figure 2. Model geometry representing (**a**) the rock domain with topography, (**b**) the rock + air domains in transparency with the ellipsoidal source beneath the resurgent dome, (**c**) quasi-static bulk modulus, (**d**) dynamic bulk modulus, (**e**) density, (**f**) Poisson's ratio.

We explore three different crust configurations. A homogeneous elastic domain with flat stress-free top surface (labeled HF) representing the average altitude of the area (~2300 m a.s.l.), a homogeneous elastic domain with topography (labeled HT), and a fully heterogeneous elastic domain with topography (labeled HeT). The topographic surface is generated by using the STRM digital elevation model (DEM) from the USGS Earth Explorer [35], resampled at 600 m resolution. Material heterogeneities (density, bulk modulus, shear modulus and Poisson's ratio; Figure 2c–f) are obtained from pressure (V_P) and shear (V_S) wave velocity distributions. Shear wave velocities V_S are from [36] while pressure velocities V_P are calculated from V_S using a V_P/V_S ratio of 1.75 for the SN block [37] and of 1.79 elsewhere, with a gradient $d(V_P/V_S)$ of 3% [33,38]. V_P, V_S velocities are converted into Poisson's ratio (ν), density (ρ) and dynamic Young's Modulus (E) using the equations in [39]:

$$\nu = 0.5 \times \left[\left(\frac{V_P}{V_S} \right)^2 - 2 \right] / \left[\left(\frac{V_P}{V_S} \right)^2 - 1 \right] \tag{1}$$

$$\rho = 1.6612 V_P - 0.4721 V_P^2 + 0.067 V_P^3 - 0.0043 V_P^4 + 0.000106 V_P^5 \tag{2}$$

$$E = \frac{V_P^2 \rho (1 + v)(1 - 2v)}{(1 - v)} \tag{3}$$

However, to properly represent the medium strain rate in a quasi-static condition, we need to refer to quasi-static mechanical properties. Laboratory tests [40,41] show, in fact, that for lithostatic pressures in the range 1–3 kbar (3.8 to 11.5 km depth), the ratio between quasi-static and dynamic bulk modulus K_s/K_d for granite is different from 1 and can vary between 0.5 (at 0.09 kbar–0.4 km depth) to 0.9 (at 3 kbar–11.5 km depth). For the range 0–3 kbar of lithostatic pressure (equivalent to the distance between the top surface and 11.5 km depth), we calculate the dynamic bulk modulus from V_P, V_S values and multiply it by the K_s/K_d ratio values from [40] at the corresponding lithostatic pressure (depth) level to estimate the quasi-static bulk modulus. The relation between quasi-static and dynamic mechanical properties is empirical and depends on several factors including stress state and stress history [40,41], however our approach leads to a better characterization of the material response with respect to what can be obtained using pure dynamical properties. An interpolation function guarantees a smooth transition between different lithostatic pressure levels. At a greater depth, where the lithostatic pressure is higher than 3 kbar, we assume a K_s/K_d ratio of 1. From the quasi-static bulk modulus, we can calculate the quasi-static shear modulus. Since Poisson's ratio is not expected to change significantly [42], we can retain its dynamic value. Crust properties are summarized in Table 1 and represented in Figure 2c–f.

Table 1. Material property parameters used for the models.

Material Parameter	Homogeneous Rock Domain	Heterogeneous Rock Domain	Air Domain [1]
Young's Modulus [GPa]	45	10–60	-
Bulk Modulus	31	8–45	
Density [kg/m^3]	2800	2450–3200	1
Poisson's ratio	0.26	0.25–0.27	-
Shear modulus	18	4–24	-
β_c [2] [GPa^{-1}]	0.049	see Section 2.4	-

[1] Air domain fluid characteristics are not solved. [2] Compressibility of the reservoir due to medium elasticity and reservoir shape; for A $\cong$ 3, $\beta_c = 7/8\mu$ (see Section 2.4).

In terms of boundary conditions, the model bottom is fixed, the top surface is stress-free while at the lateral boundaries we apply a roller condition (no displacement in the direction normal to the boundary). An infinite element condition, set at the lateral and bottom boundaries, simulates the far-field, and guarantees that the displacement vanishes at a very far distance from the original geometric size, thus avoiding any boundary effects. We prescribe a parametrized overpressure on the boundaries of the ellipsoidal cavity. The model domain is meshed with tetrahedral elements while the source boundaries and the top surfaces are discretized with triangular elements. Automatic adaptive mesh refinement tests are carried until an optimal performance is found without further variation of the output. The mesh for the whole domain is shown in Figure 3.

Figure 3. Domain mesh (example with rock domain and topography). The top surface is refined to accurately capture the topographic relief and the deformation pattern generated by the source overpressure.

We validate the FEM model for the deformation by performing a benchmark calculation for a vertical prolate ellipsoid in a flat homogeneous domain and comparing the numerical results against the analytical solution by [43], see Appendix A.

2.3. Inverse Modeling of Deformation Data

We use the FEM model described in Section 2.2 to perform numerical inversions of EDM baseline changes and uplift (leveling-GPS). Inversions are performed in two steps.

In the first step, we jointly invert the EDM and leveling-GPS data for the period 1985–1999 [13] and for each model configuration (with/without topography or with topography and heterogeneities) we infer the best-fit source dimensions, position, and overpressure. In the second step, we keep the deformation source stable in size and location and further optimize only for the source overpressure by performing a second inversion of the leveling-GPS data for the period 1982–1999 [28]. This second step provides the source parameters needed to model the gravity changes.

Inverse modeling is performed using the Nelder-Mead solver [44] and coupling the structural mechanics and the optimization module in COMSOL Multiphysics. The link between the data to invert and the source parameters is built by setting up the objective function [45]

$$F_i = [(M_i - D_i) \times W_i]^2 \tag{4}$$

where

$$W_i = a_i / \sum_j a_j \tag{5}$$

and

$$a_i = \left| \frac{D_i}{E_i} \right| \tag{6}$$

M_i are the modeled data, D_i the observed data, E_i the observation error, W_i the weights and the index i relates to each benchmark. The inversion goal is to minimize the least weighted squares Equation (4).

The inversion of EDM and leveling-GPS datasets for the period 1985–1999, is made considering the following seven parameters of the ellipsoidal source: the semiaxes (E_a, E_b and E_c), oriented along the cartesian x, y and z axis respectively; the horizontal position along the x and y direction (E_x, E_y) with respect to a reference point located at longitude $-119°$W, latitude 37.5°N; the source vertical position (E_z), with $E_z = 0$ m at the sea level; the overpressure (ΔP) applied at the source internal boundaries. The first three parameters control the source geometry, the second three control the source position and the last one controls the source overpressure. The source is assumed to be vertical, with a plunge of 90°. For each parameter we assign an initial value (used in the first iteration), and lower and upper bounds to search for reasonable values during the computation (Table 2). Initial values and ranges of parameters are based on results from previous studies (e.g., reference [7] and references therein). In particular, because of a poor data coverage in the south caldera rim, we constrained the north-south source position E_y to fall within the resurgent dome area. The solver is set to perform a maximum number of 400 model evaluations. This threshold has been chosen by looking at the convergence rate and considering that further evaluations no longer have any significant impact on the value of the objective function.

Table 2. Initial values and ranges of the parameters used in the FEM model inversions.

Parameter Name	Initial Value [m]	Lower Bound [m]	Upper Bound [m]
E_a (x-semiaxis)	1500	500	5000
E_b (y-semiaxis)	1500	500	5000
E_c (z-semiaxis)	3000	500	5000
E_x (center x-coord) [1]	8000	4000	12,000
E_y (center y-coord) [1]	20,400	18,000	23,000
E_z (depth) [2]	5000	4000	8000
Parameter Name	**Initial Value [Pa]**	**Lower Bound [Pa]**	**Upper Bound [Pa]**
ΔP (overpressure)	7.00×10^7	5.00×10^7	1.00×10^8

[1] E_x and E_y represent the source center coordinates along the x and y direction with respect to a reference point located at longitude −119°W and latitude 37.5°N. [2] Source depth relative to the stress-free surface and not the sea level, i.e., accounting for the average elevation of LVC area (~2300 m a.s.l.).

According to [29], the inflation source could be slightly tilted with a dip angle between 91 and 105 degrees. To check whether this was the case for our source, we performed preliminary tests, including additional parameters for a source rotation of ±10 degrees around each of the three cartesian axes. Results showed that in all cases (with/without topography or heterogeneities) the optimal rotation is minimal, <1° around x and y axis and <2° degrees around the z axis. For this reason, we did not include these parameters in our inversions.

2.4. Computation of Gravity Changes

The total gravity change recorded at a benchmark during unrest episodes contains the effect of different contributions: (i) the free-air effect, due to the vertical displacement of the benchmarks at the ground surface during unrest; (ii) the water table effect, proportional to the water table level change in the area; (iii) the deformation effect, due to the coupling between elastic deformation and gravity; and (iv) the residual gravity, which depends on the density change related to the introduction of the new mass into the pressurized volume (e.g., [20–22]). Furthermore, the estimation of gravity variation is sensitive to model complexities, such as volumetric source geometry, topography, material heterogeneities and fluid compressibility (e.g., [22–27]).

The best fit parameters of the ellipsoidal source from the three different crust configurations (HF, HT, HeT; Figure 2), are used to compute the gravity change at the free surface. Following the methodology in [46], we first compute the displacement field from the previously estimated best source models (cf. Section 2.3), and we then solve the Poisson's equation relating the gravity potential (φ_g) to the change in density distribution ($\Delta\rho$) caused by the subsurface mass redistribution $\nabla^2\varphi_g = -4\pi G\Delta\rho(x, y, z)$, where G is the gravitational constant. The gravity change can be then computed as $\delta_g = -\partial \varphi_g/\partial z$. In particular, the relation between the gravity potential φ_g and the density variations can be expressed by the following contributions:

$$\nabla^2\varphi_{g1} = 4\pi G(\boldsymbol{u}\cdot\nabla\rho_0) \quad (7)$$

$$\nabla^2 \varphi_{g2\Delta V} = 4\pi G(\boldsymbol{u} \cdot \nabla \rho_0) \tag{8}$$

$$\nabla^2 \varphi_{g3} = 4\pi G(\rho_0 \nabla \cdot \boldsymbol{u}) \tag{9}$$

$$\nabla^2 \varphi_{g2V} = 4\pi G(\rho_{in}) \tag{10}$$

where $\boldsymbol{u}$ is the inflation-related displacement field, ρ_0 the embedding medium density and ρ_{in} is the density of the intruding fluid. Equation (7) gives the gravity contribution δ_{g1} due to the displacement of density boundaries in heterogeneous media, corresponding to the Bouguer correction at the surface in case of flat homogeneous models. Equation (8) gives the gravity contribution $\delta_{g2\Delta V}$ due the displacement of the source boundaries, which implies replacement of the surrounding mass. Equation (9) gives the term δ_{g3}, which considers the effect of dilatational/compressional strains in the host rock, while Equation (10) gives the term δ_{g2V} which considers the input of material (of density ρ_{in}) into the source volume [21]. Equations (7)–(9) can be used to compute the massless deformation contribution to the gravity changes while Equation (10) represents the contribution due to the source mass change.

To numerically solve the Poisson's equations, we modify the model geometry by adding an additional domain with same size of the rock domain (Figure 2b), but made of air (assuming E = 1 Pa, ρ = 1 kg/m^3, ν = 0.25; Table 1). Furthermore, solving for all contributions to the gravity potential requires the embedded source to be a domain and not a cavity, as done during the inversion of displacements. Poisson's Equations (7) and (8) are solved on the stress-free surface and on the source boundaries, respectively. Poisson's Equation (9) is solved on the domains surrounding the source, and (10) is solved on the source domain.

We validate the FEM model by performing a benchmark calculation for a vertical prolate ellipsoid in a flat homogeneous domain and comparing the numerical results to the analytical solutions by [47] (see Appendix A).

When estimating the gravity changes due to reservoir inflation, it is important to consider that the volume change accommodating for the input of new mass could arise, not only from the expansion of the source wall that deforms the surrounding medium, but also from the compression of the material stored in the reservoir (e.g., [25,48–50]). The relation between the actual volume of the mass intrusion, ΔV_m, and the volume change from the inversion of deformation data (geodetic volume, cf. Section 3.2), ΔV, is (e.g., [48,51])

$$\Delta V_m = \Delta V \times \left(1 + \frac{\beta_m}{\beta_c}\right) = \Delta V \times r_V \tag{11}$$

where $\beta_m = \frac{1}{K_m}$ is the compressibility of the material stored in the reservoir, $\beta_c = \frac{1}{K_c}$ is the compressibility of the reservoir due to medium elasticity and reservoir shape, K_m and K_c are the bulk moduli, r_V is the volume ratio, and β_m is a function of several parameters, like pressure, gas volume fraction, temperature, phenocryst content and source depth (e.g., Table 3 from [52]). Finite element calculations of reservoir compressibility β_c as a function of the source geometric aspect ratio $A = \frac{E_c}{\sqrt{E_a E_b}}$ indicates that in our case $\beta_c \approx \frac{7}{8\mu}$, where μ is the shear modulus (see Figure 5 in [53], and Tables 1 and 3). β_c can also be computed as [48]:

$$\beta_c = \left(\frac{1}{V}\right)\left(\frac{\Delta V}{\Delta P}\right) \tag{12}$$

where ΔP is the overpressure, and V is the source volume before the application of the overpressure (Table 4). Finally, the density corrected for the effect of compressibility (ρ_{cmp}) can be computed as:

$$\rho_{cmp} = \rho_{in}\left(\frac{1}{r_V}\right) \tag{13}$$

Table 3. Best fit source parameters, and associated uncertainties estimates σ, obtained from the joint inversion of EDM + Leveling data for the period 1985–1999 and optimal overpressure obtained from the inversion of leveling data for the period 1982–1999. HF: homogeneous flat crust; HT: homogeneous crust with topography; HeT: heterogeneous crust with topography.

Model	E_a (m)		E_b (m)		E_c (m)		A		E_x (m)		E_y (m)		E_z [1] (m)		ΔP 1985–1999 (MPa)		ΔP 1982–1999 (MPa)	
		σ		σ		σ		σ		σ		σ		σ		σ		σ
HF	1726	149	1491	141	4553	605	2.8	0.8	9145	374	18,006	921	7674	634	69	2	88	2
HT	1865	162	1556	140	4136	419	2.4	0.6	9077	397	18,009	919	7610	628	67	2	85	2
HeT	1217	146	1133	150	3032	254	2.6	0.6	8958	487	18,000	1039	7519	521	65	2	84	2

[1] Note that here we indicate the source depth with respect to the stress-free surface and not the sea level, i.e., accounting for the average elevation of LVC area (~2300 m a.s.l.).

Table 4. Density values of the intrusion, and associated uncertainties estimates σ, obtained from the inversion of residual gravity for 1982–1999.

Model	V [km^3]		ΔV [km^3]		ΔP [MPa]		β_c [GPa^{-1}]		β_m [GPa^{-1}]		r_V		ρ_{in} [kg/m^3]	ρ_{cmp} [kg/m^3]	
		σ		σ		σ	(*)	σ		σ		σ			σ
HF	49	9	0.21	0.04	88	2	0.049	0.014	0.026	0.001	1.53	0.15	2670	1741	172
HT	50	8	0.21	0.04	85	2	0.049	0.012	0.026	0.001	1.53	0.13	2720	1782	151
HeT	17	3	0.21	0.04	84	2	0.147	0.037	0.026	0.001	1.18	0.05	2184	1856	72

(*) Equation (12).

3. Results

3.1. Deformation: Best Fit Source

We find that the three different crust configurations (homogeneous, flat elastic half-space HF; homogeneous elastic domain with topography, HT; heterogeneous elastic domain with topography, HeT) give similar results for the position (E_x, E_y), depth (E_z), geometric aspect ratio (A), and pressure change (ΔP) of the source (Table 3).

The nonlinearity of the inverse problem makes the evaluation of uncertainties difficult; nonlinear error propagation is a difficult problem to address, COMSOL does not have a feature that allows extraction of the covariance matrix, and the model covariance matrix may not give a good estimate of the uncertainties [54]. A solution could be to employ a Monte Carlo method. Unfortunately, this method requires each model to be run thousands of times. We employ the result from the inversions (350 to 400 runs) to mimic a Monte Carlo method and obtain an estimate of the uncertainties of the source parameters. We then propagate the errors to the density results (see Appendix B).

The source is moved about 1 km eastwards (E_x), 2.4 km southwards (E_y) and 600 m deeper (E_z), with respect to its starting position and starting depth. No major differences can be seen between a homogeneous flat crust (HF), homogeneous crust with topography (HT) or heterogeneous crust with topography (HeT). The source size is similar for the HF and HT crust models, while we can observe in the HeT crust model a significant reduction of about 1/3 of all semiaxes. The source shapes for each crust model before and after inversion are showed in Figure 4. Although the main source parameters are similar for the three different models (Table 3), material heterogeneities make a difference, especially in the estimate of the absolute volume of the source (see Table 4).

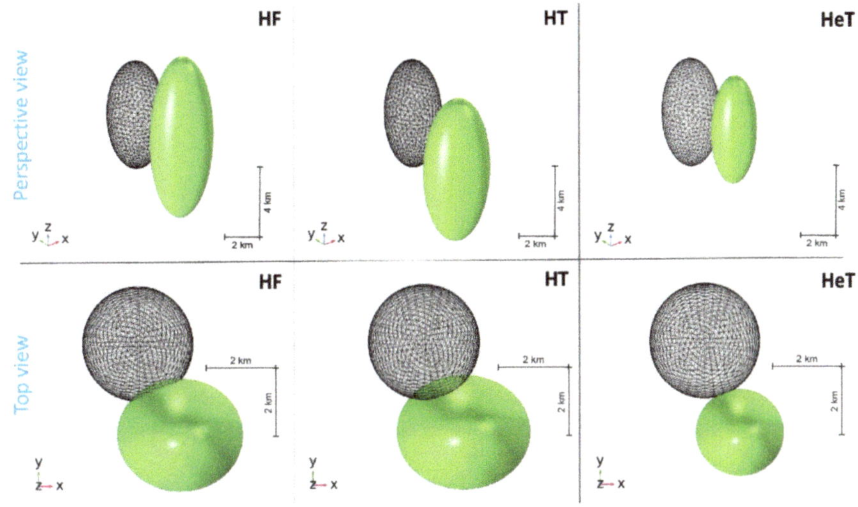

Figure 4. Perspective views and top views of the source shape before (black wireframe) and after (green shaded) the joint inversion of EDM and leveling data for the period 1985–1999. HF: homogeneous flat crust; HT: homogeneous crust with topography; HeT: heterogeneous crust with topography.

Figure 5 shows the total displacement (combination of vertical and horizontal displacements) at the free surface for each model configuration. For the HF and HT cases we can clearly observe two lobes with higher displacement northwards and southwards of the source with the topography playing a minor damping role and a slight clockwise rotation of the northern lobe. When heterogeneities are introduced (HeT), the southern lobe disappears while the northern lobe further rotates clockwise and shows an area of maximum total displacement.

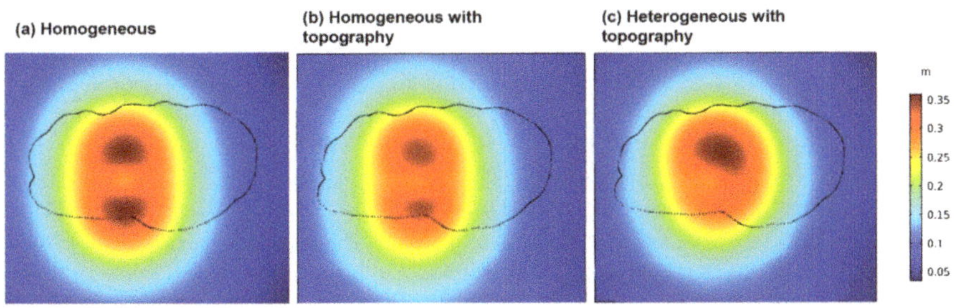

Figure 5. Total displacement at the free surface (top view) from the joint inversion of EDM and leveling data for the period 1985–1999. Black dotted line represents the caldera border. (**a**) Homogeneous model without topography. (**b**) Homogeneous model with topography. (**c**) Heterogeneous model with topography. The color scale (0.05–0.35 m) is the same for the three panels.

3.2. Fit to Deformation Data

Figures 6 and 7 compare the modeled and observed values for horizontal (EDM baseline length changes) and vertical displacements (uplift) for 1985–1999, obtained from the numerical joint inversions of EDM and leveling data. Observed values and numerical results for the baseline changes and for the leveling data over the period 1985–1999 are reported in Tables S1 and S2 in Supplementary Materials.

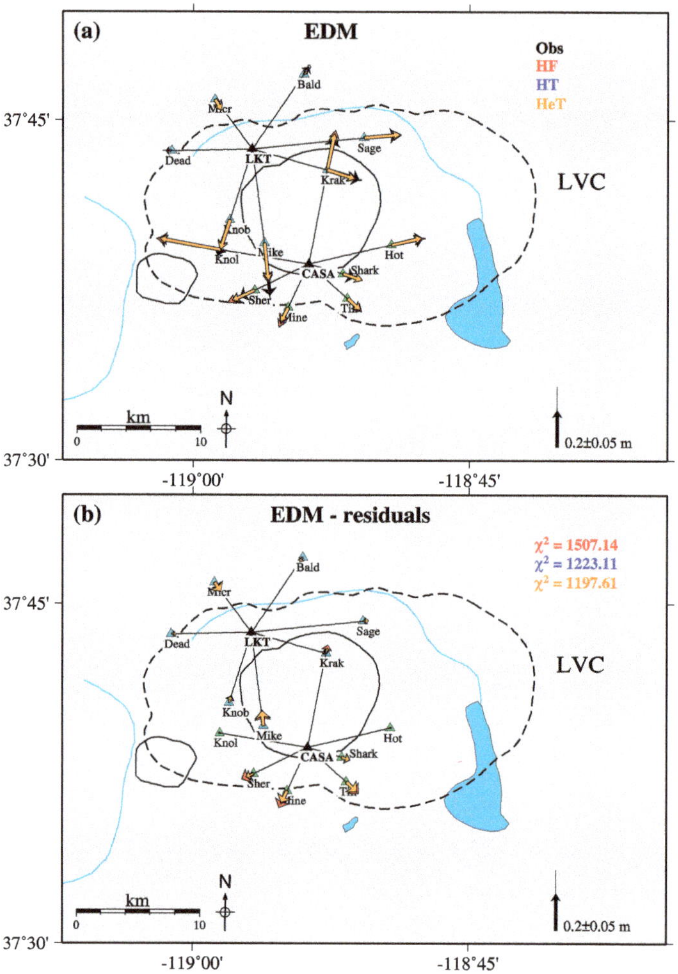

Figure 6. (**a**) Comparison between observed (black arrows with thin solid line as error) and modeled (colored arrows corresponding to different model configurations) EDM baseline length changes. (**b**) Corresponding residuals between modeled and observed baseline length changes. Obs: observations; HF: homogeneous flat crust; HT: homogeneous crust with topography; HeT: heterogeneous crust with topography. χ^2 values for each model are shown on top right.

Results for EDM show a good agreement between models and observations (Figure 6a) except for LKT-MIKE baseline which is slightly underestimated in all three crustal models. However, the difference between the models' results and the measurement for LKT-MIKE is within the data uncertainty (thin solid black lines). From the EDM residuals (Figure 6b) we can observe that the fit improves when we add topography (χ^2 decrease by 19% from HF to HT, red and blue arrows) and material heterogeneities (χ^2 decrease by 21% from HF to HeT, red and yellow arrows).

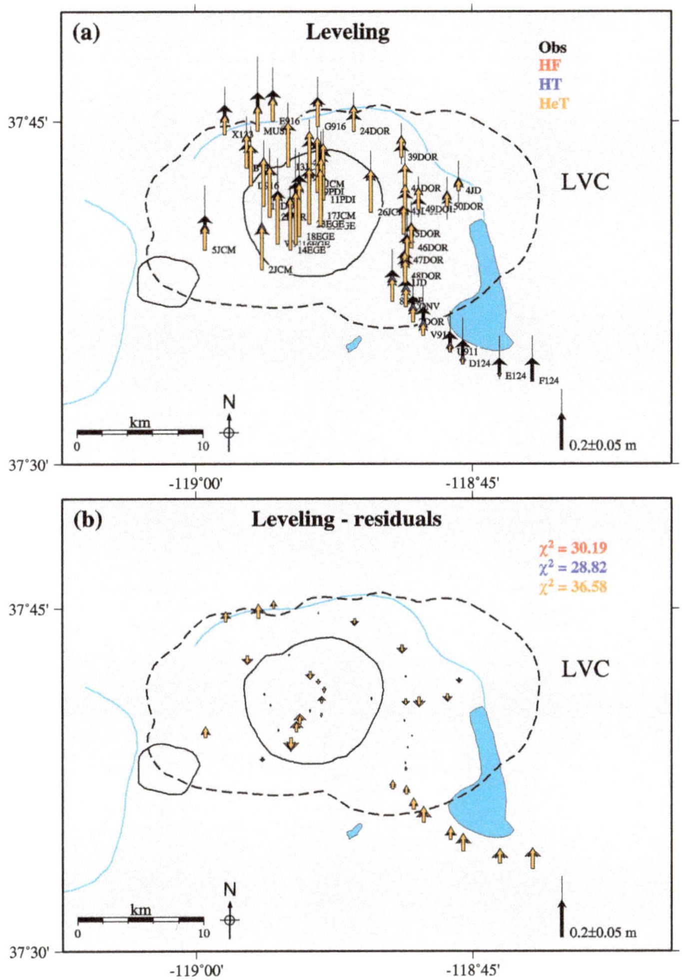

Figure 7. (**a**) Comparison between observed (black arrows with thin solid line as error) and modeled (colored arrows corresponding to different model configurations) uplift for the period 1985–1999. (**b**) Corresponding residuals between modeled and observed uplift. Obs: observations; HF: homogeneous flat crust; HT: homogeneous crust with topography; HeT: heterogeneous crust with topography. χ^2 values for each model are shown on top right.

The inversion results for the leveling data (Figure 7a) show a good agreement between the observed and modeled data at the benchmarks located inside the caldera border (black dashed line). However, the model underestimates the observed uplift by 5–10 cm at the benchmarks located outside the southeastern (SE) caldera border (near Crowley Lake) and by 2–5 cm for the benchmarks located at the northwestern (NW) caldera border (Figure 8). This discrepancy influences the χ^2 value (Figure 7b), which slightly decreases when topography is included (5% decrease in χ^2 from HF to HT) but increases when we add the material heterogeneities (16% increase in χ^2 from HF to HeT). This is probably because, in the HeT model, the material outside the caldera area is stiffer than the material inside it (Figure 2).

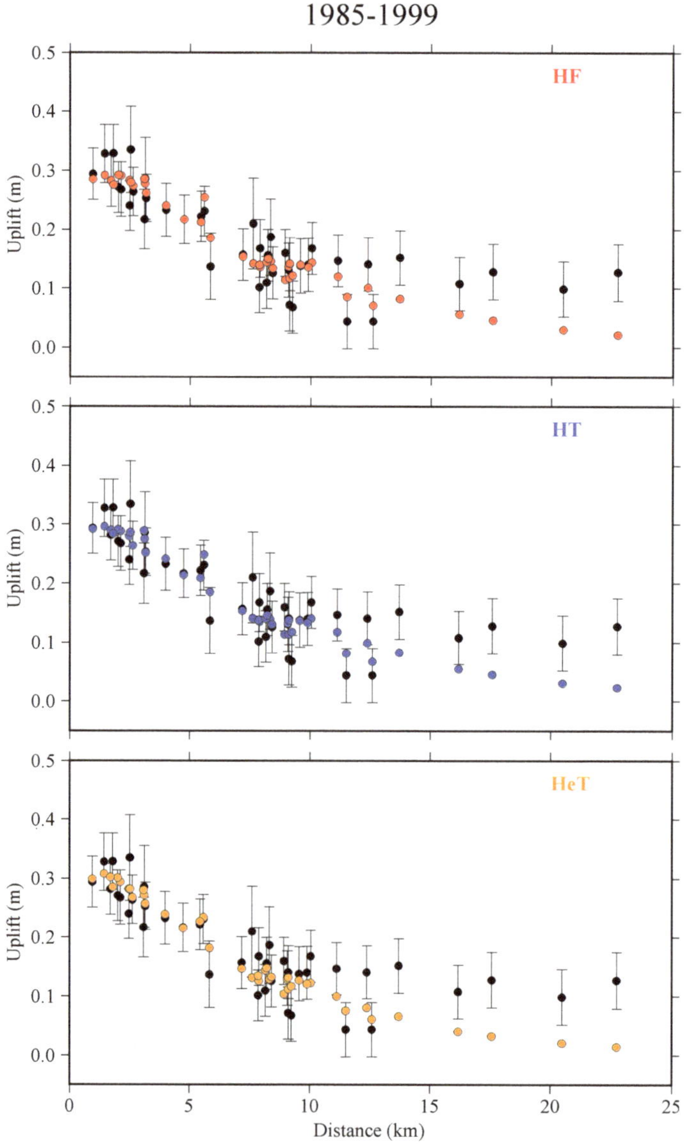

Figure 8. Comparison between observed (black circles with 1-sigma error bars) and modeled (colored circles corresponding to different model configurations) uplift at leveling benchmarks for the data 1985–1999 ordered according to the horizontal distance from the resurgent dome center. HF: homogeneous flat crust; HT: homogeneous crust with topography; HeT: heterogeneous crust with topography.

Figures 9 and 10 compare the observed leveling data over the period 1982–1999 and the correspondent model results. The latter were obtained by further optimizing the source overpressure using the leveling data 1982–1999, while keeping the same source location and size from the joint inversion of EDM and leveling data from 1985–1999 (c.f. Section 2.3). In this case, we reach a good agreement between the models and uplift for all three crustal models, with residuals within the observation error (Figures 9b and 10). Some discrepancy is still observed for the benchmarks southeast of the caldera border, since the uplift in this area is probably controlled by the deformation along the Sierra Nevada block [55]. In this case, the χ^2 value is reduced by 36% when we add topography (from HF to HT) and by a further 10% when we also add the heterogeneities (46% total decrease χ^2 in from HF to HeT). Observed values and numerical results for the leveling data over the period 1982–1999 are reported in Table S4, Supplementary Materials.

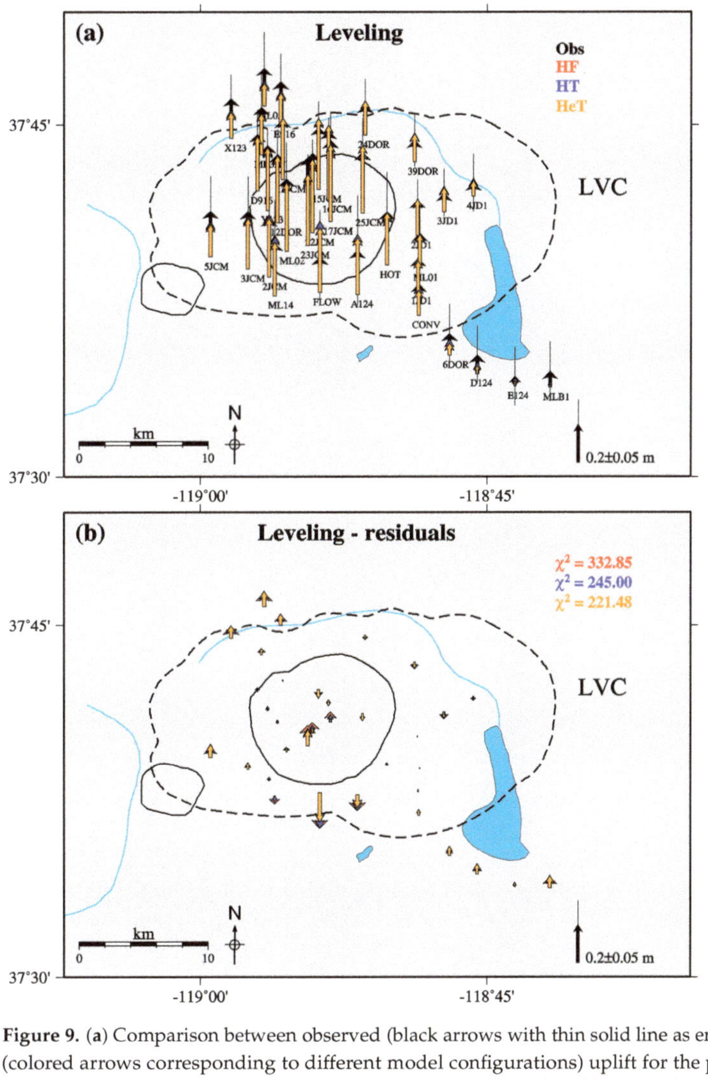

Figure 9. (**a**) Comparison between observed (black arrows with thin solid line as error) and modeled (colored arrows corresponding to different model configurations) uplift for the period 1982–1999. (**b**) Corresponding residuals between modeled and observed uplift. Obs: Observations; HF: homogeneous flat crust; HT: homogeneous crust with topography; HeT: heterogeneous crust with topography. χ^2 values for each model are shown on top right.

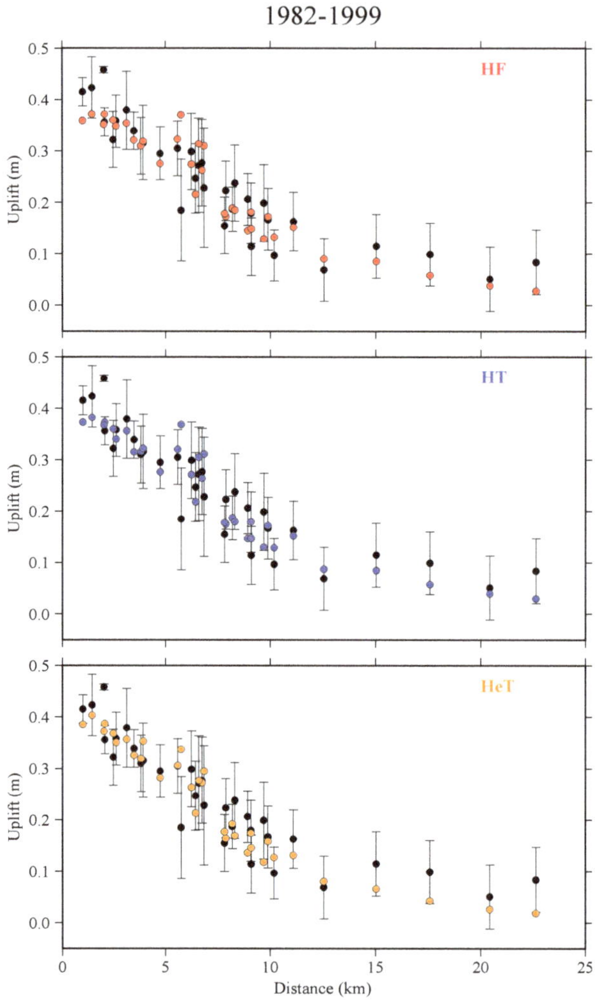

Figure 10. Comparison between observed (black circles with 1-sigma error bars) and modeled (colored circles corresponding to different model configurations) uplift at leveling benchmarks for the data in 1982–1999 ordered according to the horizontal distance from the resurgent dome center. HF: homogeneous flat crust; HT: homogeneous crust with topography; HeT: heterogeneous crust with topography.

The large residuals observed for three benchmarks close to the center of the resurgent dome are from the exploitation of the hydrothermal aquifers by the local geothermal power plant [13]. Other discrepancies are because of motion along faults in the caldera South Moat [4] or the Sierra Block (e.g., Figures 7 and 9) [8,55]. The heterogeneous model (HeT) can better fit the uplift for 1982–1999 than the two homogeneous models (HF and HT; Figure 9).

3.3. Density of the Intrusion

The observed gravity change, after free-air and water-table correction, shows a positive anomaly centered on the resurgent dome, with peak amplitude of about 60 μGal ([28] and black circles in Figure 11), that suggests mass intrusion into the sub-caldera crust beneath the resurgent dome.

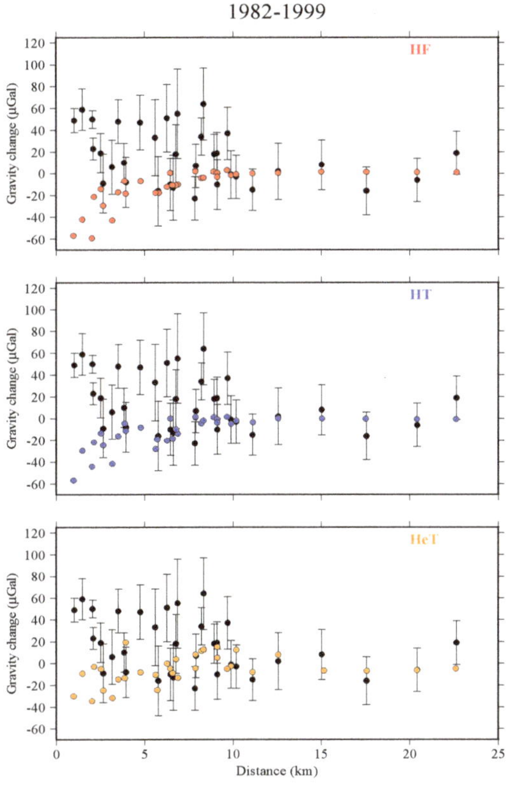

Figure 11. Comparison between observed gravity changes (1982–1999, Table S4, Supplementary Materials) after the removal of free air and water table effect (black circles with 1-sigma error bars) and the total deformation contribution to gravity changes ($\delta_{g1} + \delta_{g2\Delta V} + \delta_{g3}$) from the solution of Equations (7)–(9) corresponding to different model configurations (colored circles), ordered according to the horizontal distance from the resurgent dome center. HF: homogeneous crust, red circle; HT: homogeneous crust with topography, blue circle; HeT: heterogeneous crust with topography.

The estimate of the density of the intrusion requires three steps. First, we calculate the gravity changes associated with the deformation of the source and of the surrounding crust ($\delta_{g1} + \delta_{g2\Delta V} + \delta_{g3}$) by solving (7)–(9), the so-called "deformation effects", see Figure 11. The gravity variations due to deformation effects are substantial in the near-field of the source location, with the highest values at the source-tip location (up to -60 μGal, i.e., comparable, in magnitude, to the observed gravity change after free-air and water-table correction) and decaying to magnitudes <10 μGal at a distance of ~10 km.

We then subtract the deformation effects from the observed gravity changes in order to calculate the residual gravity (black circle with error bars in Figure 12). It is worth noting that the observed gravity changes had been previously corrected for water table noise and the free-air effect (Table S4, Supplementary Materials) (details in [28]). The residual gravity, δ_{g2V}, depends on the mass change accompanying the deformation. Following the methodology described in Section 2.3, we solve the inverse problem with Poisson's Equation (10) to obtain the density value for the intruding fluid which best matches the observed residual gravity change δ_{g2V}. Numerical results for modeled residual gravity change agree within the measurement errors with most of the residual gravity observations

(Figure 12). Figure 12 shows a good fit to the observed residual gravity for the FEM model of a homogeneous crust with topography (HT). Adding additional information about heterogeneities in the crust does not significantly improve the fit.

Table 4 shows the resulting density values of the intrusion, assuming that the mass change is due to either an incompressible (ρ_{in}) or compressible (ρ_{cmp}) fluid intrusion. The relation between the density of incompressible (ρ_{in}) or compressible (ρ_{cmp}) fluid intrusion is given in Equation (13). Reference [53] and Equation (12) allow for computing the compressibility β_c of the crust surrounding the intrusion from quasi-static elastic properties (Table 1) and the results of the FEM models (Table 4).

The density of the intrusion depends on magma and reservoir compressibility—Equations (11)–(13). According to [56], the isothermal compressibility for a rhyolite, with a volume fraction of water between 0% and 30%, is 0.026 ± 0.0011 GPa^{-1}. We estimated a crust compressibility of 0.147 ± 0.037 GPa^{-1} for the heterogeneous model. Using these values, we obtained a density of 1856 ± 72 kg/m^3. This density is consistent with a rhyolite melt (no crystals) with 20% to 30% of dissolved hydrothermal fluids.

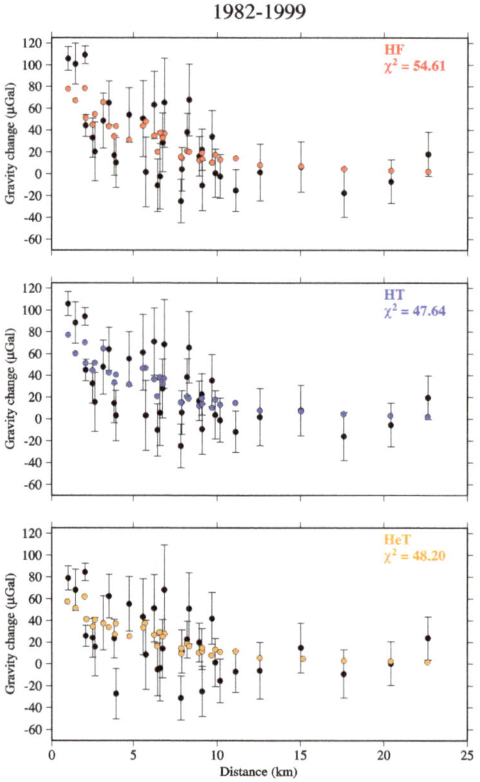

Figure 12. Comparison between observed (black circles with 1-sigma error bars) and modeled (colored circles corresponding to different model configurations) residual gravity, δ_{g2V}. The misfit value χ^2 for each case is also indicated. The corresponding best fit values for density, under the assumption of incompressible magma, are shown in Table 4. HF: homogeneous crust, red circle; HT: homogeneous crust with topography, blue circle; HeT: heterogeneous crust with topography, orange circle.

4. Summary

We build a 3D finite element model to investigate the source of observed displacements and gravity changes at Long Valley Caldera for 1982–1999. Using the geodetic data available in Long Valley Caldera GIS Database (https://doi.org/10.3133/ds81, accessed on 15 February 2021)—EDM baseline change, uplift, and gravity change measurements—we explore different model configurations starting from a flat and fully homogeneous domain and then adding additional complexities, such as topography and medium heterogeneities. Limits on the coverage of the existing geodetic monitoring network, ambiguities on the interpretation of subsurface distribution of the crust elastic properties, and the nature of non-linear inversion make our models and solutions non-unique. We can improve the bounds on the parameters of the deformation source by employing an appropriate modeling approach.

Since the joint inversion of horizontal and vertical deformation better constrain the geometry of the deformation source, we first invert EDM and leveling data for the period 1985–1999 to infer the size and location of an ellipsoidal source under the caldera. We then optimize the estimate of the source volume and mass change by performing a second set of inversions of leveling and gravity data for 1982–1999. This is the period with the best signal to noise ratio for gravity data.

One advantage of our approach is, not only the inclusion of topography, but also of full heterogeneities (3D). The influence of mechanical heterogeneities in LVC has been considered in other works [30,57–59], but while past works relied on simplified models in terms of geometry (e.g., 2D) or the material heterogeneities distribution (e.g., only vertical), in this study, similarly to [8], we implement both lateral and vertical material heterogeneities. Furthermore, we account for the difference between static and dynamic mechanical properties, since the use of a dynamic bulk modulus for the overall domain would overestimate the medium rigidity at shallow depths (lithostatic pressure < 3 kbar).

To estimate the subsurface mass change of the deformation source, we first estimate the so called "deformation effects" given by (7)–(9); see Figure 11. We then subtract the "deformation effects" from the gravity changes to obtain the residual gravity, since the residual gravity, δ_{g2V}, depends on the mass change accompanying the deformation. Finally, we solve the inverse problem with Poisson's Equation (10) to obtain the density value for the intruding fluid which best matches the observed residual gravity change (Figure 12).

The density of the intrusion will change if the fluid is either incompressible (ρ_{in}) or compressible (ρ_{cmp}), since the density of the intrusion depends on the ratio β_m/β_c (Equation (13), Table 4). We can estimate β_c either from the shear modulus and source aspect ratio (Table 1; [53]) or from the absolute volume of the source inferred from our numerical analysis (Table 4). Both approaches calculate the same value of β_c for a homogeneous medium. We find the second approach more appropriate for our heterogeneous model.

5. Conclusions

Gravity data are usually noisier than deformation data (e.g., [28]) but are essential for estimating the density of intrusion, because changes in the gravity potential are related to the changes in density distribution caused by the subsurface mass redistribution. Without gravity data, we cannot obtain information about the nature of the deformation. In this specific case, the major ambiguity is not coming from the errors in the gravity data but from the uncertainty about the appropriate value of magma compressibility. Reference [56] present experimental values of the isothermal compressibility of rhyolite, andesite, and basalt glasses as a function of the volume fraction of water (see Table 4 and Figure 5 in [56]). We assume here the compressibility values for rhyolite, i.e., the main component of erupted magma in LVC [18]. According to [56], the isothermal compressibility for a rhyolite, with a volume fraction of water between 0% and 30%, is 0.026 ± 0.001 GPa^{-1}. Using our inversion results for source parameters, for the heterogeneous case we estimated a reservoir compressibility of 0.147 ± 0.037 GPa^{-1} (Table 4). We therefore obtained a density of 1856 ± 72 kg/m^3. This density is consistent with a rhyolite melt with 20% to 30% of dissolved hydrothermal fluids.

Supplementary Materials: The following are available online at https://www.mdpi.com/article/10.3390/rs13204054/s1, Table S1: Horizontal deformation for the period 1985–1999 (two-color EDM data), Table S2: Vertical deformation (uplift) for the period 1985–1999, Table S3: Vertical deformation (uplift) for the period 1982–1999 (leveling data), Table S4: Gravity data (1982–1999).

Author Contributions: Conceptualization, F.P. and F.S.; methodology, F.P. and F.S.; software, M.B.; validation, F.P. and F.S.; formal analysis, F.P.; investigation, F.P. and F.S.; resources, M.B.; data curation, M.B.; writing—original draft preparation, F.P. and F.S.; writing—review and editing, all authors; visualization, M.B.; supervision, M.B.; project administration, M.B.; funding acquisition, M.B. All authors have read and agreed to the published version of the manuscript.

Funding: This research was funded by the—U.S. Geological Survey, Volcano Science Center, and by Sapienza—University of Rome, Piccoli Progetti Universitari 2020.

Institutional Review Board Statement: Not applicable.

Informed Consent Statement: Not applicable.

Data Availability Statement: Gravity and deformation data employed in this paper are available on-line from the Long Valley Caldera GIS Database ([13]; https://doi.org/10.3133/ds81, accessed on 15 February 2021), and as Supplementary Materials.

Acknowledgments: The authors wish to thank James Hickey (Bristol University) and Gilda Currenti (INGV) for the discussion about the model implementation. Mehdi Nikkoo (GFZ) for supporting the benchmark studies with analytical solutions and for the helpful conversations. The authors thank the Reviewers for the thorough and insightful comments which helped to improve the manuscript and Emily Montgomery-Brown (USGS) for internal review of the manuscript. Any use of trade, firm, or product names is for descriptive purposes only and does not imply endorsement by the U.S. Government.

Conflicts of Interest: The authors declare no conflict of interest.

Appendix A Validation of Numerical vs. Analytical Solution

To assess the accuracy of the finite element calculations, the absence of edge effects and ensure that the geometry and mesh adopted yield sufficient sensitivity, we validated the FEM, the numerical solution, for a vertical prolate ellipsoid in a homogeneous model without topography against the analytical solutions from [43]. For this comparison, we used a prolate ellipsoid with horizontal semiaxes a = b = 1.5 km and vertical semiaxis c = 3 km. The ellipsoid depth is 7 km and an outward (inflating) overpressure of 70 MPa is applied. We first compare the horizontal and vertical components of displacement on the stress-free surface on ±40 km EW profile (Figure A1a). Successively, we calculate the gravity changes due to the source deformation and to an intruding fluid with a density of 2700 kg/m^3 using the method described in the main text (Section 2.4) and compare them with the analytical solutions from [47] (Figure A1b,c). For the gravity contribution from the mass change (δ_{g2V}), we also checked with the solutions from [60]. We observe a good agreement between the analytical and FEM solutions for both the displacements and the gravity components. Minor differences visible over the first 5 km distance from the source are inside the ranges of uncertainty, which are ~1 mm for horizontal displacements, ~1 cm for vertical displacements and ~10 μGal for gravity changes.

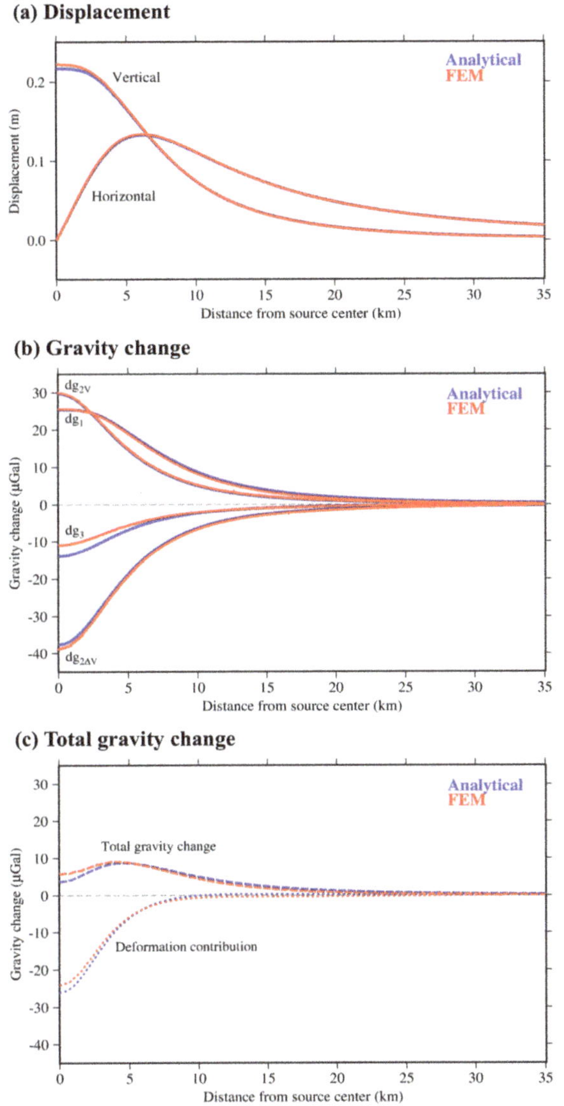

Figure A1. Benchmark test. Comparison between analytical (blue) and numerical (red) solutions for an ellipsoidal source in a homogeneous space with no topography. All solutions are shown along a radial distance centered at the source horizontal position. (**a**) Comparison of horizontal and vertical displacement at the free surface. (**b**) Comparison between the various components of gravity changes at the surface. (**c**) Total gravity change obtained by summing all the components shown in (**b**) and the gravity change only due to the contribution of deformation ($\delta_{g1} + \delta_{g2\Delta V} + \delta_{g3}$).

Appendix B Error Propagation

$$A = \frac{E_c}{\sqrt{E_a E_b}} \quad \sigma_A^2 = 4\left[\frac{\sigma_{E_c}^2}{E_a E_b} + \frac{(E_b E_c)^2}{4(E_a E_b)^3}\sigma_{E_a}^2 + \frac{(E_a E_c)^2}{(E_a E_b)^3}\sigma_{E_b}^2\right] \quad \text{(A1)}$$

$$V = \frac{4}{3}\pi E_a E_b E_c$$
$$\sigma_V^2 = \frac{16}{9}\pi^2\left[(E_b E_c)^2\sigma_{E_a}^2 + (E_a E_c)^2\sigma_{E_b}^2 + (E_a E_b)^2\sigma_{E_c}^2\right] \quad (A2)$$

$$\Delta V \sim \frac{\Delta P}{\mu}V \quad \sigma_{\Delta V}^2 \sim \frac{1}{\mu^2}\left[V^2\sigma_{\Delta P}^2 + \Delta P^2\sigma_V^2\right] \quad (A3)$$

$$\beta_c = \frac{1}{V}\frac{\Delta V}{\Delta P} \quad \sigma_\beta^2 = \left[\left(\frac{\Delta V}{V^2\Delta P}\right)^2\sigma_V^2 + \left(\frac{1}{V\Delta P}\right)^2\sigma_{\Delta V}^2 + \left(\frac{\Delta V}{V\Delta P^2}\right)^2\sigma_{\Delta P}^2\right] \quad (A4)$$

$$r_v = 1 + \frac{\beta_m}{\beta_c} \quad \sigma_r^2 = \left[\frac{\sigma_{\beta m}^2}{\beta_c^2} + \frac{\beta_m^2}{\beta_c^4}\sigma_{\beta_c}^2\right] \quad (A5)$$

$$\rho_{cmp} = \frac{\rho_{in}}{r_v} \quad \sigma_\rho = \frac{\rho_{in}}{r_v^2}\sigma_r \quad (A6)$$

References

1. Hill, D.P.; Montgomery-Brown, E.-K.; Shelly, D.R.; Flinders, A.F.; Prejean, S. Post-1978 tumescence at Long Valley Caldera, California: A geophysical perspective. *J. Volcanol. Geotherm. Res.* **2020**, *400*, 106900. [CrossRef]
2. Tiampo, K.F.; Rundle, J.B.; Fernandez, J.; Langbein, J.O. Spherical and ellipsoidal volcanic sources at Long Valley caldera, California, using a genetic algorithm inversion technique. *J. Volcanol. Geotherm. Res.* **2000**, *102*, 189–206. [CrossRef]
3. Newman, A.V.; Dixon, T.H.; Ofoegbu, G.I.; Dixon, J.E. Geodetic and seismic constraints on recent activity at Long Valley Caldera, California: Evidence for viscoelastic rheology. *J. Volcanol. Geotherm. Res.* **2001**, *105*, 183–206. [CrossRef]
4. Langbein, J.O. Deformation of the Long Valley Caldera, California: Inferences from measurements from 1988 to 2001. *J. Volcanol. Geotherm. Res.* **2003**, *127*, 247–267. [CrossRef]
5. Newman, A.V.; Dixon, T.H.; Gourmelen, N. A four-dimensional viscoelastic deformation model for Long Valley Caldera, California, between 1995 and 2000. *J. Volcanol. Geotherm. Res.* **2006**, *150*, 244–269. [CrossRef]
6. Liu, Z.; Dong, D.; Lundgren, P. Constraints on time-dependent volcanic source models at Long Valley Caldera from 1996 to 2009 using InSAR and geodetic measurements. *Geophys. J. Int.* **2011**, *187*, 1283–1300. [CrossRef]
7. Montgomery-Brown, E.K.; Wicks, C.W.; Cervelli, P.F.; Langbein, J.O.; Svarc, J.L.; Shelly, D.R. Renewed inflation of Long Valley Caldera, California (2011 to 2014). *Geophys. Res. Lett.* **2015**, *42*, 5250–5257. [CrossRef]
8. Silverii, F.; Pulvirenti, F.; Montgomery-Brown, E.K.; Neely, W.R. The 2011-2019 Long Valley Caldera inflation: New insights from separation of superimposed geodetic signals and 3D modeling. *Earth Planet. Sci. Lett.* **2021**, *569*, 117055. [CrossRef]
9. Savage, J.C.; Clark, M.M. Magmatic resurgence in Long Valley Caldera, California: Possible Cause of the 1980 Mammoth Lakes Earthquakes. *Science* **1982**, *217*, 531–533. [CrossRef] [PubMed]
10. Savage, J.C.; Cockerham, R.; Estrem, J.E.; Moore, L.R. Deformation near the Long Valley Caldera, eastern California, 1982-1986. *J. Geophys. Res.* **1987**, *92*, 2721–2746. [CrossRef]
11. Langbein, J.; Dzurisin, D.; Marshall, G.; Stein, R.; Rundle, J. Shallow and peripheral volcanic sources of inflation revealed by modeling two-color geodimeter and leveling data from Long Valley Caldera, California, 1988-1992. *J. Geophys. Res.* **1995**, *100*, 12487–12495. [CrossRef]
12. Battaglia, M.; Roberts, C.; Segall, P. Magma intrusion beneath Long Valley caldera confirmed by temporal changes in gravity. *Science* **1999**, *285*, 2119–2122. [CrossRef] [PubMed]
13. Battaglia, M.; Segall, P.; Murray, J.; Cervelli, P.; Langbein, J. The mechanics of unrest at Long Valley Caldera, California: 1. Modeling the geometry of the source using GPS, leveling and 2-color EDM data. *J. Volcanol. Geotherm. Res.* **2003**, *16*, 170. [CrossRef]
14. Fialko, Y.; Simons, M.; Khazan, Y. Finite source modelling of magmatic unrest in Socorro, New Mexico, and Long Valley, California. *Geophys. J. Intern.* **2001**, *146*, 191–200. [CrossRef]
15. Feng, L.; Newman, A.V. Constraints on continued episodic inflation at Long Valley Caldera, based on seismic and geodetic observations. *J. Geophys. Res.* **2009**, *114*, 6240. [CrossRef]
16. Ji, K.H.; Herring, T.A.; Llenos, A.L. Near real-time monitoring of volcanic surface deformation from GPS measurements at Long Valley Caldera, California. *Geophys. Res. Let.* **2013**, *40*, 1054–1058. [CrossRef]
17. Langbein, J.; Hill, D.P.; Parker, T.N.; Wilkinson, S.K. An episode of reinflation of the Long Valley Caldera, eastern California, 1989–1991. *J. Geophys. Res.* **1993**, *98*, 15851–15870. [CrossRef]
18. Hildreth, W. Fluid-driven uplift at Long Valley Caldera, California: Geologic perspectives. *J. Volcanol. Geotherm. Res.* **2017**, *341*, 269–286. [CrossRef]
19. Carbone, D.; Currenti, G.; Del Negro, C. Elastic model for the gravity and elevation changes before the 2001 eruption of Etna volcano. *Bull. Volcanol.* **2007**, *69*, 553–562. [CrossRef]
20. Walsh, J.B.; Rice, J.R. Local changes in gravity resulting from deformation. *J. Geophys. Res.* **1979**, *84*, 156–170. [CrossRef]

21. Bonafede, M.; Mazzanti, M. Modelling gravity variations consistent with ground deformation in the Campi Flegrei caldera (Italy). *J. Volcanol. Geotherm. Res.* **1998**, *81*, 137–157. [CrossRef]
22. Currenti, G.; Del Negro, C.; Ganci, G. Modelling of ground deformation and gravity fields using finite element method: An application to Etna volcano. *Geophys. J. Int.* **2007**, *169*, 775–786. [CrossRef]
23. Amoruso, A.; Crescentini, L.; Berrino, G. Simultaneous inversion of deformation and gravity changes in a horizontally layered half-space: Evidences for magma intrusion during the 1982-1984 unrest at Campi Flegrei caldera (Italy). *Earth Planet. Sc. Lett.* **2008**, *272*, 181–188. [CrossRef]
24. Crescentini, L.; Amoruso, A. Effects of crustal layering on the inversion of deformation and gravity data in volcanic areas: An application to the Campi Flegrei caldera, Italy. *Geophys. Res. Lett.* **2007**, *34*, 9919. [CrossRef]
25. Currenti, G. Numerical evidence enabling reconciliation of gravity and height changes in volcanic areas. *Geophys. J. Int.* **2014**, *197*, 164–173. [CrossRef]
26. Trasatti, E.; Bonafede, M. Gravity changes due to overpressure sources in 3D heterogeneous media: Application to Campi Flegrei caldera, Italy. *Ann. Geophys.* **2008**, *51*, 119–133. [CrossRef]
27. Trasatti, E.; Bonafede, M.; Ferrari, C.; Giunchi, C.; Berrino, G. On deformation sources in volcanic areas: Modeling the Campi-Flegrei (Italy) 1982-84 unrest, Earth planet. *Sci. Lett.* **2011**, *306*, 175–185. [CrossRef]
28. Battaglia, M.; Segall, P.; Roberts, C. The mechanics of unrest at Long Valley Caldera, California: 2. Constraining the nature of the source using geodetic and micro-gravity data. *J. Volcanol. Geotherm. Res.* **2003**, *127*, 219–245. [CrossRef]
29. Battaglia, M.; Hill, D.P. Analytical modeling of gravity changes and crustal deformation at volcanoes: The Long Valley caldera, California, case study. *Tectonophysics* **2009**, *471*, 45–57. [CrossRef]
30. Battaglia, M.; Segall, P. The Interpretation of Gravity Changes and Crustal Deformation in Active Volcanic Areas. *Pure Appl. Geophys.* **2004**, *161*, 1453–1467. [CrossRef]
31. Tizzani, P.; Battaglia, M.; Zeni, G.; Atzori, S.; Berardino, P.; Lanari, R. Uplift and magma intrusion at Long Valley caldera from InSAR and gravity measurements. *Geology* **2009**, *37*, 63–66. [CrossRef]
32. Roberts, C.; Jachens, R.; Morin, R. High-precision Stations for Monitoring Gravity Changes in Long Valley Caldera, California. U.S. Geological Survey Open-File Report 88-0050; 1988. Available online: https://pubs.usgs.gov/of/1988/0050/report.pdf (accessed on 15 February 2021).
33. Frassetto, A.M.; Zandt, G.; Gilbert, H.; Owens, T.J.; Jones, C.H. Structure of the Sierra Nevada from receiver functions and implications for lithospheric foundering. *Geosphere* **2011**, *7*, 898–921. [CrossRef]
34. Ruppert, S.; Fliedner, M.M.; Zandt, G. Thin crust and active upper mantle beneath the Southern Sierra Nevada in the western United States. *Tectonophysics* **1998**, *286*, 237–252. [CrossRef]
35. Farr, T.G.; Rosen, P.A.; Caro, E.; Crippen, R.; Duren, R.; Hensley, S.; Kobrick, M.; Paller, M.; Rodriguez, E.; Roth, L.; et al. The shuttle radar topography mission. *Rev. Geophys.* **2007**, *45*, 183. [CrossRef]
36. Flinders, A.F.; Shelly, D.R.; Dawson, P.B.; Hill, D.P.; Tripoli, B.; Shen, Y. Seismic evidence for significant melt beneath the Long Valley Caldera, California, USA. *Geology* **2018**, *46*, 799–802. [CrossRef]
37. Montgomery-Brown, E.; Shelly, D.; Hsieh, P. Snowmelt-triggered earthquake swarms at the margin of Long Valley Caldera, California. *Geophys. Res. Lett.* **2019**, *46*, 3698–3705. [CrossRef]
38. Bernardino, M.V.; Jones, C.H.; Levandowski, W.; Bastow, I.; Owens, T.J.; Gilbert, H. A multicomponent Isabella anomaly: Resolving the physical state of the Sierra Nevada upper mantle from Vp/Vs anisotropy tomography. *Geosphere* **2019**, *15*, 2018–2042. [CrossRef]
39. Brocher, T.M. Empirical relations between elastic wavespeeds and density in the Earth's crust. *Bull. Seismol. Soc. Am.* **2005**, *95*, 2081–2092. [CrossRef]
40. Cheng, C.H.; Johnston, D.H. Dynamic and static moduli. *Geophys. Res. Lett.* **1981**, *8*, 39–42. [CrossRef]
41. Ciccotti, M.; Mulargia, F. Differences between static and dynamic elastic moduli of a typical seismogenic rock. *Geophys. J. Int.* **2004**, *157*, 474–477. [CrossRef]
42. Li, H.B.; Zhao, J.; Li, T.J. Triaxial compression tests on a granite at different strain rates and confining pressures. *Int. J. Rock Mech. Min. Sci.* **1999**, *36*, 1057–1063. [CrossRef]
43. Yang, X.; Davis, P.M.; Dieterich, J.H. Deformation from inflation of a dipping finite prolate spheroid in an Elastic Half-Space as a model for volcanic stressing. *J. Geophys. Res.* **1988**, *93*, 4249–4257. [CrossRef]
44. Nelder, J.A.; Mead, R. A simplex method for function minimization. *Comput. J.* **1965**, *7*, 308–313. [CrossRef]
45. Hickey, J.; Gottsmann, J.; Mothes, P. Estimating volcanic deformation source parameters with a finite element inversion: The 2001–2002 unrest at Cotopaxi volcano. *Ecuad. J. Geophys. Res. Solid Earth* **2015**, *120*, 1731. [CrossRef]
46. Currenti, G.; Del Negro, C.; Ganci, G. Finite element modeling of ground deformation and gravity field at Mt Etna. *Ann. Geophys.* **2008**, *51*, 105–119. [CrossRef]
47. Nikkhoo, M.; Rivalta, E. Analytical solutions for gravity changes caused by triaxial volumetric sources. *Earth Space Sci. Open Arch.* **2021**, *2*, 7706. [CrossRef]
48. Rivalta, E.; Segall, P. Magma compressibility and the missing source for some dike intrusions. *Geophys. Res. Lett.* **2008**, *35*, L04306. [CrossRef]
49. Bonafede, M.; Ferrari, C. Analytical models of deformation and residual gravity changes due to a Mogi source in a viscoelastic medium. *Tectonophysics* **2009**, *471*, 4–13. [CrossRef]

50. Battaglia, M.; Lisowski, M.; Dzurisin, D.; Poland, M.P.; Schilling, S.; Diefenbach, A.; Wynn, J. Mass addition at Mount St. Helens, Washington, inferred from repeated gravity surveys. *J. Geophys. Res.* **2018**, *123*, 1856–1874. [CrossRef]
51. Mastin, L.G.; Lisowski, M.; Roeloffs, E.; Beeler, N. Improved constraints on the estimated size and volatile content of the Mount St. Helens magma system from the 2004–2008 history of dome growth and deformation. *Geophys. Res. Lett.* **2009**, *36*, L20304. [CrossRef]
52. Miyagi, I.; Kita, N.; Morishita, Y. The geochemical and petrological characteristics of prenatal caldera volcano: A case of the newly formed small dacitic caldera, Hijiori, Northeast Japan. *Contrib. Miner. Pet.* **2017**, *172*, 1–17. [CrossRef]
53. Anderson, K.; Segall, P. Physics-based models of ground deformation and extrusion rate at effusively erupting volcanoes. *J. Geophys. Res.* **2011**, *116*, 7939. [CrossRef]
54. Snieder, R. The role of nonlinearity in inverse problems. *Inverse Probl.* **1998**, *14*, 387. [CrossRef]
55. Cervelli, P.F.; Langbein, J.O.; Perkins, J.P.; Svarc, J.L.; Owen, S.E. Campaign GPS Measurements from 2000-2010 in the Sierra Block South of Long Valley Caldera, CA, USA. In Proceedings of the American Geophysical Union Fall Meeting, San Francisco, CA, USA, 13–17 December 2010.
56. Malfait, W.J.; Sanchez-Valle, C.; Ardia, P.; Médard, E.; Lerch, P. Amorphous Materials: Properties, Structure, and Durability: Compositional dependent compressibility of dissolved water in silicate glasses. *Am. Mineral.* **2011**, *96*, 1402–1409. [CrossRef]
57. Long, S.M.; Grosfils, E.B. Modeling the effect of layered volcanic material on magma reservoir failure and associated deformation, with application to Long Valley caldera, California. *J. Volcanol. Geoth. Res.* **2009**, *186*, 349–360. [CrossRef]
58. Magni, V.; Battaglia, M.; Tizzani, P.; Manconi, A.; Walter, T. Axial symmetric crustal deformation model for Long Valley Caldera, California. In Proceedings of the COMSOL Conference 2008, Hannover, Germany, 4–6 November 2008.
59. Pepe, S.; Tizzani, P.; Manconi, A. Numerical inversion of surface deformation at Long Valley Caldera (California) by using 3D mechanical models. In Proceedings of the COMSOL Conference 2010, Paris, France, 17–19 November 2010.
60. Clark, D.A.; Saul, S.J.; Emerson, D.W. Magnetic and gravity anomalies of a triaxial ellipsoid. *Explor. Geophys.* **1986**, *17*, 189–200. [CrossRef]

remote sensing

MDPI

Article

Plume Height Time-Series Retrieval Using Shadow in Single Spatial Resolution Satellite Images

Sophie Pailot-Bonnétat [1,*], Andrew J. L. Harris [1], Sonia Calvari [2], Marcello De Michele [3] and Lucia Gurioli [1]

[1] Laboratoire Magmas et Volcans, Université Clermont Auvergne, CNRS, IRD, OPGC, 63000 Clermont-Ferrand, France; andrew.harris@uca.fr (A.J.L.H.); lucia.gurioli@uca.fr (L.G.)

[2] Osservatorio Etneo—ezione di Catania, Istituto Nazionale di Geofisica e Vulcanologia, Piazza Roma 2, 95125 Catania, Italy; sonia.calvari@ingv.it

[3] BRGM, Bureau de Recherches Géologiques et Minières, 3 Avenue Claude Guillemin, F-45060 Orléans, France; m.demichele@brgm.fr

* Correspondence: sophie.pailot_bonnetat@uca.fr; Tel.: +33-(0)631358090

Received: 13 October 2020; Accepted: 30 November 2020; Published: 3 December 2020

Abstract: Volcanic plume height is a key parameter in retrieving plume ascent and dispersal dynamics, as well as eruption intensity; all of which are crucial for assessing hazards to aircraft operations. One way to retrieve cloud height is the shadow technique. This uses shadows cast on the ground and the sun geometry to calculate cloud height. This technique has, however, not been frequently used, especially not with high-spatial resolution (30 m pixel) satellite data. On 26 October 2013, Mt Etna (Sicily, Italy) produced a lava fountain feeding an ash plume that drifted SW and through the approach routes to Catania international airport. We compared the proximal plume height time-series obtained from fixed monitoring cameras with data retrieved from a Landsat-8 Operational Land Imager image, with results being in good agreement. The application of the shadow technique to a single high-spatial resolution image allowed us to fully document the ascent and dispersion history of the plume–cloud system. We managed to do this over a distance of 60 km and a time period of 50 min, with a precision of a few seconds and vertical error on plume altitude of ±200 m. We converted height with distance to height with time using the plume dispersion velocity, defining a bent-over plume that settled to a neutral buoyancy level with distance. Potentially, the shadow technique defined here allows downwind plume height profiles and mass discharge rate time series to be built over distances of up to 260 km and periods of 24 h, depending on vent location in the image, wind speed, and direction.

Keywords: Etna volcano; lava fountain; paroxysmal explosive eruptions; ash plume height; Landsat 8 satellite images; mass discharge rate time-series

1. Introduction

Volcanic ash clouds injected into the atmosphere represent a threat to population health and aircraft operations [1]. This is a recurrent problem, and despite the improved monitoring of volcanic clouds, aircraft encounters and incidents occur regularly (e.g., [1–5]). Damage to aircraft, traffic rerouting and airport traffic disruption can cause millions of dollars in losses for airline companies and any related businesses [6]. Even explosive eruptions at volcanoes located in unpopulated areas represent a threat because the associated ash clouds can be transported thousands of kilometers away from their sources [7]. In volcanic hazard assessment, the retrieval of cloud height is a key parameter because it can be linked to dispersion [8,9], and is used to derive other, higher-level parameters that allow better classification of the eruption and understanding of the associated eruption dynamics, such as the Volcanic Explosive Index (VEI) [10] and mass discharge rate (MDR, i.e., mass flux per unit time in kg s^{-1}) [8].

The synoptic view offered by satellites is especially useful in observing large-scale phenomena such as volcanic clouds [11,12]. Other advantages of satellite data include their free or cheap availability, rapid (near-real-time) dissemination, synoptic view and global coverage [7]. Moreover, many volcanoes that experience low-frequency but high-magnitude and -intensity—Vulcanian through Plinian—eruptions [13,14], do not possess ground-based monitoring networks so that sometimes satellite data are the only available data to detect, track and study an eruption [11,12,15]. Satellites have thus been used for an increasing number of studies focused on detecting volcanic clouds and tracking their dispersal since the 1970s. As part of this, several methods that use satellite sensor-based data have been developed to retrieve cloud heights (as reviewed in Supplementary Materials Table S1). One of them is the height-from-shadow technique, whereby the length of the shadow cast by the cloud onto the ground is measured, and a "simple" trigonometric relationship with the sun geometry delivers the cloud height [16]. However, the shadow technique has not frequently been used for volcanological applications, especially using high-spatial resolution (30 m pixel) data, such as LANDSAT's MSS, TM, ETM+ and OLI, Terra's ASTER and ESA's Sentinel (cf. Supplementary Materials Table S2). The height-from-shadow method does have some issues, such as the assumptions that need to be made regarding cloud shape or plume top topography [17]. It is also only applicable during daytime when shadows are present. It is, however, unclear why such high-spatial resolution data have not been used to more frequently retrieve heights for volcanic clouds. This is especially so when considering problems associated with the commonly used height-from-temperature or brightness temperature method, which suffers from problems such as plume transparency or ice formation [7,18–21], and which are more typically applied to low-spatial resolution (1–4 km pixel) data from sensors such as AVHRR, MODIS, GOES, and SEVIRI. The shadow technique also has the advantage of not requiring stereo-imaging or temperature measurement, nor does it require assumptions regarding cloud thermal and/or emissive and transmissive properties. In this study, we thus develop and validate a method that allows the shadow technique to be used to (i) retrieve cloud height in high-spatial resolution data, and (ii) convert the downwind plume heights to at-vent MDR time series.

A distinction needs to be made between volcanic plumes, which are connected to the source (rooted to the vent), and volcanic clouds, which have become detached from the source [22]. The former is characterized by a dominant vertical component, the latter by a dominant horizontal component. Eruptive columns can be divided from bottom to top into three regions. The first and lowermost is the gas-thrust region immediately above the vent where the plume rises as a momentum-driven jet [8,23,24]. The second is the convective region, where the density of the plume becomes less than that of the surrounding atmosphere, atmospheric mixing occurs and ascent by buoyancy takes over [8,23]. The third is the umbrella region, where the densities of the plume and the atmosphere become equal and the cloud spreads laterally, defining the neutral buoyancy level (NBL) [23,25]. Initially, though, the maximum plume height will be higher than the NBL because of the inertia of the particles arriving at the top of the column to lead to overshooting [23]. Thus, while the dispersal of volcanic clouds depends on their interaction with atmospheric motions (mostly wind speed and direction), plume dynamics will also depend on atmospheric stratification through the entire height of their ascent, variation in wind speed with height, as well as "source" conditions, notably MDR and air entrainment rates [8,22]. Depending on the eruption intensity (MDR) and wind strength (speed), we can find a continuum of plume behaviors. The end members are umbrella clouds, which are strong plumes that rise above the vent and spread horizontally around the neutral buoyancy height, and bent-over or weak plumes [26] that are significantly distorted downwind and may not reach the NBL before being sheared-off downwind [8].

For this study, a cloud-free Landsat-8 Operational Land Imager (OLI) image of the 26 October 2013 lava fountaining event at Mount Etna (Sicily, Italy) was selected. This target has the advantage of possessing excellent ground-truth data, as are available from the continuously monitoring camera network operated by the Istituto Nazionale di Geofisica e Vulcanologia (INGV—Osservatorio Etneo, Catania, Italy). This consists of a network of thermal and visible cameras (Figure 1), which have been previously applied to plume tracking [27] and which are routinely used for monitoring. Their specifications are detailed in Ref. [28], where the GPS-stamped image frames allow for precisely timed event detection. In processing the Landsat 8 data, to produce plume or cloud height profiles versus distance from the vent, several parameters such as plume top topography, shape, ground topography and satellite viewing geometry need to be taken into account [cf. 17]. This is done here through developing a methodology to extract height with distance using the cloud shadow. We convert these profiles into times series using cloud dispersal velocities obtained from the INGV camera network. This allows us to plot plume height over the vent for one hour prior the satellite image acquisition and to derive an MDR time series. This permits us to (i) describe the effects of settling to the NBL, (ii) assess the effect of wind on the bending of a fire fountain, and (iii) derive physical and empirical relations for the ascent and dispersal of this fountain-fed plume–cloud system.

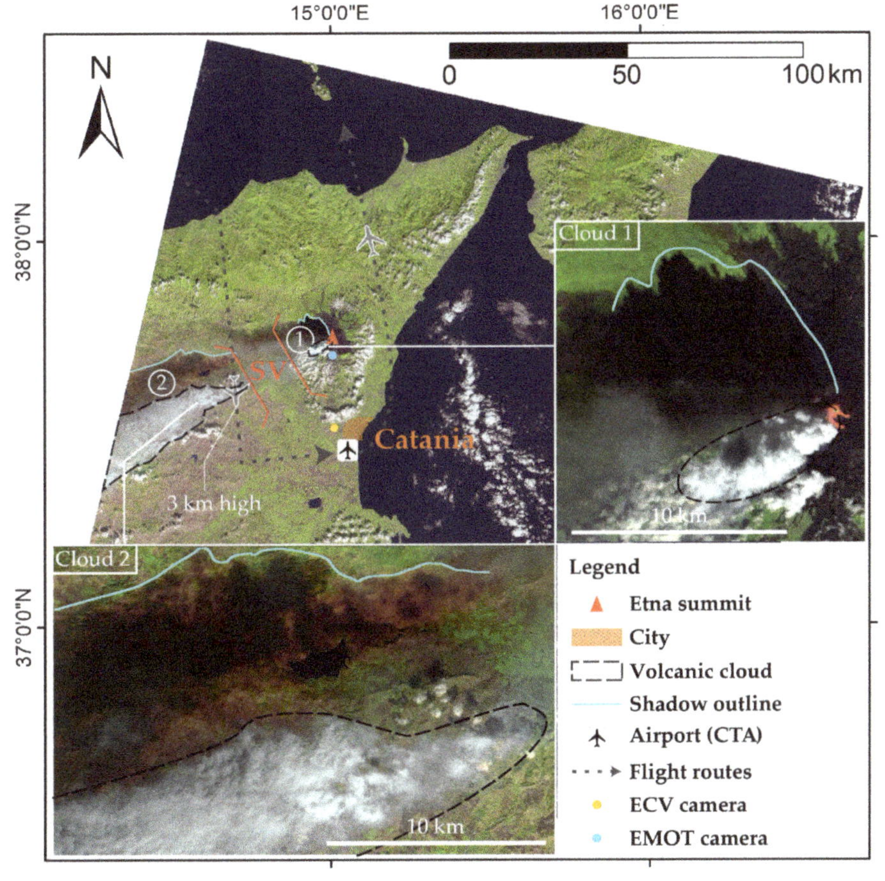

Figure 1. Ortho-rectified Landsat 8 OLI image of Mt. Etna acquired on 26 October 2013 at 09:37 UTC (courtesy of National Aeronautics and Space Administration (NASA)/United States Geological Survey (USGS)). Flight routes and altitudes for air traffic approaching Catania Fontarossa airport are taken from www.flightradar24.com. The two cameras from the INGV—Osservatorio Etneo camera network used in this paper are marked. A zoom on each cloud is given as inset. SV = Simeto Valley; CTA = Catania Fontanarossa airport.

2. Case Study: OLI Imagery of the 25–26 October 2013 Fountaining at Etna

Mount Etna is characterized by Strombolian and lava fountaining activity, and frequent effusive eruptions (e.g., [29,30]). The climaxes of lava fountaining at Etna, which gave rise to ash plumes and clouds, are termed "paroxysmal episodes" [28]. The fall out of lapilli and ash represents a frequent hazard to the population residing on the flanks of Etna [31], which is around one million [32]. The threat to Catania airport is also a persistent hazard [5] where, for example, between 2011 and 2015 alone there were 49 lava fountain events [28], and 150 between 1990 and 2015 [33]. Fountaining and Strombolian, as well as Vulcanian, activity can be sourced from flank vents feeding effusive activity, as well as from the summit craters. Etna hosts five active summit craters: Bocca Nuova (BN), Voragine (VOR), North-East Crater (NEC), South-East Crater (SEC) and the New South-East Crater (NSEC) [28,34]. Of these, SEC has become the most common locus of lava fountain activity since its formation in the 1970s (e.g., [28,35]). The NSEC developed from a pit on the east flank of the SEC that formed in 2011 and fed fountaining episodes between 2011 and 2015 [29,30]. It is one of these events, that of 25–26 October 2013, that we focus on here.

2.1. The 25–26 October 2013 Eruption

The 25–26 October 2013 eruption occurred after a six-month-long pause in activity [36]. It was marked by activity from several craters at the same time, with Strombolian explosions starting on 25 October at the NSEC. Its intensity increased during the night, and a lava fountain began around ~01:30 (hh:mm; all times are Coordinated Universal Time, UTC) on 26 October [36] feeding a light-toned ash plume. Lava flowed out of the NSEC at ~03:15 [36] and the NEC also started to emit a dark-gray ash plume at ~06:20 that dispersed at a lower altitude than the NSEC plume. At the same time, occasional explosions occurred at BN. The lava fountain at NSEC lasted until around 10:00, reaching an average height of 430 m [28]. Its plume consisted initially mostly of gas, but the amount of pyroclastics increased with the increase in intensity, causing tephra fallout. The plume was, though, rapidly bent-over toward the west southwest by the wind [37].

A satellite image from Landsat 8's OLI sensor was acquired at 09:37 on 26 October around the end of the lava fountain episode (Figure 1). The two clouds from the NSEC and the NEC are distinguishable from the tonal difference and are visibly being blown toward the WSW by the wind. The light-toned cloud from the NSEC seems to be separated into two parts, the first extending 8 km from the vent; then a second, after a gap of 22 km, being at least 30 km long (it running off of the western edge of the image). The two parts will be referred to, hereafter, as cloud 1 for the near vent part and cloud 2 for the distal part. Cloud heights for this event are available for validation purposed from the INGV camera network (Figure 1), and have been derived from this same image using an independent, parallax-based method, i.e., the Plume Elevation Model (PEM) of De Michele et al. (2019) [38], thus also allowing inter-method comparison.

2.2. Data

Since the method developed here needs a clear view of the shadow cast by a plume, or cloud, on the ground, only satellite images taken during daylight can be considered. The meteorological cloud cover needs to be as little as possible to ensure the best visibility, even though shadows cast over meteorological clouds can be used, provided that the meteorological cloud top height is known or can be calculated at the same time. The volcanic cloud must also be sufficiently opaque so as to provide a clearly defined shadow. However, even if the cloud is not dense or is partially transparent, the method can be applied as long as a shadow is visible. This is an advantage over the brightness temperature method or the PEM procedure, where transparency or lack of optical thickness can cause plume height underestimations [7,38].

The LANDSAT 8 OLI image was downloaded directly from the Earthdata Search engine (https://search.earthdata.nasa.gov/search), which provides access to NASA's Earth Observing System Data and Information System (EOSDIS) services. The product available is a level 1 product, which is terrain-corrected. The pre-processing of such level 1 products includes geo-referencing (alignment of imagery to its correct geographic location) and ortho-rectification (correction for the effects of relief and view direction on pixel location) to ensure the exact positioning of the image [39]. For a cloud, though, this is not ideal, as geo-referencing can introduce distortions when a cloud is projected onto the surface [40]. LANDSAT 8's OLI and Thermal Infrared Sensor (TIRS) provides 11 different bands of data at a spatial resolution of 15–100 m (Table 1).

Table 1. LANDSAT 8's OLI and TIRS instrument characteristics.

Bands	Wavelength (μm)	Spatial Resolution (m)	Measure
Band 1—Coastal aerosol	0.43–0.45	30	TOA/SR
Band 2—Blue	0.45–0.51	30	TOA/SR
Band 3—Green	0.53–0.59	30	TOA/SR
Band 4—Red	0.64–0.67	30	TOA/SR
Band 5—NIR	0.85–0.88	30	TOA/SR
Band 6—SWIR 1	1.57–1.65	30	TOA/SR
Band 7—SWIR 2	2.11–2.29	30	TOA/SR
Band 8—Panchromatic	0.50–0.68	15	TOA
Band 9—Cirrus	1.36–1.38	30	TOA
Band 10—Thermal Infrared (TIRS) 1	10.6–11.19	100	Temperature
Band 11—Thermal Infrared (TIRS) 2	11.50–12.51	100	Temperature

NIR = near infrared; SR = surface reflectance; SWIR = shortwave infrared; TIRS = thermal infrared; TOA = top of atmosphere reflectance.

3. Methodology

3.1. Height-from-Shadow Time Series from a Single High-Spatial Resolution Image

The height-from-shadow technique is based on the fact that the plume and cloud casts a shadow on the ground. If the length of the shadow and the angle between the ground and the sunray direction is known, then basic geometry delivers an approximation of the plume height [17,19,41–46]. We here lay out a six-step methodology that begins with the definition of the fundamental Sun–Earth astronomical relationships that underpin the methodology, and ends with the generation of a cloud height time series.

3.1.1. Sun–Cloud Geometric Relations

The Sun–Earth astronomical relationships from Iqbal (1983) [47] were used to calculate the sun elevation and sun azimuth at the time of image acquisition, as converted to true solar time by taking into account the latitude and longitude of the summit of Etna. To facilitate the length measurements, the North-oriented image was rotated clockwise by the sun azimuth angle (Ψ) using cubic convolution. As a result, the sunrays are orientated in a vertical direction, so that shadow length along a correct orientation can be made along easy-to-define vertical lines. (Figure 2) A second line, called the intersection line, was composed of as many segments as there were direction changes in the cloud, and was drawn following the center of clouds 1 and 2. This line represents the position of the cloud height profile that will be created. The image was contrast-enhanced to maximize the intensity of the shadow darkness and the contrast with the background. As shown in Figure 2, a shadow length measurement (SLM) was carried out for each peak and trough in the enhanced shadow outline (i.e., at each point of variation, the shadow being the same between each turn-around point). Each selected point on the shadow outline was then linked to its matching point on the intersection line, following a vertical line, which now (following the rotation) represents the sun azimuth direction (Figure 2). The measurements were made in pixels and converted to distance (in m) using the pixel size calculated to take into account off-nadir and Earth curvature distortion following Harris (2013) [48].

Finally, each shadow length measurement was linked to a distance from the vent using the geographical coordinate of each pixel and taking into account Earth curvature.

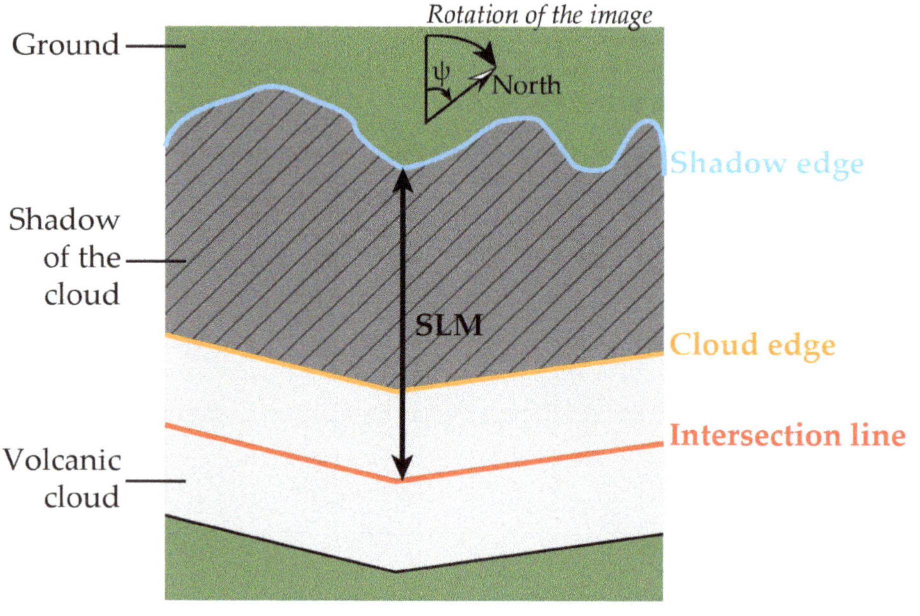

Figure 2. Schematic satellite view of the rotated plume/shadow system and parameter definition.

3.1.2. Viewing and Shadow Geometry

Whether the entire length of the shadow is visible or not depends on the viewing geometry of the satellite with respect to the sun geometry (Figure 3a). Taking into account the distance between the Sun and the Earth relative to the size of the object that casts a shadow on the Earth's surface, the sunrays are assumed to be parallel lines. If the entire length of the shadow is visible, then a simple geometry is valid. However, if there is no area on the ground illuminated by the sun between the cloud and its shadow, then the distance needed for the calculation is missing a part of the shadow which is hidden behind the cloud (Figure 3b). In the case of a hidden shadow on a horizontal surface, the following correction can be applied:

$$\textit{Plume altitude (hidden shadow case)} = D' \tan\alpha + \frac{(D' \tan\alpha)^2}{D' \tan\sigma - D' \tan\alpha} \tag{1}$$

where α is the sun elevation and σ the satellite scan angle. In our case, there was no hidden part, so Equation (1) did not need to be applied.

Four cases need to be considered depending on whether shadows are also apparent on the cloud top or not. In the first case, there are no shadows on the plume top, meaning that the plume top is relatively flat (Figure 4a). In the second case, there are shadows on the plume top, meaning that two different heights can be measured: the cloud hedge height, and the variation in height between the cloud edge and structures within the cloud (Figure 4b). In the third case, the shadow extends to the edge of the cloud, meaning only the cloud top height can be measured (Figure 4c). The final case is one intermediate between cases i and iii (Figure 4d). For the imaged case, the high-spatial resolution allowed us to determine visually that cloud 1 belongs to cases ii, iii and iv, and cloud 2 belongs to case i. As a result, our height measurements use the length of the cloud shadow on the ground (D) to obtain the height of the cloud edge (H), and the lengths of shadows cast on the cloud top (d) to obtain variations in the height (h) across the cloud depending on the case identified (Figure 4). The case that needed to be applied to any point in the cloud was determined by visual inspection of the imagery.

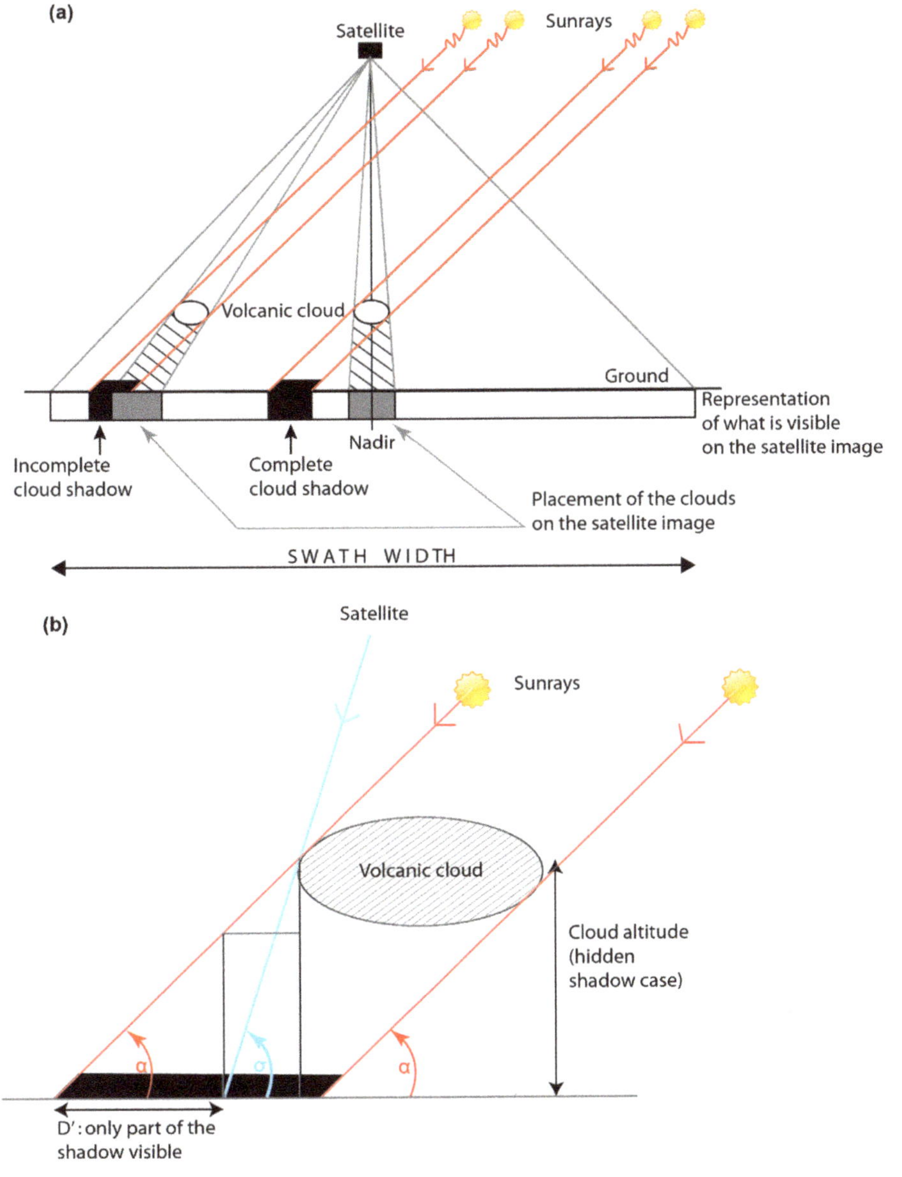

Figure 3. Satellite and Sun viewing geometry responsible for a "missing" shadow portion. (**a**) General geometry. (**b**) Detailed geometry for the case where a portion of the shadow is hidden.

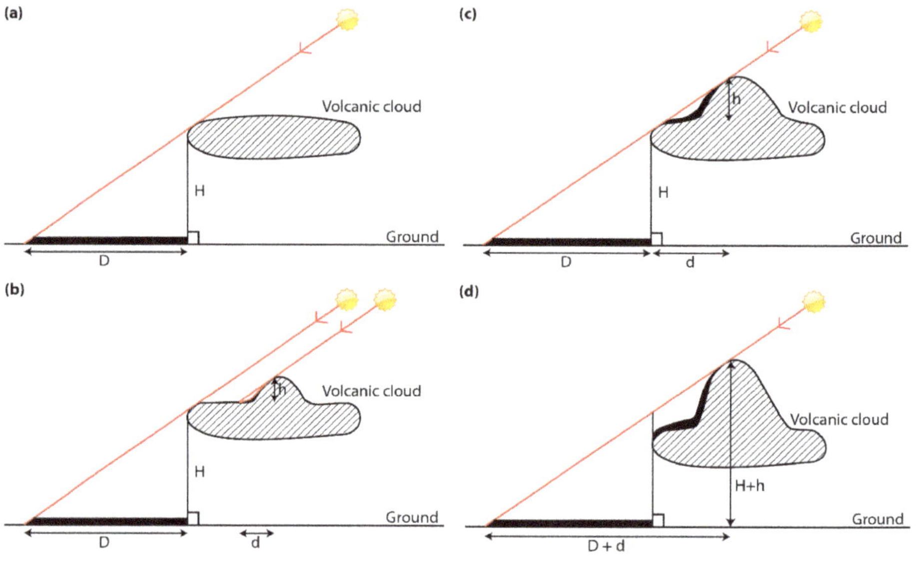

Figure 4. Plume/cloud height measurement at nadir case depending on the shadow distribution. (**a**) Case i: flat plume. (**b**) Case ii: plume with topography where measurement of H and h is possible. (**c**) Case iii: plume with topography where measurement of H and h is possible. (**d**) Case iv: intermediate case: only the measurement of H+h is possible.

3.1.3. Height Calculations and Corrections

The simple trigonometric relationship between sun elevation, shadow length and cloud height uses a distance projected onto a horizontal plane. However, the shadow length visible on the satellite image is modified by the slope of the terrain onto which it is projected. Two two-dimensional cases were considered. For case 1, the shadow is cast onto a surface tilted in the same direction as the sunrays (Figure 5a) and in the opposite direction in case 2 (Figure 5b). The plume altitude for each case can be calculated following:

$$Plume\ altitude\ (case\ 1) = \tan\alpha\left(\frac{H1}{\tan(\sin^{-1}(\frac{H1}{SLM}))}\right) + AI \tag{2}$$

$$Plume\ altitude\ (case\ 2) = \tan\alpha\left(\frac{H1}{\tan(\sin^{-1}(\frac{H1}{SLM}))} + \frac{H1}{\tan\alpha}\right) + AI \tag{3}$$

where $H1$ is the elevation difference between the cloud shadow edge and its matching point in the intersection line, SLM is the shadow length measurement, and AI is the altitude of the point on the intersection line. This last value is added to the above-surface height measurement to obtain the plume altitude in meters above sea level (m a.s.l.), rather than meters above the ground surface (Figure 5b). The sign of $H1$ determines the case: if $H1$ is negative, then the geometry can be described by case 1; if positive, by case 2. The ground surface altitudes were obtained using Google Earth Pro with a precision of ±5 m.

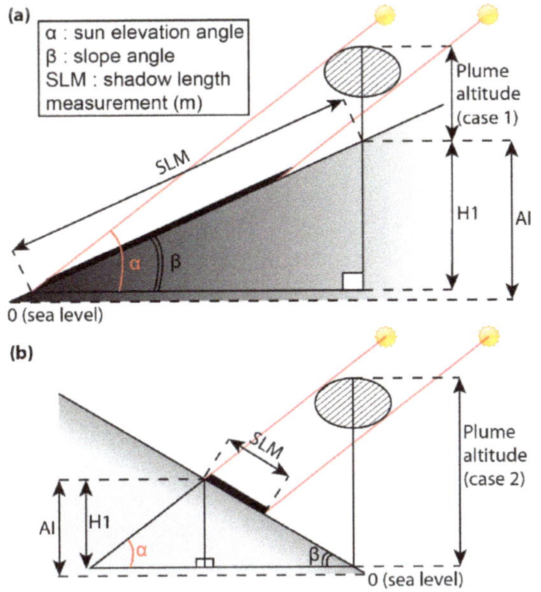

Figure 5. Effect of topography on shadow length measurement. (**a**) Case 1: surface tilted in the same direction as the sunray direction. (**b**) Case 2: surface tilted in the opposite direction to the sun ray direction.

Next, we need to correct for the effect of the ortho-rectification, where the raw data are projected onto a digital elevation model. While this insures that all ground points are located correctly and image distortion is removed [39], it also projects airborne objects (i.e., clouds) from their position above the ground onto the topography so that they consequently become distorted. As a result, the shadow edge line is at its correct location on level 1 data, but the cloud edges and cloud-top features are not (Figure 6a). The cloud appears, instead, as a flat feature that is draped across the topography. This affects the shadow length measurements. We can obtain the shadow length measurement in geometrically corrected level 1 (SLM_1) data, but we need the shadow length measurement in reality (SLM_r), i.e., corrected for geometric distortion, if we are to correctly calculate the actual cloud height, h_p (Figure 6b). In the case of a projection onto a plane in the direction of the sunrays, the problem can be resolved in two dimensions (Figure 6b). In Figure 6b, $H1$ is the altitude difference between plume edge and the shadow edge as projected onto the topography in the ortho-rectified image. $H2$ is the altitude difference between the cloud edge as located on the ortho-rectified image and the intersection with the sunrays above this point on the ground. This, effectively, projects the position of the cloud edge from its flattened position on the ground vertically upwards onto its actual above-ground position. As a result, we define parameter x, which is the altitude difference between the cloud edge as placed on the level 1 image and the actual location of the cloud edge in the air (Figure 6b). We thus first need to calculate SLM_r (Equation (4)), which then allows us to estimate h_p using Equation (5):

$$SLM_r = \frac{SLM_1 * \left[H1 + H2 + \left(\tan\alpha * \frac{H2*SLM_1}{\tan\sigma - H1 - H2}\right)\right]}{H1 + H2} \quad (4)$$

$$h_p = SLM_r - \mathrm{H1} - x \quad (5)$$

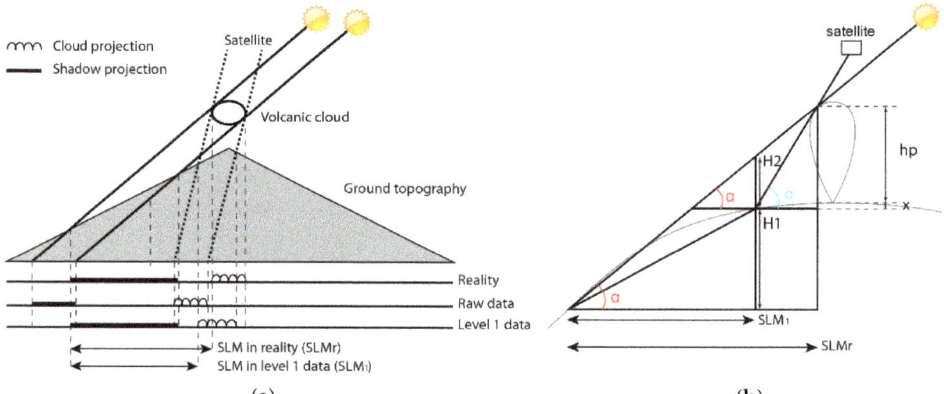

Figure 6. (**a**) Comparison of cloud and shadow projections between raw and post-processed (geometrically corrected, level 1) data. (**b**) Geometry for the resolution of the ortho-rectification issue in two dimensions.

Glaze et al. (1999) argued that one of the issues of the shadow technique is that the sunrays, most of the time, do not hit the cloud at the very top, but instead may intersect with the cloud a little below its top [17]. This results in an underestimation of the cloud height. In such a case, the cloud maximum height, H_{max}, can, however, be calculated by assuming a spherical geometry for a convective ash puff, as drawn by Sparks et al. (1997) [8]. This spherical geometry is described in Figure 7, and in such a case the height h_s to add to the plume altitude of Equation (5) in order to obtain the maximum cloud height is given by:

$$h_s = 2\sin(\frac{\alpha}{2})\frac{d_w}{2}\cos(\frac{180-\alpha}{2}) \tag{6}$$

where d_w is the cloud width measured with the same method as the *SLM*. Consequently:

$$H_{max} = plume\ altitude\ (case\ 1\ or\ 2) + h_s \tag{7}$$

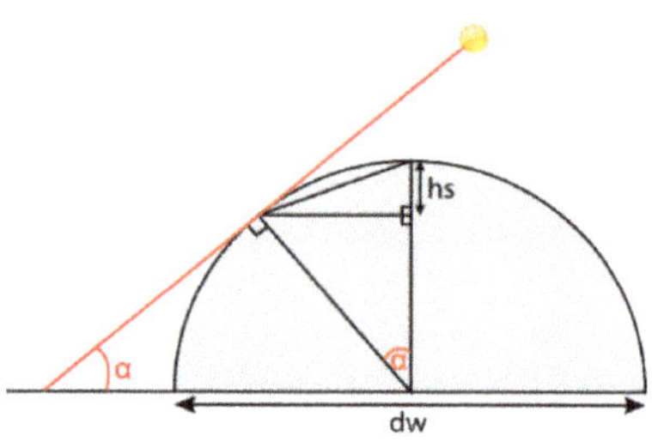

Figure 7. Spherical plume top geometry correction applied to estimate the maximum cloud height; a case to be used when sunrays intersect with the plume not at the very top of the plume, but a little below.

3.1.4. Error Calculation

A few error sources can be neglected because of the relatively low scan angles of Low Earth Orbit (LEO) sensors such as Landsat [49]. This minimizes problems due to pixel distortion, overlap, rotation and spherical shape [48]. The first major component of the possible errors is radiation smearing. Indeed, the spectral radiance arriving at the sensor is not 100% exclusive to each pixel. The spatial distribution of radiance across the pixel is described by the point spread function [48]. This has a bell shape, and for TM pixels spreads into one neighboring pixel [50]. The effect is especially strong when the contrasts between radiances are great, such as between dark shadow and light-colored ground. Consequently, spectral radiance from shadow pixels contaminates two to three neighboring pixels, causing the limit to blur by ±2 or 3 pixels. Moreover, the rotation performed in order to facilitate the shadow length measurements involves a restructuring of the pixels in relation to each other using cubic convolution. This operation is known to change the gray-scale value of some pixels, and can result in the insertion of "fill" pixels [51], possibly modifying the pixels selected during measurements. This effect is estimated to also affect an area of ±1 pixels. We can obtain the shadow length measurement error by multiplying the number of pixels by the pixel size previously calculated, and then the cloud height error by trigonometry using the sun azimuth angle. A summary of the resulting cloud height error estimations is given in Table 2. Cloud 2's altitude error is greater than cloud 1 because of its greater transparency, which made it more difficult to select edge and shadow pixels. H_{max} is the parameter with the greatest error because its calculation used two more pixel selections (for the plume width).

Table 2. Plume height error estimations.

Cloud 1		Cloud 2
Cloud altitude error	H_{max} error	Cloud altitude error
±140 m	±210 m	±185 m

We note that our solution for the effect of ortho-rectification is only for a 2D case, when the problem is three-dimensional. One solution would be the application of the Fmask algorithm developed by Zhu and Woodcock (2011) [52], and its upgrade for application to mountainous areas MFmask [40]. These algorithms are designed to automatically detect cloud and cloud shadow and match them according to their shapes. One of their inputs, though, is the cloud base height, which is needed to produce a double projection and increase the match between cloud and cloud shadow. To do this, Zhu and Woodcock (2011) used the temperature method to extract height [52]. This introduces further uncertainty inherent in applying the height-from-temperature method, which we strive to avoid here. Nevertheless, given the topography, sun angles and satellite-viewing geometry for the case considered here, we estimate only a slight underestimation of shadow length (by no more than 10%) for these low cloud heights. Error will, though, increase with cloud height, thus becoming more of a concern for higher (sub-Plinian and Plinian) clouds.

3.1.5. Cloud Dispersal Velocity Calculation

The ground-based video camera footage was used to calculate the cloud dispersal velocity. Frames from these data are stamped with GPS times, so that timings are accurate to less than a second. Every 10 min, two consecutive camera frames were selected (separated by less than 30 s). A point in each frame was chosen based on the same recognizable plume top shape and the number of pixels between the two points was counted. The points were chosen to be the highest and farthest from the vent possible in order to avoid the cloud dispersal velocity being affected by the jet velocity. Now velocity can be obtained from the time difference between the two frames and the distance moved. This gave a mean dispersal velocity of 18.5 m s^{-1}, with a range of 12.6 to 26.7 m s^{-1} (Table 3). This is in agreement with the value used by De Michele et al. (2019) [38] and the wind speed profile from the National Center for Environmental Prediction (NCEP)/National Center for Atmospheric Research (NCAR) at 12:00 UTC (i.e., 18 m s^{-1} at 10,000 m a.s.l.).

3.1.6. Generation of Cloud Height Time Series

Each cloud height calculated along the intersection line was linked to an emission time using the distance from the vent (d_{VIx}) and the cloud dispersal velocity (V). To do this, the time of release of the cloud at any given point can be obtained by subtracting the time the cloud has taken to travel from the vent to the point (=d_{VIx}/V) from the time of image acquisition (i.e., 09:37:47 UTC). For cloud 1, the cloud dispersal velocity used was that calculated between 9:30 and 9:40. Since the proximal edge of cloud 1 was calculated to have been emitted at 9:32:58 and the image was acquired just before 09:38, only one cloud dispersal velocity was available (Table 1). For cloud 2, the cloud dispersal velocity was changed whenever the velocity for each 10-min period of cloud dispersal velocity calculation from the ECV camera changed (Table 1). The time series generated in this way from a single, static, high-spatial resolution image gives us the possibility to track plume height for several hours prior to image acquisition, with a temporal resolution of (in our case) down to 1 s over around one hour.

Table 3. Cloud dispersal velocities calculated with ECV camera footage.

Time Range	Cloud Dispersal Velocity (m/s) (This Study)
8:40–8:50	17.4
8:50–9:00	19.6
9:00–9:10	16.7
9:10–9:20	21.5
9:20–9:30	26.2
9:30–9:40	26.7

3.2. Cloud Height Validation

3.2.1. Cloud Height from the INGV-OE ECV Camera

The ground-based camera that we use is named after its location label (i.e., ECV) and is located 26.7 km to the South of the NSEC (see [28] for location and further camera details). Scollo et al. (2014) have developed a method to retrieve plume height from images acquired by the camera located at ECV [53], which involves placing a grid showing a succession of heights over the image. The ECV camera can capture videos of Etna's eruption columns up to a height of 9 km a.s.l, with a height calculation error of ±500 m [53]. Scollo et al. (2014) provided two examples of grids applied to two previous events [53]. These were modified in order to fit the 26 October 2013 event (Figure 8). The wind on 26 October was blowing the plume in the WSW direction while the two example grids were fitted for plumes spreading in the ESE direction. Consequently, the grid was horizontally mirrored and overlain onto the camera frames (Figure 8). The plume height was then measured as the maximum apparent height on the frame.

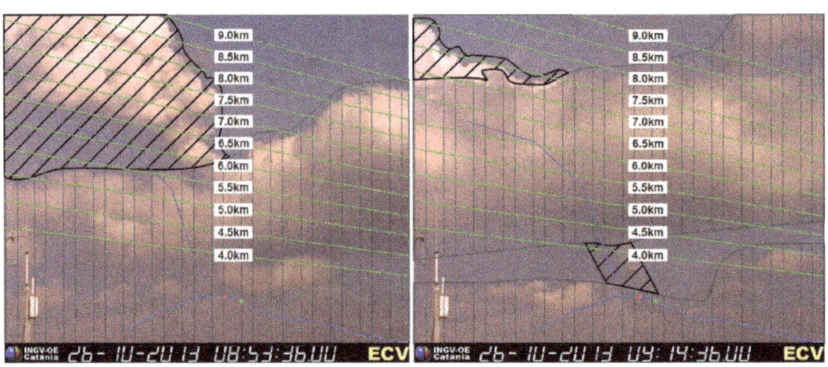

Figure 8. Plume height retrieval from ECV camera frames (at 08:53:36 and 09:14:36) using Scollo et al.'s (2014) [53] grid (green lines). The blue lines are Etna's outline and the direction of the plume ascent (wind direction dependent). The meteorological and volcanic clouds are highlighted respectively in grey and black hatching.

The fountaining activity started to produce a plume at around 02:45, and the sun rose around 05:00. Consequently, being a visible camera, ECV was not usable for the first two hours of the eruption. However, once enhanced, all frames after 04:30 had sufficient luminosity to allow the plume top to be viewed. As such, we were able to derive a cloud height time series with a temporal resolution of one frame per minute between 04:30 and 09:37 for validation of the OLI-derived time series.

3.2.2. Cloud Height from PEM

De Michele et al. (2019) created a method to extract the cloud top height from ortho-rectified Landsat 8 OLI images [38]. This method takes advantage of the physical offset between the multispectral and the panchromatic bands of the push broom sensor. This results in a short time lag (0.52 s) between acquisitions of each band. This is equivalent to the use of the stereoscopic parallax introduced by Prata and Turner (1997) [54] to take advantage of the two viewing angles provided by the Along-Track Scanning Radiometer (ATSR), but using only one single satellite pass. The Plume (top) Elevation Model (PEM) method was introduced by De Michele et al. (2016) [55] and applied to the plume from the 2014 Holuhraun (Iceland) eruption using raw (level 0) OLI data [55], and has since been developed for standard (level 1) OLI data for the same 26 October 2013 Etna lava fountain event considered here [38]. The method produces a digital elevation model for the plume top (a PEM). A new PEM for Etna's 26 October 2013 cloud was created here to correct for the ground topography. Cloud height was then retrieved at the same points used by the shadow technique. The PEM calculations, however, use a pixel window of 48 pixels and, as the window is moved across the image in steps of 24 pixels, the pixel window covers half of its neighboring window so that values are smoothed. Consequently, the spatial resolution decreases compared to the original image (with an equivalent pixel size of 240 m), and plume heights are smoothed.

4. Results

From here on, our cloud measurements are referred to as altitude if given in terms of their position above sea level or height if above ground level. For cloud 1, 84 height measurements were made over the 7.7 km of its downwind extent (Figure 9a). There is a gap of 22 km where no cloud is apparent. Then, cloud 2, for which we have 62 height measurements, extends 19.5 km to the image edge (Figure 9a).

4.1. Cloud Height Variation with Distance

Near-vent, cloud 1 ascends at a rate of 530 m of ascent per 100 m of downwind travel to reach an altitude of 7.6–8.0 km a.s.l. (4.4–4.8 km above the vent) after 1.47 km. Thereafter, ascent is less steep (90 m of ascent per 100 m of downwind travel), so that the cloud reaches a peak altitude of 10–10.4 km a.s.l. after 4.15 km. This peak altitude is equivalent to a 6.8–7.2 km high cloud. Thereafter, the cloud descends to 7.2–7.6 km a.s.l. at 7.7 km. We assume that these trends with distance reflect plume overshooting followed by adjustment (settling) to the neutral buoyancy level (cf. [41]). There is some variation in altitude after the peak at 4.15 km. For example, there are local peaks of 9–9.4 and 8.4–8.8 km at 4.8 and 6.6 km, respectively, and a low of 7.6–8 km at 5.5 km. The shape of the profile in Figure 9a is typical of a bent-over plume, where the peak point (which will be vertically over the vent in no-wind or high-MDR conditions [28,56]) has been blown 4.15 km downwind for a bent-over angle of 67°.

Cloud 2 begins 30 km away from the vent. Overall, cloud 2 ascends in altitude from 7.8–8.2 km a.s.l. at the closest point to the vent to 9.3–9.7 km a.s.l. at the image edge (59 km from the vent). It thus ascends at a rate of around 50 m per kilometer. Cloud 2 shows less frequent and less marked variations than Cloud 1, but there is a low between 33.4 and 36.2 km where the cloud altitude decreases by 700 m to a height of 7.4–7.6 km (Figure 9a).

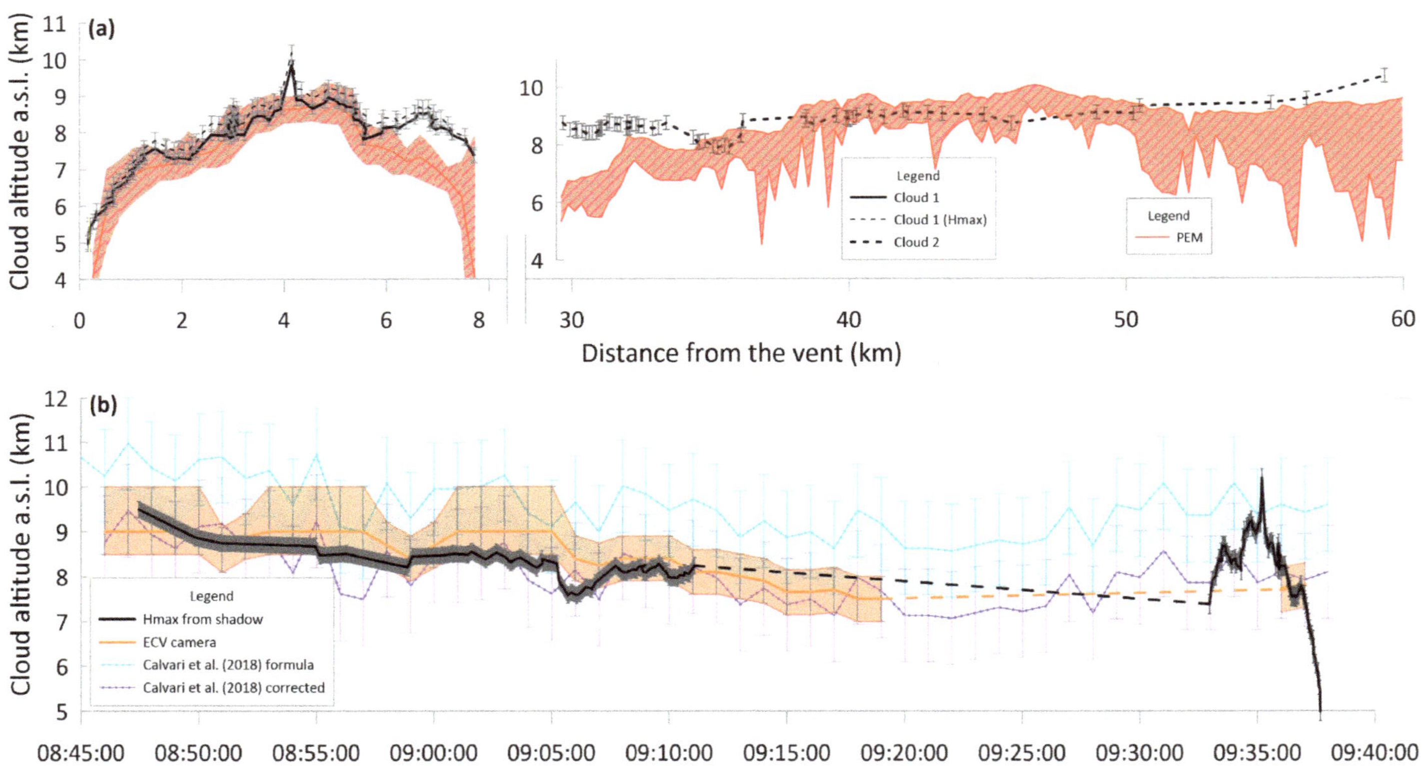

Figure 9. (**a**) Comparison between cloud 1 and cloud 2's altitude with distance from the vent obtained using the shadow technique and PEM methods. (**b**) Comparison of cloud altitude time series obtained from the OLI images using the shadow technique, ECV camera and from applying Calvari et al.'s (2018) MDR versus height relation [28]. Dashed lines are extrapolated across times for which there are no data under the assumption of plume height stability under stable MDR conditions and sustained lava fountain activity.

4.2. Cloud Height Time Series

Figure 9b displays the time series of plume altitude above the vent. In distance plots, the first events chronologically are furthest from the vent and the most recent events are closest to the vent. Thus, the temporal profile is a mirror of the distance profile. Empirically, MDR and plume height have been linked with increasing MDR, resulting in increasing plume height (e.g., [8,57]). This means that, assuming that the height fluctuations measured down the axis of the cloud represent variations at the source, the general increase in cloud altitude with distance down cloud 2 can be explained by a decreasing MDR with time, where the low point would be a short period of lower MDR. That is, in general, the oldest and furthest travelled part of the cloud was emitted at a higher MDR than the youngest and nearest (to the vent) part. For this case, we can trace the evolution of the plume emission and cloud dispersal back in time for 50 min prior to the acquisition of the satellite image. While the imaged portion of cloud 2 was erupted between 08:47 and 09:11, cloud 1 emission began at 09:33 and emission was continuing at the time of image acquisition at 9:37. Given an optimally placed event and low wind speeds, we can potentially derive a time series with a duration of more than 24 h. For example, the leading edge of plume with a source in the lowermost SW corner of a 180 × 185 km OLI image will, in a 3 m s^{-1} wind blowing to the NE, take 24 h to traverse the image.

4.3. Comparison with PEM and ECV Camera

In Figure 9a we compare the cloud heights from the shadow with the cloud heights generated by the PEM method. There is a good agreement between these two datasets for cloud 1, except for the more proximal part where the plume begins to become transparent. There is another difference between the two height estimations at the beginning of cloud 2 for the same reason. Cloud transparency thus seems to be responsible of underestimations of between 25% and 50% in the height using the PEM method in these two regions. In addition, the PEM does not pick up the overshooting region (the highest portion of cloud 1) because the PEM method sub-samples the pixels in the image, thus smoothing the height profile. For cloud 2, there is a great deal of variation in the PEM measurements because the measurements have been made at the border of the cloud. This is a mixed pixel effect, where pixels become mixed in two senses: vertically and horizontally. In the vertical sense, the signal from a semi-transparent cloud and that from the underlying ground become mixed. In the horizontal sense, where a pixel is only partially filled by the cloud, the pixel-integrated spectral-radiance is a product of the portion of the pixel occupied by the cloud and ground occupying the remainder of the pixel (Marsh, 1981). We can also have cases of vertical and horizontal mixing (a semi-transparent cloud that partially fills a pixel). These effects result in a much greater error than for the shadow method. The PEM underestimates the height if the pixel selected is too close to the ground, or overestimates the height if the pixel selected is closer to the center of the cloud.

The cloud altitudes from the shadow technique are also in line with those of the ECV camera (Figure 9b), meaning that the plume dispersion velocities are most likely correct, and that the satellite-retrieved altitudes appear valid. Unfortunately, the weather became cloudy from around 09:20 onwards, blocking the ECV camera view of the plume top until 09:36. The satellite-retrieved cloud altitudes are, however, able to fill this gap—Figure 9b shows a generally declining plume altitude, from around 9.3–9.7 to 7.8–8.2 km a.s.l. above the vent between 08:45 and 09:10. This is a total decline of 1500 m in 25 min, for a rate of 60 m of decline per minute. There is then a rapid waxing period between 09:33 and 09:35, when the plume picked up from around 7.3–7.7 to 10–10.4 km a.s.l. This is an increase of 2600 m in 2 min, for a rate of ascent of 1300 m per minute, or around 22 m s^{-1}.

5. Discussion

5.1. Cloud Separation: Shut-Down in Activity?

Between 8 and 30 km, we see an apparent gap in the cloud in the OLI image (Figure 1). The question is: does this relate to and signify a pause in activity? Given the cloud dispersal velocity, any such pause would have lasted for 20 min beginning at 09:13 and ending at 09:33 (Figure 9b). Cloud cover did not allow plume top height observations from the ECV camera between 09:20 and 09:36, but a lava fountain was visible from the near-vent EMOT camera at this time. The lava fountain height decreased by half between 09:12 and 09:30. Concomitantly, the plume altitude viewed by the ECV camera decreased by 500 m between 09:11 and 09:19. The difference in plume behavior is visible on Figure 8; at 9:14, the plume is much lower than at 8:53. There was then a plume altitude increase between 09:33 and 09:35, but at no point did the eruptive activity stop. This decrease in plume altitude coincides with a decrease in levels of sulfur dioxide and ash emissions (cf. Figure 4 of [58]). The short-lived (22 min long) decrease in lava fountain intensity resulted in a plume height decrease but also a decrease in the quantity of erupted material. When contrast-enhanced, there is in fact material visible in the OLI image in the air between the two clouds. However, it appears as a haze, and is not optically thick enough to produce a shadow. Thus, the gap between the two clouds, although being coincident with a decrease in activity, does not mark a shut down. The gap in the volcanic cloud is also coincident with a zone that lacks meteorological clouds (Figure 1). This coincides with the valley of the Simeto river, where elevations fall away to around 400 m a.s.l., compared with 1000 m a.s.l. in the mountains to the west. The two volcanic clouds are also light-toned, indicating that they were rich in water vapor. We thus suggest that the atmospheric conditions across the Simeto valley prevented both meteorological and water-rich (ash-poor) volcanic clouds from condensing in this area, enhancing the "disappearance" of the cloud during the decrease in activity. This apparent absence of the volcanic cloud raises concerns regarding the detection of volcanic material by satellite in this particular area, which happens to be the flight corridor for aircraft coming into the Catania Fontanarossa airport (Figure 1).

5.2. Settling of Cloud Height to Neutral Buoyancy Level

The height values obtained from the shadow technique and the ECV camera images are in good agreement for the proximal cloud, there being a difference of just 20–200 m (Figure 9b). Due to wind effects, the position at which the cloud reaches its maximum height is offset by 4 km to the SW (Figure 9a). Between 4 and 8 km, the cloud settles back to the level of neutral buoyancy, which is around 1 km below the overshoot height. For the case considered here, settling is systematic and follows the trend:

$$Cloud\ altitude\ (m) = -0.415 * Distance\ from\ the\ vent\ (m) + 11.093.\ \left(r^2 = 0.68\right) \tag{8}$$

It is important to note that this relation lumps all of the complexities in the plume settling dynamics into a single, simple relation, where our objective is simply to derive a trend for plume height with distance for this case where a simple, systematic trend can be seen.

Calvari et al. (2018) proposed a formula linking lava fountain, *H* (*lava fountain*), and associated plume, *H* (*plume*), heights (in m) [28]:

$$H(plume) = 5.26\ H(lava\ fountain) + 6.830 \tag{9}$$

This relation was empirically derived by considering data from 20 lava fountaining events at the NSEC between 2011 and 2013, but did not include the event of the case study given here. The values calculated with Equation (8) are systematically higher than the heights obtained by the shadow method (Figure 9b). However, our measurements are mostly for the height of neutral buoyancy and not maximum plume height. Thus, the 800–2200 m difference in Figure 9b between the result of Equation (9) and the measurement of plume height from the shadow may be due to plume settling. This therefore likely gives the difference between maximum plume height, *H* (*plume*), and neutral buoyancy height, *H* (*NB*), so that:

$$H(NB) = H(plume) - average\ offset = 5.26\ H(lava\ fountain) + 5.130.\ \left(r^2 = 0.50\right) \tag{10}$$

Here, the average offset is the median value for the differences between Equation (9)'s output and the plume height at all down-wind measurement points. This relation is supported by the fact that the maximum plume height measured from the image is, in fact, in very good agreement with that expected from the relation of Calvari et al. (2018) [28] (Figure 9b). Thereafter, the results of Equation (9) and the measurement of cloud height diverge downwind due to plume settling.

5.3. Mass Flux Fluctuation Effect

Assuming constant rates of air entrainment (cf. [59]), an empirical relation that enables the estimation of the *MDR* required to drive a plume to height *H* (*plume*) is that of Woodhouse et al. (2013) [60]:

$$H(plume) = 0.318 * MDR^{0.253} \tag{11}$$

where *H* (*plume*) is measured in kilometers and *MDR* is the mass flux measured in kilograms per second. However, this relation is designed for sustained plumes associated with Plinian eruptions [60], and not for lava fountains, which have different convective properties. During lava fountaining events, magmas can ascend the conduit to erupt as a spray of magma clots and gas at the surface [8,61]. A variable—but usually minor when compared with Plinian plumes—amount of fine pyroclasts is produced to drive a convective plume above the fountain [8]. Most of the heat produced during lava fountains is held by the coarse ejecta, which fall from the plume rapidly [8]. As a result, only a small percentage of the thermal energy flux is conveyed to the plume, so that volcanic plumes above fountaining vents are usually weaker and less high than those fed at comparable MDR during Vulcanian and sub-Plinian eruptions [8]. Moreover, Equation (11) requires maximum plume height and not the neutral buoyancy height calculated in most of the points in this study. We thus need to adapt Equation (11) for a jet-fed thermal, using the height corrected for the settling effect following Equation (10). Calvari et al. (2018) provide a time series of volumetric discharge rates (VDR) for this event [28]. We used the relation of Woodhouse et al. (2013) [60] to estimate the MDR from each of our *H* (*plume*) and corrected the height using Equation (11), then converted them to VDR using an assumed tephra density of 2650 kg m^{-3} dense rock equivalent (DRE) [28] and compared them (Figure 10).

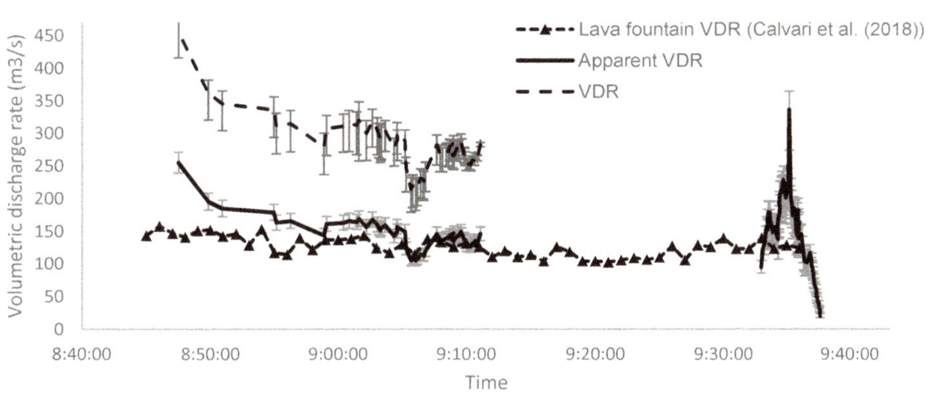

Figure 10. Comparison between cloud volumetric discharge rate (VDR) estimated from H (plume) using Equation (10) from Woodhouse et al. (2013) [60] and measurements of Calvari et al. (2018) [28].

Following Andronico et al. (2018) [37], only 4% of the erupted magma volume DRE accounts for the distal tephra fallout (i.e., the cloud), and 23% for the proximal tephra fallout (the lava fountain). This means that the mass discharge rate for particles going into the cloud should be around 17% of that for those involved in feeding the lava fountain. Considering that the VDR calculated from Calvari et al. (2018) fluctuates between 100 and 150 m^3 s^{-1} [28], the DRE flux (for the particulate portion of the mixture) should be around 17–26 $m^3\ s^{-1}$. Instead, in Figure 10, the DRE flux fluctuates between 250 and 450 $m^3\ s^{-1}$. However, the difference is systematic, allowing us to multiply the relation of Woodhouse et al. (2013) [60] by a factor of 0.065 ± 0.005 to take into account the differing dynamics and mass partitioning between a Plinian and a lava fountain plume. The following corrected equation was only fitted to our single event, and applies to the volume flux of solid particles (as DRE) and not the whole cloud (mixture of particles, gas and entrained air):

$$H(plume) = 0.318 * (VDR * 0.065 \pm 0.005)^{0.253} \tag{12}$$

5.4. Wind Effects on the Plume above the Vent

The effects of the wind on the plume are visible in the cloud 1 profile (Figure 9), because the crosswind speed is higher than the plume rise velocity [25]. At the acquisition time, the plume dispersal velocity was 12.5 m s^{-1}. However, there seem to be two dynamics within the ascending column (Figure 11). It first ascends with a slope of 530 m per 100 m of downwind travel to reach an altitude of 7.6–8 km a.s.l. (4.4–4.8 km above the vent) for a downwind bending 1.47 km. Then, the ascent is less steep (90 m per 100 m of downwind travel) until the cloud reaches its peak altitude of 10–10.4 km a.s.l. (7840 m above the vent) after 4.15 km. The first part of the ascending column has an angle of 72° and the second part an angle of 51° (Figure 11).

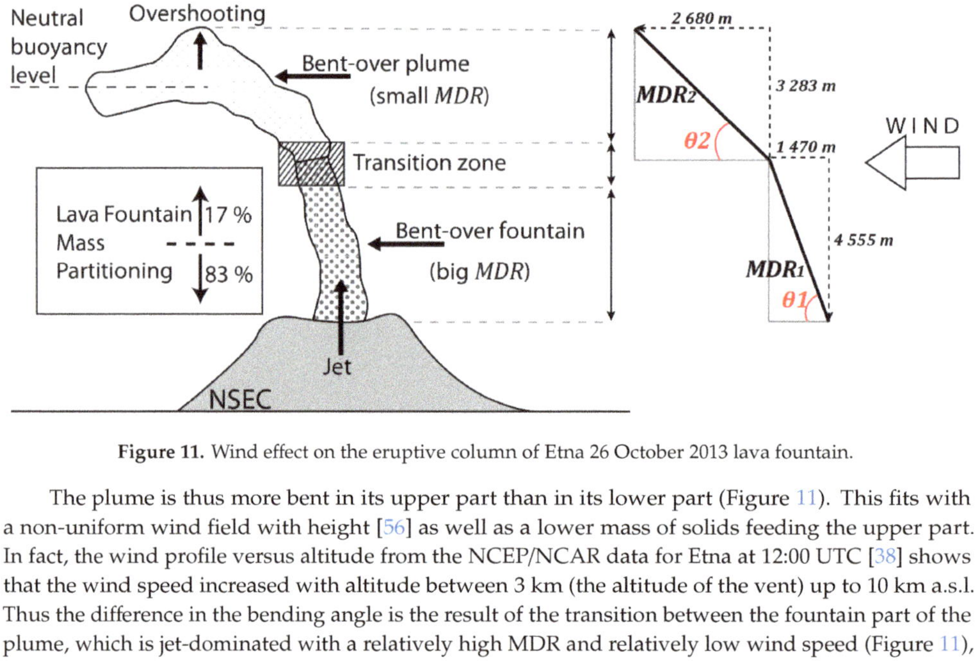

Figure 11. Wind effect on the eruptive column of Etna 26 October 2013 lava fountain.

The plume is thus more bent in its upper part than in its lower part (Figure 11). This fits with a non-uniform wind field with height [56] as well as a lower mass of solids feeding the upper part. In fact, the wind profile versus altitude from the NCEP/NCAR data for Etna at 12:00 UTC [38] shows that the wind speed increased with altitude between 3 km (the altitude of the vent) up to 10 km a.s.l. Thus the difference in the bending angle is the result of the transition between the fountain part of the plume, which is jet-dominated with a relatively high MDR and relatively low wind speed (Figure 11), and the upper tephra-rich part where the wind speed is higher, where the MDR is lower, and where more air is incorporated inside the plume [8,56]. The upper part has transitioned to a buoyant phase, where the air entrainment rates can be significantly higher for buoyant thermals fed by a jet in such "weak" (Strombolian and Hawaiian) plume cases [22].

5.5. Plume/Cloud Ascent and Dispersion Dynamics: Summary

The 26 October 2013 fountain event at Etna produced a plume that, above the vent, was a bent-over rooted thermal. A decrease in mass flux between the lower fountain-fed portion of the plume and the upper portion, in tandem with the increasing wind speed with altitude, resulted in an increase in the degree of bending 2.5 km above the vent. This represents the transition between the jet of large clasts that fall back to the ground (thereby removing mass and thermal energy from the system) and the plume of tephra that continues to ascend by convection (Figure 11). In such a scenario, the upper part is bent over more easily than the lower part. Ascent rates were 22 m s^{-1} up to a maximum height of 10–10.4 km above (but offset by 4.2 km downwind from) the vent. Thereafter, overshooting tephra settled to a neutral buoyancy level at around 7.8–8.2 km a.s.l. after 7.7 km (Figure 12). A lack of condensation in the water-rich plume meant that there was an apparent gap in the cloud between 8 and 30 km. However, thereafter an apparent increase in cloud height implied higher mass fluxes in the 50-min period prior to image acquisition (Figure 12).

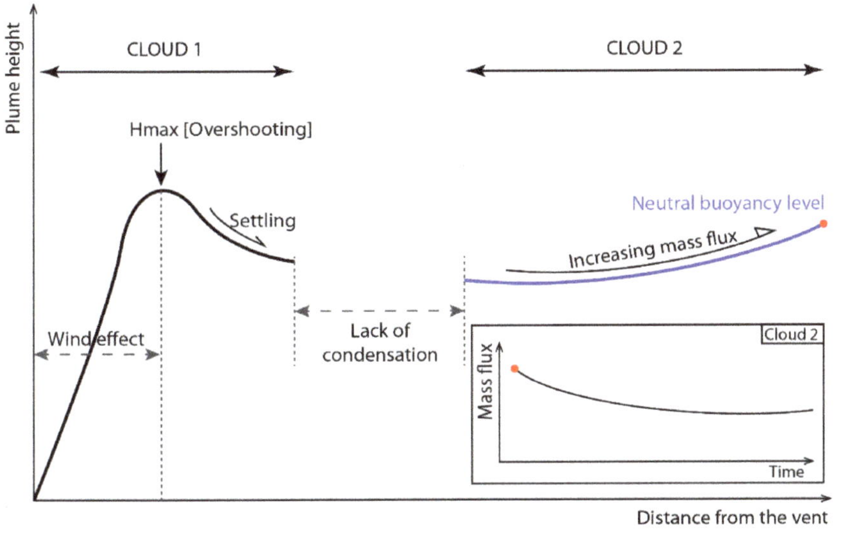

Figure 12. Summary of the dispersion dynamics of the 26 October 2013 Etna fountain-fed plume and cloud system. The red dot indicates the visible end of cloud 2 at the image edge.

6. Conclusions

The cloud-height-from-shadow technique set up and validated in this study allowed the retrieval of a cloud altitude time-series from a single LANDSAT 8 OLI image, allowing us to document the ascent and dispersion history of a plume–cloud system emitted during a fountaining event at Etna volcano. The high-spatial resolution of the LANDSAT 8 product allowed us to detail and quantify cloud and plume dynamics over a distance of 60 km and over a time period of 50 min, with the precision of a few seconds and vertical error on plume altitude of ±200 m. Potentially, our method allows downwind plume height profiles and MDR time series to be built over distances of up to 260 km and periods of 24 h, depending on source location in the image, wind speed and direction. The results were found to be in good agreement with measurements of plume height from ground-based cameras, as well as the PEM method of De Michele et al. (2019) [38].

The data set derived here for cloud/plume altitude versus distance and time allowed a number of empirical relations to be derived and refined for a fountain-fed plume. These include refinement for relations between settling and distance, maximum plume height and level of neutral buoyancy, and MDR and plume height. These relations are, however, strictly calibrated for the 26 October 2013 event, but do suggest peculiarities in the dynamics and mass partitioning between different segments of a fountain-fed plume when compared with Vulcanian and Plinian plumes. However, whether the empirical, best-fitting relations given here can be viewed as a general model for fountain-fed plumes requires the collection of data for further events so as to build a model based on a robust statistical approach. We thus advocate the targeting of high-spatial resolution satellite sensors to fountaining events, especially those where ancillary ground-based thermal camera videos are available (e.g., Etna, Hawaii, and Piton de la Fournaise) to follow up on, and check, the relations implied here.

The time series generated here uses a single, static, high-spatial resolution image, allowing time series with a resolution of seconds to be built from a single image. The temporal resolution depends on the distance between pixels selected, whereby an increase in the number of points where measurements are made will increase the processing time. Potentially, though, the measurement can be made pixel-by-pixel, allowing a measurement to be made every few tens of meters which, in high wind speed conditions, will convert to a measurement every few seconds. Such time series have, to date, typically been the domain of sensors mounted on geostationary satellites with spatial resolutions of 1–4 km, and temporal resolutions of 15 min (e.g., [16,21,42]). The method thus represents a great increase in both the spatial and temporal resolutions of the record, providing data essential for constraining and running models for plume ascent and dispersal cf. [62]. There are, however, a few limits to the application of this method. First, it can only apply to daytime events, as we required the presence of a shadow. Second, the conversion to a time series requires reliable data on wind speed or cloud dispersal velocity. Because these vary with time, height and distance, a single value might not give reliable results. Finally, such explosive events (being short, just a few hours in duration) are difficult to capture given the typical 16-day return period of LEO satellites carrying high spatial resolution sensors needed for the method applied here. This argues for more use of pointing capabilities, such as that represented by the ASTER Urgent Response Protocol [63], or constellations, wherein currently the Landsat-ASTER-Sentinel satellites represent just such a network.

Supplementary Materials: The following are available online at http://www.mdpi.com/2072-4292/12/23/3951/s1, Table S1: Reference list for volcanological applications of satellite data to volcanic plumes. Table S2: Sun-Earth geometry calculations.

Author Contributions: Conceptualization, A.J.L.H. and L.G.; methodology, S.P.-B., A.J.L.H., S.C. and M.D.M.; validation, A.J.L.H. and S.C.; formal analysis, S.P.-B., S.C. and M.D.M.; investigation, S.P.-B., A.J.L.H. and S.C.; resources, S.P.-B., A.J.L.H., S.C. and, M.D.M.; data curation, S.P.-B.; writing—original draft preparation, S.P.-B.; writing—review and editing, A.J.L.H., S.C., M.D.M. and L.G.; visualization, S.P.-B.; supervision, A.J.L.H.; project administration, L.G.; funding acquisition, L.G. All authors have read and agreed to the published version of the manuscript.

Funding: This research was funded by CNES-TOSCA (Terre Solide), grant number 10 3703 "Integration of sample return data and remote sensing for advanced understanding of volcanic ash formation and dispersion" (PI: Lucia Gurioli).

Acknowledgments: The authors would like to thank the INGV-OE personnel for the cameras network maintenance. This manuscript benefited greatly from the valuable input on satellite viewing geometry and orthorectification of Daniel Raucoules (BRGM Orléans). We are grateful to Mike Ramsey and Diego Coppola who assisted us in satellite data collection. We owe much to Nathan Roberts from USGS support staff and the "calibration/validation team" for details on Landsat's acquisition time information. We wish to thank Stefano Mannini for his input on satellite data processing.

Conflicts of Interest: The authors declare no conflict of interest.

References

1. Casadevall, T.J. *Volcanic Ash and Aviation Safety; Proceedings of the First International Symposium on Volcanic Ash and Aviation Safety*; U.S. Geological Survey Bulletin: Washington, DC, USA, 1994.
2. Kienle, J.; Shaw, G.E. Plume dynamics, thermal energy and long-distance transport of vulcanian eruption clouds from Augustine Volcano, Alaska. *J. Volcanol. Geotherm. Res.* **1979**, *6*, 139–164. [CrossRef]
3. Hanstrum, B.N.; Watson, A.S. A case study of two eruptions of Mount Galunggung and an investigation of volcanic eruption cloud characteristics using remote sensing techniques. *Autralian Meteorol. Mag.* **1983**, *31*, 171–177.
4. Aloisi, M.; D'Agostino, M.; Dean, K.G.; Mostaccio, A.; Neri, G. Satellite analysis and PUFF simulation of the eruptive cloud generated by the Mount Etna paroxysm of 22 July 1998. *J. Geophys. Res. Solid Earth* **2002**, *107*, ECV 9-1–ECV 9-12. [CrossRef]

5. Guffanti, M.; Mayberry, G.C.; Casadevall, T.J.; Wunderman, R. Volcanic hazards to airports. *Nat. Hazards* **2009**, *51*, 287–302. [CrossRef]
6. Alemanno, A. *Governing Disasters—The Challenges of Emergency Risk Regulation*; Edward Elgar Publishing Limited: Cheltenham, UK, 2011; ISBN 9780857935731.
7. Oppenheimer, C. Volcanological applications of meteorological satellites. *Int. J. Remote Sens.* **1998**, *19*, 2829–2864. [CrossRef]
8. Sparks, R.S.J.; Bursik, M.; Carey, S.; Gilbert, J.S.; Glaze, L.S.; Sigurdsson, H.; Woods, A.W. *Volcanic Plumes*; Wiley: Chichester, UK, 1997; ISBN 9780471939016.
9. Bonadonna, C.; Folch, A.; Loughlin, S.; Puempel, H. Future developments in modelling and monitoring of volcanic ash clouds: Outcomes from the first IAVCEI-WMO workshop on Ash Dispersal Forecast and Civil Aviation. *Bull. Volcanol.* **2012**, *74*, 1–10. [CrossRef]
10. Newhall, C.G.; Self, S. The volcanic explosivity index (VEI) an estimate of explosive magnitude for historical volcanism. *J. Geophys. Res.* **1982**, *87*, 1231. [CrossRef]
11. Francis, P.W. Infra-red techniques for volcano monitoring and prediction-A review. *J. Geol. Soc. Lond.* **1979**, *136*, 355–359. [CrossRef]
12. Francis, P.W.; Oppenheimer, C. Applications of satellite remote sensing techniques to volcanology. In *Understanding the Terrestrial Environment—The Role of Observation from Space*; Paylor & Francis: London, UK, 1992; pp. 37–52.
13. Francis, P.W.; Oppenheimer, C. *Volcanoes*, 2nd ed.; Oxford University Press: Oxford, UK, 2003; ISBN 9780199254699.
14. Pyle, D.M. Sizes of Volcanic Eruptions. In *The Encyclopedia of Volcanoes*; Elsevier: London, UK, 2015; pp. 257–264.
15. Mouginis-Mark, P.J.; Pieri, D.C.; Francis, P.W.; Wilson, L.; Self, S.; Rose, W.I.; Wood, C.A. Remote sensing of volcanos and volcanic terrains. *EOS Trans. Am. Geophys. Union* **1989**, *70*, 1567–1575. [CrossRef]
16. Glaze, L.S.; Francis, P.W.; Self, S.; Rothery, D.A. The 16 September 1986 eruption of Lascar volcano, north Chile: Satellite investigations. *Bull. Volcanol.* **1989**, *51*, 149–160. [CrossRef]
17. Glaze, L.S.; Wilson, L.; Mouginis-Mark, P.J. Volcanic eruption plume top topography and heights as determined from photoclinometric analysis of satellite data. *J. Geophys. Res. Solid Earth* **1999**, *104*, 2989–3001. [CrossRef]
18. Woods, A.W.; Self, S. Thermal disequilibrium at the top of volcanic clouds and its effect on estimates of the column height. *Nature* **1992**, *355*, 628–630. [CrossRef]
19. Holasek, R.E.; Self, S.; Woods, A.W. Satellite observations and interpretation of the 1991 Mount Pinatubo eruption plumes. *J. Geophys. Res. B Solid Earth* **1996**, *101*, 27635–27655. [CrossRef]
20. Zakšek, K.; Hort, M.; Zaletelj, J.; Langmann, B. Monitoring volcanic ash cloud top height through simultaneous retrieval of optical data from polar orbiting and geostationary satellites. *Atmos. Chem. Phys.* **2013**, *13*, 2589–2606. [CrossRef]
21. Marchese, F.; Falconieri, A.; Pergola, N.; Tramutoli, V. A retrospective analysis of the Shinmoedake (Japan) eruption of 26-27 January 2011 by means of Japanese geostationary satellite data. *J. Volcanol. Geotherm. Res.* **2014**, *269*, 1–13. [CrossRef]
22. Patrick, M.R. Dynamics of Strombolian ash plumes from thermal video: Motion, morphology, and air entrainment. *J. Geophys. Res.* **2007**, *112*, B06202. [CrossRef]
23. Carey, S.; Bursik, M. Volcanic Plumes. In *The Encyclopedia of Volcanoes*; Elsevier: London, UK, 2015; pp. 571–585.
24. Bonaccorso, A.; Calvari, S. A new approach to investigate an eruptive paroxysmal sequence using camera and strainmeter networks: Lessons from the 3–5 December 2015 activity at Etna volcano. *Earth Planet. Sci. Lett.* **2017**, *475*, 231–241. [CrossRef]
25. Carey, S.; Sparks, R.S.J. Quantitative models of the fallout and dispersal of tephra from volcanic eruption columns. *Bull. Volcanol.* **1986**, *48*, 109–125. [CrossRef]
26. Slawson, P.R.; Csanady, G.T. On the mean path of buoyant, bent-over chimney plumes. *J. Fluid Mech.* **1967**, *28*, 311. [CrossRef]
27. Andò, B.; Pecora, E. An advanced video-based system for monitoring active volcanoes. *Comput. Geosci.* **2006**, *32*, 85–91. [CrossRef]

28. Calvari, S.; Cannavò, F.; Bonaccorso, A.; Spampinato, L.; Pellegrino, A.G. Paroxysmal Explosions, Lava Fountains and Ash Plumes at Etna Volcano: Eruptive Processes and Hazard Implications. *Front. Earth Sci.* **2018**, *6*. [CrossRef]
29. Behncke, B.; Branca, S.; Corsaro, R.A.; De Beni, E.; Miraglia, L.; Proietti, C. The 2011–2012 summit activity of Mount Etna: Birth, growth and products of the new SE crater. *J. Volcanol. Geotherm. Res.* **2014**, *270*, 10–21. [CrossRef]
30. De Beni, E.; Behncke, B.; Branca, S.; Nicolosi, I.; Carluccio, R.; D'Ajello Caracciolo, F.; Chiappini, M. The continuing story of Etna's New Southeast Crater (2012–2014): Evolution and volume calculations based on field surveys and aerophotogrammetry. *J. Volcanol. Geotherm. Res.* **2015**, *303*, 175–186. [CrossRef]
31. Coltelli, M.; Del Carlo, P.; Vezzoli, L. Discovery of a Plinian basaltic eruption of Roman age at Etna volcano, Italy. *Geology* **1998**, *26*, 1095. [CrossRef]
32. Istituto Nazionale di Statistica-Roma, I. Demografia in Cifre. Available online: http://demo.istat.it/ (accessed on 1 October 2020).
33. Vulpiani, G.; Ripepe, M.; Valade, S. Mass discharge rate retrieval combining weather radar and thermal camera observations. *J. Geophys. Res. Solid Earth* **2016**, *121*, 1–17. [CrossRef]
34. Del Carlo, P.; Vezzoli, L.; Coltelli, M. Last 100 ka tephrostratigraphic record of Mount Etna. *GMS* **2004**, *143*, 77–89.
35. Corradini, S.; Guerrieri, L.; Lombardo, V.; Merucci, L.; Musacchio, M.; Prestifilippo, M.; Scollo, S.; Silvestri, M.; Spata, G.; Stelitano, D. Proximal Monitoring of the 2011–2015 Etna Lava Fountains Using MSG-SEVIRI Data. *Geosciences* **2018**, *8*, 140. [CrossRef]
36. Greco, F.; Currenti, G.; Palano, M.; Pepe, A.; Pepe, S. Evidence of a shallow persistent magmatic reservoir from joint inversion of gravity and ground deformation data: The 25–26 October 2013 Etna lava fountaining event. *Geophys. Res. Lett.* **2016**, *43*, 3246–3253. [CrossRef]
37. Andronico, D.; Behncke, B.; De Beni, E.; Cristaldi, A.; Scollo, S.; Lopez, M.; Lo Castro, M.D. Magma Budget From Lava and Tephra Volumes Erupted during the 25-26 October 2013 Lava Fountain at Mt Etna. *Front. Earth Sci.* **2018**, *6*. [CrossRef]
38. De Michele, M.; Raucoules, D.; Corradini, S.; Merucci, L.; Salerno, G.; Sellitto, P.; Carboni, E. Volcanic cloud top height estimation using the plume elevation model procedure applied to orthorectified Landsat 8 data. test case: 26 October 2013 Mt. Etna eruption. *Remote Sens.* **2019**, *11*, 785. [CrossRef]
39. Young, N.E.; Anderson, R.S.; Chignell, S.M.; Vorster, A.G.; Lawrence, R.; Evangelista, P.H. A survival guide to Landsat preprocessing. *Ecology* **2017**, *98*, 920–932. [CrossRef]
40. Qiu, S.; He, B.; Zhu, Z.; Liao, Z.; Quan, X. Improving Fmask cloud and cloud shadow detection in mountainous area for Landsats 4–8 images. *Remote Sens. Environ.* **2017**, *199*, 107–119. [CrossRef]
41. Dean, K.G.; Bowling, S.A.; Shaw, G.E.; Tanaka, H. Satellite analyses of movement and characteristics of the Redoubt Volcano plume, January 8, 1990. *J. Volcanol. Geotherm. Res.* **1994**, *62*, 339–352. [CrossRef]
42. Holasek, R.E.; Self, S. GOES weather satellite observations and measurements of the May 18, 1980, Mount St. Helens eruption. *J. Geophys. Res. Solid Earth* **1995**, *100*, 8469–8487. [CrossRef]
43. Kinoshita, K. Observation of flow and dispersion of volcanic clouds from Mt. Sakurajima. *Atmos. Environ.* **1996**, *30*, 2831–2837. [CrossRef]
44. Denniss, A.M.; Harris, A.J.L.; Rothery, D.A.; Francis, P.W.; Carlton, R.W.T. Satellite observations of the April 1993 eruption of Lascar volcano. *Int. J. Remote Sens.* **1998**, *19*, 801–821. [CrossRef]
45. Prata, A.J.; Grant, I.F. Retrieval of microphysical and morphological properties of volcanic ash plumes from satellite data: Application to Mt Ruapehu, New Zealand. *Q. J. R. Meteorol. Soc.* **2001**, *127*, 2153–2179. [CrossRef]
46. Tupper, A.; Carn, S.A.; Davey, J.; Kamada, Y.; Potts, R.; Prata, F.; Tokuno, M. An evaluation of volcanic cloud detection techniques during recent significant eruptions in the western 'Ring of Fire'. *Remote Sens. Environ.* **2004**, *91*, 27–46. [CrossRef]
47. Iqbal, M. Sun–Earth Astronomical Relationships. In *An Introduction to Solar Radiation*; Elsevier: London, UK, 1983; pp. 1–28. ISBN 9780123737502.
48. Harris, A.J.L. *Thermal Remote Sensing of Active Volcanoes*; Cambridge University Press: Cambridge, UK, 2013; ISBN 9781139029346.
49. Mouginis-Mark, P.J.; Domergue-Schmidt, N. Acquisition of satellite data for volcano studies. *Geophys. Monogr. Ser.* **2000**, *116*, 9–24. [CrossRef]

50. Markham, B.L. Characterization of the Landsat sensors' spatial responses. *IEEE Trans. Geosci. Remote Sens.* **1985**, *GE-23*, 864–875. [CrossRef]
51. Oppenheimer, C.; Francis, P.W.; Rothery, D.A.; Carlton, R.W.T.; Glaze, L.S. Infrared image analysis of volcanic thermal features: Lascar Volcano, Chile, 1984-1992. *J. Geophys. Res.* **1993**, *98*, 4269–4286. [CrossRef]
52. Zhu, Z.; Woodcock, C.E. Object-based cloud and cloud shadow detection in Landsat imagery. *Remote Sens. Environ.* **2012**, *118*, 83–94. [CrossRef]
53. Scollo, S.; Prestifilippo, M.; Pecora, E.; Corradini, S.; Merucci, L.; Spata, G.; Coltelli, M. Eruption column height estimation of the 2011-2013 Etna lava fountains. *Ann. Geophys.* **2014**, *57*. [CrossRef]
54. Prata, A.J.; Turner, P.J. Cloud-top height determination using ATSR data. *Remote Sens. Environ.* **1997**, *59*, 1–13. [CrossRef]
55. De Michele, M.; Raucoules, D.; Arason, Þ. Volcanic Plume Elevation Model and its velocity derived from Landsat 8. *Remote Sens. Environ.* **2016**, *176*, 219–224. [CrossRef]
56. Bursik, M. Effect of wind on the rise height of volcanic plumes. *Geophys. Res. Lett.* **2001**, *28*, 3621–3624. [CrossRef]
57. Mastin, L.G.; Guffanti, M.; Servranckx, R.; Webley, P.W.; Barsotti, S.; Dean, K.G.; Durant, A.; Ewert, J.W.; Neri, A.; Rose, W.I.; et al. A multidisciplinary effort to assign realistic source parameters to models of volcanic ash-cloud transport and dispersion during eruptions. *J. Volcanol. Geotherm. Res.* **2009**, *186*, 10–21. [CrossRef]
58. Sellitto, P.; Di Sarra, A.; Corradini, S.; Boichu, M.; Herbin, H.; Dubuisson, P.; Sèze, G.; Meloni, D.; Monteleone, F.; Merucci, L.; et al. Synergistic use of Lagrangian dispersion and radiative transfer modelling with satellite and surface remote sensing measurements for the investigation of volcanic plumes: The Mount Etna eruption of 25-27 October 2013. *Atmos. Chem. Phys.* **2016**, *16*, 6841–6861. [CrossRef]
59. Suzuki, Y.J.; Koyaguchi, T. Effects of wind on entrainment efficiency in volcanic plumes. *J. Geophys. Res. Solid Earth* **2015**, *120*, 6122–6140. [CrossRef]
60. Woodhouse, M.J.; Hogg, A.J.; Phillips, J.C.; Sparks, R.S.J. Interaction between volcanic plumes and wind during the 2010 Eyjafjallajökull eruption, Iceland. *J. Geophys. Res. Solid Earth* **2013**, *118*, 92–109. [CrossRef]
61. Allard, P.; Burton, M.; Muré, F. Spectroscopic evidence for a lava fountain driven by previously accumulated magmatic gas. *Nature* **2005**, *433*, 407–410. [CrossRef]
62. Costa, A.; Suzuki, Y.J.; Cerminara, M.; Devenish, B.J.; Ongaro, T.E.; Herzog, M.; Van Eaton, A.R.; Denby, L.C.; Bursik, M.; de' Michieli Vitturi, M.; et al. Results of the eruptive column model inter-comparison study. *J. Volcanol. Geotherm. Res.* **2016**, *326*, 2–25. [CrossRef]
63. Ramsey, M.S. Synergistic use of satellite thermal detection and science: A decadal perspective using ASTER. *Geol. Soc. Lond. Spec. Publ.* **2016**, *426*, 115–136. [CrossRef]

remote sensing

MDPI

Article

The Extensive Parameters as a Tool to Monitoring the Volcanic Activity: The Case Study of Vulcano Island (Italy)

Salvatore Inguaggiato [1,*], Fabio Vita [1], Iole Serena Diliberto [1], Agnes Mazot [2], Lorenzo Calderone [1], Andrea Mastrolia [1] and Marco Corrao [3]

[1] Istituto Nazionale di Geofisica e Vulcanologia, Sezione di Palermo, Palermo Via Ugo La Malfa, 90146 Palermo, Italy; fabio.vita@ingv.it (F.V.); iole.diliberto@ingv.it (I.S.D.); lorenzo.calderone@ingv.it (L.C.); andrea.mastrolia@ingv.it (A.M.)

[2] GNS Science Wairakei Research Centre, Wairakei, 114 Karetoto Road, Taupo Private Bag 2000, Taupo 3352, New Zealand; a.mazot@gns.cri.nz

[3] Geologist, Capo d'Orlando (Me) 98071, Italy; corrao@conteco.it

* Correspondence: salvatore.inguaggiato@ingv.it; Tel.: +39-091-6809-435

Abstract: On Vulcano Island (Italy), many geochemical crises have occurred during the last 130 years of solfataric activity. The main crises occurred in 1978–1980, 1988–1991, 1996, 2004–2007, 2009–2010 and the ongoing 2021 anomalous degassing activity. These crises have been characterized by early signals of resuming degassing activity, measurable by the increase of volatiles and energy output emitted from the summit areas of the active cone, and particularly by increases of gas/water ratios in the fumarolic area at the summit. In any case, a direct rather than linear correspondence has been observed among the observed increase in the fluid output, seismic release and ground deformation, and is still a subject of study. We present here the results obtained by the long-term monitoring (over 13 years of observations) of three extensive parameters: the SO_2 flux monitored in the volcanic plume, the soil CO_2 flux and the local heat flux, monitored in the mild thermal anomaly located to the east of the high-temperature fumarole. The time variations of these parameters showed cyclicity in the volcanic degassing and a general increase in the trend in the last period. In particular, we focused on the changes in the mass and energy output registered in the period of June–December 2021, to offer in near-real-time the first evaluation of the level and duration of the actual exhalative crisis affecting Vulcano Island. In this last event, a clear change in degassing style was recorded for the volatiles emitted by the magma. For example, the flux of diffused CO_2 from the soils reached the maximum never-before-recorded value of 34,000 g m^{-2} d^{-1} and the flux of SO_2 of the plume emitted by the fumarolic field on the summit crater area reached values higher than 200 t d^{-1}. The interpretation of the behavior of this volcanic system, resulting from the detailed analyses of these continuous monitoring data, will complete the framework of observations and help in defining and possibly forecasting the next evolution of the actual exhaling crisis.

Keywords: SO_2 flux; CO_2 flux; heat flux; Vulcano Island; geochemical crisis; extensive parameters

Citation: Inguaggiato, S.; Vita, F.; Diliberto, I.S.; Mazot, A.; Calderone, L.; Mastrolia, A.; Corrao, M. The Extensive Parameters as a Tool to Monitoring the Volcanic Activity: The Case Study of Vulcano Island (Italy). *Remote Sens.* **2022**, *14*, 1283. https://doi.org/10.3390/rs14051283

Academic Editors: Zhong Lu and David Gomez-Ortiz

Received: 18 January 2022
Accepted: 1 March 2022
Published: 5 March 2022

1. Introduction

Volatiles degassing from volcanic systems is a peculiar and useful tool for monitoring the volcanic activity by characterizing the geochemistry of shallow plumbing systems, and by highlighting the changes of the volcanic outgassing in order to forecast the next evolution of the system.

Many scientists have carried out investigations to identify the main active degassing structures that are present on volcanic systems by means of the volatile degassing reaching the shallow hydrothermal plumbing systems, either dissolved in groundwater, bubbling out of surface water, outpoured by the ground as diffuse degassing or released through fumaroles and volcanic plume [1].

The fluid phase expanding towards the surface to respond to the physical and chemical gradients allow for the tracking and interpreting of the energy and mass exchange between the magmatic source, the shallow plumbing system and the atmosphere.

Many volcano observatories have been established throughout the world, and the network of observable phenomena include some geochemical parameters selected among the extensive and intensive variables.

The extensive parameters, like CO_2, SO_2 and heat fluxes from soil and plume, can be monitored continuously by remote sensing, thanks to the technological evolution that refined the optical instruments and to the active research, verified and constrained by the direct measurements (and continuous monitoring) of spy parameters carried out in some select locations.

Such a geochemical tool has been successfully applied to Vulcano Island, the southernmost of the islands of the Aeolian archipelago, characterized by solfataric activity in the last few centuries after the last eruption occurred in 1888–1890 [2–12]. The solfataric activity is mainly concentrated in the top part of the island in the crater of the Fossa and the Bay of Levante (Figure 1). Furthermore, anomalous diffuse degassing from the soils is present in the Palizzi area and in part of the inhabited center of Vulcano.

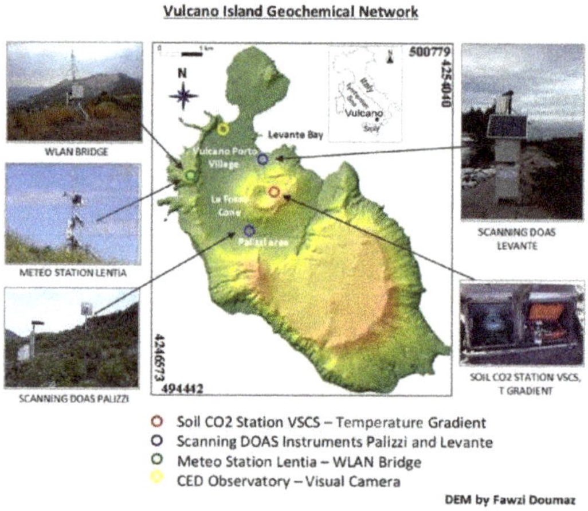

Figure 1. Location map of monitoring stations supplying the discussed datasets with pictures of each measurement system: W-LAN bridge; scanning DOAS Levante and Palizzi for SO_2 plume flux monitoring; VSCS station for CO_2 diffuse flux and temperature monitoring.; Meteo station Lentia.

During the last 130 years of observed normal solfataric activity, many geochemical crises occurred and were recognized. In particular, these crises have been characterized by clear signals of resuming degassing activity with increases of volatiles and energy output emitted from summit areas of Vulcano Island and by increases of gas/water ratio of the main high-temperature fumaroles located in the La Fossa Crater. The main crises observed occurred in 1978–1980 [13], 1988–1991 [14–18], 1996 [19], 2004–2007 [20–22], 2009–2010 [5,7] and the ongoing 2021 anomalous degassing activity.

Many other studies, carried out in different volcanic areas, already highlighted that the diffuse degassing of CO_2 and the associated thermal output cannot be considered a negligible component in the energy and mass balance of solfataric activity [1,23–31].

Moreover, many different geochemical investigations were carried out to estimate the CO_2 output of discharged fluids from Vulcano Island [3,6,32,33]. It was also discovered that the summit area accounts for more than 90% of total CO_2 discharged from the island [6] and is a suitable site for installing a geochemical monitoring station to investigate volcanic activity. Therefore, an automated soil CO_2 fluxes monitoring station was installed in

September 2007 within the active summit crater of La Fossa, beyond the fumarolic areas [5] to investigate, in near-real-time, any changes in the level of solfataric activity.

Other investigations about the energy output diffused from the soils have been carried out by [20,34] that introduced the theoretical background for the evaluation and monitoring of diffuse thermal release. Such areas have been defined as steam-heated soils (SHS), or sub-fumaroles, and are worthy of attention in the scientific investigations of volcanic activity because they can highlight the turning condition from background and anomalous exhaling activity by monitoring the volcanic activity at a safe distance from the main conduits [35–40].

The best site for monitoring the changes in the heat release from a volcanic system is where the condensation zone is just below the monitored profile because the simple temperature measurements in the ground track the vertical shifts of the massive convective front and the correlated changes in the diffuse heat flux. Figure 2 shows a schematic section of the steam-heated soil in the sub-fumaroles zone. In the same Figure, on the left is the linear temperature distribution within the shallow ground, accounting for the diffusive heat transfer.

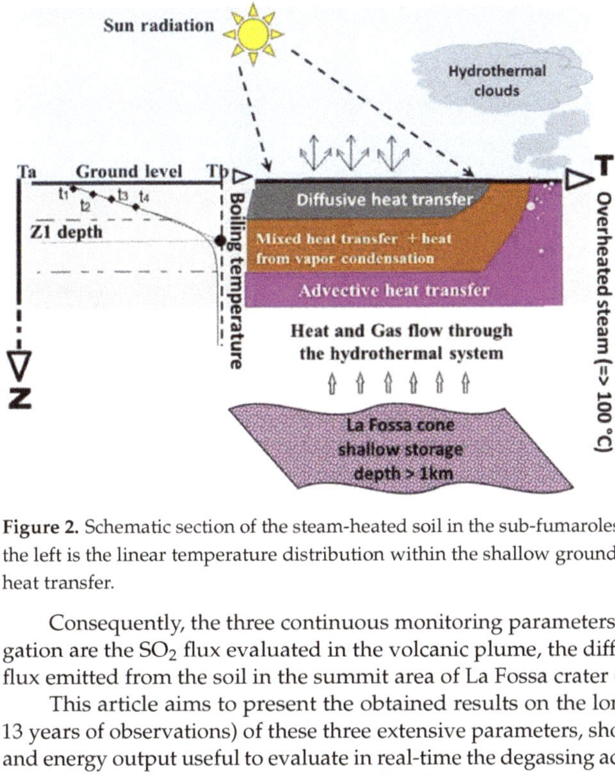

Figure 2. Schematic section of the steam-heated soil in the sub-fumaroles zone. In the same Figure, on the left is the linear temperature distribution within the shallow ground, accounting for the diffusive heat transfer.

Consequently, the three continuous monitoring parameters considered in this investigation are the SO_2 flux evaluated in the volcanic plume, the diffuse CO_2 flux and the heat flux emitted from the soil in the summit area of La Fossa crater (Figure 1; [6,7,20]).

This article aims to present the obtained results on the long-term monitoring (over 13 years of observations) of these three extensive parameters, showing changes in the mass and energy output useful to evaluate in real-time the degassing activity level of this volcanic system. Moreover, we analyzed in detail the last strong ongoing degassing that occurred in Vulcano Island in the period June–December 2021, in order to show the evolution of the actual exhaling crisis.

2. Geological and Volcanological Background

Vulcano Island is one of the most active volcanoes of the Tyrrhenian Sea, the exposed summit of a volcanic edifice sited in the southern-most sector of the Aeolian archipelago (Figure 3). The Aeolian Archipelago is part of a volcanic arc located in a complex geodynamic setting. The complexity of this regional sector in the southern Tyrrenium sea is due to three different tectonically active areas resulting from the Neogene–Quaternary Africa–Eurasia convergence with extension zones oriented NE–SW and NNW–SSE in west

Calabria and NE Sicily, respectively, an E–W oriented belt of compression affecting the southern Tyrrhenian Sea above western and central Sicily and the NNW–SSE trending right-lateral Aeolian–Tindari–Letojanni fault system in between [40–44] (Ruch. J. et al., 2016; Pecerillo et al., 2006, 2013; De Astis 2013; Neri et al., 2005; D'Agostino and Selvaggi, 2004). The whole volcanic edifice forming the island of Vulcano rises from the local sea floor (about 1000 m b.s.l.) up to the maximum height of 499 m a.s.l. at Monte Aria [43–47]. The Island consists of overlapping volcanic edifices whose formation began at 120 ka [43,48,49]. The volcanic system comprises the Primordial Vulcano edifice (120–100 ka), dissected by the development of Il Piano caldera that lasted until around 100 ka before the present and subsequently by the La Fossa caldera (dating back from 80 to 15 ka before the present) [48–50]. In the last 15 ka, new eruptive vents were activated in the northern sector of the island, developing in a N–S direction. The two most recent volcanic structures are the La Fossa cone, the active volcanic center developing the stratocone that started to erupt around 5.5 ka b.p. [51], and Vulcanello, the small volcanic peninsula in the northern part of La Fossa caldera. The Vulcanello peninsula, showing three coalescent crater rims, is composed of a lava platform and three partially overlapping scoria cones aligned NE–SW along the northern ring fault of La Fossa caldera [52].

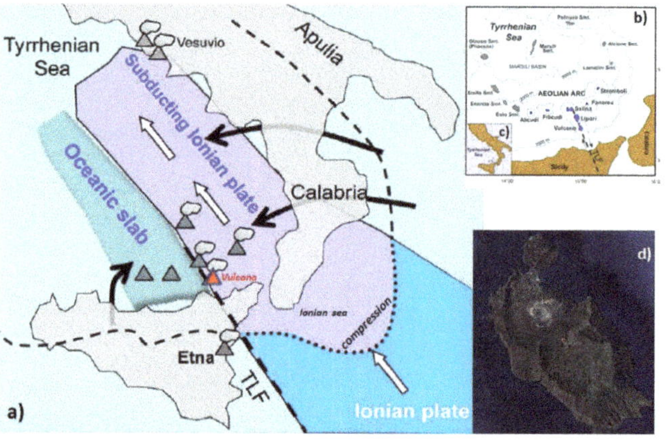

Figure 3. (**a**) Geodynamic setting; (**b**) Aeolian Archipelago sketch map; (**c**) South Italy map, modified from Peccerillo et al. 2013; (**d**) Vulcano Island image from Google Earth.

The presence of a magmatic chamber below the La Fossa cone is required to account for the time variations of diffuse degassing registered in the area and the whole volcanic edifice forming the island of Vulcano rises from the local sea floor (about 1000 m b.s.l.) up to the maximum height of 499 m a.s.l. at Monte Aria [45]. The Island consists of overlapping volcanic edifices whose formation began at 120 ka [43,48,49]. The volcanic system comprises the Primordial Vulcano edifice (120–100 ka), dissected by the development of Il Piano caldera that lasted until around 100 ka before the present and subsequently by the La Fossa caldera (dating back from 80 to 15 ka before the present [48,50]. In the last 15 ka, new eruptive vents were activated in the northern sector of the island, developing in a N–S direction. The two most recent volcanic structures are the La Fossa cone, the active volcanic center developing the stratocone that started to erupt around 5.5 ka b.p., [51] and Vulcanello, the small volcanic peninsula in the northern part of La Fossa caldera. The Vulcanello peninsula, showing three coalescent crater rims, is composed of a lava platform and three partially overlapping scoria cones aligned NE–SW along the northern ring fault of La Fossa caldera [52]. The presence of a magmatic chamber below the La Fossa cone is required to account for the time variations of diffuse degassing registered in the area and is also suggested by the high geothermal gradients measured in the Agip boreholes [53]. Different geophysical and geochemical evidence accounts for the evaluated depths of magma storage. These two different approaches agree with indicating a multibaric and multi-thermal plumbing system during the entire life of Vulcano, and in identifying the different magma storages at about 20 km of depth at 13–8 km and 5–3 km, as well as the shallowest storage system zone, indicatively located about 1–2 km beneath the La Fossa cone [42]. The magmatic intrusion located 1–2 km of depth, and the magmatic storage at about 5–3 km, have been strongly influencing the shallow hydrothermal system, feeding the fumarolic field of the Grancratere of La Fossa, during the historical observation period.

3. Materials and Methods

3.1. Plume SO_2 Fluxes, Network and Methodology of the UV Scanning DOAS Network on Vulcano Island

The remote sensing technique of passive DOAS (differential optical absorption spectroscopy), [54,55] allows the quantifying of different volcanic gases within the columns emitted from active volcanoes by collecting the spectra in the ultraviolet region (UV) in order to supply indirect measurements of magmatic volatiles [7,56–58]. The DOAS method is based on the principles of absorption spectroscopy (Bouger–Beer-Lambert law) and is used for the quantification of different trace gases' concentrations (e.g., SO_2, NO_2, BrO).

In the framework of the "Network for Observation of Volcanic and Atmospheric Change" (NOVAC), a worldwide network of permanent scanning DOAS instruments was installed on 19 volcanoes around the world, to measure in real-time the emitting SO_2 fluxes for volcanic gas monitoring [59]. The NOVAC network was established with a project funded by the European Union in the period 2005–2010, thereafter, it continued to be maintained and expanded, thanks to resources supplied from the volcano observatories initially involved in the active project. The NOVAC community currently counts about 160 stations on 47 volcanoes in different regions of the world, now including Iceland, Philippines, Indonesia, Papua New Guinea and Montserrat [60]. The present work shows the updated results of the ground-based scanning DOAS network installed at Vulcano Island Italy in 2008 (Figures 1 and 4). The results presented here are the first example of SO_2 continuous monitoring, performed in a closed conduit volcano with solfataric activities and relatively low gas emissions, like the active cone of Vulcano Island. On the active cone of Vulcano Island, these long-term measurements of SO_2 flux have provided relevant information about the reference level of volcanic activity; they would contribute to evaluating the number of volcanic gases released into the atmosphere by this natural source. In 2015, the monitoring network had been implemented, thanks to the Vulcamed project, with the installation of a new UV-scanning DOAS station on the NW side of the island. This implemented configuration allowed the tracking of plumes for more than 80% of days during each solar cycle (one year). All the data acquired from each measurement station are telemetered in real-time to reference the Vulcano Observatory (in this case to the local INGV volcanological center Carapezza) through a wireless system. From the Vulcano Observatory, the data are transmitted via the internet in near-real-time to the Istituto Nazionale di Geofisica e Vulcanologia (INGV, Palermo); moreover, both the raw and the analyzed data have been hosted in Gothenburg and mirrored in Brussels and Heidelberg [61].

The NOVAC scanning DOAS instruments provide a direct method for measuring the total SO_2 flux from Vulcano Island and for sharing data with a worldwide scientific community involved in geophysical and environmental research. Details of the instrument and operation routines are described in [59,60,62]. The system consists of a single spectrometer from the Ocean Optics Company (S2000 spectrograph), an embedded personal computer, a global positioning system receiver, a timer, an optical fiber and a telescope.

The spectrometer used (S2000 from the Ocean Optics Company) a 2400 lines/mm grating combined with a 50 nm slit, which provided an optical resolution of ca. 0.6 nm over a wavelength range of 280 nm to 390 nm.

The telescope automatically moves from horizon to horizon using a stepper motor. The control unit is running on a Linux operating system and includes a serial, USB 2 and

Ethernet communication ports. Serial ports are used for communication with the spectrometer and control of the scanner's stepper motor. The USB port can be used for powering the spectrometer, while the Ethernet port is usually used for data transfer to radio modems. A digital thermometer allows the recording of the internal temperature, a voltmeter checks the battery voltage and a GPS antenna records the position and working time.

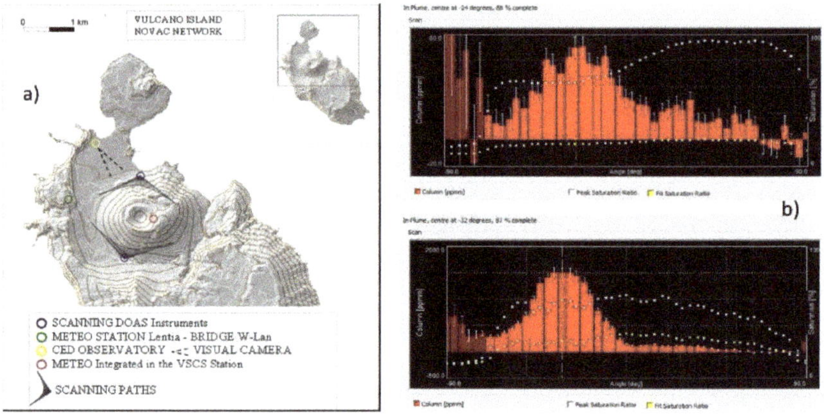

Figure 4. (**a**) UV scanning DOAS and conic scanning path; (**b**) two examples of the distribution of the plume on two measurements made under conditions of different degassing activities. The graphs were produced by the NOVAC software and show the positions of the telescope during a single measurement, from 90° to −90° (X axis; horizon to horizon) for a total of 50 positions for one complete measure, and the column density of SO_2 (Y axis; ppmm). The white dots represent the peak intensities of the light, and the red bars the column density for each position of the telescope during the measurement. (**a**) Measurement during 2020 (normal activity, ca. 14 t d^{-1}). (**b**) Measurement on 12 October 2021 (increased activity, ca. 140 t d^{-1}).

The telescope consists of a single plane-convex quartz lens with a diameter of 25.4mm and a focal length of 7.5 cm, as well as a Hoya (U330) UV filter that reduces intensity of light with wavelengths longer than 360 nm. A stepper motor automatically moves the telescope from horizon to horizon (90° to −90°). At every elevation angle, 15 spectra are collected for one measurement. Every full measurement cycle contained fifty-three spectra: one Zenit sky spectrum, one dark current spectrum, one offset spectrum, and fifty measurement spectra. It usually takes about 10 min, depending on the light conditions of the sky. The plume is scanned along a conical surface rather than a vertical one (Figure 4)—this configuration significantly increases the wind direction range that can be covered by a single scanning instrument, and at the same time reduces problems related to multiple scattering effects [60]. The telescope is coupled to one (single-beam) quartz optical fiber with a diameter of 600 μm. This combination gives a field of view of 8 mrad. The optical fiber is coupled with the entrance slit of the spectrometer, which has a width of 50 μm and height of 1 mm. The NOVAC instruments are powered by an array of 12V batteries and solar panels. A timer is added to interrupt operation of the instrument at night and reset the instrument in the morning.

The spectra are analyzed according to the DOAS procedures, based on the Beer–Lambert Law, to analyze the collected data and evaluate the flux [63,64]. Each spectrum is first corrected for the electronic noise (offset and dark current) and the optical density is obtained by dividing each spectrum measurement by a reference spectrum (i.e., a spectrum taken outside the plume), and then a logarithm is taken of the result of this calculation. A polynomial fit is used to take into account any broadband extinction structure caused by broadband absorption of trace gases [55,64]. A high-pass filter is applied to the absorption cross-sections to use only the remaining high-frequency structure, which is unique for any trace gas, and which can therefore be used to determine its abundance. Shifts and squeezes are allowed for the absorption cross-sections to compensate for any small shift, caused mainly by variations in the temperatures of the spectrometer and the detector unit [7].

3.2. Environmental Parameters Network (Lentia and La Fossa Crater)

The main weather station, used to elaborate the data supplied by the UV scanning DOAS, was installed on Lentia Hill (Figure 1) at an altitude comparable to that of the Fossa crater (350 m a.s.l.). This location ensured the same exposure to the main wind directions, and the absence of any physical barrier between the sensor positions and the fumarole field of La Fossa cone was the condition to register the best local weather variable fitting to the DOAS measurements to extrapolate the real Flux of SO_2 and other associated species. The weather station "Lentia" engineered by Davis was the model Vantage Pro2 and provided the following variables: wind speed and direction, rainfall, air temperature, relative humidity, infrared radiation and UV radiation (Figure 1). The communication with the remote server was ensured by a wireless system. Acquired data passing via the Vulcano Observatory arrived at Palermo data center acquisition (INGV) in real-time, passing via the Vulcano Observatory.

A second meteorological station was integrated into the VSCS soil CO_2 fluxes measurement station, located in the summit area of the La Fossa Crater. The anemometer used in this case is Wind-Sonic Gill, a 2-axis ultrasonic wind sensor, which provides data on wind speed (0–60 m/s) and wind direction (0−359°). This ultrasonic wind sensor is ideal for use in the adverse conditions found in the crater area of Vulcano, having no moving parts to jam, break or wear out and being robust and resistant to corrosion. The SHT75 thermo-hygrometer allows the measurement of relative humidity in the air with a range of Rh% 0–100% and air temperature with a temperature range of T −40 to 100 C ° (Figure 1).

The wind speed has been utilized inside of a NOVAC program algorithm [64–66] to calculate the flux by the concentration of the molecules in the cross-section, perpendicular to the direction of propagation of the plume (Figure 3). Local wind monitoring is crucial because any erroneous estimation of wind speed and direction of propagation results in underestimations or overestimations of the flux measurement [1]. Here, we have also tested the wind data provided by global model from European Centre for Medium-Range Weather Forecasts (ECMWF) or the National Oceanic and Atmospheric Administration (NOAA) that have been considered in other volcanic assets. We found that the global modeling of weather parameters does not fit well the local dispersion of the small volcanic plume, generated by the active cone of La Fossa, due to the actual condition affecting the dispersion of gases (e.g., low altitude of the source, small distance from the populated areas) and the position of the Vulcano Porto Village, laying very close to the main fumaroles. For these reasons, we generally used the wind data from two very local weather stations (Lentia and VSCS in Figure 1) to better evaluate the SO_2 flux and better constrain the input data requested for tracking the local dispersion of noxious gases, like the sulfur dioxide of volcanic origin.

Furthermore, a UV-visible camera (MOBOTIX M25 Res. 1280 × 720 HD) was installed in 2016 on the roof of the Carapezza volcanological center in the direction of the volcano summit and a daily time-lapse is recorded with frames every 5 min (Figure 5). This camera has a dual purpose, to directly observe the degassing activity and to have direct feedback on the plume emitted from the solfataric area of the Fossa crater. This allows us to have a view of both direction and height of the plume, useful for processing the data of the slant column density of SO_2 to be multiplied by the wind speed. Moreover, the visual information acquired by the UV-visible camera allows us to choose, which is the UV-scanning-DOAS station to utilize for the calculation of the output of SO_2.

3.3. Soil CO_2 Fluxes, Methodology

The most sensible site for monitoring the summit degassing and evaluating the volcanic activity level is the summit crater area, which generally releases over 90% of the total

CO_2 degassing from Vulcano Island [6]. For this reason, in September 2007, an automated soil CO_2 monitoring station (VCSCS) was installed within the active summit crater of La Fossa beyond the fumarolic areas (Figure 1; the station is manufactured by West Systems ltd., [67]. Carbon dioxide was measured with a Dräger Polytron IR spectrometer, which operates in the range of 0–9999 ppm (precision of ± 5 ppm); environmental parameters (wind direction and speed, soil and atmosphere temperatures, atmospheric pressure and soil and atmosphere relative humidity) are acquired at the same time and in the same place [5]. The near-real-time measurement of CO_2 fluxes (VCSCS) is carried out on an hourly basis and the data are either stored in loco or transmitted directly to the INGV–Palermo geochemical monitoring center via internet utilizing WLAN/rooter service [5]. In case of failure in the transmission system, the local storage prevents the loss of data and the time series can be updated by in situ downloading.

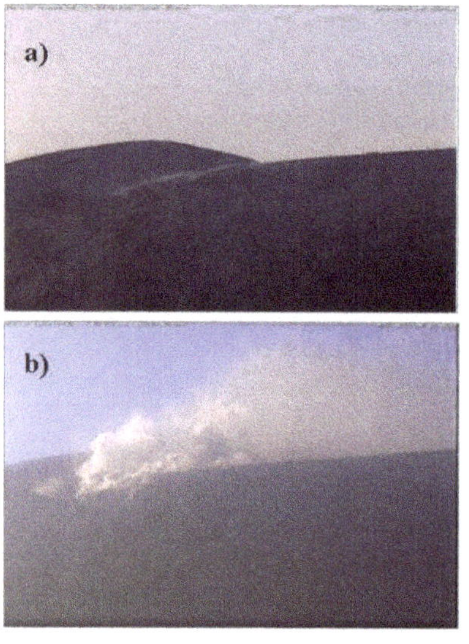

Figure 5. Photos of the Vulcano crater obtained by video monitoring. (**a**) View of the fluid emission from the main fumaroles field during the period of background solfataric activity degassing; (**b**) view of the fluid emission from the main fumaroles field during the anomalous degassing of 2021 crisis.

3.4. Soil CO_2 Output, Methodology and Campaigns

The last survey of diffuse soil CO_2 flux, in order of time, was carried out on 30 September 2021, in coincidence with the maximum CO_2 emission (34,000 g m^2 d^{-1}) from soils, recorded by the VCSCS station. The survey was performed with the West Systems equipment on 164 points located in the summit area of the active cone of La Fossa, exactly in the same target area of the 2007 survey [6] outside of the main fumarole field. These CO_2 flux data have been plotted on a normal probability plot [68] to assess the anomaly threshold for soil CO_2 flux and differentiate among different populations of soil CO_2 flux, following a Graphical Statistical Approach (GSA). Soil CO_2 flux data were processed for surface mapping using the sequential Gaussian simulation [69,70] (sGs), and this stochastic simulation allowed the best interpolation of the sets of data and evaluation of the output of diffuse CO_2 flux associated with the standard deviation of the total CO_2 output. The soil diffuse CO_2 output has been extended to cover the crater, except the fumarole area, and has been compared to the results of the 2007 survey [6].

3.5. Thermal Gradient and Heat Fluxes

The dataset of ground temperatures consists of temperature measurements in the porous ground (loose volcanic breccia) recorded on the west of the high-temperature fumarole vents located on the northern slope of the active cone of La Fossa caldera. The temperature data were recorded at the VSCS station (Figure 1) on a vertical line of four PT1000 sensors (temperature range of −40 °C to 150 °C, accuracy ±0.2 °C, resolution ±0.1°C), the distance between each sensor was 0.15 m, and thus the total depth of the monitored profile was 0.60 m. The temperature data were recorded hourly.

The measured profile is located a short distance (2 m) from the accumulation chamber for the diffuse CO_2 flux monitoring in order to monitor both the heat and gas flux at the same time and in the same diffuse degassing zone (Figure 2D). Therefore, temperature data in the shallowest layer of ground have been stored in the same server at the same time as CO_2 flux data diffusely released there.

The diffuse gas emission is essentially CO_2 because the main component of the hydrothermal advection, which is water vapor, has in this case condensed just below the conductive layer, and all the other condensable species are trapped in the liquid phase within the saturated ground. Therefore, temperature data in the shallowest layer of ground have been stored in the same server at the same time as CO_2 flux data diffusely released there.

These areas of mild thermal anomaly were named sub-fumarole area by [70,71] or steam heated soil (SHS) by other authors [1]. The method for measuring the heat flux from a shallow profile of temperature in volcanic areas was developed by [70] and further applied in continuous monitoring mode by other authors such as [72,73]. At the best condition, the monitored profile is almost dry (low humidity and absence of a liquid phase and steam) and the condensation zone of the steam ascending from the deep magmatic source stands at a short distance from the bottom of the profile along the z-axis [73]. In this almost-dry condition, the continuous monitoring of the diffuse heat flux associated with the thermal grounds has been ensured for long-term acquisition and avoided frequent interpolations due to missing data, or not-ideal conditions of the site.

The main issue in managing temperature data from the SHS monitoring stations is that the result is strongly site-sensitive, so the thermal monitoring profile must be located in steam-heated soils, along active faults, eruptive fractures or close to active fumaroles conduits, to promptly highlight magmatic impulses or new changes of trends. The heat flux monitored on the earth's surface is indeed the result of the combined effect of forces of opposite origin—the endogenous geothermal source, generated by buried magma on one side and the exogenous radiative source of sun on the other side. Another issue is the high sensitivity of the shallow ground to the natural changes of external conditions because rainfall, strong winds and sun radiations strongly perturb the ground temperature of the shallow layer of ground. The synchronous monitoring of atmospheric variables at the local scale and the cross-correlation with other volcanic variables, independently monitored, are strongly recommended in order to avoid erroneous interpretation of the temperature trends. In general, the best location for the temperature monitoring profile is at same distance from a fumarole vent, where the steam advection degrades from the convective to the diffusive transfer, and therefore the temperature distribution within the ground (thermal gradient and depth of the massive convective front) becomes the main variable to be directly correlated to the local heat flux. The previous monitoring experiences [33,73] proved that the best location for the temperature monitoring profile is at same distance from fumarole vents, where the steam advection degrading from the convective to the diffusive transfer is the best location for the temperature monitoring profile. In this case, the temperature distribution within the ground (thermal gradient and depth of the massive convective front) becomes the main variable directly correlating with the local heat flux. In the SHS, the main changes of thermal gradient result in changes to the conductive heat flow, but also in changes in the depth of the massive convective front, reflecting any contraction or expansion of the convective zone manifested by the neighboring fumaroles.

3.5.1. Depth Limit of the Conductive Layer (Z1)

The depth limit of the conductive layer (Z1) was calculated based on the temperature gradient following the hypothesis that the monitored profile was dry and the conductive heat transfer law dominated. The profiles showing a linear fit coefficient (R, calculated on 4 contemporary temperatures) lower than 0.99 indicated not ideal condition. The local temperature variation/depth range (Dt/Dz) slope, measured along the shallow ground profile to a maximum depth of 0.6 m, indicated the temperature distribution in the uppermost ground layer when the heat transport was essentially conductive. Extrapolation of each linear slope to the local boiling point (100 °C; 1 atm) indicated the Z1 at the VSCS station at the recording time. Thus, by extrapolating the linear profiles of temperature to the boiling point, we can evaluate the local changes in Z1 and also track the vertical shifts of the massive convective front throughout the monitoring period by following the time variations of Z1. Figure 6 gives an example of the linear thermal gradient (red line and equation) used to evaluate the depth limit of the conductive layer (Z1), that is the shallowest depth of the convective front, with the logarithm equation also fitting the air temperature and the resulting boiling point (blue curve and equation). R^2 is the correlation value for the correspondent regression. The temperature values in the example of Figure 6 X were recorded in January 2019, the 1° at H 12:00. When the heat release increases the convective front moves toward the interface between air, the local condition and the site could turn from sub-fumarole to fumarole condition. In this case, the linear regression coefficient results get lower than 0.990 and the thermal gradient is low because the diffuse heat flux becomes just a minor component of the heat release; more undetected variables should be investigated to quantify the real heat flux leaving the ground.

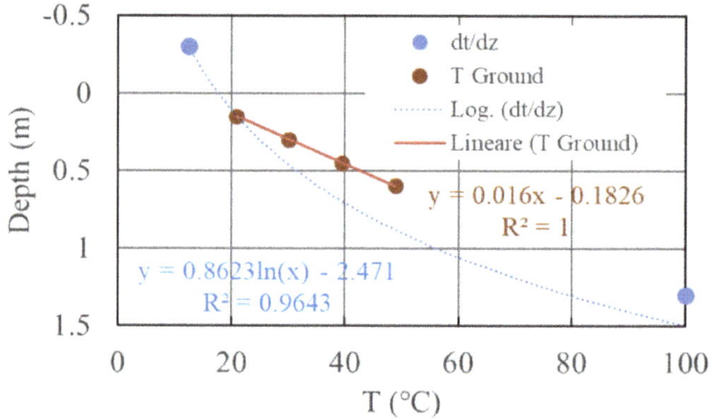

Figure 6. Example of a ground profile showing the thermal gradient in dry ground condition (brown circles) and air temperature (blue circle) recorded at the VSCS station. The brown line and equation show the linear gradient extrapolated to the depth limit of the conductive layer (Z1); the blue curve and equation shows the logarithm equation also fitting the air temperature. R^2 is the correlation value for the correspondent regression.

3.5.2. Heat Flux of the Steam Heated Ground (SHS-Flux)

The heat flux of the steam heated ground (SHS-Flux) is calculated in the superficial ground zone (above the Z1 level), using the following heat flux equation simplified by [74]:

$$\text{SHS-Flux} = \lambda(t4 - t2)/(zt4 - zt2) \quad (1)$$

where t4 and t2 are the temperatures, measured along the shallow vertical profiles at different depths (zt4 = −0.6 m; zt2= −0.3 m) and λ is the thermal conductivity, which, in the dry ground conditions requested for SHS flux evaluations, is assumed to be constant for a specific site (in this case we applied the value $\lambda = 0.8\ \text{W}\cdot\text{m}^{-1}\cdot\text{K}^{-1}$, in accordance with [37–74]). In order to minimize the sun radiation effect, the SHS-Flux was calculated based on the temperature gradient, excluding the uppermost sensor of temperature (T1 at depth of 0.15 m). When/if the temperature of the sensors in the ground increases to more than 70 °C, the temperature gradient is no more linear and the convective component of the heat flux becomes dominant. In this case, steam is the main component in the porous ground, and the monitored layer loses the ideal dry condition requested for the total conductive heat transfer. When the ground temperature of the monitored profile is higher than 70 °C, the SHS monitoring system can still supply the real start and end time of the increased heat release, but in this instance, the evaluated heat flux is highly underestimated, not including the convective heat transfer, that is—temporary—the main component of the local heat flux.

4. Results

4.1. SO_2 Plume Fluxes

The SO_2 plume fluxes monitored from 2008 to 2021 showed values between 7 and 248 t d^{-1} (Figure 7), with an average value of 25 t d^{-1}. In the following discussion, this last value, plus the standard deviation, will be considered the actual threshold value (25 +/− 21 t d^{-1}) to indicate the background convective outgassing, sourced from the high-temperature fumarole field of the La Fossa cone.

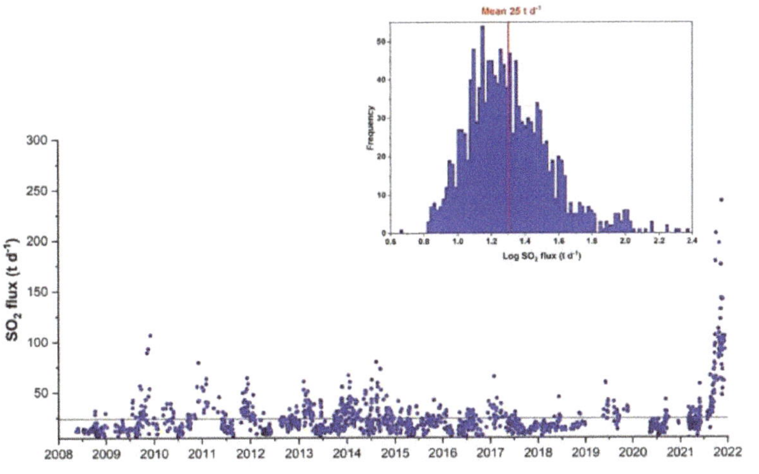

Figure 7. Daily average of SO_2 fluxes from the volcanic plume of the 2008–2021 period. Histogram of the SO_2 fluxes data with a mean value of 25 t d^{-1}.

Two major peaks in outgassing were observed in detail in September–December 2009, respectively (Inguaggiato et al. 2012), which reached values of SO_2 fluxes of around 100 t d^{-1}; and the ongoing stronger increases occurred in September–December 2021 with flux values that had abundantly exceeded 200 t d^{-1}, up to a maximum value of 248 t d^{-1} to date (one order of magnitude over the mean value of the last 13 years). To date, the last period of anomalous outgassing is still ongoing (31 December 2021).

In addition, several smaller increases with peaks of 75, 60, 50 and 40 t d^{-1} respectively occurred over the long observation period. During 2015, on the other hand, lower degassing values were recorded for the whole year, and the SO_2 flux values slightly exceeded the average value of 25 t d^{-1}.

4.2. Soil CO_2 Fluxes

The fluxes of CO_2 from the soils measured at the VSCS station represented the time variations of diffuse gas emissions from the summit of the active cone of the island of Vulcano in the period 2007–2021. The monitoring station showed values between 100 and 34,000 g m^{-2} d^{-1}, with an average value of 1637 g m^{-2} d^{-1} (Figure 8). The two largest increases in the diffuse CO_2 fluxes from the ground occurred respectively in the period September–December 2009 [5], with values of 14,000 g m^{-2} d^{-1}, and from September to

December 2021, with the highest values of 34,000 g m^{-2} d^{-1}. After the peak occurred in 2009, a general decreasing trend, with minor peaks, was observed until the 2015 period, when the lowest fluxes (around 100 g m^{-2} d^{-1}) were observed. Thereafter, starting from 2016 onwards, a reversal in the diffuse degassing trend has been observed, with a series of increasing peaks in 2016, 2017, 2018 and 2019. However, in these previous years, the maximum values of soil CO_2 fluxes never exceeded 5000 g m^{-2} d^{-1}. Starting from 2020 onwards, the VSCS station showed a main positive trend of diffuse gas emissions with flux values increasing from 500 g m^{-2} d^{-1} to 34,000 g m^{-2} d^{-1}, recorded on 30 September 2021. This value, reached in September, represented an increase of 20 times higher with respect to the average values evaluated in 13 years of observation.

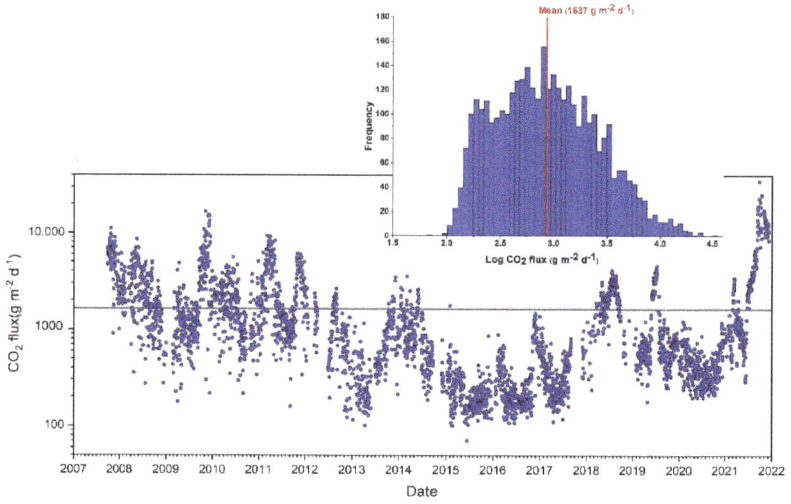

Figure 8. Daily average (24 measurements/day) of CO_2 fluxes of VSCS station of the 2007–2021 period. Histogram of the CO_2 fluxes data with a mean value of 1637 g m^{-2} d^{-1}.

4.3. Soil CO_2 Output

On 30 September 2021, the survey of CO_2 fluxes in a target area of La Fossa caldera has been carried out to detect the ongoing strong increases of soil CO_2 fluxes recorded in the summit area by the VCSCS station. This survey started in coincidence with the maximum CO_2 fluxes recorded in the summit area by the VCSCS station, (34,000 g m^{-2} d^{-1}). The total output of CO_2 degassing from the soils in this area has been estimated as 1548 t d^{-1} (see Table 1 for sGs statistics and reference surfaces). This CO_2 flux value was around 16 times higher than the 2007 campaigns (92 t d^{-1}; Figure 4; [6]), although they were performed on the same target area of comparable extension (Table 1). In both the surveys, the targeted area excluded the sector interested by the main fumarole release (Figure 9); moreover, the last survey confirmed CO_2 fluxes by the VCSCS station as representative of the whole summit degassing of the crater area of this active cone.

Table 1. Summit area CO_2 flux output (t d^{-1}).

Date	Number of Samples	Area (m^2)	Total CO_2 Output (t/d)	Standard Deviation (t/d)	Total CO_2 Output (t/d) Area of 440,755 m^2
Sep-2021	164	88,900	312	7.1	1547
Sep-2007	244	70,575	15	0.6	94

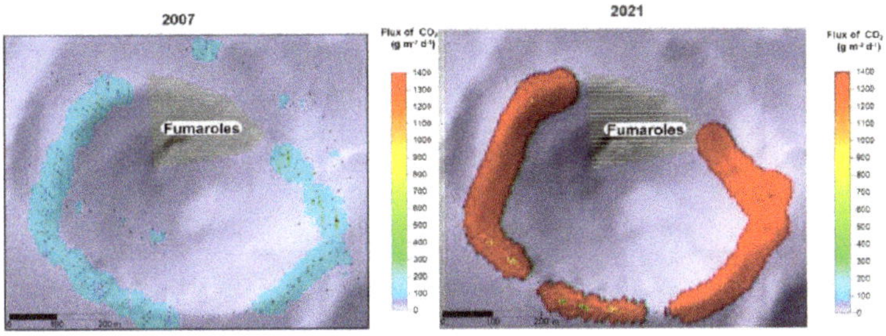

Figure 9. Summit soil degassing area maps relative to 2007 and 2021; the brown field shows the main fumarole release. The distribution of CO_2 flux over the summit area and estimates of the total CO_2 discharge were derived by sequential Gaussian simulation (sGs).

The probability plot resulting from the flux measured in the last survey (Figure 10a and Tables 1 and 2) shows a three-modal distribution, consisting of three lognormal populations named A, B and C, with a partial overlapping of the samples in (A) and (B). Based on the GSA technique proposed by [69], the mean CO_2 flux and the 95% confidence interval of the mean [74] are for the population (A) 606 g m^{-2} d^{-1} (336–1679 g m^{-2} d^{-1}); for the population (B) 1343 g m^{-2} d^{-1} (1150–1629 g m^{-2} d^{-1}); and for the population (C) 13,518 g m^{-2} d^{-1} (12,816–14,789 g m^{-2} d^{-1}).

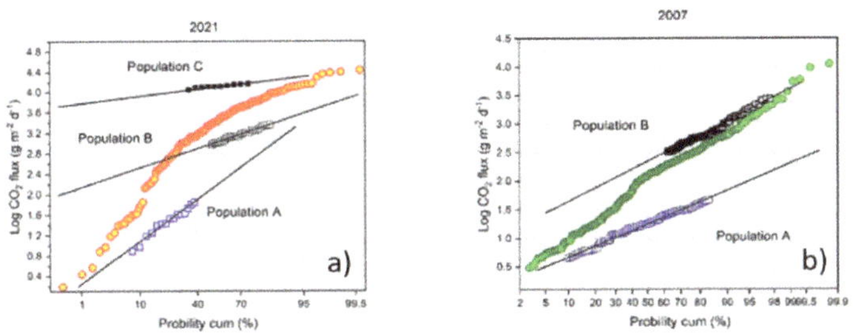

Figure 10. (**a**) Log CO_2 fluxes vs the cumulated probability (%) of the 2007 survey; (**b**) Log CO_2 fluxes vs the cumulated probability (%) of the 2021 survey. Both surveys were performed at the summit area.

Table 2. Proportions of each population with a mean CO_2 flux (in g m^{-2} d^{-1}) and the corresponding 90% confidence intervals obtained by statistical graphical approach.

	Crater 2021		
Population of CO_2 Flux	**Mean Flux of CO_2 (g m^{-2} d^{-1})**	**90% Confidence Interval (g m^{-2} d^{-1})**	**Proportion (%)**
A	606	336–1679	29
B	1343	1150–1629	57
C	13,518	12,816–14,789	14
	Crater 2007		
A	30	28–34	40
B	455	414–512	60

The results of the GSA for the data measured in September 2021 highlights two new populations characterized by high CO_2 flux that were never registered before (Figure 10b and Tables 1 and 2) [5], suggesting a new and more active degassing level, possibly of magmatic origin.

4.4. Thermal Monitoring

The thermal monitoring of diffuse heat flux from the marginal zone of the high-temperature fumarole field of the active cone of La Fossa Caldera began in April 2018. For more than three years, we registered the hourly variation of temperature in the section of the ground crossed by diffuse gas emission. Both the depth limit of the conductive layer (z1 depth) and the diffuse heat flux resulted in the hourly monitored shallow temperature gradient. The period from June 2018 to June 2019 was characterized in general by background solfataric activity, as can be confirmed by comparing the general report of volcanic surveillance. For this reason, the average value of the z1 depth, extrapolated during the same period, represents the reference value of the depth limit of steam condensation during the background degassing. The time variations of Z1 depth (Figure 9) show that the z1 depth has been generally deeper than −0.9m, except in July 2018, during summer 2019 (from the end of May to 8 August) and from 26 June to the end of 2021 (Figure 11). From 28 September to 11 October, the Z1 depth rose (see Figure 10) and the monitored profile included a narrow level interested by mixed heat transfer (above a depth of 0.4m, both conductive and convective) while the deepest part was essentially convective (below a depth of 0.4 m), and two levels reached the same maximum temperature of 93 °C. The shallowest depth limit of the conductive layer was registered on 11 October 2021 and rose to −0.4m below the ground level.

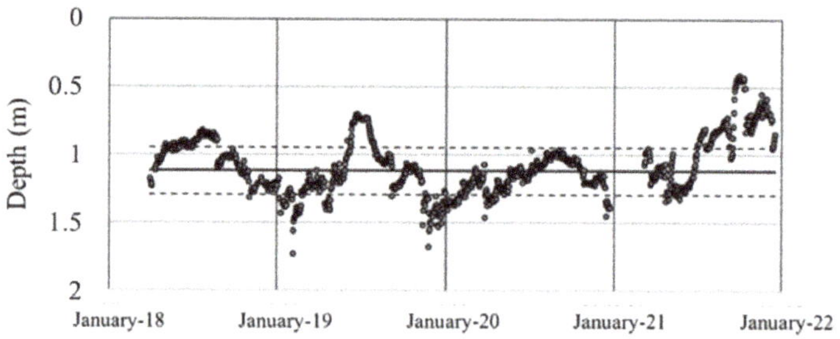

Figure 11. Time variation of the depth limit of the conductive layer recorded at VSCS station. The dashed lines indicate the standard deviation around the average value, evaluated during the period of background diffuse degassing (June 2018–June 2019).

This long-term monitoring of the thermal anomaly in this diffuse degassing zone also suggested the statistic threshold value for normal SHS flux values representing the background activity (Figure 12). The heat flux included in the range of values from 43 to 50 w × m^{-2} (average value plus standard deviation evaluated during the first year, June 2018–2019) indicates the normal background oscillation of heat flux emitted from the ground by the diffusive layer, while all the heat flux values higher than the average, plus the standard deviation, indicate periods of anomalous heat release. The most intense impulses of diffuse heat flux have been recorded in the year 2021, with the longest and most intense heat flux release starting after 28 August. From 20 September to 14 October, the station showed values higher than 65 w × m^2, but these values highly underestimated the local heat output, because the convective heat transfer (undetermined by the SHS monitoring station) has been the dominant form of heat transfer along the profile after 17 September 2021. Indeed, from 28 September to 10 October 2021, we observed the highest temperature values ever recorded at this site, and two levels of the recorded profile reached the buffered temperature of 93 °C. This thermal record is our reference to indicate the acme of the actual exhalative crisis by measuring the effect directly (i.e., by contact sensors) on the thermal anomaly located in the summit area at La Fossa caldera. According to the VSCS monitoring station, the thermal effect of the volcanic unrest has been evident from the end of June 2021 and has remained persistent to the end of the year 2021, as indicated by the rate of increase of the ground temperatures, registered again after any external perturbation occurred in the last three months. However, starting on the night of 29 September 2021, very strong winds and intense rainfall events highly perturbed the soil gas emission, often causing sudden decreases in the shallow ground temperatures.

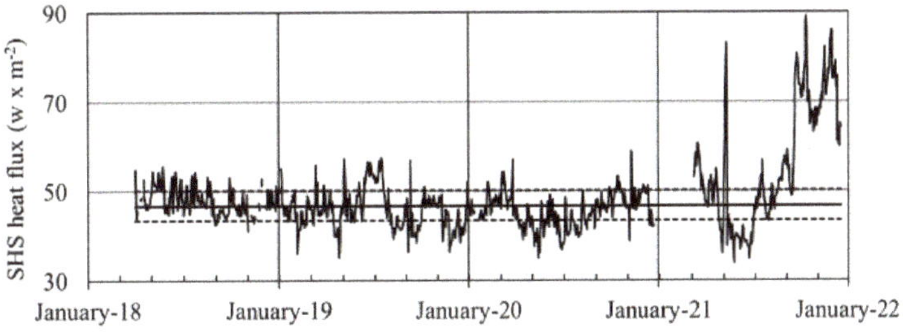

Figure 12. Time variation of diffuse heat flux recorded in the sub-fumarole area since April 2018. The grey area includes the average value and standard deviation evaluated during the period of background diffuse degassing (June 2018–June 2019).

5. Discussion

The plumbing system of this volcanic island is fed by deep magmatic input and is modulated by the overlying hydrothermal system [5,9,22] which is able to absorb or release its volatile compounds as a function of the dynamic input/output energy balance. The volatiles fluxes monitored on the summit area of the La Fossa cone during the solfataric activity represent the main surface manifestation of the shallow plumbing system.

In particular, the time variations of both SO_2 and CO_2 fluxes, tracked by the respective monitoring stations since 2008, have shown two major increases of solfataric activity, highlighted by evident anomalies both in the convective and in the conductive transfer of gases occurring in 2009 and 2021. Furthermore, in the long-term, the diffuse gas emissions showed that a decreasing trend characterized the period from 2009 to 2015, while an opposite, increasing trend has been registered from 2016 to 2021. Taking into account the diffuse degassing, we can divide the entire observation period (2007–2021) into two sub-periods, 2008–2014 (Figure 13) and 2016–2021 (Figure 14) for both recorded parameters (SO_2 and CO_2, to better understand the degassing processes in progress.

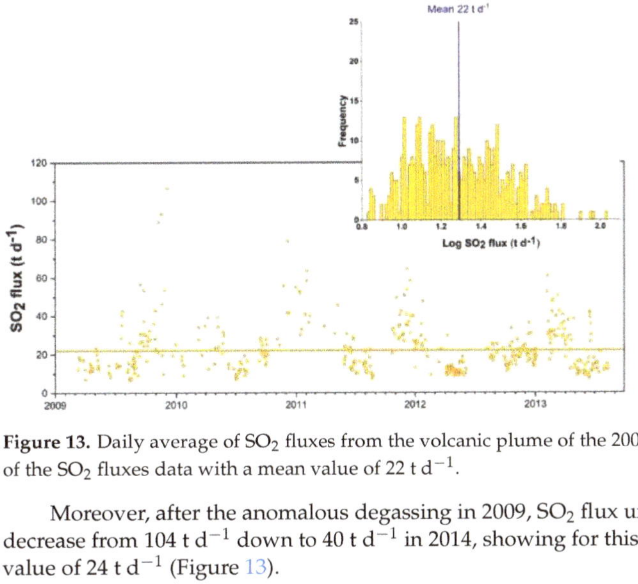

Figure 13. Daily average of SO_2 fluxes from the volcanic plume of the 2008–2014 period. Histogram of the SO_2 fluxes data with a mean value of 22 t d^{-1}.

Moreover, after the anomalous degassing in 2009, SO_2 flux underwent a progressive decrease from 104 t d^{-1} down to 40 t d^{-1} in 2014, showing for this sub-period the average value of 24 t d^{-1} (Figure 13).

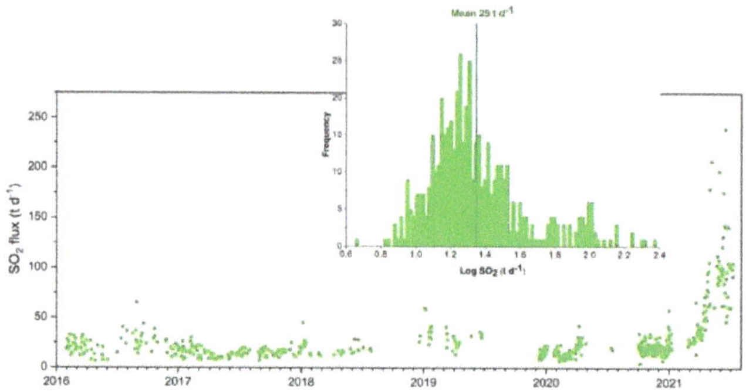

Figure 14. Daily average of SO_2 fluxes from the volcanic plume of the 2016–2021 period. Histogram of the SO_2 fluxes data with a mean value of 29 t d^{-1}.

The second sub-period, from 2016 to 2021, has been characterized by the average value of 29 t d^{-1}, a generally more sustained outgassing, and several small anomalies in 2016, 2018 and 2019, showing values exceeding 50 t d^{-1}. Finally, the strongest increase in the convective degassing has begun in April 2021 and reached values up to 250 t d^{-1}, an order of magnitude higher than the average value in September–October 2021 (Figure 14).

The behavior of the two sub-periods is better visualized by GSA, as shown in Figure 15 and in Table 3. In detail, the 2008–2014 period was characterized by a unimodal distribution of degassing families, while the 2016–2021 period has been characterized by the bimodal distribution with two different degassing families with frequencies of 86% and 14 %, respectively.

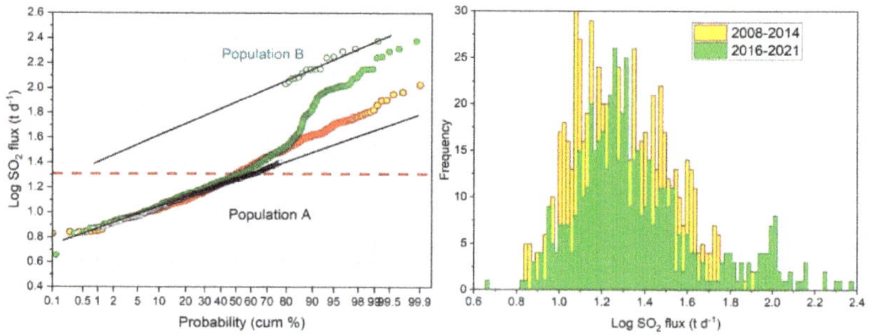

Figure 15. Partition of the SO_2 flux measurements recorded daily from the active DOAS monitoring system, according to the method of Graphical Statistical Approach proposed by [50].

Table 3. Results of the partition shown in Figure 14 of the SO_2 flux measurements recorded daily from the active DOAS monitoring system, based on the method of Graphical Statistical Approach proposed by [68].

Population of SO_2 Flux	Mean Flux of SO_2 (t d^{-1})	90% Confidence Interval (t d^{-1})	Proportion (%)
A	19.5	18.8–20.4	86
B	85	77–95	14

The monitoring station for the diffuse degassing of CO_2 highlighted a clearer discontinuity between the two sub-periods, also highlighted by GSA on the SO_2 flux data. The first sub-period from 2008 to 2014 (Figure 16) is characterized by a major anomaly in 2009

(11,000 g m^{-2} d^{-1}), but also by an average value of approximately 1720 g m^{-2} d^{-1} and by a series of decreasing peaks of CO_2 flux values reaching 6000, 4000 and 2500 g m^{-2} d^{-1}, respectively, in the period 2010–2012. Finally, in 2013–2015, the CO_2 flux diffused from the ground at the summit station VSCS reached the absolute minimum value of about 100 g m^{-2} d^{-1}, while the second sub-period from 2016 to 2021 (Figure 17) showed a series of increasing CO_2 flux peaks at 2000, 4000 and 5000 g m^{-2} d^{-1}, respectively, in 2016, 2018 and 2019, corresponding to subsequent CO_2 pulses. Finally, the second half of 2020 was characterized by a growing trend, which increased almost continuously until September–October 2021, reaching the highest recorded values of 34,000 g m^{-2} d^{-1}.

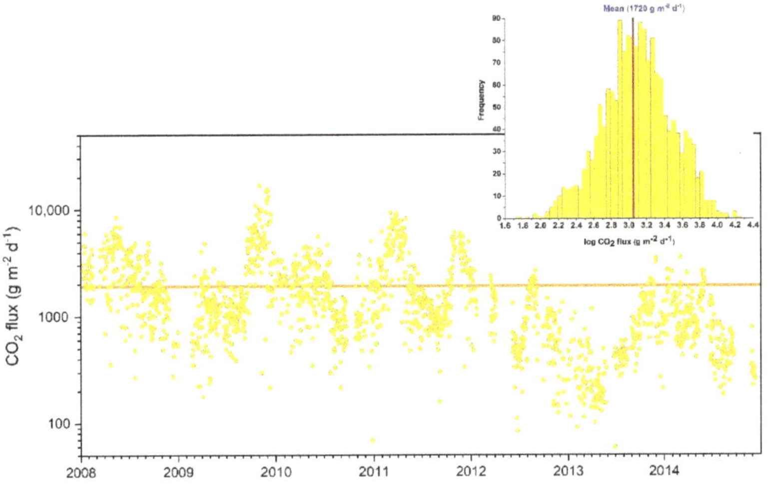

Figure 16. Daily average (24 measurements/day) of CO_2 fluxes of VSCS station of the 2008–2014 period. Histogram of the CO_2 fluxes data with a mean value of 1720 g m^{-2} d^{-1}, reported on the horizontal yellow line.

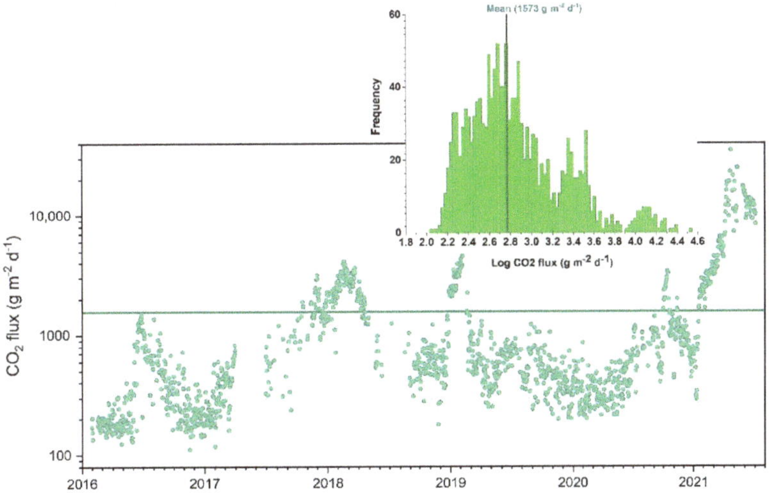

Figure 17. Daily average (24 measurements/day) of CO_2 fluxes of VSCS station of the 2016–2021 period. Histogram of the CO_2 fluxes data with a mean value of 1573 g m^{-2} d^{-1}, reported on the horizontal green line.

The cumulated probability plot, applied to the CO_2 flux data, show that the 2007–2014 period is characterized by a unimodal distribution of degassing family, while the 2016–2021 period is characterized by a three-modal distribution with different degassing families respectively showing frequencies of 10%, 85 and 5 %, with the respective mean CO_2 flux values diffused from the ground of 180, 663 and 8155 g m^{-2} d^{-1} (Figures 14–16 and Table 4).

Table 4. Results of the partition shown in Figure 18 of the diffuse CO_2 flux recorded daily at the VSCS station, based on the method of Graphical Statistical Approach proposed by [68].

Population of CO_2 Flux	Mean Flux of CO_2 (g m^{-2} d^{-1})	90% Confidence Interval (g m^{-2} d^{-1})	Proportion (%)
A	180	177–184	10
B	663	635–693	85
C	8155	7676–8726	5

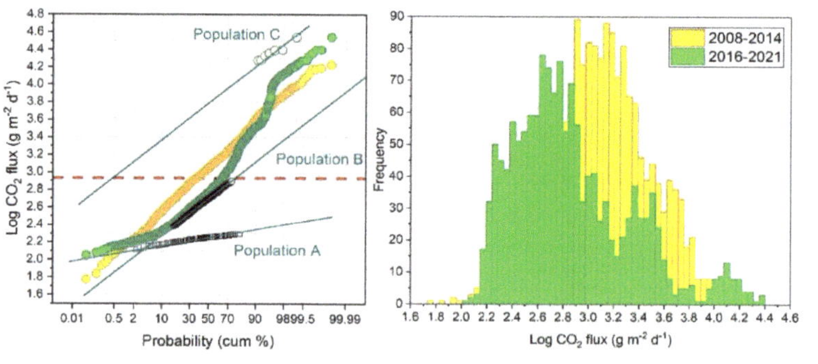

Figure 18. Partition of the diffuse CO_2 flux from the ground recorded daily from the VSCS monitoring system, according to the method of Graphical Statistical Approach proposed by [68].

Figure 19a shows the strong increase, registered during 2021 on both fluxes degassing rate (daily) and SO_2 fluxes in September 2021, with 3 t d^{-2} rate of SO_2 increasing daily and over 100 t d^{-2} plume SO_2 flux.

Figure 19b shows the strong increase, registered during 2021 on both fluxes degassing rate (daily) and CO_2 fluxes in September 2021, with 1000 g m^2 d^{-2} rate of CO_2 increasing daily and over 10,000 g m^2 d^{-1} diffuse CO_2 flux. However, the anomalous diffuse degassing started in June–July 2021, with a minimum rate of about 60 g m^2 d^{-2} and fluxes of 3000 g m^2 d^{-1}. In the following period, from October to December 2021, a decreasing rate of CO_2 fluxes has been recorded; even if the monthly CO_2 fluxes remained around 10,000 g m^2 d^{-1}, that is one order of magnitude higher than the average value recorded in the last 5 years (500 g m^2 d^{-1}).

Focusing on the last year of heat flux monitoring (Figure 20), the diffuse heat flux diffused from the ground oscillated within the background range of values from April to July, but in May and October, the heat flux showed the greatest dispersion of daily values. Such great dispersions reflect the condition of an altered equilibrium in the outgassing, but only a multi-parameter correlation could confirm this hypothesis. The anomalous heat flux resulting at VSCS during May does not show a positive correlation with the other two extensive parameters discussed here (CO_2 flux diffused from soil and SO_2 flux released in the volcanic plume emitted by the high-temperature fumaroles). Afterwards, the heat flux at the VSCS station began increasing in June 2021, but the strongest increasing rates were observed in September and November. The ideal condition requested for this simplified thermal monitoring system was lost for about two weeks (28 September–10 October), when the conductive transfer became the minor component in the local heat balance and the convective transfer was undetermined. For this reason, the heat flux based only on the temperature gradient resulted in a high underestimation, temporarily interrupting the direct correlation with the general trend of the anomalous gas emissions from the volcanic system.

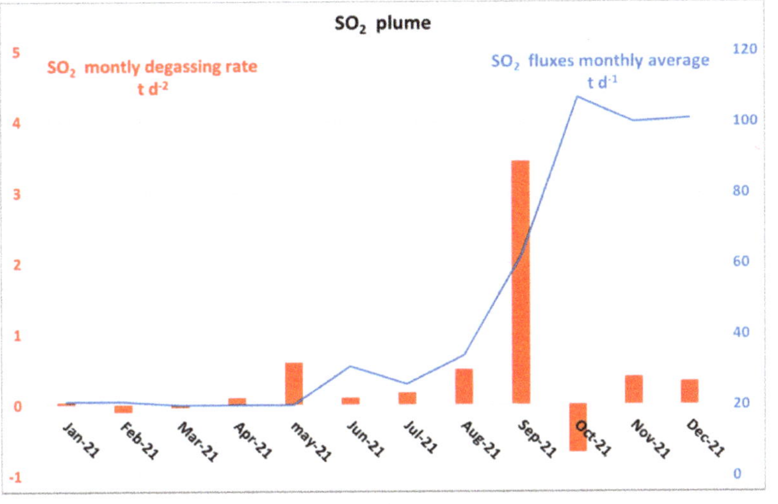

(a)

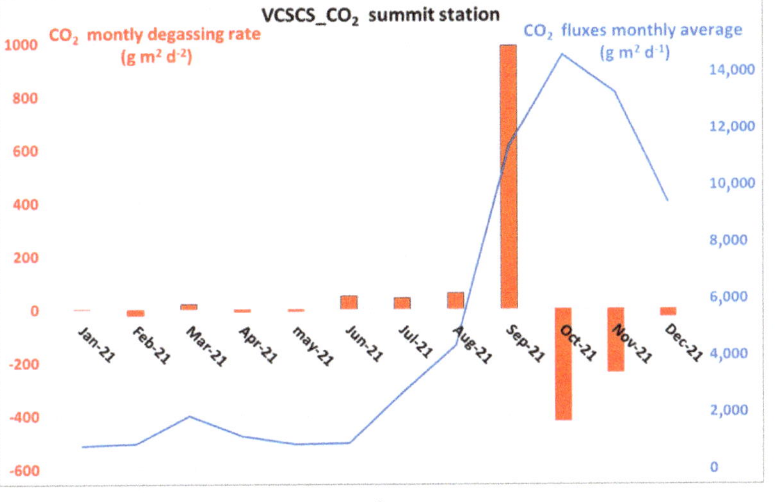

(b)

Figure 19. (**a**) SO_2 plume degassing, monthly SO_2 degassing rate (t d^{-2}) and SO_2 fluxes monthly average (t d^{-1}) of 2021 period. Red bars indicate the increasing degassing rate; blue bars indicate the decreasing degassing rate. The brown line represents the monthly average of SO2 fluxes. (**b**) VCSCS summit station, monthly CO_2 degassing rate and CO_2 fluxes monthly average of 2021 period. Red bars indicate the increasing degassing rate; blue bars indicate the decreasing degassing rate. The brown line represents the monthly average of CO_2 fluxes.

During the second half of 2021, a clear change in degassing style was recorded for the volatiles emitted by the magma in the summit area, such as the flux of diffused CO_2 from the soils and the flux of the SO_2 of the plume emitted by the fumarolic field. In particular, the volatiles released from the magma stored 2–3 km below the Vulcano edifice, rose towards the surface and interacted with the shallow hydrothermal system (Figure 21). During the background level of activity, groundwater receive, modulate and partially buffer, the fluids released from the magmatic bottom source produced a volatile scrubbing process that modulated the normal shallow solfataric degassing activity. This condition determined a uni-modal distribution in the degassing style, as we observed from 2007 to 2015 (Figure 15). Differently, during a geochemical crisis, the groundwaters were not able to modulate or buffer the new increased volatiles fluxes anymore. In this case, the monitoring stations showed that the heat and CO_2 fluxes in the diffuse degassing areas increased over the statistical thresholds of the background, and the DOAS network revealed more than one order of magnitude increase of SO_2 flux from the fumaroles field. This is what we observed during the last period, and these conditions determine a tri-modal distribution of degassing families (Figure 18).

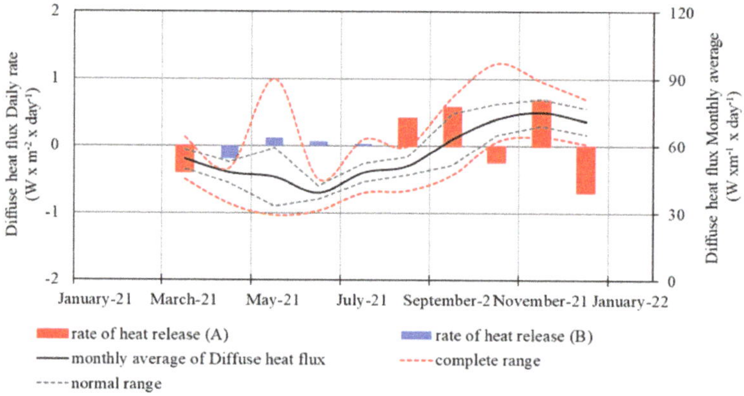

Figure 20. Time variation of diffuse heat flux from the summit station VSCS (red curve) and daily rate of heat release evaluated on a monthly base during the year 2021 (red bars for anomalous degassing average, blue bars for the background average degassing). The red border indicates the entire range of values and the grey border indicates the statistical "normal range" of values from the hourly record, resulting from a unimodal distribution around the average values. The highest values of diffuse heat fluxes occurred seldom in the months of May and much more frequently in October 2021.

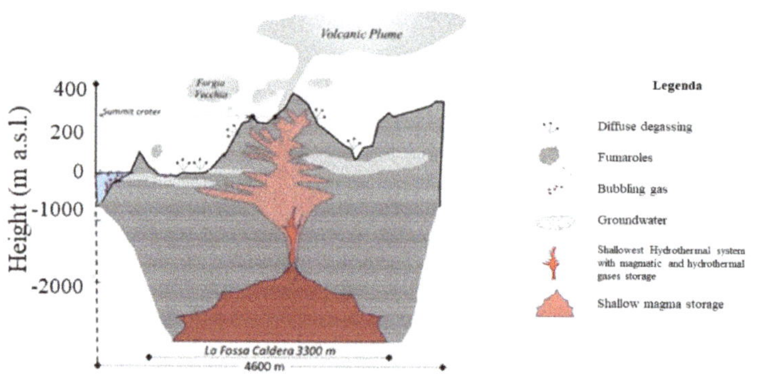

Figure 21. NW–NE profile of the Caldera La Fossa running from Mt. Lentia, Vulcano Piano. The ratio between height and length of this profile is H/L = 1.5; the total length is 4.6 km; and La Fossa Caldera occupies 3.3 km along this section. The schematic section shows the model of solfataric release in background condition (modified from [5,9,55] Inguaggiato et al., 2012; Federico et al., 2010; Fusillo et al., 2015).

6. Conclusions

The long-term monitoring of the hydrothermal release from the summit area of the active cone of La Fossa highlighted the huge increase of energy and mass output occurring in the second half of 2021 on the Island of Vulcano. The correlation among these three different and independent geochemical parameters allowed the quantification of timing and the intensity of the actual volcanic crisis, updating the obtained results and evaluations in almost real-time. The three remotely controlled monitoring systems discussed here are tracking the evolution of exhaling activity without exposing the research units involved in the maintenance of the network to the growing gas hazard.

This geochemical monitoring method, based on the selection of a few independent variables, which are closely related to the energy and mass flux escaping from the deep magmatic system, is still sustainable, even when the severe environmental conditions in the main fumaroles area become ever more adverse and the visitors are forbidden to reach the summit zone of the active cone due to the increase of solfataric activity.

Thanks to the long-term time series of monitoring data, these authors defined the local thresholds of the background for the exhaling activity by means of simple but robust statistical evaluations based on several thousands of validated measures to show the behavior of each monitored variable: SO_2 flux plume, diffuse CO_2 flux and diffuse heat flux.

The monitoring station of these fluid emanations are extremely site-sensible, but the location selected here ensures correlated evidence of massive steam releases from the summit of the La Fossa cone by means of independent acquisitions by remote sensing flux evaluations in the volcanic plume, CO_2 diffused flux and thermal anomalies.

The long-term data comparisons show that the simple thermal monitoring of the shallow ground is useful to track the time variation of the extensive flux of hydrothermal origin, thanks to the direct relationship between the temperature of the ground and the flow of steam and hot gas rising through the main fractures and fumarole conduits.

In conclusion, the extensive parameters indicate that the increases in mass and energy output in the summit area of the La Fossa crater began in June 2021. Then, a sharp variation starting in September 2021 brought both the SO_2 flux in the plume, as well as the CO_2 and the heat fluxes from the diffuse degassing zone to their respective actual absolute maximum values. In particular, we underline that the plume SO_2 flux, soil CO_2 fluxes from the VCSCS station and soil CO_2 output, all measured with different techniques but all related to the outgassing activity, showed sharp increases in September 2021, with similar factors (e.g., 10, 20 and 16 times higher respectively) respect to their background degassing values.

After these sharp increasing fluxes, we observed from October to December a halt of the growing rate of the extensive parameter with a slight decrease, followed by a stabilization of the phenomenon on new levels, an order of magnitude higher than the background evaluated during the previous 13 years of observation. On the base of the degassing sketch model supplied so far, we strongly suggest following up this monitoring activity and comparing observable parameters (geophysical and geochemical ones) as far as possible to capture—in useful time—any variation of this state of activity.

Finally, the long-term monitoring of this volcanic system confirmed that this last crisis had been far more intense than the previous ones because it falls well over the normal oscillation of the hydrothermal system. To date, the degassing activity is still very intense, and the system seems to have reached a new equilibrium state characterized by high fluid pressures in the shallowest storage zone.

Many different processes may alter this new equilibrium condition, possibly resulting in the destabilization of this new dynamic degassing balance between the input/output of volatiles through the shallow hydrothermal systems.

Tracking the trending variation of extensive fluxes, we could follow the evolution of this active system towards different scenarios: the opening and/or reactivation of faults, corresponding an increase in surface permeability caused by seismic activity, coupled with a decrease in the deep input, could bring the pressure of the surface plumbing system back to pre-crisis values, while a new deep input of volatiles from magmatic storage, occurring in this new energetic steady state, could trigger the next unrest, leading to the next phreatic and/or magmatic explosive activity.

The prediction of the occurrence of a phreatic eruption and the identification of possible precursors has been a debated topic for many years. This debate has led to the identification of the actions to be adopted in order to characterize and monitor a volcanic system [75] and identify a modus operandi able to evaluate the variation of the state of volcanic activity [1,76]. The presence of a modern, multidisciplinary and efficient monitoring system is a fundamental requirement to achieve useful results for the possible prediction of paroxysmal events. The island of Vulcano possesses all these requisites and the evolution of the volcanic activity in progress will certainly provide useful information to enrich the international case history, better linking the observed geochemical variations of the hydrothermal system with changes in volcanic activity. Furthermore, it is important and desirable for the data presented in this work to be compared with the geophysical data acquired from the other monitoring networks installed on the island (accounting for seismicity, deformation, tremor, etc.) to formulate a more complete dynamic model, including the deep and shallow plumbing systems of Vulcano Island.

Author Contributions: Conceptualization, S.I. and F.V.; methodology, F.V., I.S.D., L.C., M.C. and A.M. (Andrea Mastrolia); validation, F.V., I.S.D. and L.C.; investigation, S.I.; data curation, L.C., F.V. and S.I.; writing—original draft preparation, S.I., I.S.D. and A.M. (Agnes Mazot); writing—review and editing, S.I. and I.S.D. All authors have read and agreed to the published version of the manuscript.

Funding: This research was funded by the INGV-DPCN (Italian National Institute of Geophysics and Volcanology—Italian National Department for Civil Protection) volcanic surveillance program of Vulcano island, ObFu 0304.010. Moreover, this investigation was partially funded by the TORS project in the framework of institutional INGV projects "Ricerca Libera" ObFu 9999.549; and Pianeta Dinamico Task V2, ObFu 1020.010.

Data Availability Statement: Not Applicable.

Acknowledgments: The authors wish to thank their colleagues at the Istituto Nazionale di Geofisica e Vulcanologia of Palermo for their help in acquiring and processing data and for their support in field logistics. We also thank Fawzi Dumaz for the DEM map of Vulcano Island.

Conflicts of Interest: The authors declare no conflict of interest.

References

1. Inguaggiato, S.; Diliberto, I.S.; Federico, C.; Paonita, A.; Vita, F. Review of the evolution of geochemical monitoring, networks and methodologies applied to the volcanoes of the Aeolian Arc (Italy). *Earth Sci. Rev.* **2018**, *176*, 241–276. [CrossRef]
2. Diliberto, I.S. Cyclic Behavior in the Fumaroles Output Detected by Direct Measurement of Temperature of the Ground. *Eng. Proc.* **2021**, *5*, 47. [CrossRef]
3. Camarda, M.; De Gregorio, S.; Capasso, G.; Di Martino, R.M.R.; Gurrieri, S.; Prano, V. The monitoring of natural soil CO_2 emissions: Issues and perspectives. *Earth Sci. Rev.* **2019**, *198*, 102928. [CrossRef]
4. Capasso, G.; Favara, R.; Inguaggiato, S. Chemical features and isotopic composition of gaseous manifestations on Vulcano Island, Aeolian Islands, Italy: An interpretative model of fluid circulation. *Geochim. Cosmochim. Acta* **1997**, *61*, 3425–3440. [CrossRef]
5. Inguaggiato, S.; Calderone, L.; Inguaggiato, C.; Mazot, A.; Morici, S.; Vita, F. Long-time variation of soil CO_2 fluxes at summit crater of Vulcano (Italy). *Bull. Volcanol.* **2012**, *74*, 1859–1863. [CrossRef]
6. Inguaggiato, S.; Mazot, A.; Diliberto, I.S.; Inguaggiato, C.; Madonia, P.; Rouwet, D.; Vita, F. Total CO_2 output from Vulcano island (Aeolian Islands, Italy). *Geochem. Geophys. Geosys.* **2012**, *13*, 1–19. [CrossRef]
7. Vita, F.; Inguaggiato, S.; Bobrowski, N.; Calderone, L.; Galle, B.; Parello, F. Continuous SO_2 flux measurements for Vulcano Island, Italy. *Ann. Geophys.* **2012**, *55*, 2. [CrossRef]
8. Paonita, A.; Federico, C.; Bonfanti, P.; Capasso, G.; Inguaggiato, S.; Italiano, F.; Madonia, P.; Pecoraino, G.; Sortino, F. The episodic and abrupt geochemical changes at La Fossa fumaroles (Vulcano Island, Italy) and related constraints on the dynamics, structure, and compositions of the magmatic system. *Geochim. Cosmochim. Acta.* **2013**, *120*, 158–178. [CrossRef]
9. Federico, C.; Capasso, G.; Paonita, A.; Favara, R. Effects of steam-heating processes on a stratified volcanic aquifer: Stable isotopes and dissolved gases in thermal waters of Vulcano Island (Aeolian archipelago). *J. Volcanol. Geotherm. Res.* **2010**, *192*, 178–190. [CrossRef]
10. Aiuppa, A.; Burton, M.; Mure, F.; Inguaggiato, S. Intercomparison of volcanic gas monitoring methodologies performed on Vulcano Island, Italy. *Geophys. Res. Lett.* **2004**, *31*, L02610. [CrossRef]

11. Aiuppa, A.; Inguaggiato, S.; McGonigle, A.J.S.; O'Dwyer, M.; Oppenheimer, C.; Padgett, M.J.; Rouwet, D.; Valenza, M. H2S fluxes from Mt. Etna, Stromboli, and Vulcano (Italy) and implications for the sulfur budget at volcanoes. *Geochim. Cosmochim. Acta* **2004**, *69*, 1861–1871. [CrossRef]
12. O'Dwyer, M.; Padgett, M.J.; McGonigle, A.J.S.; Oppenheimer, C.; Inguaggiato, S. Real-time measurement of volcanic H_2S and SO_2 concentrations by UV spectroscopy. *Geopys. Res. Lett.* **2003**, *30*, 1652. [CrossRef]
13. Carapezza, M.; Nuccio, P.M.; Valenza, M. Genesis and evolution of the fumaroles of vulcano (Aeolian Islands, Italy): A geochemical model. *Bull. Volcanol.* **1981**, *44*, 547–563. [CrossRef]
14. Barberi, F.; Neri, G.; Valenza, M.; Villari, L. 1987–1990 unrest at Vulcano. *Acta Vulcanol.* **1991**, *1*, 95–106.
15. Capasso, G.; Inguaggiato, S.; Nuccio, P.M.; Pecoraino, G.; Sortino, F. Compositional evolution of the fumarolic gases emitted at the crater of Vulcano during 1987–1993. In Proceedings of the Independent Publishers Group World Organization of Volcano Observatories Meeting, Guadalupe, CA, USA, 13–17 December 1993; pp. 109–126.
16. Cioni, R.; D'Amore, F. A genetic model for the crater fumaroles of Vulcano island (Sicily, Italy). *Geothermics* **1984**, *13*, 375–384. [CrossRef]
17. Chiodini, G.; Cioni, R.; Falsaperla, S.; Guidi, M.; Marini, L.; Montalto, A. Geochemical and seismological investigations at Vulcano (Aeolian Islands) during 1978–1989. *J. Geophys. Res.* **1992**, *97*, 11025–11032. [CrossRef]
18. Badalamenti, B.; Chiodini, G.; Cioni, R.; Favara, R.; Francofonte, S.; Gurrieri, S.; Hauser, S.; Inguaggiato, S.; Italiano, F.; Magro, G.; et al. Special field workshop at Vulcano (Aeolian Islands) during summer 1988. Geochemical results. *Acta Vulcanol.* **1991**, *1*, 223–227.
19. Capasso, G.; Favara, R.; Francofonte, S.; Inguaggiato, S. Chemical and isotopic variations in fumarolic discharge and thermal waters at Vulcano Island (Aeolian Island, Italy) during 1996: Evidence of resumed volcanic activity. *J. Volcanol. Geotherm. Res.* **1999**, *88*, 167–175. [CrossRef]
20. Diliberto, I.S. Long-term monitoring on a closed-conduit volcano: A 25-year long time-series of temperatures recorded at La Fossa cone (Vulcano Island). *J. Volcanol. Geotherm. Res.* **2017**, *346*, 151–160. [CrossRef]
21. Cannata, A.; Diliberto, I.S.; Alparone, S.; Gambino, S.; Gresta, S.; Liotta, M.; Madonia, P.; Milluzzo, V.; Aliotta, M.; Montalto, P. Multiparametric Approach in Investigating Volcano-Hydrothermal Systems: The Case Study of Vulcano (Aeolian Islands, Italy). *Pure Appl. Geophys.* **2012**, *169*, 167–182. [CrossRef]
22. Capasso, G.; Federico, C.; Madonia, P.; Paonita, A. Response of the shallow aquifer of the volcano-hydrothermal system during the recent crises at Vulcano Island (Aeolian Archipelago, Italy). *J. Volcanol. Geotherm. Res.* **2014**, *273*, 70–80. [CrossRef]
23. Girona, T.; Costa, F.; Newhall, C.; Taisne, B. On depressurization of volcanic magma reservoirs by passive degassing. *J. Geophys. Res. Solid Earth* **2014**, *119*, 8667–8687. [CrossRef]
24. Maldonado, L.F.M.; Inguaggiato, S.; Jaramillo, M.T.; Garzon Valencia, G.; Mazot, A. Volatiles and energy released by Puracè volcano. *Bull. Volcanol.* **2017**, *79*, 84. [CrossRef]
25. Inguaggiato, S.; Vita, F.; Cangemi, M.; Mazot, A.; Sollami, A.; Calderone, L.; Morici, S.; Jacome Paz, M.P. Stromboli volcanic activity variations inferred by fluids geochemistry observations: Sixteen years of continuous soil CO_2 fluxes monitoring (2000–2015). *Chem. Geol.* **2017**, *469*, 69–84. [CrossRef]
26. Inguaggiato, S.; Cardellini, C.; Taran, Y.; Kalacheva, E. The CO_2 flux from hydrothermal systems of the Karymsky volcanic Centre, Kamchatka. *J. Volcanol. Geotherm. Res.* **2017**, *346*, 1–9. [CrossRef]
27. Inguaggiato, S.; Vita, F.; Rouwet, D.; Bobrowski, N.; Morici, S.; Sollami, A. Geochemical evidence of the renewal of volcanic activity inferred from CO_2 soil and SO_2 plume fluxes: The 2007 Stromboli eruption (Italy). *Bull. Volcanol.* **2011**, *73*, 443–456. [CrossRef]
28. Inguaggiato, S.; Vita, F.; Cangemi, M.; Calderone, L. Increasing Summit Degassing at the Stromboli Volcano and Relationships with Volcanic Activity (2016–2018). *Geosciences* **2019**, *9*, 176. [CrossRef]
29. Inguaggiato, S.; Vita, F.; Cangemi, M.; Calderone, L. Changes in CO_2 Soil Degassing Style as a Possible Precursor to Volcanic Activity: The 2019 Case of Stromboli Paroxysmal Eruptions. *Appl. Sci.* **2020**, *10*, 4757. [CrossRef]
30. Inguaggiato, S.; Vita, F.; Cangemi, M.; Inguaggiato, C.; Calderone, L. The Monitoring of CO_2 Soil Degassing as Indicator of Increasing Volcanic Activity: The Paroxysmal Activity at Stromboli Volcano in 2019–2021. *Geosciences* **2021**, *11*, 169. [CrossRef]
31. Jacome Paz, M.P.; Taran, Y.; Inguaggiato, S.; Collard, N. CO_2 flux and chemistry of El Chichon crater lake (Mexico) in the period 2013–2015: Evidence for the enhanced volcano activity. *Geophys. Res. Lett.* **2016**, *43*, 127–134. [CrossRef]
32. Diliberto, I.S.; Gurrieri, S.; Valenza, M. Relationships between diffuse CO_2 emissions and volcanic activity on the island of Vulcano (Aeolian Islands, Italy) during the period 1984–1994. *Bull. Volcanol.* **2002**, *64*, 219–228. [CrossRef]
33. Granieri, D.; Carapezza, M.L.; Chiodini, G.; Avino, R.; Caliro, S.; Ranaldi, M.; Tarchini, L. Correlated increase in CO_2 fumarolic content and diffuse emission from La Fossa crater (Vulcano, Italy): Evidence of volcanic unrest or increasing gas release from a stationary deep magma body? *Geophys. Res. Lett.* **2006**, *33*, L13316. [CrossRef]
34. Carapezza, M.L.; Granieri, D. CO_2 soil flux at Vulcano (Italy): Comparison of active and passive methods and application to the identification of actively degassing structure. *Appl. Geochem.* **2004**, *19*, 73–88. [CrossRef]
35. Chiodini, G.; Granieri, D.; Avino, R.; Caliro, S.; Costa, A.; Werner, C. Carbon dioxide diffuse degassing and estimation of heat release from volcanic and hydrothermal systems. *J. Geophys. Res.* **2005**, *110*, B08204. [CrossRef]
36. Aubert, M.; Diliberto, S.; Finizola, A.; Chebli, Y. Double origin of hydrothermal convective flux variations in the Fossa of Vulcano (Italy). *Bull. Volcanol.* **2008**, *70*, 743–751. [CrossRef]

37. Aubert, M.; Alparone, S. Variation d'origine sismique du flux de chaleur convectif dans La Fossa de Vulcano (Italie). *Comptes Rendus L'academie Sci. Ser. II A Earth Planet Sci.* **2000**, *330*, 603–610. [CrossRef]
38. Diliberto, I.S. Long-term variations of fumaroles temperatures on Vulcano Island (Italy). *Ann. Geophys.* **2011**, *54*, 2011. [CrossRef]
39. Diliberto, I.S. Time series analysis of high temperature fumaroles monitored on the island of Vulcano (Aeolian Archipelago, italy). *J. Volcanol. Geotherm. Res.* **2013**, *264*, 150–163. [CrossRef]
40. Ruch, J.; Vezzoli, L.; De Rosa, R.; Di Lorenzo, R.; Acocella, V. Magmatic control along a strike—Slip volcanic arc: The central Aeolian arc (Italy). *Tectonics* **2016**, *35*, 407–424. [CrossRef]
41. Peccerillo, A.; De Astis, G.; Faraone, D.; Forni, F.; Frezzotti, M.L. Compositional variations of magmas in the Aeolian arc: Implications for petrogenesis and geodynamics. *Geol. Soc. Lond. Mem.* **2013**, *37*, 491–510. [CrossRef]
42. Peccerillo, A.; Frezzotti, M.L.; De Astis, G.; Ventura, G. Modeling the magma plumbing system of Vulcano (Aeolian Islands, Italy) by integrated fluid-inclusion geobarometry, petrology, and geophysics. *Geology* **2006**, *34*, 17–20. [CrossRef]
43. De Astis, G.; Lucchi, F.; Dellino, P.; La Volpe, L.; Tranne, C.A.; Frezzotti, M.L.; Peccerillo, A. Geology, volcanic history and petrology of Vulcano (central Aeolian archipelago). *Geol. Soc. Lond. Mem.* **2013**, *37*, 281–349. [CrossRef]
44. Neri, G.; Barberi, B.; Oliva, G.; Orecchio, B. Spatial variation of seismogenic stress orientations in Sicily, south Italy. *Phys. Earth Planet. Inter.* **2005**, *148*, 175–191. [CrossRef]
45. Goes, S.; Giardini, D.; Jenny, S.; Hollenstein, C.; Kahle, H.-G.; Geiger, A. A recent reorganization in the south-central Mediterranean. *Earth Planet. Sci. Lett.* **2004**, *226*, 335–345. [CrossRef]
46. Billi, A.; Barberi, G.; Faccenna, C.; Neri, G.; Pepe, F.; Sulli, A. Tectonics and seismicity of the Tindary Fault System, southern Italy: Crustal deformations at the transition between ongoing contractional and extensional domains located above the edge of a subducting slab. *Tectonics* **2006**, *25*, TC2006. [CrossRef]
47. D'Agostino, N.; Selvaggi, G. Crustal motion along the Eurasia-Nubia plate boundary in the Calabrian Arc and Sicily and active extension in the Messina Straits from GPS measurements. *J. Geophys. Res.* **2004**, *109*, B11402. [CrossRef]
48. Keller, J. The island of Vulcano. *Rend. Soc. Italy Miner. Petr.* **1980**, *36*, 369–414.
49. Lanza, R.; Zanella, E. Paleomagnetic secular variation at Vulcano (Aeolian islands) during the last 135 kyr. *Earth Planet. Sci. Lett.* **2003**, *213*, 321–336. [CrossRef]
50. Gioncada, A.; Sbrana, A. "La Fossa Caldera", Vulcano: Inferences from deep drillings. *Acta Vulcanol.* **1991**, *1*, 115–126.
51. Frazzetta, G.; Gillot, P.Y.; La Volpe, L.; Sheridan, M.F. Volcanic hazards at Fossa of Vulcano: Data from the last 6000 years. *Bull. Volcanol.* **1984**, *47*, 105–124. [CrossRef]
52. Ventura, G.; Vilardo, G.; Milano, G.; Pino, N.A. Relationships among crustal structure volcanism and strike-slip tectonics in the Lipari-Vulcano volcanic complex (Aeolian Islands, Southern Tyrrhenian Sea, Italy). *Phys. Earth Planet Inter.* **1999**, *116*, 31–52. [CrossRef]
53. Fusillo, R.; Di Traglia, F.; Gioncada, A.; Pistolesi, M.; Wallace, P.J.; Rosi, M. Deciphering post-caldera volcanism: Insight into the Vulcanello (Island of Vulcano, Southern Italy) eruptive activity based on geological and petrological constraints. *Bull. Volcanol.* **2015**, *77*, 1–23. [CrossRef]
54. Platt, U.; Stutz, J. *Differential Optical Absorption Spectroscopy Principles and Applications*; Physics of Earth and Space Environments; Springer: Berlin/Heidelberg, Germany, 2008; p. 597. ISBN 978-3-540-75776-4.
55. Platt, U. Differential optical absorption spectroscopy (DOAS). In *Air Monitoring by Spectroscopic Techniques*; Chemical Analysis Series; Sigrist, M.W., Ed.; John Wiley & Sons Inc.: Hoboken, NJ, USA, 1994; Volume 127, pp. 27–83.
56. Edmonds, M.; Herd, H.; Galle, B.; Oppenheimer, C. Automated, high time-resolution measurements of SO_2 flux at Soufriere Hills Volcano, Montserrat. *West Indies Bull. Volcanol.* **2003**, *65*, 578–586. [CrossRef]
57. Kern, C.; Masias, P.; Apaza, F.; Reath, K.A.; Platt, U. Remote measurement of high preeruptive water vapor emissions at Sabancaya volcano by passive differential optical absorption spectroscopy. *J. Geophys. Res. Solid Earth* **2017**, *122*, 3540–3564. [CrossRef]
58. Vita, F.; Kern, C.; Inguaggiato, S. Development of a portable active long-path differential optical absorption spectroscopy system for volcanic gas measurements. *J. Sens. Sens. Syst.* **2014**, *3*, 355–367. [CrossRef]
59. Galle, B.; Johansson, M.; Rivera, C.; Zhang, Y.; Kihlman, M.; Kern, C.; Lehmann, T.; Platt, U.; Arellano, S.; Hidalgo, S. Network for Observation of Volcanic and Atmospheric Change (NOVAC): A global network for volcanic gas monitoring -Network layout and instrument description. *J. Geophys. Res.* **2010**, *115*, D05304. [CrossRef]
60. Arellano, S.; Galle, B.; Apaza, F.; Avard, G.; Barrington, C.; Bobrowski, N.; Bucarey, C.; Burbano, V.; Burton, M.; Chacon, Z.; et al. Synoptic analysis of a decade of daily measurements of SO_2 emission in the troposphere from volcanoes of the global ground-based Network for Observation of Volcanic and Atmospheric Change. *Earth Syst. Sci. Data* **2021**, *13*, 1167–1188. [CrossRef]
61. Vita, F.; Arellano, S.; Inguaggiato, S.; Galle, B. SO_2 Flux of-VULCANO-Volcano from the NOVAC Data-Base, [Data Set], v.001, The NOVAC Database 2020. Available online: https://doi.org/10.17196/novac.vulcano.001 (accessed on 30 November 2021). [CrossRef]
62. Galle, B.; Delgado, H.; Garzon, G.; Vogel, L.; Platt, U. *NOVAC Project Final Report, EU-FP6*; Chalmers University of Technology: Gothenburg, Sweden, 2011; p. 100.
63. IUPAC. *Compendium of Chemical Terminology*, 2nd ed.; The "Gold, Book"; McNaught, A.D., Wilkinson, A., Eds.; Blackwell Scientific Publications: Oxford, UK, 1997; Online version (2019-) created by Chalk, S.J.; ISBN 0-9678550-9-8. [CrossRef]
64. Platt, U.; Perner, D. Measurements of Atmospheric Trace Gases by Long Path Differential UV/Visible Absorption Spectroscopy. In *Optical and Laser Remote Sensing*; Killinger, D.A., Mooradien, A., Eds.; Springer Verlag: New York, NY, USA, 1983; pp. 95–105.

65. Johansson, M. *NOVACProject/NovacPPP: NovacPPP (Post Processing Program)*; Version 2.0.0; Zenodo: Geneve, Switzerland, 2021. [CrossRef]
66. Johansson, M.; Galle, B.; Zhang, Y.; Rivera, C. The dual-beam mini-DOAS technique–measurements of volcanic gas emission, plume height and plume speed with a single instrument. *Bull. Volcanol.* **2009**, *71*, 747–751. [CrossRef]
67. Chiodini, G.; Cioni, R.; Guidi, M.; Raco, B.; Marini, L. Soil CO_2 flux measurements in volcanic and geothermal areas. *Appl. Geochem.* **1998**, *13*, 543–552. [CrossRef]
68. Sinclair, A.J. Selection of threshold values in geochemical data using probability graphs. *J. Geochem. Explor.* **1974**, *3*, 129–149. [CrossRef]
69. Deutsch, C.; Journel, A. *GSLIB: Geostatistical Software and User's Guide*, 2nd ed.; Oxford University Press: New York, NY, USA, 1998; 369p.
70. Aubert, M. Practical evaluation of steady heat discharge from dormant active volcanoes: Case study of Vulcarolo fissure (Mount Etna, Italy). *J. Volcanol. Geotherm. Res.* **1999**, *92*, 413–429. [CrossRef]
71. Aubert, M.; Alparone, S. Hydrothermal convective flux variation related to a seismo-tectonic crisis in the Fossa of Vulcano (Italy). *CR Geosci.* **2000**, *330*, 603–610.
72. Gaudin, D.; Ricci, T.; Finizola, A.; Delcher, E.; Alparone, S.; Barde-Cabusson, S.; Brothelande, E.; Di Gangi, F.; Gambino, S.; Inguaggiato, S.; et al. Heat flux-based strategies for the thermal monitoring of sub-fumarolic areas: Examples from Vulcano and La Soufriere de Guadeloupe. *J. Volcanol. Geotherm. Res.* **2017**, *343*, 122–134. [CrossRef]
73. Diliberto, I.S.; Gagliano Candela, E.; Morici, S.; Pecoraino, G.; Bellomo, S.; Bitetto, M.; Longo, M. Changes in heat release by hydrothermal circulation monitored during an eruptive cycle on Mount Etna (Italy). *Bull. Volcanol.* **2018**, *80*, 4. [CrossRef]
74. David, M. *Geostatistical Ore Reserve Estimation*; Developments in Geomathematics; Elsevier: New York, NY, USA, 1977; Volume 2, 363p.
75. Barberi, F.; Bertagnini, A.; Landi, P.; Principe, C. A review on phreatic eruptions and their precursors. *J. Volcanol. Geotherm. Res.* **1992**, *52*, 231–246. [CrossRef]
76. Selva, J.; Bonadonna, C.; Branca, S.; De Astis, G.; Gambino, S.; Paonita, A.; Ricciardi, A. Multiple hazards and paths to eruptions: A review of the volcanic system of Vulcano (Aeolian Islands, Italy). *Earth Sci. Rev.* **2020**, *207*, 103186. [CrossRef]

remote sensing

MDPI

Article

Integration of DInSAR Time Series and GNSS Data for Continuous Volcanic Deformation Monitoring and Eruption Early Warning Applications

Brianna Corsa, Magali Barba-Sevilla, Kristy Tiampo * and Charles Meertens

Department of Geological Sciences, The Collaborative Institute of Research in Environmental Sciences (CIRES), University of Colorado Boulder, Boulder, CO 80309, USA; Brianna.Corsa@colorado.edu (B.C.); magali.barba@colorado.edu (M.B.-S.); charles.meertens@colorado.edu (C.M.)

* Correspondence: kristy.tiampo@colorado.edu

Abstract: With approximately 800 million people globally living within 100 km of a volcano, it is essential that we build a reliable observation system capable of delivering early warnings to potentially impacted nearby populations. Global Navigation Satellite System (GNSS) and satellite Synthetic Aperture Radar (SAR) document comprehensive ground motions or ruptures near, and at, the Earth's surface and may be used to detect and analyze natural hazard phenomena. These datasets may also be combined to improve the accuracy of deformation results. Here, we prepare a differential interferometric SAR (DInSAR) time series and integrate it with GNSS data to create a fused dataset with enhanced accuracy of 3D ground motions over Hawaii island from November 2015 to April 2021. We present a comparison of the raw datasets against the fused time series and give a detailed account of observed ground deformation leading to the May 2018 and December 2020 volcanic eruptions. Our results provide important new estimates of the spatial and temporal dynamics of the 2018 Kilauea volcanic eruption. The methodology presented here can be easily repeated over any region of interest where an SAR scene overlaps with GNSS data. The results will contribute to diverse geophysical studies, including but not limited to the classification of precursory movements leading to major eruptions and the advancement of early warning systems.

Keywords: integrated DInSAR and GNSS time series; geodetic dataset; volcanic deformation; early warning applications; natural hazards

Citation: Corsa, B.; Barba-Sevilla, M.; Tiampo, K.; Meertens, C. Integration of DInSAR Time Series and GNSS Data for Continuous Volcanic Deformation Monitoring and Eruption Early Warning Applications. *Remote Sens.* **2022**, *14*, 784. https://doi.org/10.3390/rs14030784

Academic Editors: Sonia Calvari, Eugenio Sansosti, Alessandro Bonaccorso, Annalisa Cappello and Flora Giudicepietro

Received: 1 January 2022
Accepted: 26 January 2022
Published: 8 February 2022

1. Introduction

Volcanic eruptions, earthquakes, and tsunamis occur over numerous spatial and temporal scales. Although these phenomena are often studied individually, there is frequently interconnectivity between disaster types. For example, concentrated swarms of earthquakes, elevated readings of gas emission, and increased ground motion over volcanic regions may indicate an impending eruption [1–8]. Most active volcanoes around the world are monitored using geodetic data sets such as Synthetic Aperture Radar (SAR) and Global Navigation Satellite System (GNSS) data, in conjunction with other ground-based instruments, with the goal of providing early warning for major eruptions and reducing risk to nearby populations or infrastructure [9]. While several studies have attempted to forecast or model potential volcano hazards using remote sensing techniques [2,10–16], there is currently no single framework in place that simultaneously consolidates geodetic data from multiple sensors, freely provides scientists with near real-time continuous time series products and is capable of distinguishing and broadcasting geophysical events.

The GeoScience CyberInfrastructure Framework (GeoSCIFramework or GSF) project aims to improve intermediate-to-short term forecasts of catastrophic natural hazard events, allowing researchers to instantly detect phenomena and reveal more suppressed, long-term motions of Earth's surface at unprecedented spatial and temporal resolutions. These goals will be accomplished by applying big data analytics and training machine learning algorithms to recognize patterns across various data signals during noteworthy events. When complete, the system will be capable of processing and delivering large streams of near real-time data from a mix of Differential Interferometric SAR (DInSAR) imagery, GNSS, and other geodetic-related sensors, as well as seismic, gas emission, and thermal data.

DInSAR quantifies line-of-sight (LOS) ground deformation with mm-cm precision, and GNSS data delivers precise point positioning and timing data to determine exact location and deformation measurements, also with mm-cm scale precision. Furthermore, DInSAR processing can be combined with GNSS data to obtain 3D ground surface motions [17–20]. Together, these time series produce high resolution, sub-centimeter precision measurements of ground deformation over large swaths of Earth's surface with dense spatiotemporal coverage, which provides scientists with a greater understanding of crustal or shallow subsurface dynamics over volcanic regions.

We focus on generating an automated DInSAR time series processing routine that is integrated with GNSS data into a unified deformation field to provide more constrained deformation rates and vector measurements related to volcanic activity. We process Sentinel-1A/B SAR data into time series over Hawaii from November 2015 to April 2021 and integrate those results with GNSS data at various station positions (Figure 1). The DInSAR + GNSS integrated time series can be used to describe the full extent of ground motions through time with decreased uncertainty in three directions of motion (east-west, north-south, and up-down). We present the unified DInSAR and GNSS time series and compare them to the original datasets.

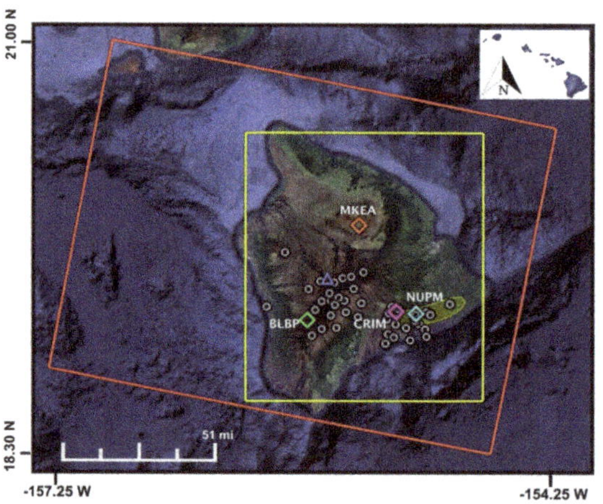

Figure 1. Region of study over the Big Island of Hawaii. The red outline shows the extent of the SAR scenes used for this study, Path 87 Frame 526, downloaded from the Alaskan Satellite Facility Vertex portal [21]. The yellow box shows the cropped outline of each interferogram used when generating time series. Grey circles and colored diamonds indicate GNSS station locations, which were used to create kriging interpolated GNSS maps in Section 2.3. Twenty-four-hour final solution GNSS time series data, from stations listed in Table 1 and aligned to the local, fixed, Pacific Plate reference frame were obtained through the Nevada Geodetic Laboratory (NGL), University of Nevada Reno (http://geodesy.unr.edu/ (accessed on 1 November 2021)). Stations are maintained by the USGS HVO [22,23], and data are archived and distributed by the UNAVCO GAGE facility. We take a closer look at the time series over the diamond GNSS locations in Section 3.2, where the blue diamond is the NUPM GNSS station, the purple diamond corresponds to the CRIM station, orange represents the MKEA station, and green is the BLBP station. The blue triangle shows the location of the Mauna Loa volcano summit, and the yellow polygon indicates where the East Rift Zone is located. Background image taken from Google Earth/Data SIO, NOAA, U.S. Navy, NGA, GEBCO/Data LDEO-Columbia, NSF, NOAA/ Imagery Date: 13 December 2015.

Our region of interest is over the Big Island of Hawaii. Unlike most volcanic systems, Hawaii experiences frequent eruptive activity and provides an opportunity for scientists to record extensive observations over multiple events. Kilauea volcano has erupted 34 times since 1952 [24]. Two volcanic eruptions are captured within our time series between November 2015 and April 2021. The first eruption occurred in May 2018, when the lava lake within the Halema'uma'u crater (collocated with CRIM station, Figure 1) drained following an intrusion into, and subsequent eruption from, the Pu'u 'Ō'ō crater (collocated with NUPM station, Figure 1) and Kilauea's lower East Rift Zone (ERZ) (Yellow polygon, Figure 1). Sudden changes in lava lake levels, increased micro-seismicity around Kilauea's summit, and deformation along the NE strike of the ERZ were all precursory indications reported by the Hawaiian Volcano Observatory (HVO) that an impending eruption might occur [10,25]. According to [26], an obstruction in the magma plumbing system at Pu'u 'Ō'ō volcano caused widespread pressurization in the volcano, driving magma into the lower, eastern flank. Activity over the ERZ decelerated by September 2018 and remained quiet until December 2020, when a summit eruption that continued through May 2021 refilled the lava lake within the Halema'uma'u crater [24]. Although we observe this activity within the separate datasets, DInSAR and GNSS, as shown in Section 2, by integrating the two geodetic datasets together, we recover a 3-D spatial map, instead of 1-D LOS motion, with the resolution of the DInSAR images, that includes time series for motion in the east-west, north-south, and up direction, at each location. In addition, there is a significant improvement in the accuracy of each component of motion (east, north, up) relative to either the DInSAR or GNSS data alone, as presented in Sections 3 and 4.

2. Materials and Methods

The automated DInSAR processing routine was separated into three different components. The first section (Stage 1) was built based on GMTSAR source code [27,28] to process single look complex (SLC) DInSAR satellite imagery into interferograms using the small baseline subset (SBAS) method [29] performed in parallel. The second phase (Stage 2) applies the New SBAS (NSBAS) inversion [30] method to the GMTSAR interferograms of Stage 1 and generates the DInSAR time series and the cumulative deformation map. Finally, the third component (Stage 3) produces integrated 3D displacements using the LOS deformation from Stage 2, the geometry of the SAR acquisition, and precise, 3D vector positioning measurements [12,18–20]. The routine can process the fused data at a single pixel, which is delivered as a plotted time series, or as an interpolated displacement map over a larger region, created from an array of available GNSS stations within the extent of the SAR scene.

2.1. Data

For this study, 250 descending Sentinel-1A/B SLC images were acquired between November 2015 to April 2021 along Path 87 Scene 526 (Figure 1; Supplementary Material List S1) through the Alaskan Satellite Facility (ASF) Vertex portal [21]. Twenty-four-hour final solution GNSS data was managed by the USGS HVO and archived by the UNAVCO GAGE facility; processed time series were generated by and distributed through NGL [22,23]. Data from 48 GNSS stations over Hawaii (Table 1, Figure 1) were obtained over the same period of time, decimated to match the sampling rate of the InSAR time series, and used to create the interpolated map for this study. We used the ordinary kriging interpolation algorithm [31–33] supported by an exponential distribution model to construct standard variograms from the 48 GNSS station data.

While GNSS is known for its high precision in the horizontal directions (east and north), estimates of vertical motion have a larger uncertainty [34,35]. On the other hand, with an incidence angle range of 18.3° to 46.8°, DInSAR sensors are most sensitive to vertical displacements and can help to improve ground velocity estimates in the up direction [20,35]. Our integrated results provide a better representation, and therefore, a better understanding of the volcanic deformation pattern and subsurface-surface behavior through time. The workflow presented here can be easily repeated or applied to other locations where GNSS data falls within an SAR scene footprint.

Table 1. GNSSS stations over Hawaii Island used in this study for time series integration with DInSAR data. Bolded station corresponds to colored diamonds in Figure 1. GNSS data were obtained through the NGL, and stations were maintained by the USGS HVO [22,23].

Station Name:	Latitude (°N)	Longitude (°W)	Elevation (m)	Station Name:	Latitude (°N)	Longitude (°W)	Elevation (m)
AHUP	19.379	−155.266	1104.881	KULE	19.249	−155.323	57.839
AINP	19.373	−155.458	1567.881	MANE	19.339	−155.273	996.466
ALAL	19.381	−155.592	3203.593	MKAI	19.356	−155.176	892.897
ALEP	19.541	−155.644	2922.262	**MKEA**	**19.801**	**−155.456**	**3754.657**
ANIP	19.396	−155.517	2599.215	MLCC	19.563	−155.491	2886.947
APNT	19.264	−155.202	42.009	MLES	19.464	−155.553	3841.48
BLBP	**19.355**	**−155.711**	**2664.265**	MLRD	19.556	−155.533	3082.687
BYRL	19.412	−155.26	1099.085	MLSP	19.451	−155.592	4078.4
CNPK	19.392	−155.306	1123.818	MMAU	19.374	−155.178	949.575
CRIM	**19.395**	**−155.274**	**1147.6**	MOKP	19.485	−155.599	4132.709
ELEP	19.45	−155.525	3378.14	NPOC	19.393	−155.11	809.836
GOPM	19.322	−155.222	759.313	**NUPM**	**19.385**	**−155.175**	**933.27**
HLNA	19.293	−155.31	698.278	OUTL	19.387	−155.281	1103.498
HOLE	19.315	−155.128	408.431	PAT3	19.43	−155.572	3831.48
JCUZ	19.384	−155.102	826.863	PHAN	19.447	−155.638	3700.613
JOKA	19.434	−155.004	482.625	PIIK	19.322	−155.564	2308.363
KAEP	19.281	−155.121	38.147	PMAU	19.677	−155.818	2033.189
KAMO	19.395	−155.122	781.432	PUH2	19.421	−155.908	50.715
KAON	19.278	−155.282	288.305	PUKA	19.506	−155.479	3026.304
KFAP	19.438	−155.441	2073.534	RADF	19.584	−155.431	2414.046
KHKU	19.317	−155.637	2641.483	SLPC	19.407	−155.67	3141.234
KNNE	19.286	−155.686	2468.357	STEP	19.536	−155.575	3419.067
KOSM	19.363	−155.316	990.363	TOUO	19.504	−155.703	2535.406
KTPM	19.341	−155.16	783.049	UWEV	19.421	−155.291	1257.633

2.2. Building Interferograms

DInSAR processing is challenging and complex in that several improvements, including orbital, topographic, and atmospheric corrections, must be applied to isolate the deformation signal. As we consider an automated system that will be streaming continuous Sentinel-1A/B data over Hawaii Island, we must decide how these corrections are applied and systematized within the processing routine. This first component requires several input parameters and files to run, including the SAR images, the corresponding orbital correction files, a Digital Elevation Model (DEM), and the number of allowed days between potential interferometric image pairs.

For the time series presented in this paper, we used only precise orbital corrections on each Sentinel-1 A/B SLC image. These precise files, which were also obtained through the ASF Vertex portal [21], are not available to users until two weeks after the SAR image is acquired by the Sentinel satellite. Thus, for the real-time automated system, and due to the Sentinel-1A/B satellites' repeat period of 6–12 days, the most recent SAR image acquisitions over the region of interest were first processed using real-time orbital corrections and then updated with the precise corrections two weeks later. We thoroughly review the effects of real-time vs. precise orbital corrections in [9] and determine that the use of real-time orbits is sufficient for early-warning applications. For the topographic correction, we used a 30 m resolution SRTM DEM that completely covers the SAR image footprint.

After the orbital and topographic corrections are applied, the complex interferogram is formed. We allowed the algorithm to pair any two images that were obtained within 35 days of each other. From the 250 descending SLC images acquired between November 2015 and April 2021, 671 interferometric image pairs were successfully generated (Supplementary Material List S2). For each pair, interferometric products of phase, coherence, phase gradient, and LOS displacement were constructed in both radar and geographical coordinates (Figure 2). The code utilizes the Snaphu phase unwrapping method [36] and geocodes all interferometric output products. These outputs are formatted as GMT-compatible grids, which easily convert to GeoTIFF or other GDAL-compatible raster drivers.

2.3. DInSAR Time Series Generation

The second component of the automated system uses the GIAnT software [37] to construct LOS displacement maps and time series. In GIAnT, unwrapped interferograms are converted to units of millimeters before any processing takes place. The software assembles the input data, the unwrapped interferograms, coherence maps, masks, and metadata, into binary files and a composite HDF5 file that is compatible with GIAnT processing.

The user can choose between several atmospheric model corrections, including the European Centre for Medium-Range Weather Forecasts (ECMWF) model, which is built into the software so that the atmospheric files are pulled and applied automatedly. Currently, ECMWF is in the midst of migrating its data to an updated server, rendering the files temporarily unavailable. Therefore, for this study, the Generic Atmospheric Correction Online Service (GACOS) atmospheric model correction was applied. The GACOS model integrates the ECMWF atmospheric files with continuous GNSS tropospheric delay estimates, which currently must be manually ordered through the online portal [38–41]. The goal is to incorporate it into automated processing in the future.

The user can also choose between different time series inversion methods, available from GIAnT, to estimate filtered time series. The SBAS and NSBAS techniques, which are implemented here, are applicable when the differential interferograms have a small spatial baseline, a characteristic of Sentinel-1A/B data [29,30]. In particular, the NSBAS technique also estimates DEM errors and compensates for pixels that have missing observations. This inversion method estimates the LOS phase change of each pixel independently using a linear system.

For Stage 1, the automated scripts ran using high-performance computing nodes on the SUMMIT supercomputer located at the University of Colorado, Boulder, for approximately eight hours to produce 694 interferograms between November 2015 and April 2021. The NSBAS technique applies inversion on the 694 input interferograms to generate a time series with a cumulative displacement measurement for every date, or time-slice, included within the data. In this case, the time series produced 275 time-slices (See Figure 3). The atmospheric correction and inversion process in Stage 2 took approximately two hours to complete. Once the historical data has been processed, however, adding a new acquisition to the time series takes on the order of a couple of hours to fully acquire and process the necessary data, then output all the updated DInSAR-related products and time series.

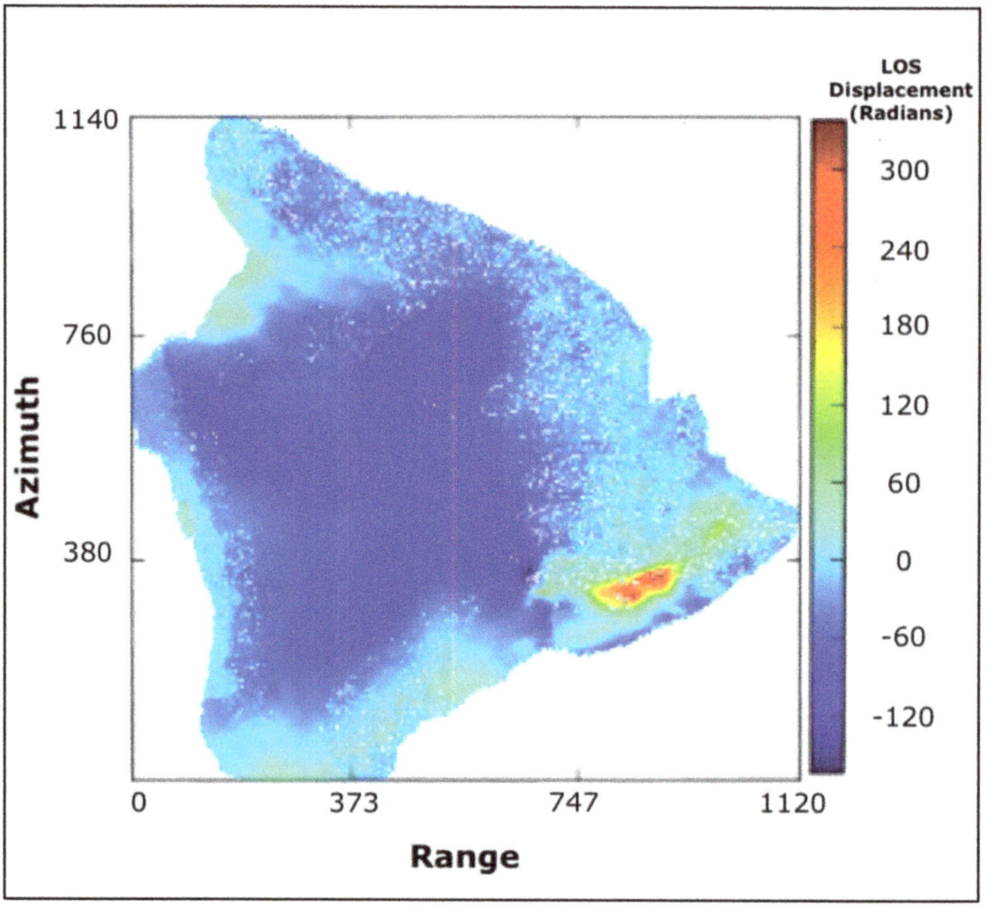

Figure 2. Example of an unwrapped LOS phase interferogram between 23 April 2018 and 4 June 2018, over the Big Island of Hawaii (Path 87 Frame 526) in units of radians from Stage 1 of the automated processing routine, based on GMTSAR. Here, warm colors and positive values are concentrated along the ERZ and represent an increase in slant range, corresponding to ground motion away from the satellite over this time. Cool colors and negative values represent a decrease in slant range, which means the ground moved towards the satellite. This color convention is reversed when the units of radians are converted to millimeters of deformation.

2.4. Integration of Geodetic Datasets

We integrate our DInSAR time series with GNSS data from 48 stations over Hawaii to produce 3D high-resolution cumulative displacement maps with corresponding errors. The 24 h final GNSS solutions are provided in three components (east, north, up) and aligned to the local, fixed Pacific plate reference frame to minimize linear trends of the tectonic plate motion. It is worth noting that NGL also provides solutions aligned to the International GNSS Service-14 (IGS14) reference frame, which is based on the International Terrestrial Reference Frame (ITRF) and holds a no-net-rotation (NNR) [42]. We present a comparison of the results from both reference frames in the Supplementary Material (Figures S1–S3).

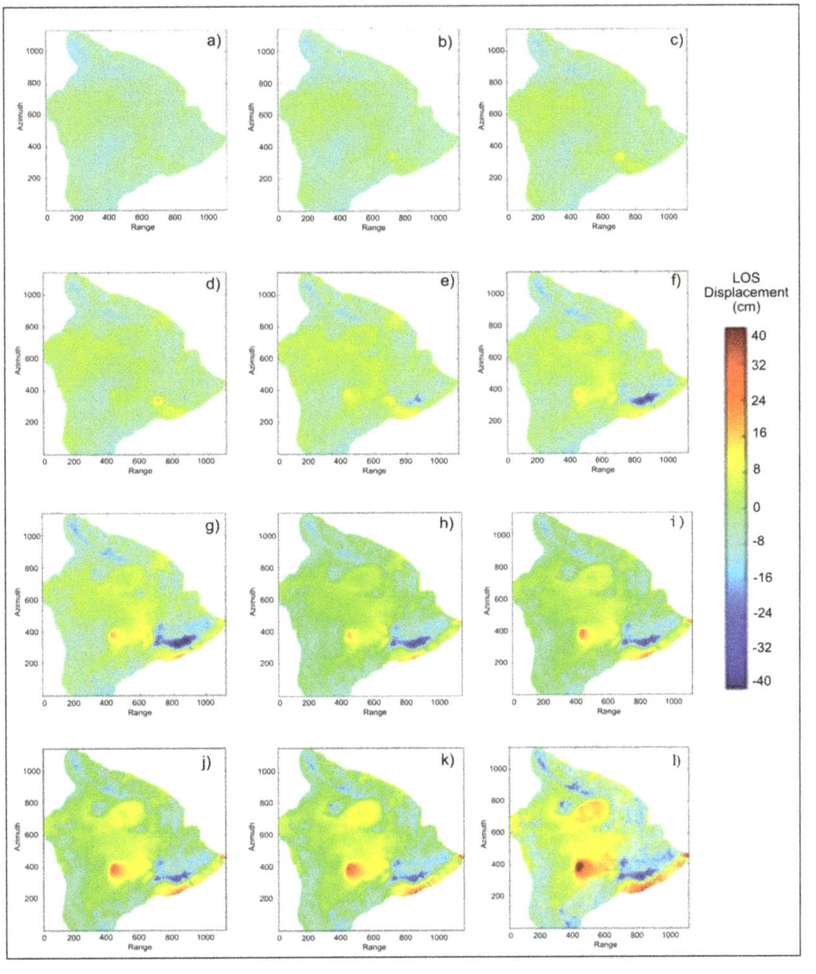

Figure 3. Cumulative LOS displacement DInSAR time series results for Sentinel-1A/B Path 87 Frame 526 data over the Big Island of Hawaii from November 2015 to April 2021. These are twelve of the possible 275 time-steps from the 5.5-year-long time series. Each submap corresponds to the total deformation between (**a**) 11 November 2015 and (**b**) 09 May 2016; (**c**) 05 November 2016; (**d**) 04 May 2017; (**e**) 18 November 2017; (**f**) 05 May 2018; (**g**) 07 November 2018; (**h**) 06 May 2019; (**i**) 14 November 2019; (**j**) 12 May 2020; (**k**) 08 November 2020; and (**l**) 13 April 2021. Once the phase is converted to units of millimeters, the sign convention in GIAnT changes. Here, warm, positive colors represent regions of uplift and cool; negative colors correspond to subsidence.

First, we temporally subset the GNSS data from 48 stations to extract the data corresponding to the interval of the DInSAR time series. For each time step, we used the ordinary kriging algorithm and the 3D displacements from the 48 stations to compute a variogram. The spatial covariance of the GNSS data determines the structure of the variogram, and the weights were calculated from the data using an exponential distribution model to interpolate undefined points across the spatial field [31,32]. The kriging algorithm outputs three cumulative displacement maps (east, north, and up) with the same discretization and geocoding as the DInSAR output from Section 2.2 (Figure 4), [12,18,19,43–45].

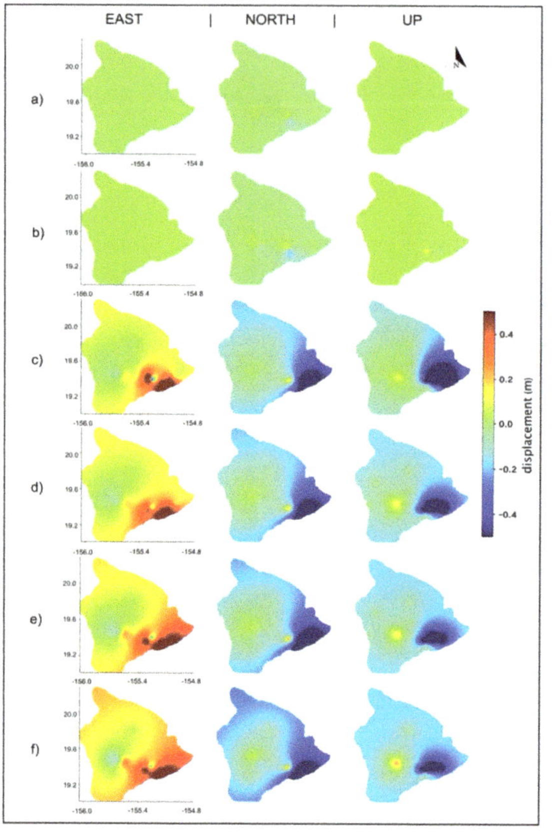

Figure 4. Interpolated GNSS displacement maps generated using 48 Hawaiian GNSS stations and the ordinary kriging algorithm. Each submap corresponds to the total deformation between 11 November 2015 and (**a**) 23 December 2016; (**b**) 18 November 2017; (**c**) 28 June 2018; (**d**) 14 November 2019; (**e**) 12 May 2020; and (**f**) 13 April 2021.

To combine all the geodetic displacement data into a single time series product, we expand on the DInSAR+GNSS velocity integration method developed by [12]. The method uses a Bayesian statistical modeling approach in conjunction with Markov random field (MRF) theory to combine datasets with varying spatio-temporal extents. Integration of DInSAR and GNSS high-resolution, 3D surface displacement measurements was achieved by minimization of the energy function related to the corresponding Gibbs random field (GRF) distribution, in which the joint probability density of the variables are positive [12,18,19,33]. The energy function is as follows:

$$U(b/a) = \sum_{i=1}^{N} \frac{(b_i - a_i)^2}{2\sigma_i^2} \tag{1}$$

where $U\left(\frac{b}{a}\right)$ is the likelihood energy, b is the observation with uncertainty σ, a is the unknown parameter, and N is the number of observations or pixels within the acquisition. The resulting adaptation of Equation (1) to our geodetic displacement datasets is:

$$U\left(d^{LOS}, d_x^{GPS}, d_y^{GPS}, d_z^{GPS}/d_x, d_y, d_z\right) = \sum_{i=1}^{N} C^{LOS}\left(d^{LOS} - S_x^{LOS}d_x - S_y^{LOS}d_y - S_z^{LOS}d_z\right)^2 + C_x^{GPS}\left(d_x^{GPS} - d_x\right)^2 + C_y^{GPS}\left(d_y^{GPS} - d_y\right)^2 + C_z^{GPS}\left(d_z^{GPS} - d_z\right)^2 \quad (2)$$

with coefficients:

$$C^{LOS} = \frac{1}{2(\sigma^{LOS})^2}, \quad C_x^{GPS} = \frac{1}{2(\sigma_x^{GPS})^2}, C_y^{GPS} = \frac{1}{2(\sigma_y^{GPS})^2}, C_z^{GPS} = \frac{1}{2(\sigma_z^{GPS})^2} \quad (3)$$

where σ is the standard deviation for the measurements, d^{LOS} is the cumulative LOS displacement, $\left[S_x^{LOS}, S_y^{LOS}, S_z^{LOS}\right]$ are the unit vectors pointing from the ground to the satellite, and $\left[d_x^{GPS}, d_y^{GPS}, d_z^{GPS}\right]$ are the 3D displacements from the kriging interpolated GNSS data. The remaining equation details are outlined in the Supplementary Material (Equations S1–S8).

Uncertainties associated with the 3D displacement maps and the plotted time series are automatically generated with the corresponding products. These are integrated using the same methodology as the observations and are a combination of the raw GNSS data uncertainties and the LOS DInSAR uncertainties (Equations S6–S8 in Supplementary Material). For the 3D displacement maps, an additional error is introduced when we interpolate data from the 48 GNSS stations into a variogram using the ordinary kriging algorithm. We present the error analysis for that integration in the Supplementary Material (Figures S4 and S5).

3. Results

3.1. Cumulative Deformation Maps

The automated processing system can provide the assimilated DInSAR+GNSS time series to users in multiple ways. Over large areas, the code produced 3D high-resolution cumulative displacement maps and the corresponding uncertainties for each individual date or time-step of the series. Figure 5 shows the evolution of 3D surface motions captured in our time series, revealing the long-term deformation response to the May 2018 and December 2020 eruptions. The pixel resolution of the fused products was 100 m, the same as the DInSAR results.

Slight inflation (~10 cm) over the Kilauea volcano (purple diamond, Figures 1 and 5) was observed in the fused DInSAR+GNSS results as early as November 2016, nearly 1.5 years before the eruption took place (Figure 5a). By August 2017 (Figure 5b), cumulative inflation over Kilauea reached +26 cm. This magma intrusion is further supported by the pattern observed over the same region in the north-component subplot, in which the southern flank of Kilauea moved approximately 15 cm further south, and the northern flank moved 8 cm further north.

Upward motion over Mauna Loa volcano (blue triangle, Figures 1 and 5) also began to increase in August 2017. By February 2018 (Figure 5c), this inflation pattern was recognizable over Mauna Loa, where the ground reached +16 cm of uplift and was obvious over the Kilauea volcano after reaching +30 cm of total uplift. Over the following months, inflation at Mauna Loa began to slow, while it continued to increase at a similar/opposite rate over Kilauea. This interaction between the two volcanic systems has been studied before [46–49] and provides further evidence of upper mantle links and magma transportation.

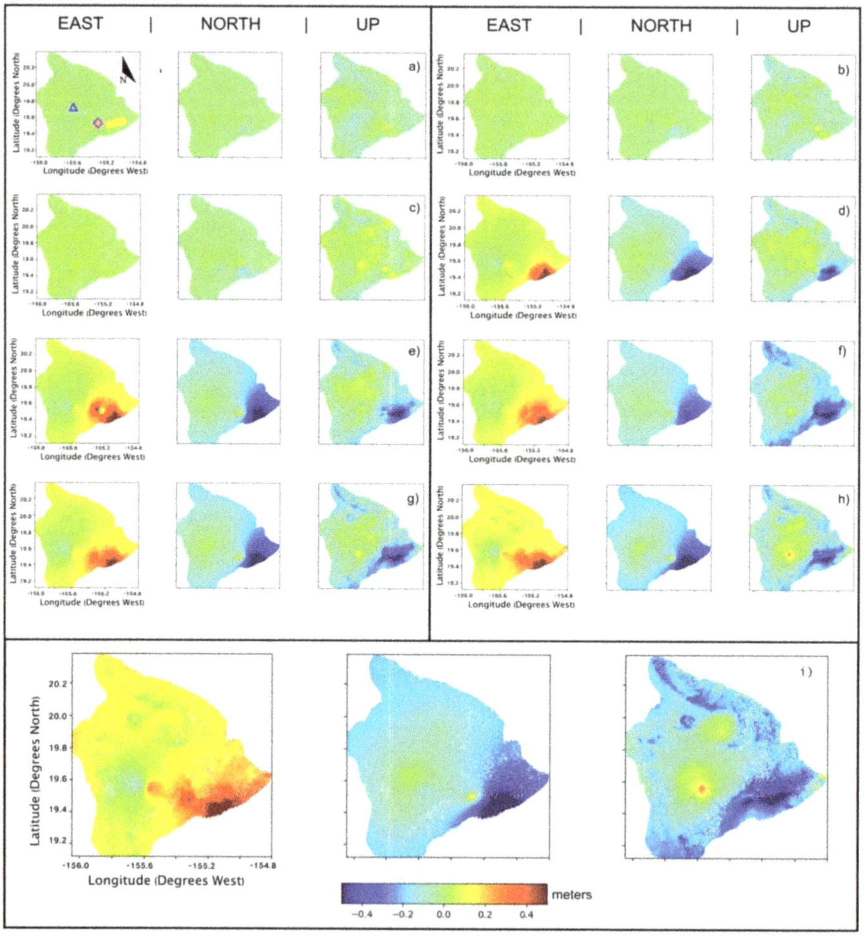

Figure 5. Integrated DInSAR+GNSS 3D cumulative displacement maps from November 2015 to April 2021 in east, north, and up components of motion. Each submap corresponds to the total deformation between 22 November 2015, and (**a**) 05 November 2016; (**b**) 14 August 2017; (**c**) 22 February 2018; (**d**) 05 May 2018; (**e**) 28 June 2018; (**f**) 31 December 2018; (**g**) 14 November 2019; (**h**) 14 December 2020; (**i**) 13 April 2021. Subplots (**a–c**) are pre-eruption (which occurred 04 May 2018). In subplot a, the blue triangle corresponds to the summit of Mauna Loa, the purple diamond represents the summit of Kilauea, and the yellow polygon overlays the ERZ.

On 30 April 2018, the Pu'u 'Ō'ō eruptive vent collapsed [48–50], and on May 4, a Mw 6.9 earthquake occurred on the south flank of Kilauea, initiating the 2018 volcanic eruption. Our integrated results from 5 May 2018 (Figure 5d) recorded the 43 cm southeast ground rupture of the earthquake. Next, the sudden evacuation of subsurface material resulted in a rapid −10 cm ground deflation at Kilauea's crater and −80 cm over the ERZ near Pu'u 'Ō'ō volcano (yellow polygon, Figures 1 and 5a).

As this eruptive activity began over the ERZ, the ground at Mauna Loa inflated in the east-west directions from a continued subsurface magma injection. This divergent, horizontal pattern became more obvious by June 2018 (Figure 5e), after Mauna Loa experienced an additional +3 cm of uplift (a net total of 13 cm). A closer look at deformation over the Kilauea crater in June 2018 indicated additional, sudden ruptures that moved the localized ground 25 cm to the west and in the vertical direction, Kilauea crater subsided another −9 cm. This deformation was likely due to the accelerated down-drop of Kilauea's caldera combined with 62 summit collapse events between May 16 and August 2 [48,49]. HVO reported that some collapse events released energy equivalent to a Mw 4.7 to 5.4 earthquake; our results captured these details.

Another interesting feature includes the three zones of subsidence radially surrounding Kilauea's summit and the Halema'uma'u crater. These deflated regions are spatially separated by ~120 degrees. The two zones to the north and the southwest both experienced between −35 and −45 cm subsidence. The southeast leg of the three-pronged sinking feature fed into the ERZ, which has subsided more than −1.17 m to date. No active surface lava was observed after 4 September 2018 [48,49], but the ground surface over the ERZ continued to deflate through December of that year (Figure 5f), reaching a cumulative minimum of −1.32 m.

Soon after, the cumulative subsidence from the ERZ began to shrink, indicating that during the recovery from the 2018 eruption, the subsurface chambers may have begun refilling with magma. In November 2019 (Figure 5g), the east-west inflation pattern over Mauna Loa became more prominent, and uplift reached a total of +21 cm. The north-south component revealed that the Kilauea crater also recovered 17 cm of motion to the north (against the dominant southeastern motion of the ERZ and the coastline just south of the region) and regained +26 cm in vertical surface height.

Over the next year leading to the December 2020 summit eruption, another +11 cm of upward motion occurred at Mauna Loa (net total of +32 cm), and +14 cm occurred along the ERZ (net total of −83 cm) (Figure 5h). Immediately following the summit eruption, the ground height at Kilauea summit measured a total of +20 cm, while the region surrounding the crater subsided once more, decreasing the total ground height by −30 cm. In mid-April 2021, when our time series ended (Figure 5I), a cumulative total of +21 cm of uplift occurred at the Kilauea crater, and +46 cm of uplift had occurred at Mauna Loa in response to the 2018 eruption and 2020 summit eruptions. Kilauea's surrounding volcanic flank and the ERZ experienced −50 cm and −1.0 m of subsidence, respectively.

These products allowed us to observe the total cumulative deformation pattern in the east, north, and up directions over the entire island from November 2015 to April 2021 with improved accuracy and detail. The 3D cumulative displacement maps provided new information regarding the pre-, during-, and post-eruption phases of the Hawaiian volcanic system at unprecedented spatial scales and revealed surface effects from magma movement and seismic activity leading to two different types of eruptions. For example, Figure 6 converts the integrated results and raw GNSS data from 3D to 1D, LOS deformation to match the raw DInSAR results, for easy comparison of all final results across the three datasets. While all three datasets visually complement one another, the results from integrating the GNSS data with the DInSAR data, Figure 6A, showed a more constrained uplift pattern than the DInSAR or GNSS results alone. Figure 7 compares the integrated results with the kriged GNSS data, illustrating the additional information provided by incorporating the DInSAR data. Uncertainty estimates for each component of motion at each time step are provided and analyzed in the Supplementary Material (Figures S6–S10).

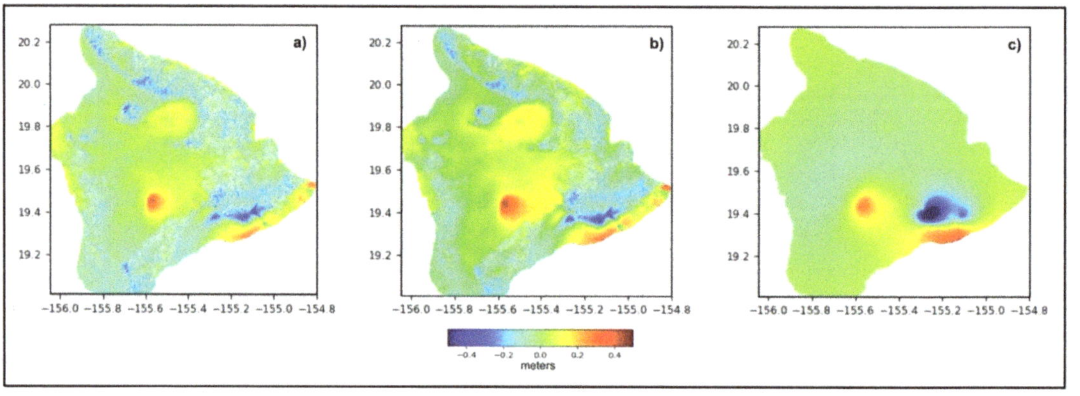

Figure 6. Comparison of (**a**) the final time-step of the integrated time series converted to LOS with (**b**) the final cumulative LOS DInSAR scene and (**c**) the final cumulative GNSS interpolated map, also converted to LOS.

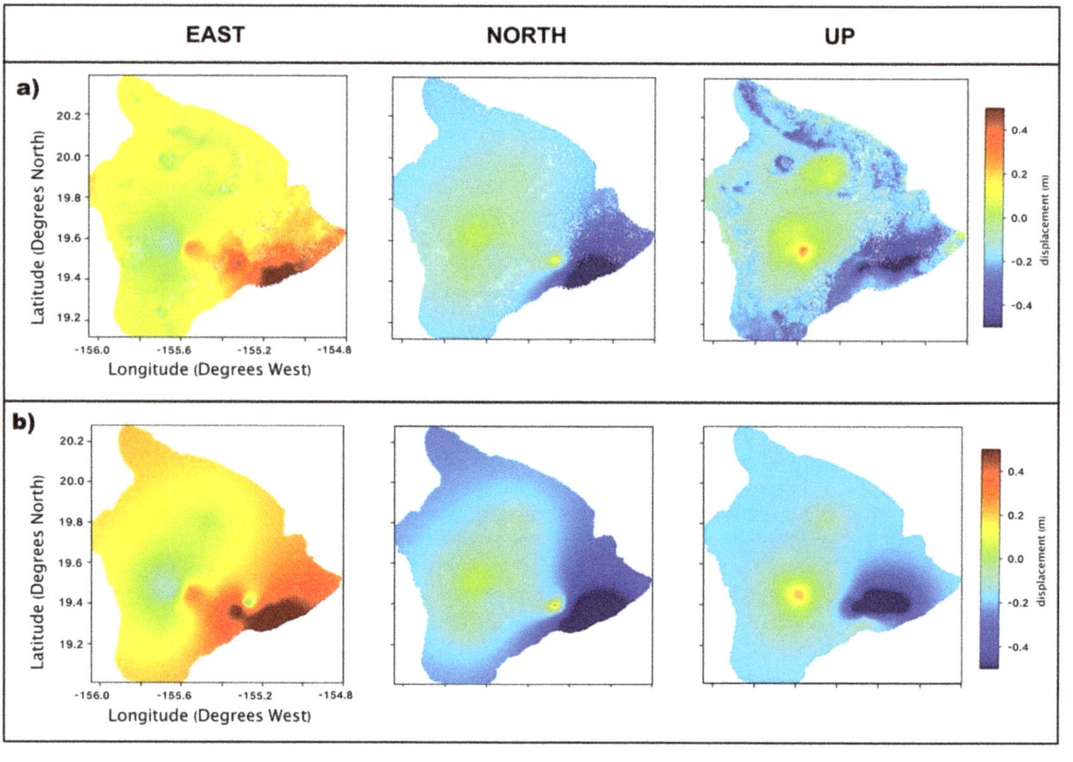

Figure 7. Comparison of (**a**) the final time-step from the integrated time series with (**b**) the final time-steps of the GNSS variograms interpolated with kriging in the east, north, and up directions of motion.

3.2. Plotted Time Series

In addition to generating the assimilated displacement maps, we present time series at four locations within the image where the GNSS stations overlap with a corresponding

DInSAR pixel coordinate. These stations are strategically positioned across the island and include both active and relatively stable regions. Figures 8–11 present the DInSAR+GNSS integrated time series at CRIM, NUPM, MKEA, and BLBP GNSS stations, respectively. We compare each component of motion from the integrated dataset with the raw GNSS time series and LOS DInSAR time series. These graphs depict the ground movement of a single point at every time-step of the series and record the surface response of that specific location to the two volcanic eruptions. The selected GNSS stations monitor a mixture of the most and least active regions of Hawaii and span a large spatial extent of the island. Similar to the 3D displacement maps, these individual time series provide researchers with a comprehensive study of surface deformation through time. By comparing the integrated product against the original GNSS time series, we find that information in the horizontal direction is slightly improved and information in the vertical direction is significantly enhanced once fused with LOS DInSAR measurements. The plotted time series provide additional insight into the temporal behavior at any given location.

The CRIM station (Figure 8) is located at the summit of Hawaii's most active volcano, Kilauea, along the southern rim of the Halema'uma'u crater (purple diamond, Figure 1). At the CRIM station, the integrated results align nicely with the raw GNSS data. Over the period of time leading up to the 2018 eruption (November 2015 through May 2018), the ground at the CRIM station experienced 15.56 cm of horizontal motion to the southeast and 22 cm of uplift. The May 4 Mw 6.9 earthquake jolted the ground 11.4 cm further southeast and uplifted suddenly by 2 cm before volcanic activity dominated the signal. From the eruption itself, the ground moved 1.65 m northwest and subsided almost 2 m before it began a period of steady recovery at the beginning of 2019. Between January 2019 and December 2020, the CRIM station recovered 41 cm of south-southeast motion and uplifted approximately 20 cm from its lowest surface height. Finally, we observed another smaller jolt of ground motion 10.44 cm northeast and 10 cm down in response to the 20 December 2020 summit eruption.

The NUPM station (Figure 9) is positioned at the Pu'u 'Ō'ō volcano (blue diamond, Figure 1), which connects with the ERZ. We present plotted time series from this location to show the sudden displacement from the May 4 Mw 6.9 earthquake. The ground at NUPM stayed relatively stable between November 2015 and May 2018, leading up to the volcanic eruption, with ~4.5 cm southeast motion and stable (+/−1 cm) fluctuations in the vertical direction. Suddenly, the signal broke 58.83 cm southeast and 28 cm downward. Volcanic activity took over the signal and continued over the next couple of months, adding 20.88 cm of horizontal motion to the southeast (net total of 79.71 cm) and another 38 cm of subsidence (net total of 66 cm) before entering a state of ground recovery. Leading to the December 2020 summit eruption, the ground at the NUPM station recovered ~8.95 cm to the northwest and regained 17 cm of uplift. The summit eruption was slightly detectable at this location, having recorded 2 cm of motion to the northeast.

The MKEA station (Figure 10) lies along the southeastern flank of the dormant volcano, Mauna Kea (orange diamond, Figure 1). Data from November 2015 to January 2020 at the MKEA station was associated with the highest uncertainty values throughout this study. In the horizontal directions, the integrated dataset agreed nicely with the GNSS data; however, in the vertical orientation, the data trended more with the DInSAR time series results between 2015 and 2019. In January 2020, HVO reported an instrumental dome replacement, after which the variability in the integrated signal tightened, yet still deviated slightly (~1 cm) from the raw GNSS data. This time series provides an important example of the GNSS dataset with the highest error used in this study and demonstrates how combing the GNSS data with DInSAR also keeps a system of checks and balances of the integrated system. Even with high uncertainty, we could distinguish a 2.5 cm trend southeastward during the 2018 eruption.

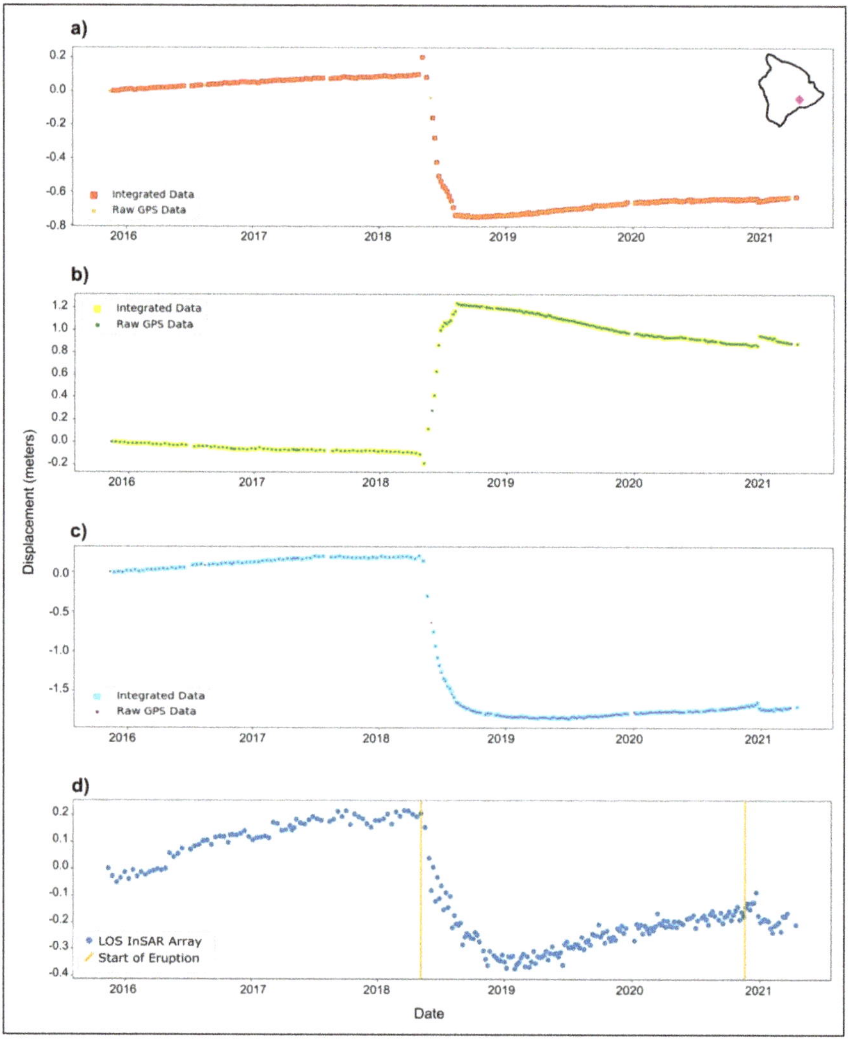

Figure 8. Integrated results compared to original, raw GNSS time series in the (**a**) east-, (**b**) north-, and (**c**) up-components of motion at the CRIM GNSS station (19.395°N, −155.274°W). (**d**) DInSAR LOS time series at the same pixel, over CRIM station. Yellow vertical lines indicate the May 2018 and December 2020 volcanic eruptions. The inset in subfigure **a** shows location of CRIM station in Hawaii.

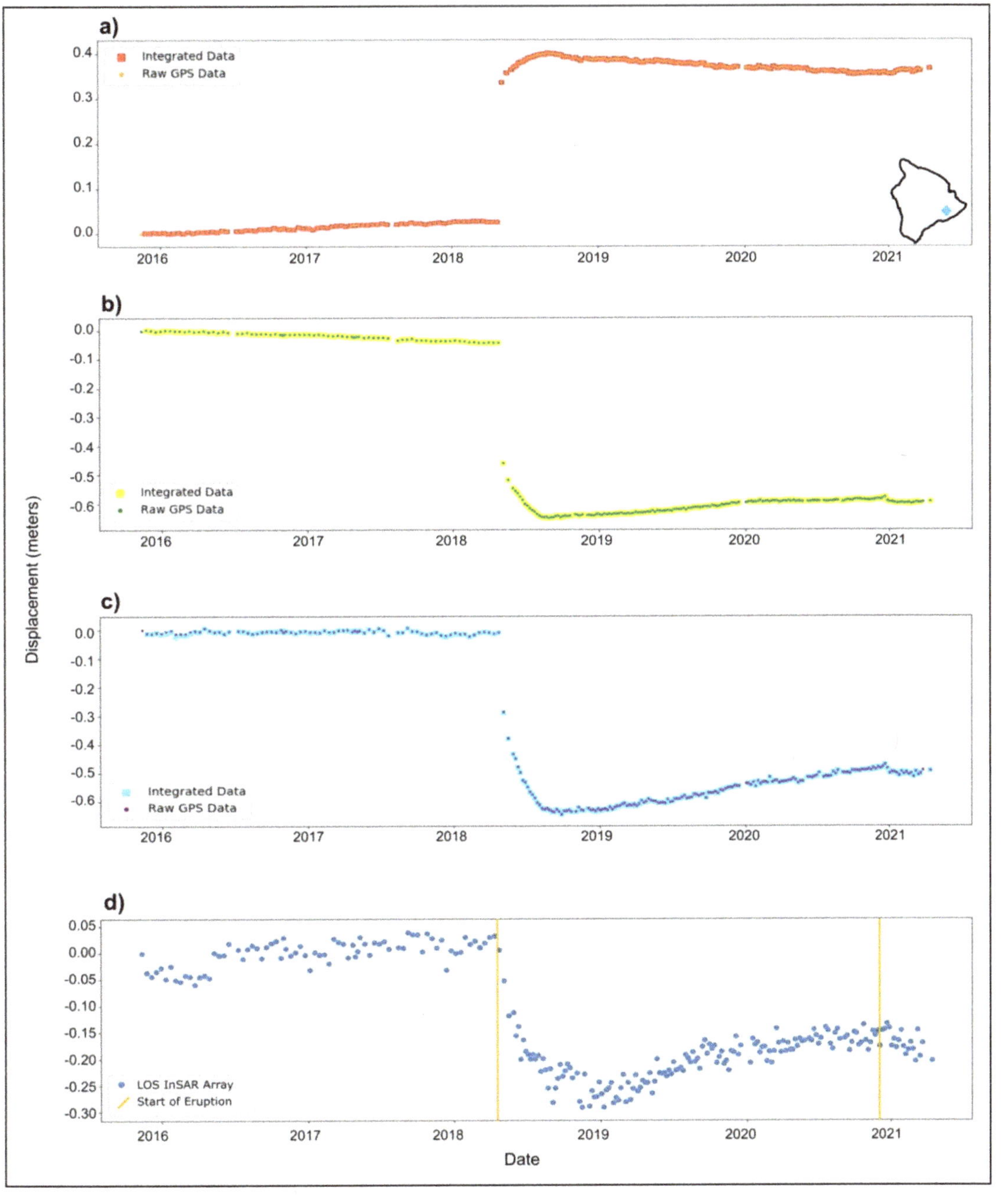

Figure 9. Integrated results compared to original, raw GNSS time series in the (**a**) east-, (**b**) north-, and (**c**) up-components of motion at NUPM GNSS station (19.385°N, −155.175°W). (**d**) DInSAR LOS time series at the same pixel, over NUPM station. Results clearly distinguish the Mw 6.9 earthquake rupture in 2018 and continued motion due to volcanic activity. Yellow lines are as in Figure 8, above. The inset in subfigure (**a**) shows location of NUPM station in Hawaii.

Figure 10. Integrated results compared to original, raw GNSS time series in the (**a**) east-, (**b**) north-, and (**c**) up-components of motion at MKEA GNSS station (19.801°N, −155.456°W). Motion in the east-west and north-south directions are slightly more constrained, while motion in the up-down direction is significantly transformed after combining the DInSAR and GNSS datasets together. (**d**) DInSAR LOS time series at the same pixel, MKEA station. Yellow lines are as in Figure 8, above. The inset in subfigure **a** shows location of MKEA station in Hawaii.

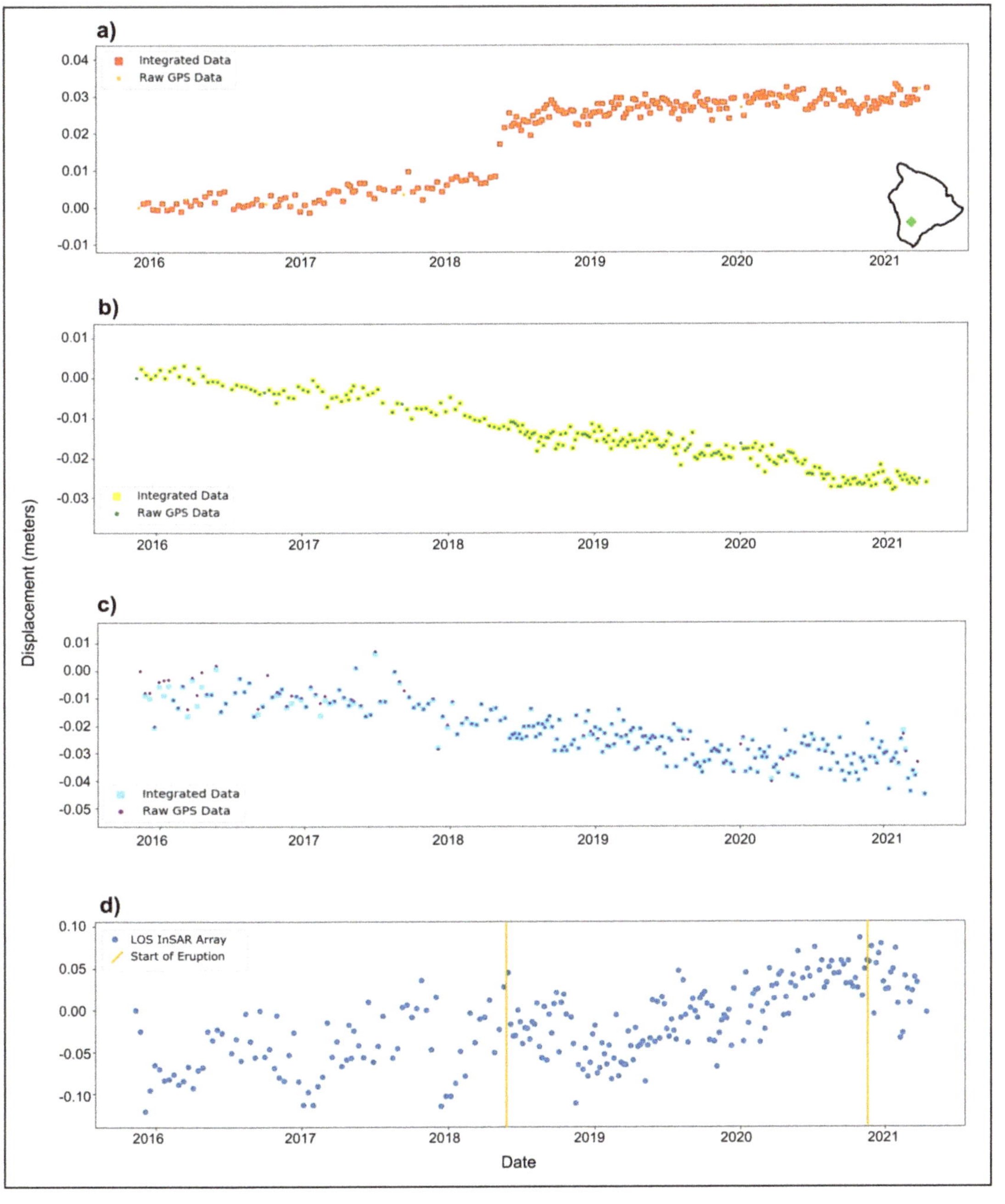

Figure 11. Integrated results compared to original, raw GNSS time series in the (**a**) east-, (**b**) north-, and (**c**) up-components of motion at BLBP GNSS station (19.355°N, –155.711°W). (**d**) DInSAR LOS time series at the same pixel, BLBP station. Yellow lines are as in Figure 8, above. The inset in subfigure **a** shows location of BLBP station in Hawaii.

The scatter observed in the DInSAR data, and the integrated results in the up direction in Figure 10 are partially due to the fact that the GPS data is at a single, isolated point, while the DInSAR results average the value of deformation over a 100 × 100 m^2 pixel. The DInSAR values may also contain small amounts of residual tropospheric noise. Examination of the MKEA station (Figure 10), which is associated with the highest uncertainty measurements from the array of 48 GNSS stations, illustrates how the integrated dataset deviates significantly from the GNSS data in the up direction. Motions in the east-west and north-south directions are slightly more constrained, and motion in the up-down direction is significantly transformed after combining the DInSAR and GNSS datasets together. The integration algorithm weighs the DInSAR time series more than the GNSS data based on the errors of the individual datasets (see Supplemental Material).

Finally, the BLBP GNSS site (Figure 11) is situated over an area of comparatively stable ground along the southwestern part of the island, where relatively low activity was maintained over the duration of our time series (green diamond, Figure 1). Leading up to the 2018 volcanic eruption, the ground at BLBP experienced approximately 1 cm of eastward motion, while remaining within ± 4 mm of motion in the north-south direction. At the time of the 2018 eruption, the BLBP station moved ~12 cm to the east, remained steady in its north-south component of motion, and unclearly moved in the up-down direction. Nevertheless, over the entire time series, we detected that the ground at the BLBP station moved a total of ~3 cm to the southeast, and ~4 cm of subsidence occurred between November 2015 and April 2021.

4. Discussion

The 2018 event at Kilauea volcano was the largest caldera collapse and most effusive volcanic eruption in Hawaii within the past 200 years [26,48–51]. Combining high-quality geodetic datasets, such as DInSAR and GNSS, resulted in detailed observations and precise measurements regarding the development and evolution of volcanic events.

The median uncertainty associated with the LOS DInSAR dataset was 6.60 mm, and the average was 6.54 mm. The maximum uncertainty associated with our LOS DInSAR displacement maps came from the final cumulative time slice and was equal to 1.03 cm. This uncertainty considers errors estimated from the DEM, orbital, and atmospheric corrections applied during processing.

The median errors associated with the raw GNSS data in the east, north, and up directions were 0.81 mm, 0.74 mm, and 3.27 mm, respectively. The corresponding maximum uncertainties associated with the raw GNSS data in the east, north, and up directions were 7.63 mm, 3.28 mm, and 9.65 mm. For direct comparison with the LOS DInSAR data, the converted LOS vector component of the raw GNSS data took on a maximum uncertainty of 1.27 cm.

Integrated uncertainties were combined from the raw DInSAR and GNSS sigma values using the same methodology as the real data (see Supplementary Material Equations S1–S8). Maximum integrated uncertainty in the LOS component of motion was 7.93 mm. When generating the 3D displacement maps, however, the error introduced from using the kriging algorithm and interpolating over undefined points in the spatial field must also be included, and in doing so, resulting in increased maximum uncertainties in the east, north, and up directions of 7.07 cm, 5.82 cm, and 5.71 cm, respectively.

At a single pixel, where kriging interpolation was not applied, the accuracy of our deformation measurements improved. For example, at the corresponding pixel where the CRIM GNSS station is situated, the median uncertainties for the raw GNSS time series data in the east, north, and up directions of motion were 0.77 mm, 0.70 mm, and 3.06 mm, respectively. The raw LOS DInSAR time series at this specific location had a median uncertainty of 1.99 mm. By integrating the two geodetic datasets together, the fused dataset had an improved median uncertainty of 0.76 mm, 0.70 mm, and 1.95 mm in the east, north, and up directions. Further analyses of the errors associated with each dataset are available in the Supplementary Material.

Many other studies have utilized geodetic remote sensing techniques to better understand magmatic systems and volcanic eruptive history. Our raw interferometric products and LOS displacement values are consistent with those produced by the USGS HVO and European Space Agency, also processed with GMTSAR software with Snaphu phase unwrapping [52]. Furthermore, our results extend further evidence of the 62 summit collapse events between May 16 and August 2, reported by HVO. Another study by [50] used raw GNSS and DInSAR measurements from Hawaii as input for multi-reservoir conceptual models to quantitatively constrain the hydraulic connectivity between magmatic systems. By exploiting high-quality geodetic data, they learned that the Halema'uma'u magmatic reservoir is distinct from the South Caldera reservoir and that the ERZ is being fed simultaneously by both chambers. Our LOS inflation and deflation patterns over Kilauea and the ERZ agreed with their observations derived from descending interferograms from November 2018 through March 2019. Lundgren et al. [51] also used airborne InSAR measurements from the Glacier and Ice Surface Topography Interferometer (GLISTIN-A) instrument over Hawaii to measure the bulk volume of subaerial lava flows between 18 May and 15 September 2018. They found that 0.593 +/−0.011 km^3 sourced from the Lower ERZ and −0.836 +/−0.002 km^3 of material resulted from the summit collapse. Finally, a study at Piton de la Fournaise, France, used four interferograms to determine the displacement source and a temporal inversion of GNSS data to describe the dynamics of magmatic propagation [53].

These examples show the broad, interdisciplinary applications that come from a mix of DInSAR and GNSS monitoring over volcanic regions and further support the relevance and benefits of this study. A more accurate, fused dataset creates potential for improved volumetric analysis during large events, enhanced volcanic source modeling and mapping of magma chamber geometries or subsurface transport channels, and may provide a better means to forecast initial eruption sites or potential lava flow pathways along the surface.

5. Conclusions and Upcoming Work

Geodetic datasets such as DInSAR and GNSS time series contain valuable information for earth scientists. These data are particularly useful in analyzing the long-term evolution of volcanic systems. When integrated together, they deliver improved, more constrained estimates of 3D motions of the Earth's surface. Inflation estimates such as that detected in our integrated time series prior to both eruptions are useful for hazard response purposes and applications and underline the importance of deformation monitoring in volcanic regions. The methods performed here can be applied to other regions in the world and can be easily adapted to different volcanic systems.

The integrated 3D displacement maps and associated time series provided new estimates and details of the spatial and temporal dynamics of the 2018 Hawaiian eruption. In conjunction with seismic, tide gauge, gas emission, and thermal datasets, the individual InSAR and GNSS time series and the combined results presented here will be streamed through machine learning algorithms capable of identifying pixels or clusters of pixels that exhibit anomalous movement. This will facilitate the detection and analysis of precursor motion and can be used to inform early warning systems. Eventually, the system will run continuously, providing researchers with large streams of historic or near real-time data from an array of geodetic sensors.

With a continuous, scalable, near real-time data streaming architecture in place, the combined data will contribute to the early detection of major geophysical events such as volcanic eruptions and improve our understanding of the fundamental processes involved over the locations of interest. The fused dataset can be used to improve volcanic modeling systems, using the long-term surface response to map the subsurface magmatic plumbing system. The results presented here also have additional detailed insight into earthquake dynamics, and further research may lead to the distinction or isolation of certain geophysical phenomena within the signal. Finally, the application of machine learning algorithms to these datasets will improve our understanding of the connections between these volcanoes or how activity at one location may have implications for potential activity at another and be critical inputs for early warning applications. These results will provide researchers with better methods in real-time or near real-time to monitor volcanic regions and evaluate resulting surface deformations with millimeter precision, ultimately advancing scientific discovery and benefitting efforts to further protect human life and property.

Supplementary Materials: The following are available online at https://www.mdpi.com/article/10.3390/rs14030784/s1, Supplementary **List S1**: 250 descending Sentinel-1 A/B SLC image IDs used in this study from Path 87 Frame 526 between 23 November 2015–13 April 2021; Supplementary **List S2**: List of 694 (671 successful) interferometric image date pairs crated from Supplementary List S1 using GMT5SAR software and allowing a maximum of 35 days between images; **Figures S1–S3**: Comparison of integrated, plotted time series results in the ITRF-14 reference frame for CRIM, MKEA, and BLBP GNSS stations, respectively; **Equations S1–S8**: Integrating Geodetic Datasets and Inversion Processes; **Figure S4**: Raw GNSS uncertainty results; **Figure S5:** Variance associated with the Kriging interpolation used to generate the GPS displacement maps; **Figure S6:** Uncertainties associated with the cumulative, integrated DInSAR+GNSS results; **Figures S7–S10**: Integrated uncertainty for the plotted time series at the overlapping pixel location for CRIM, NUPM, MKEA, BLBP GNSS stations, respectively.

Author Contributions: Conceptualization, B.C., M.B.-S., K.T.; methodology, B.C., M.B.-S., K.T.; software, B.C., M.B.-S.; validation, B.C., M.B.-S., K.T., C.M.; formal analysis, B.C., M.B.-S., K.T., C.M.; investigation, B.C.; resources, B.C., M.B.-S., K.T., C.M.; data curation, B.C.; writing—original draft preparation, B.C.; writing—review and editing, B.C., K.T.; visualization, B.C.; supervision, B.C., M.B.-S., K.T., C.M.; project administration, K.T., C.M.; funding acquisition, K.T., C.M. All authors have read and agreed to the published version of the manuscript.

Funding: This research was funded by NSF Grant No. 1835566, Grant No. 1835791, Collaborative Research:Framework:Data:NSCI:HDR:GeoSCIFramework:Scalable Real-Time Streaming Analytics and Machine Learning for Geoscience and Hazards Research.

Institutional Review Board Statement: Not applicable.

Informed Consent Statement: Not applicable.

Data Availability Statement: GNSS processed time series are freely available and downloadable through the Nevada Geodetic Laboratory (http://geodesy.unr.edu/ (accessed on 3 April 2021)). Data stations are maintained by the USGS HVO [23,24], and data are archived and distributed by the UNAVCO GAGE facility. Sentinel-1 A/B DInSAR imagery is also freely available and downloadable through the Alaska Satellite Facility Vertex Portal [22] (https://search.asf.alaska.edu/#/ (accessed on 15 April 2021)). In the future, all products in this paper and produced as part of the project will be available under the GeoSciFramework project website (https://www.unavco.org/projects/major-projects/earthcube/geosciframework/geosciframework.html (accessed on 5 August 2021)). For DInSAR Data used in this study: Copernicus Sentinel data [2015–2021]. Retrieved from ASF DAAC [acquired 01 February 2021], processed by ESA. For Atmospheric Data used in this study: GACOS is supported by NERC through the center for seismic, volcanic, and tectonic observation and modeling (comet, Ref.: come30001) and LICs (Ref. Ne/k010794/1) and esa-most dragon-4 (Ref. 32244). We recognize the European Centre for medium-range weather forecasts (ECMWF) for its high-resolution tropospheric products. For GNSS Stations used in this study: Station MKEA is operated by the Jet Propulsion Laboratory and UNAVCO with NASA funding. Data are freely available from the NASA Crustal Dynamics Date Information System (CDDIS). For the following stations: ANIP, ALAL, APNT, HLNA, KAON, KHKU, KNNE, MKEA, MLES, MLRD, PAT3, PHAN, PIIK, PUH2, PUKA, RADF, and SLPC, citation: OBSERVER: Benjamin Brooks, AGENCY: Pacific GPS Facility, U of Hawaii/and Blewit et al. (2018). The following information obtained from UNAVCO's online DOI system: (https://www.unavco.org/data/doi/search/search.html) (accessed on 4 April 2021). Hawaii GPS Network-AHUP-Ahua P.S. Start: 30-May-1999, End: 16-May-2019 Segall, Paul, Miklius, Asta, 1999. Hawaii GPS Network-AHUP-Ahua P.S., UNAVCO, GPS/GNSS Observations Dataset, https://doi.org/10.7283/T5ST7MW0) (accessed on 4 April 2021). Hawaii GPS Network-AINP-Ainapo P.S. Start: 08-Oct-2002, End: 16-May-2019, Miklius, Asta, 2002, Hawaii GPS Network-AINP-Ainapo P.S., UNAVCO, GPS/GNSS Observations Dataset, https://doi.org/10.7283/T5348HMF) (accessed on 4 April 2021). Hawaii GPS Network-ALEP-ALEA Permanent P.S. Start: 15-Jul-2004, End:

15-May-2019, Miklius, Asta, 2004, Hawaii GPS Network-ALEP-ALEA Permanent P.S., UNAVCO, GPS/GNSS Observations Dataset, https://doi.org/10.7283/T52805VG) (accessed on 4 April 2021). Hawaii GPS Network-BLBP-Bluebird P.S., Start: 08-Jun-2004, End: 15-May-2019, Miklius, Asta, 2004, Hawaii GPS Network-BLBP-Bluebird P.S., UNAVCO, GPS/GNSS Observations Dataset, https://doi.org/10.7283/T5FF3QKS) (accessed on 4 April 2021). Hawaii GPS Network-BYRL-Byrons Ledge P.S., Start: 27-Feb-2006, End: 16-May-2019, Miklius, Asta, 2006, Hawaii GPS Network-BYRL-Byrons Ledge P.S., UNAVCO, GPS/GNSS Observations Dataset, https://doi.org/10.7283/T5NZ85W4) (accessed on 4 April 2021). Hawaii GPS Network-CNPK-Cone Peak P.S., Start: 25-Jun-2008, End: 16-May-2019, Miklius, Asta, 2008, Hawaii GPS Network-CNPK-Cone Peak P.S., UNAVCO, GPS/GNSS Observations Dataset, https://doi.org/10.7283/T5N014RM) (accessed on 4 April 2021). Hawaii GPS Network-CRIM-Crater Rim P.S., Start: 06-Jun-2008, End: 16-May-2019, Miklius, Asta, 2008, Hawaii GPS Network-CRIM-Crater Rim P.S., UNAVCO, GPS/GNSS Observations Dataset, https://doi.org/10.7283/T5RR1WGN) (accessed on 4 April 2021). Hawaii GPS Network-ELEP-Ele'ele P.S., Start: 09-Dec-2000, End: 15-May-2019, Miklius, Asta, 2000, Hawaii GPS Network-ELEP-Ele'ele P.S., UNAVCO, GPS/GNSS Observations Dataset, https://doi.org/10.7283/T5BP011P) (accessed on 4 April 2021). Hawaii GPS Network-GOPM-Goat P.S., Start: 30-May-1999, End: 16-May-2019, Segall, Paul, Miklius, Asta, 1999, Hawaii GPS Network-GOPM-Goat P.S., UNAVCO, GPS/GNSS Observations Dataset, https://doi.org/10.7283/T5P26W50) (accessed on 4 April 2021). Hawaii GPS Network-HOLE-Holei P.S., Start: 30-May-1999, End: 16-May-2019, Segall, Paul, Miklius, Asta, 1999, Hawaii GPS Network-HOLE-Holei P.S., UNAVCO, GPS/GNSS Observations Dataset, https://doi.org/10.7283/T5RV0KR5) (accessed on 4 April 2021). Hawaii GPS Network-JCUZ-Pu'u 'O'o JCUZ P.S., Start: 06-Jul-2007,End: 16-May-2019, Miklius, Asta, 2007, Hawaii GPS Network-JCUZ-Pu'u 'O'o JCUZ P.S., UNAVCO, GPS/GNSS Observations Dataset, https://doi.org/10.7283/T5416V92) (accessed on 4 April 2021). Hawaii GPS Network-JOKA-Jonika P.S., Start: 03-Oct-2007, End: 16-May-2019, Miklius, Asta, 2007, Hawaii GPS Network-JOKA-Jonika P.S., UNAVCO, GPS/GNSS Observations Dataset, https://doi.org/10.7283/T58P5XR4) (accessed on 4 April 2021). Hawaii GPS Network-KAEP-Kaena Point P.S., Start: 30-May-1999, End: 16-May-2019, Segall, Paul, Miklius, Asta, 1999, Hawaii GPS Network-KAEP-Kaena Point P.S., UNAVCO, GPS/GNSS Observations Dataset, https://doi.org/10.7283/T5J964D6) (accessed on 4 April 2021). Hawaii GPS Network-KAMO-Pu'u Kamoamoa P.S., Start: 10-Aug-2007, End: 16-May-2019, Miklius, Asta, 2007, Hawaii GPS Network-KAMO-Pu'u Kamoamoa P.S., UNAVCO, GPS/GNSS Observations Dataset, https://doi.org/10.7283/T5G73BXZ) (accessed on 4 April 2021). Hawaii GPS Network-KFAP-Kaoiki Fault P.S., Start: 16-Jul-2004, End: 16-May-2019, Miklius, Asta, 2004, Hawaii GPS Network-KFAP-Kaoiki Fault P.S., UNAVCO, GPS/GNSS Observations Dataset, https://doi.org/10.7283/T59P2ZV2) (accessed on 4 April 2021). Hawaii GPS Network-KOSM-KOSM P.S., Start: 01-Aug-1999, End: 16-May-2019, Segall, Paul, Miklius, Asta, 2000, Hawaii GPS Network-KOSM-KOSM P.S., UNAVCO, GPS/GNSS Observations Dataset, https://doi.org/10.7283/T59S1P23) (accessed on 4 April 2021). Hawaii GPS Network-KTPM-Kalapana Trail P.S., Start: 30-May-1999, End: 16-May-2019, Segall, Paul, Miklius, Asta, 1999, Hawaii GPS Network-KTPM-Kalapana Trail P.S., UNAVCO, GPS/GNSS Observations Dataset, https://doi.org/10.7283/T5DJ5CN1) (accessed on 4 April 2021). Hawaii GPS Network-MANE-Mauka Kipuka Nene P.S., Start: 30-May-1999, End: 16-May-2019, Segall, Paul, Miklius, Asta, 1999, Hawaii GPS Network-MANE-Mauka Kipuka Nene P.S., UNAVCO, GPS/GNSS Observations Dataset, https://doi.org/10.7283/T5513W86) (accessed on 4 April 2021). Hawaii GPS Network-MKAI-Makai Makaopuhi P.S., Start: 09-Jul-2010, End: 16-May-2019, Miklius, Asta, 2010, Hawaii GPS Network-MKAI-Makai Makaopuhi P.S., UNAVCO, GPS/GNSS Observations Dataset, https://doi.org/10.7283/T57S7M01) (accessed on 4 April 2021), Hawaii GPS Network-MLCC-Mauna Loa Cinder Cone P.S., Start: 02-Jul-2004, End: 15-May-2019, Miklius, Asta, 2004, Hawaii GPS Network-MLCC-Mauna Loa Cinder Cone P.S., UNAVCO, GPS/GNSS Observations Dataset, https://doi.org/10.7283/T56W989D) (accessed on 4 April 2021). Hawaii GPS Network-MLSP-Mauna Loa Summit P.S., Start: 09-Dec-2000, End: 16-May-2019, Miklius, Asta, 2000, Hawaii GPS Network-MLSP-Mauna Loa Summit P.S., UNAVCO, GPS/GNSS Observations Dataset, https://doi.org/10.7283/T5K64G9H) (accessed on 4 April 2021). Hawaii GPS Network-MMAU-Mauka Makaopuhi P.S., Start: 09-Jul-2010, End: 16-May-2019, Miklius, Asta, 2010, Hawaii GPS Network-MMAU-Mauka Makaopuhi P.S., UNAVCO, GPS/GNSS Observations Dataset, https://doi.org/10.7283/T5CF9NBG) (accessed on 4 April 2021). Hawaii GPS Network-MOKP-Moku'aweoweo P.S., Start: 09-Dec-2000, End: 16-May-2019, Miklius, Asta, 2000, Hawaii GPS Network-MOKP-Moku'aweoweo P.S., UNAVCO, GPS/GNSS Observations Dataset, https://doi.org/10.7283/T5TQ5ZS9) (accessed on 4 April 2021). Hawaii GPS Network-NPOC-North Pu'u P.S., Start: 05-Jul-2007, End: 16-May-2019, Miklius, Asta,

2007, Hawaii GPS Network-NPOC-North Pu'u P.S., UNAVCO, GPS/GNSS Observations Dataset, https://doi.org/10.7283/T5QR4VB8) (accessed on 4 April 2021). Hawaii GPS Network-NUPM-Kane Nui O Hamo P.S., Start: 30-May-1999, End: 16-May-2019, Segall, Paul, Miklius, Asta, 1999, Hawaii GPS Network-NUPM-Kane Nui O Hamo P.S., UNAVCO, GPS/GNSS Observations Dataset, https://doi.org/10.7283/T58S4MZR) (accessed on 4 April 2021). Hawaii GPS Network-OUTL-Outlet P.S., Start: 18-April-2008, End: 16-May-2019, Miklius, Asta, 2008, Hawaii GPS Network-OUTL-Outlet P.S., UNAVCO, GPS/GNSS Observations Dataset, https://doi.org/10.7283/T5WH2N7T) (accessed on 4 April 2021). Hawaii GPS Network-PMAU-Pu'u Ma'au P.S., Start: 27-Oct-2006, End: 16-May-2019, Miklius, Asta, 2006, Hawaii GPS Network-PMAU-Pu'u Ma'au P.S., UNAVCO, GPS/GNSS Observations Dataset, https://doi.org/10.7283/T5DF6PFD) (accessed on 4 April 2021). Hawaii GPS Network-STEP-Mauna Loa Observatory P.S., Start: 12-Oct-2002, End: 15-May-2019, Miklius, Asta, 2002, Hawaii GPS Network-STEP-Mauna Loa Observatory P.S., UNAVCO, GPS/GNSS Observations Dataset, https://doi.org/10.7283/T5PZ572C) (accessed on 4 April 2021). Hawaii GPS Network-TOUO-Tube Ouo P.S., Start: 10-Jul-2004, End: 15-May-2019, Miklius, Asta, 2004, Hawaii GPS Network-TOUO-Tube Ouo P.S., UNAVCO, GPS/GNSS Observations Dataset, https://doi.org/10.7283/T55X275K) (accessed on 4 April 2021), Hawaii GPS Network-UWEV-Uwekahuna Vault P.S., Start: 30-May-1999, End: 16-May-2019, Segall, Paul, Miklius, Asta, 1999, Hawaii GPS Network-UWEV-Uwekahuna Vault P.S., UNAVCO, GPS/GNSS Observations Dataset, https://doi.org/10.7283/T51C1TXB) (accessed on 4 April 2021).

Conflicts of Interest: The authors declare no conflict of interest. The funders had no role in the design of the study; in the collection, analyses, or interpretation of data; in the writing of the manuscript, or in the decision to publish the results.

References

1. Aiuppa, A.; Moretti, R.; Cinzia, F.; Giudice, G.; Gurrieri, S.; Liuzzo, M.; Papale, P.; Shinohara, H.; Valenza, M. Forecasting Etna eruptions by real-time observation of volcanic gas composition. *Geology* **2007**, *35*, 1115–1118. [CrossRef]
2. Anantrasirichai, N.; Biggs, J.; Albino, F.; Hill, P.; Bull, D. Application of machine learning to classification of volcanic deformation in routinely generated InSAR data. *J. Geophys. Res. Solid Earth* **2018**, *123*, 6592–6606. [CrossRef]
3. Bebbington, M.S. Long-term forecasting of volcanic explosivity. *Geophys. J. Int.* **2014**, *197*, 1500–1515. [CrossRef]
4. Marzocchi, W.; Bebbington, M.S. Probabilistic eruption forecasting at short- and long-time scales. *Bull. Volcanol.* **2012**, *74*, 1777–1805. [CrossRef]
5. Phillipson, G.; Sobradelo, R.; Gottsmann, J. Global volcanic unrest in the 21st century: An analysis of the first decade. *J. Volcanol. Geotherm. Res.* **2013**, *264*, 183–196. [CrossRef]
6. Potter, S.H.; Scott, B.J.; Fearnley, C.J.; Leonard, G.S.; Gregg, C.E. Challenges and benefits of standardising early warning system. In *Observing the Volcano World*; Fearnley, C.J., Bird, D.K., Haynes, K., McGuire, W.J., Jolly, G., Eds.; Advances in Volcanology (An Official Book Series of the International Association of Volcanology and Chemistry of the Earth's Interior). Springer: New York, NY, USA, 2017. [CrossRef]
7. Rouwet, D.; Sandri, L.; Marzocchi, W.; Gottsmann, J.; Selva, J.; Tonini, R.; Papale, P. Recognizing and tracking volcanic hazards related to non-magmatic unrest: A review. *J. Appl. Volcanol.* **2014**, *3*, 17. [CrossRef]
8. Stix, J. Understanding fast and slow unrest at volcanoes and implications for eruption forecasting. *Front. Earth Sci.* **2018**, *6*, 2018. [CrossRef]
9. Kelevitz, K.; Tiampo, K.F.; Corsa, B.D. Improved real-time natural hazard monitoring automated DInSAR time series. *Remote Sens.* **2021**, *13*, 867. [CrossRef]
10. Chen, K.; Smith, J.D.; Avouac, J.-P.; Liu, Z.; Song, Y.T.; Gualandi, A. Triggering of the Mw 7.2 Hawaii earthquake of 4 May 2018 by a dike intrusion. *Geophys. Res. Lett.* **2019**, *46*, 2503–2510. [CrossRef]
11. Derauw, D.; d'Oreye, N.; Jaspard, M.; Caselli, A.; Samsonov, S. Ongoing automated ground deformation monitoring of Domuyo—Laguna del Maule area (Argentina) using Sentinel-1 MSBAS time series: Methodology description and first observations for the period 2015–2020. *J. South Am. Earth Sci.* **2020**, *104*, 102850. [CrossRef]
12. Samsonov, S.; Tiampo, K. Analytical optimization of DInSAR and GPS dataset for derivation of three-dimensional surface motion. *IEEE Geosci. Remote Sens. Lett.* **2006**, *3*, 107–111. [CrossRef]
13. Samsonov, S.V.; Feng, W.; Fialko, Y. Subsidence at cerro prieto geothermal field and postseismic slip along the indiviso fault from 2011 to 2016 RADARSAT-2 DInSAR time series analysis. *Geophys. Res. Lett.* **2017**, *44*, 2716–2724. [CrossRef]
14. Tilling, R.I. Volcanic hazards and early warning. In *Encyclopedia of Complexity and Systems Science*; Meyers, R., Ed.; Springer: New York, NY, USA, 2009. [CrossRef]
15. Valade, S.; Ley, A.; Massimetti, F.; D'Hondt, O.; Laiolo, M.; Coppola, D.; Loibl, D.; Hellwich, O.; Walter, T.R. Towards global volcano monitoring using multisensor sentinel missions and artificial intelligence: The MOUNTS monitoring system. *Remote Sens.* **2019**, *11*, 1528. [CrossRef]

16. Lundgren, P.; Girona, T.; Bato, M.G.; Realmut, V.J.; Samsonov, S.; Cardona, C.; Franco, L.; Gurrola, E.; Aivazis, M. The dynamics of large silicic systems from satellite remote sensing observations: The intriguing case of Domuyo volcano, Argentina. *Sci. Rep.* **2020**, *10*, 11642. [CrossRef]
17. Ji, P.; Lv, X.; Yao, J.; Sun, G. A new method to obtain 3-D surface deformations from InSAR and GNSS data with genetic algorithm and support vector machine. *IEEE Geosci. Remote Sens. Lett.* **2022**, *19*, 1–5. [CrossRef]
18. Samsonov, S.; Tiampo, K.; Rundle, J.; Li, Z. Application of DInSAR-GPS optimization for derivation of fine scale surface motion maps of southern California. *IEEE Trans. Geosci. Remote Sens.* **2007**, *45*, 512–522. [CrossRef]
19. Samsonov, S.; Tiampo, K.; Rundle, J. Application of DInSAR-GPS optimization for derivation of three-dimensional surface motion of southern California region along the San Andreas fault. *Comput. Geosci.* **2008**, *34*, 503–514. [CrossRef]
20. Vollrath, A.; Zucca, F.; Bekaert, D.; Bonforte, A.; Guglielmino, F.; Hooper, A.J.; Stramondo, S. Decomposing DInSAR time-series into 3-D in combination with GPS in the case of low strain rates: An application to the Hyblean Plateau, Sicily, Italy. *Remote Sens.* **2017**, *9*, 33. [CrossRef]
21. Copernicus Sentinel-1 data 2015–2021, retrieved from ASF DAAC 23-04-2021, processed by ESA. Available online: https://asf.alaska (accessed on 15 April 2021).
22. Blewitt, G.; Hammond, W.C.; Kreemer, C. Harnessing the GPS data explosion for interdisciplinary science. *Eos* **2018**, *99*, 485. [CrossRef]
23. Miklius, A. Hawaii GPS Network-CNPK-Cone Peak P.S. The GAGE Facility operated by UNAVCO, Inc., GPS/GNSS Observations Dataset. 2008. Available online: https://doi.org/10.7283/T5N014RM (accessed on 5 October 2021). [CrossRef]
24. USGS.gov. December 2020–May 2021 Eruption. Available online: https://www.usgs.gov/volcanoes/kilauea/december-2020-may-2021-eruption?qt-science_support_page_related_con=0#qt-science_support_page_related_con (accessed on 5 October 2021).
25. USGS.gov. Available online: https://volcanoes.usgs.gov/volcanoes/Kilauea/ (accessed on 5 October 2021).
26. Patrick, M.R.; Houghton, B.F.; Anderson, K.R.; Poland, M.P.; Montgomery-Brown, E.; Johanson, I.; Thelen, W.; Elias, T. The cascading origin of the 2018 Kīlauea eruption and implications for future forecasting. *Nat. Commun.* **2020**, *11*, 5646. [CrossRef]
27. Sandwell, D.; Mellors, R.; Tong, X.; Wei, M.; Wessel, P. Open radar interferometry software for mapping surface deformation. *Eos Trans. AGU* **2011**, *92*, 234. [CrossRef]
28. Sandwell, D.; Mellors, R.; Tong, X.; Wei, M.; Wessel, P. *GMTSAR: An InSAR Processing System Based on Generic Mapping Tools*; UC San Diego Scripps Institution of Oceanography: La Jolla, CA, USA, 2011; Available online: http://escholarship.org/uc/item/8zq2c02m (accessed on 1 October 2021).
29. Berardino, P.; Fornaro, G.; Lanari, R.; Sansosti, E. A new algorithm for surface deformation monitoring based on small baseline differential SAR interferograms. *IEEE Trans. Geosci. Remote Sens.* **2002**, *40*, 2375–2383. [CrossRef]
30. Doin, M.P.; Guillaso, S.; Jolivet, R.; Lasserre, C.; Lodge, F.; Ducret, G. Presentation of the small baseline NSBAS processing chain on a case example: The Etna deformation monitoring from 2003 to 2010 using Envisat data. In Proceedings of the ESA FRINGE 2011 Conference, Frascati, Italy, 19–23 September 2011.
31. Zimmerman, D.; Pavlik, C.; Ruggles, A.; Armstrong, M.P. An experimental comparison of ordinary and universal kriging and inverse distance weighting. *Math. Geol.* **1999**, *31*, 375–390. [CrossRef]
32. Matheron, G. Principles of geostatistics. *Econ. Geol.* **1963**, *58*, 1246–1266. [CrossRef]
33. Moussouris, J. Gibbs and Markov random systems with constraints. *J. Stat. Phys.* **1974**, *10*, 11–33. [CrossRef]
34. Johnson, C.W.; Lau, N.; Borsa, A. An assessment of GPS velocity uncertainty in California. *Earth Space Sci.* **2020**, *8*, e2020EA001345. [CrossRef]
35. Wang, S.-Y.; Li, J.; Chen, J.; Hu, X.-G. Uncertainty assessments of load deformation from different GPS time series products, GRACE estimates and model predictions: A case study over Europe. *Remote Sens.* **2021**, *13*, 2765. [CrossRef]
36. Chen, C.W.; Zebker, H.A. Phase unwrapping for large SAR interferograms: Statistical segmentation and generalized network model. *IEEE Trans. Geosci. Remote Sens.* **2002**, *40*, 1709–1719. [CrossRef]
37. Agram, P.; Jolivet, R.; Riel, B.-V.; Simons, M.; Doin, M.; Lasserre, C.; Hetland, E.-A. GIAnT—Generic InSAR Analysis Toolbox, American Geophysical Union, Fall Meeting 2012, abstract id. G43A-0897. Available online: https://ui.adsabs.harvard.edu/abs/2012AGUFM.G43A0897A (accessed on 1 December 2021).
38. Yu, C.; Penna, N.T.; Li, Z. Generation of real-time mode high-resolution water vapor fields from GPS observations. *J. Geophys. Res. Atmos.* **2017**, *122*, 2008–2025. [CrossRef]
39. Yu, C.; Li, Z.; Penna, N.T. Interferometric synthetic aperture radar atmospheric correction using a GPS-based iterative tropospheric decomposition model. *Remote Sens. Environ.* **2018**, *204*, 109–121. [CrossRef]
40. Yu, C.; Li, Z.; Penna, N.T.; Crippa, P. Generic atmospheric correction model for Interferometric Synthetic Aperture Radar observations. *J. Geophys. Res. Solid Earth* **2018**, *123*, 9202–9222. [CrossRef]
41. Yu, C.; Li, Z.; Penna, N.T. Triggered afterslip on the southern Hikurangi subduction interface following the 2016 Kaikōura earthquake from InSAR time series with atmospheric corrections. *Remote Sens. Environ.* **2020**, *251*, 112097. [CrossRef]
42. Rebischung, P.; Schmid, R. IGS14/igs14.atx: A new framework for the IGS Products. In Proceedings of the American Geophysical Union Fall Meeting 2016, San Francisco, CA, USA, 12–16 December 2016.
43. Samsonov, S.; Dille, A.; Dewitte, O.; Kervyn, F.; d'Oreye, N. Satellite interferometry for mapping surface deformation time series in one, two and three dimensions: A new method illustrated on a slow-moving landslide. *Eng. Geol.* **2020**, *266*, 105471. [CrossRef]

44. Cressie, N. Statistics for spatial data. In *Wiley Series in Probability and Statistics*; John Wiley and Sons, Inc.: New York, NY, USA, 1991; p. 900.
45. Kitanidis, P.K. *Introduction to Geostatistcs: Applications in Hydrogeology*; Cambridge University Press: Cambridge, UK, 1997.
46. Lin, G.; Shearer, P.M.; Matoza, R.S.; Okubo, P.G.; Amelung, F. Three-dimensional seismic velocity structure of Mauna Loa and Kilauea volcanoes in Hawaii from local seismic tomography. *J. Geophys. Res. Solid Earth* **2014**, *119*, 4377–4392. [CrossRef]
47. Miklius, A.; Cervelli, P. Interaction between Kilauea and Mauna Loa. *Nature* **2003**, *421*, 229. [CrossRef]
48. Neal, C.A.; Brantley, S.R.; Antolik, L.; Babb, J.L.; Burgess, M.; Calles, K.; Cappos, M.; Chang, J.C.; Conway, S.; Desmither, L.; et al. The 2018 rift eruption and summit collapse of Kīlauea Volcano. *Science* **2019**, *363*, 367–374. [CrossRef] [PubMed]
49. USGS HVO Overview of Kīlauea Volcano's 2018 Lower East Rift Zone Eruption and Summit Collapse. 2019. Available online: https://volcanoes.usgs.gov/vsc/file_mngr/file-224/OVERVIEW_Kil2018_LERZ-Summit_June%202019.pdf, (accessed on 12 December 2021).
50. Wang, T.; Zheng, Y.; Pulvirenti, F.; Segall, P. Post-2018 caldera collapse re-inflation uniquely constrains Kīlauea's magmatic system. *J. Geophys. Res. Solid Earth* **2020**, *126*, e2021JB021803. [CrossRef]
51. Lundgren, P.R.; Bagnardi, M.; Dietterich, H. Topographic changes during the 2018 Kīlauea eruption from single-pass airborne InSAR. *Geophys. Res. Lett.* **2019**, *46*, 9554–9562. [CrossRef]
52. Smith-Konter, B.; Ward, L.; Burkhard, L.; Foster, J.; Xa, X.; Sandwell, D. 2018 Kilauea Eruption and MW 6.9 Leilani Estates Earthquake: Line of Sight Displacement Revealed by Sentinel-1 Interferometry. 2018 Kilauea InSAR. Available online: http://pgf.soest.hawaii.edu/Kilauea_insar/ (accessed on 12 December 2021).
53. Smittarello, D.; Cayol, V.; Pinel, V.; Peltier, A.; Froger, J.-L.; Ferrazzini, V. Magma propagation at Piton de la Fournaise from joint inversion of InSAR and GNSS. *J. Geophys. Res. Solid Earth* **2019**, *124*, 1361–1387. [CrossRef]

remote sensing

MDPI

Article

Detailed Cartography of Cotopaxi's 1877 Primary Lahar Deposits Obtained by Drone-Imagery and Field Surveys in the Proximal Northern Drainage

S. Daniel Andrade [1,*], Emilia Saltos [1], Valeria Nogales [1], Sebastián Cruz [1], Gareth Lee [2] and Jenni Barclay [2]

1 Instituto Geofísico, Escuela Politécnica Nacional, Ladrón de Guevara E11-253, Quito 170100, Ecuador; emilia_saltos@hotmail.com (E.S.); vnogales@igepn.edu.ec (V.N.); sebascruz@hotmail.es (S.C.)
2 School of Environmental Sciences, University of East Anglia, Norwich Research Park, Norwich NR4 7TJ, UK; g.a.lee@uea.ac.uk (G.L.); J.Barclay@uea.ac.uk (J.B.)
* Correspondence: dandrade@igepn.edu.ec

Abstract: Cotopaxi is an active volcano in Ecuador, whose eruptions are characterized by producing destructive primary lahars which represent a major risk for the country. The hazard assessment related to such lahars relies largely on the knowledge of the latest event, which occurred on 26 June 1877, for either scenario definition or simulation calibration. A detailed (1:5000 scale) cartography of the deposits belonging to that eruption has been obtained in the proximal northern drainage of Cotopaxi. The cartography was performed through a combination of geological fieldwork, as well as the analysis and interpretation of high-definition imagery obtained by drone surveys combined with the Structure from Motion technology for image processing. Such imagery included red and green visible bands, and a near-infrared band, which allowed the obtention of NDVI imagery where the primary lahar deposits were identified and cartographed with support of fieldwork data. Both data sources are mutually complementary, and the final cartography would be impossible if any of them were not available. The results obtained represent a significant advance for the level of detail with respect to previous cartographic works. Moreover, they should allow an improved calibration of the new generation of numerical models that simulate lahar flow for hazard assessment at Cotopaxi.

Keywords: Cotopaxi volcano; 1877 eruption; primary lahars; drone-imagery; geological mapping; lahar hazard assessment

Citation: Andrade, S.D.; Saltos, E.; Nogales, V.; Cruz, S.; Lee, G.; Barclay, J. Detailed Cartography of Cotopaxi's 1877 Primary Lahar Deposits Obtained by Drone-Imagery and Field Surveys in the Proximal Northern Drainage. *Remote Sens.* **2022**, *14*, 631. https://doi.org/10.3390/rs14030631

Academic Editors: Sonia Calvari, Alessandro Bonaccorso, Annalisa Cappello, Flora Giudicepietro and Eugenio Sansosti

Received: 18 November 2021
Accepted: 19 January 2022
Published: 28 January 2022

Publisher's Note: MDPI stays neutral with regard to jurisdictional claims in published maps and institutional affiliations.

1. Introduction

Cotopaxi is an ice-capped active volcano located 50 km south of Quito, the capital city of Ecuador (Figure 1). It has long been recognized as a highly hazardous volcano, especially for its capacity to produce primary lahars, which are mixtures of water and volcanic debris formed as large portions of the glacier instantaneously melt by contact with pyroclastic density currents (PDCs) during highly explosive eruptions [1–5]. Primary lahars formed in this way are characterized by volumes reaching 10^6–10^7 m^3, and peak discharges of up to 10^3–10^4 m^3/s [6,7], with the capacity to produce widespread destruction tens of kilometers away from their source, as occurred with Armero town (Colombia), during the eruption of Nevado del Ruiz volcano in 1985 [8].

In the case of Cotopaxi, such lahars have been channelized by three main drainage systems: the Pita river to the north, the Cutuchi river to the south, and the Tamboyacu river to the east, where they have historically produced extensive damage in the neighboring inhabited areas, such as Valle de los Chillos around Sangolqui town, and many Quito suburbs to the north, or Mulalo and Latacunga, to the south (Figure 1) [1,2,6].

Given the regional-scale destructive potential of Cotopaxi's primary lahars, the Instituto Geofísico of Escuela Politécnica Nacional (IG-EPN) has elaborated several hazard maps during the last five decades, with the aim of providing information for land-use planning and policymaking, as part of their official mandate to assess volcanic hazard in Ecuador [3,9–15]. Those maps have been partially or totally based on research results produced by various scientific teams (i.e., [4,6,16–21]). They have also served as the basis for vulnerability [22,23] and risk [7,24] assessments, as well as being tools for emergency response during volcanic crises [25,26].

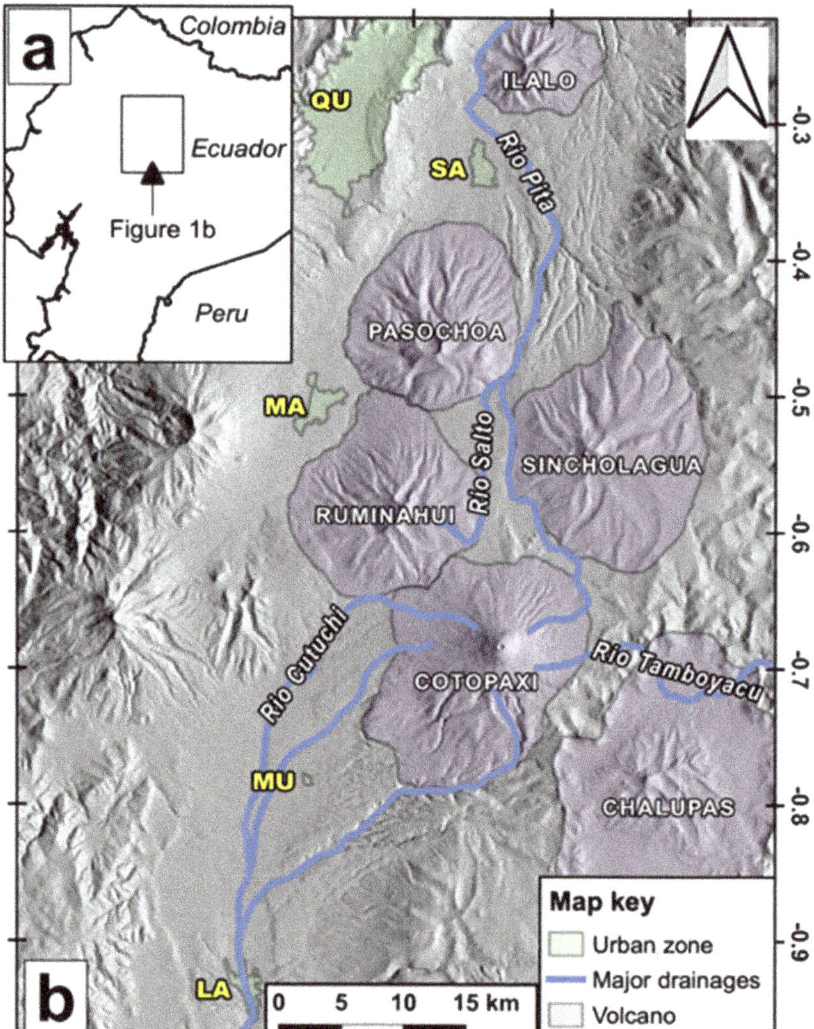

Figure 1. (**a**) Regional location of the study area and of (**b**) Cotopaxi volcano with reference to the major drainages and cities (QU: Quito; SA: Sangolquí; MA: Machachi; MU: Mulalo; LA: Latacunga). The present study is related to the proximal northern basin drained by the Río Pita and Río Salto drainages. Coordinate system in all maps is Geographic WGS84.

Because of their origin during highly explosive events, primary lahars at Cotopaxi are a relatively infrequent phenomena, averaging less than one event per century over the past 800 years [6,17]. Thus, the lahar hazard at Cotopaxi has always been assessed by deterministic approaches, which have used different numerical models calibrated with field data and corresponding lahar scenarios. Numerical models tested at Cotopaxi include 1-D hydrological [7,16,27,28], statistical [18], and more developed flow-dynamics-based codes which consider deposition, bulking and erosive processes [29,30]. Regardless of the numerical model, the field data or lahar scenario used as either entry data, boundary condition, or calibration, has been invariably derived from knowledge of the primary lahar that occurred on 26 June 1877, during the latest highly explosive eruption of the Cotopaxi volcano. Thus, lahar hazard assessments at the Cotopaxi volcano significantly rely on the knowledge related to this specific event.

It has been strongly suggested that the 1877 event may not represent the most likely nor the worst-case scenario for future eruptions of Cotopaxi [6,17]. However, it is the only one with some detailed and reliable historical accounts that help understanding the eruption evolution and its consequences [1,2]. Moreover, due to its stratigraphic position and a few facies characteristics, the lahar deposits from 1877 are the most suitable for detailed geological studies that could confidently feed and calibrate numerical models. This helps to understand why this deposit has been widely used as the standard for lahar hazard and risk assessments at Cotopaxi [7,13–15,24].

During the last decade, the combination of imagery obtained by drones with the Structure from Motion (SfM) algorithms has been established as a useful tool for geological [31–34] and volcanological [35–37] studies. Sub-decimeter resolution imagery and digital elevation models obtained with this technology have been used for the study or monitoring of lava flows [38,39], and morphological changes during eruptions [40]. Moreover, surface processes occurring at volcanoes have been studied using similar methodologies and data, such as the short- and long-term deposition/erosion processes associated with rain-triggered lahars [41–43]. In all those cases, the aim was mapping sub-metric features (scales < 1:1000) over relatively small surfaces (1–2 km^2), thus obtaining detailed measurements and assessments. A few examples also exist of this methodology being used for larger scale cartography and volcanic hazard assessments [44,45]. In many of the aforementioned cases, the combination of drone imagery with SfM is supported by complementary ground-based measurements.

In the present work, detailed cartography of the 1877 primary lahar deposit is performed in the proximal northern drainage of Cotopaxi, by means of extensive geological fieldwork and drone surveys, also supported by freely available satellite imagery. The fieldwork included identifying outcrops and lateral contacts of the 1877 lahar deposit in 172 control points and along 47 track lines. The drone surveys produced 25 cm-pixel orthomosaics obtained by SfM, composed of red, green, and near-infra-red bands covering an area of 28.4 km^2. These data were subsequently used to map the deposit contacts with the help of remote-sensing techniques (NDVI single-band images) and in correlation with field data. This survey aimed at obtaining new high-definition geological information to feed or calibrate next-generation numerical models of the 1877 primary lahar, which would allow an improved assessment of the hazards related to future eruptions of Cotopaxi volcano. The term "drone" will be preferred in this study, instead of "unmanned aerial vehicle—UAV" (or other similar), on the grounds of gender neutrality, simplicity and familiarity.

2. Context of the Proximal Northern Drainage

2.1. Morphology and Drainage Network

The study area corresponds to the plains extending at the northern, north-eastern, and eastern feet of Cotopaxi's edifice, limited by the Ruminahui and Sincholahua volcanoes to the north-west and north-east, respectively as well as by the Chalupas caldera to the east (Figure 1). These plains collect several different drainages descending through the deep gullies of Cotopaxi's middle and lower flanks, all of which are tributaries to the Rio Pita river (Figure 1). The 1877 lahar deposit, as well as several others, widely outcrop along these plains [6,17]. Given its extension, the study area has been divided into the following plains comprising specific drainages and morphologies: (1) Sindipamba; (2) Victor Punina (3) North-eastern; and (4) Eastern.

The Sindipamba plain is located to the north, just below the 4000 m level, limited by the Limpiopungo lagoon to the west and the Ingaloma hill to the east (Figure 2). At least six major gullies of Cotopaxi's north flank drain into the Sindipamba plain: the westernmost gully and two others are unnamed; Quebrada Chilcahuaycu flows at the base of the Ingaloma hill, while Quebradas Yanasacha and Yanashaco traverse the plain (Figure 2). All these drainages converge to form the Rio Salto, which flows for some ~14 km northwards, before joining the Rio Pita at the base of the Pasochoa volcano (Figures 1 and 2). The exception is Chilcahuaycu, which bends north-eastwards at the base of Ingaloma and joins the Victor Punina drainage north of the Jatunloma sector, where both flow into the Rio Pita (Figure 2).

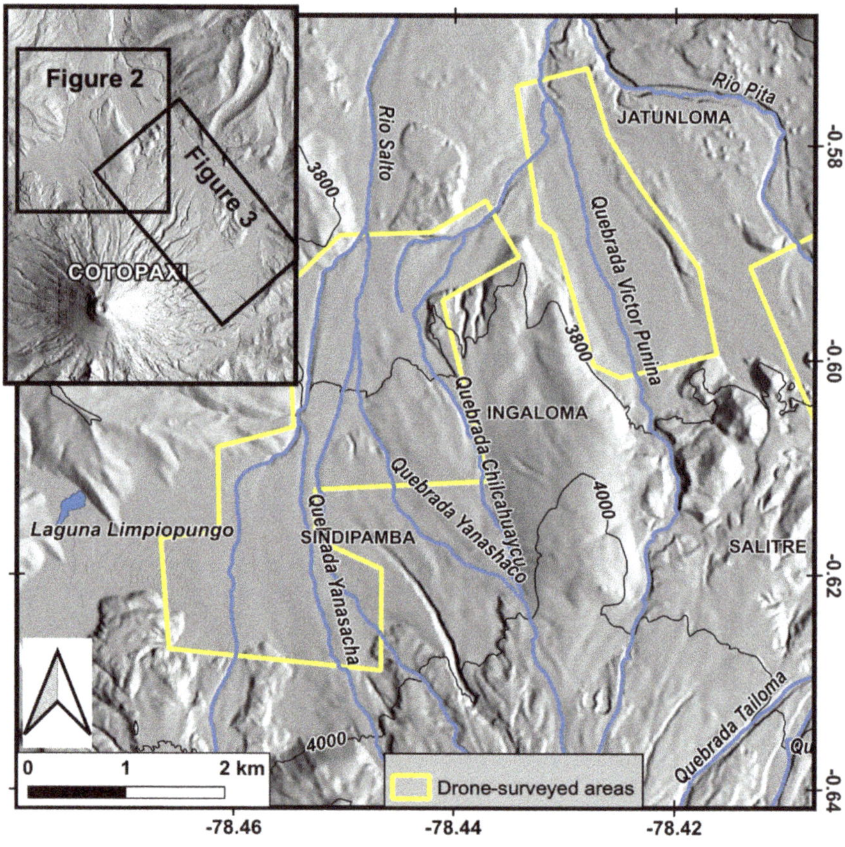

Figure 2. Location of the Sindipamba and Victor Punina plains in the northern lower flanks of Cotopaxi. The areas surveyed by drone are highlighted in yellow polygones.

The Victor Punina plain sits just east of the Ingaloma hill, below the 3800 m level, and upstream it is the only current drainage actively cutting a deep gully through the hummocks and toreva blocks that constitute the Salitre hills (Figure 2) [46]. The source zone of this drainage is the same as the Quebradas Yanashaco and Chilcahuaycu from the Sindipamba plain, and is located above the 4600 m level in the north flank of Cotopaxi (Figure 2).

The North-eastern plain extends between the Salitre and Mauca Mudadero hills, below the 4000 m level (Figure 3). Five major gullies channel drainages into this plain: Tailoma, Potrerillo, Pucarumi, Pucahuaycu, and Jatabamba, all of which have their source zones in the north-eastern flank of Cotopaxi, above the 4600 m level. Moreover, all these drainages traverse the plain and flow into the Rio Pita along the south-west foot of the Sincholahua volcano (Figures 1 and 3).

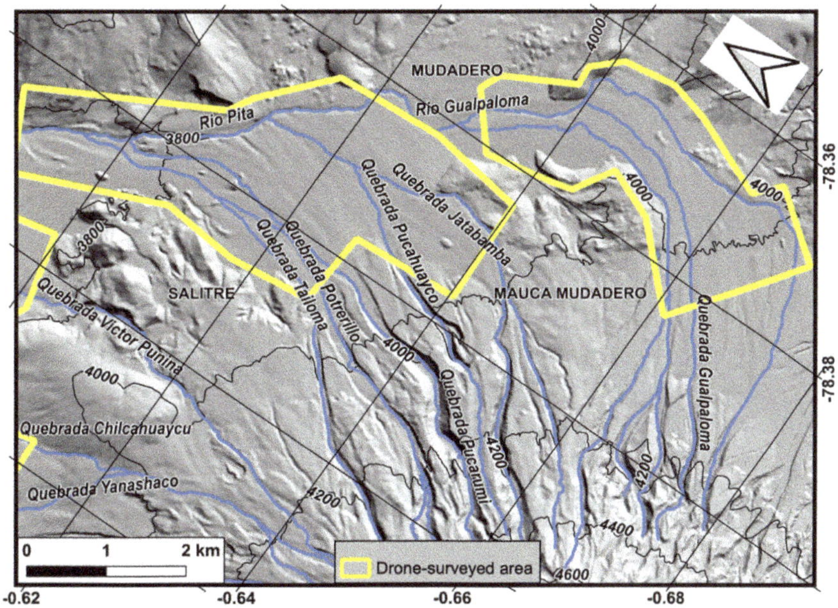

Figure 3. Location of the North-eastern and Eastern plains in the lower flanks of Cotopaxi. The areas surveyed by drone are highlighted in yellow polygons. See insert in Figure 2 for location. For improved visualization, the map has been rotated counterclockwise (north is pointing to the left).

Finally, the Eastern plain is located south of the Mauca Mudadero hills, below the 4200 m level (Figure 3). Five major gullies channel drainages into this plain, all of which are unnamed except for the Quebrada Gualpaloma, the fourth from the north. The source zone of these drainages is located in the eastern flank of Cotopaxi's edifice, above the 4600 m level. Additional gullies are further to the south in this plain, but those drainages flow into the Rio Tamboyacu towards the Amazonian basin (Figures 1 and 3).

2.2. Geological Characterization of the 1877 Deposits

The 1877 explosive eruption of Cotopaxi unfolded on the morning of June 26th, after at least three months of unrest and low-level eruptions [1,2]. The main deposits associated with this eruption were a tephra fall layer, bomb-rich pyroclastic density currents (PDCs), and the primary lahars, all of which currently outcrop around the lower flanks of Cotopaxi, though the tephra layer is less than 5 cm thick in the study area [6,17,20,47]. No lava flows occurred during this eruptive episode.

The historical accounts have been, in general, confirmed by the geological studies performed on the 1877 deposits so far. Pistolesi et al. [17] have found that the tephra deposit underlies the lahar deposit in the volcano's vicinity, and that the remains of some PDCs can be found generally in the middle and proximal lower flanks. Additionally, Mothes et al. [6] have found evidence of post-June 26th explosive activity, as described shortly after the eruption by Wolf [1], represented by a few bomb-rich PDC deposits which overlie the lahar deposit. All authors agree that both the PDC and the primary lahar are characterized by containing decimeter- to meter-sized black, cauliflower-shaped bombs. Internally, these bombs are brownish and display chilled margins and thermal fractures. However, several older primary lahar deposits (i.e., from the 1768 or 1853–1854 eruptions) also outcrop in the study area, and some contain similar black bombs [17], which may produce confusion.

A reliable field identification of the 1877 primary lahar deposit was thus critical for the present study and was based on the characteristics described above, namely, the stratigraphic position and the presence of the black, cauliflower-shaped bombs. However, two additional characteristics were recognized as useful to identify the 1877 lahar deposit. The first is related to the stratigraphic position of the deposit: as this primary lahar is the latest to have been emplaced, it is also the one whose surface has developed the least amount and diversity of vegetation ever since [48]. It was evident during field surveys that each plain of the study area displays noticeable changes in the vegetation that correlate with lateral contacts of primary lahar deposits, and that usually the least vegetated zones correspond to the 1877 deposit (Figure 4). However, as each plain is placed at different altitudes, the amount and type of vegetation also change from one to another. There are other poorly vegetated deposits in the study area (i.e., at Sindipamba and Victor Punina plains), so additional criteria are needed for a proper deposit identification.

Figure 4. Vegetation and lithic contrasts between lahar deposits belonging to the 1877 eruption and an older eruption, as observed in the North-eastern plain. Differences may appear less evident in the field when the age of the deposits are closer.

The second characteristic that helped with the identification of the 1877 deposits (either PDC or primary lahar) is a unique mechanical property of the cauliflower-shaped black bombs they contain. Regardless of the bomb size, they easily break apart in fragments through their internal thermal fractures, when slightly hit a few times with a hammer (Figure 5a). Although, externally, they look very similar, this is impossible to achieve with the black bombs present in many other deposits, which will only break through one fracture surface when strongly hit with a hammer (Figure 5b). This specific characteristic of the 1877 black bomb is so outstanding that even Theodore Wolf described it when he studied the fresh eruption deposits in September, 1877 [1]. The presence of these bombs has been considered as the primordial characteristic to identify the 1877 deposits in the field, and significant care was put into their identification during fieldwork, especially in the Sindipamba and Victor Punina plains, where other poorly vegetated deposits exist.

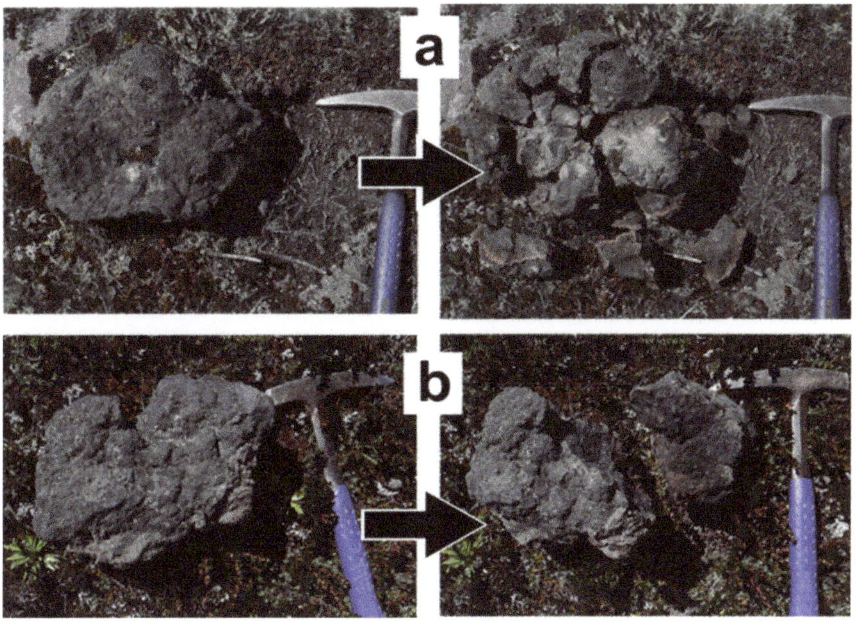

Figure 5. Contrasting mechanical behavior observed between cauliflower-shaped black bombs belonging to (**a**) the 1877 eruption, and (**b**) older eruptions. Unbroken bombs are very similar.

Finally, it was also important to differentiate between PDC and primary lahar deposits, given that both are placed in the same stratigraphical position and contain the 1877-specific black bomb. In general, the most evident differences between the two types of deposits were: (a) their surface morphologies, and (b) their surface petrographycal composition. As already noticed by previous studies [6,17], the 1877 PDC deposits display a clear decimeter- to meter-thick relief with broad and steep lateral levees and tongue-like frontal tips, which make them very visible in the landscape (Figure 6). Moreover, the PDC deposits, at least in surface, have a strong tendency to be mono-lithologic, displaying almost exclusively the decimeter- to meter-sized cauliflower-shaped black bombs. To the contrary, the primary lahar deposits display modest to low relief, with the lateral contacts usually represented by a trimline of scattered cauliflower-shaped black bombs, displaying an evident poly-lithologic composition on the surface (Figure 6).

Figure 6. A PDC deposit tops a primary lahar deposit from the 1877 eruption, as seen at Quebrada Chilcahuaycu in the Sindipamba plain (Figure 2). Notice the vegetation contrast between the lahar deposits from 1877 and from an older eruption, and the trim line formed by the specific 1877 black bombs at the lateral contact.

3. Methodologies

3.1. Geological Fieldwork

Field surveys aimed at identifying the lateral contacts and outcrops of the 1877 primary lahar and PDC deposits, were based on the criteria described above. For this, each plain of the study area was transected by foot along several paths, where two types of control points were recorded with a hand GPS-receiver: contacts and outcrops. Contact-control points were places where a clear distinction between two different deposits was possible, with one of those deposits belonging to the 1877 eruption. Most commonly, such contacts were characterized by displaying clear trim lines formed by accumulations of the black cauliflower-shaped bombs (Figures 5 and 6). Additionally, in some cases when the contacts were clearly visible and when it was judged useful, track-lines were also recorded with the GPS receiver along lateral contacts of the 1877 deposits.

On the other hand, outcrop-control points were locations where the 1877-specific black bomb (Figure 5) was found on the surface, but no contact could be identified in the field. These kind of locations were always characterized by displaying little vegetation and varying amounts of the 1877 black bombs. Most of the outcrop-control points were recorded in locations where such bombs were scarcely found in the surface.

3.2. Drone Surveying—Image Acquisition

The obtention of high-resolution imagery was central to the purpose of the present study. Given the size of the study area, this task was performed using a fixed-wing drone [49]. The device employed was a SenseFly Classic eBee drone, equipped with a Canon S110 NIR camera producing 12 Mpix frames (Table 1). The main characteristic of this sensor is the acquisition of a near-infrared (NIR) band, centered at 850 nm, in addition to the visible red and green bands. The use of this sensor was prioritized because the NIR band is known to be useful for identifying changes in the type and amount of vegetation in the surveyed areas [50–52], which is one of the key surface features that differentiates the 1877 deposits from others, as mentioned above.

Table 1. Technical features of the sensor used for image acquisition.

Resolution	12 Mpix
Ground resolution at 100 m	3.5 cm (pixel)
Sensor size	7.44 × 5.58 mm
Output formats	JPEG and/or RAW
Band centers	NIR 850 nm, Red 625 nm, Green 550 nm

The drone also bears an internal GPS receiver which inputs a geotag to each of the frames obtained by the sensor during acquisition flights. This information is later used during image processing with SfM to provide a geoposition for the images. However, given the size of the study area and the target scale for the cartography (1:5000), no ground control points were used to produce an absolute georeferencing for the imagery. Instead, as described further below, an accuracy test was performed in the processed images before using them for cartography.

The drone surveys were planned using the proprietary software eMotion 3 (which also pilots the eBee drone while flying) by creating mission blocks covering the study area and establishing an overlap of 60–70% between the individual frames in the lateral and longitudinal direction. The flights were performed at an average altitude of 160 m above the ground level, from where the sensor obtained individual frames with an average ground-resolution of ~5 cm/pixel. More than one drone flight was usually required to complete a mission block. The weather conditions during the flights were very variable, with temperatures ranging from 5 °C to 20 °C, and average wind speeds from 2.1 m/s to 9.7 m/s (flying with average windspeeds higher than 10 m/s is not recommended by the drone manufacturer). As a result, reflectance measurements by the drone sensor were affected by variable sunlight conditions between and during flights.

3.3. Image Processing and Cartography

The groups of individual frames obtained for each mission block were processed to obtain orthomosaic images, by the use of commercial software based on the Structure from Motion (SfM) algorithms and methodologies [31,53]. Such orthomosaics had three spectral bands (i.e., NIR, red and green) and ground resolutions ranging from 5 to 7 cm/pixel, but all were resampled to a 25 cm/pixel ground resolution for additional processing prior to contact identification and mapping.

As mentioned above, one key feature of the 1877 deposits is their vegetation composition, which is usually different from other deposits outcropping in the study area (Figures 4 and 6) [48]. Then, the Normal Difference Vegetation Index (NDVI) was applied to the orthomosaics with the intent to highlight the regions covered with the 1877 deposits from others [51,52,54]. The NDVI value was obtained with the following equation:

$$\mathrm{NDVI} = \frac{NIR - R}{NIR + R} \tag{1}$$

where *NIR* is the pixel value of the near-infrared band and *R* is the pixel value of the red band. The results are new single-band raster layers with pixel values between −1.0 and 1.0 which could be interpreted as a proxy for vegetation composition [54]. The obtained imagery showed that in general the lower NDVI values represent zones dominated by rock or bare soil, while the higher values represent bushes, grassland, and thicker vegetation. For mapping purposes, the final screen displays of the NDVI images were enhanced with additional contrast and both the maximum and minimum values of the layers were sometimes cut off using their respective cumulative counts. The whole drone-NDVI imagery is provided as supplementary material to this study.

Most of the mission blocks were planned to cover juxtaposed areas in order to assess the effect of the lack of ground controlled georeferencing in the cartographic quality of the orthomosaic images. After comparing the apparent location of point objects present in two adjacent images, it was observed that the differences between the objects were consistently lower than 2 m (Figure 7a,b). This means that the orthomosaic images may be used for cartography with relative georeferencing down to scales of ~1:2500. In addition, the orthomosaics were compared in the same way with freely available Google Earth imagery in order to assess the absolute georeferencing [55]. In this case, the differences raised to up to 4.5 m, implying that cartography at 1:5000 scale was still possible without an absolute georeferencing from the orthomosaics (Figure 7c).

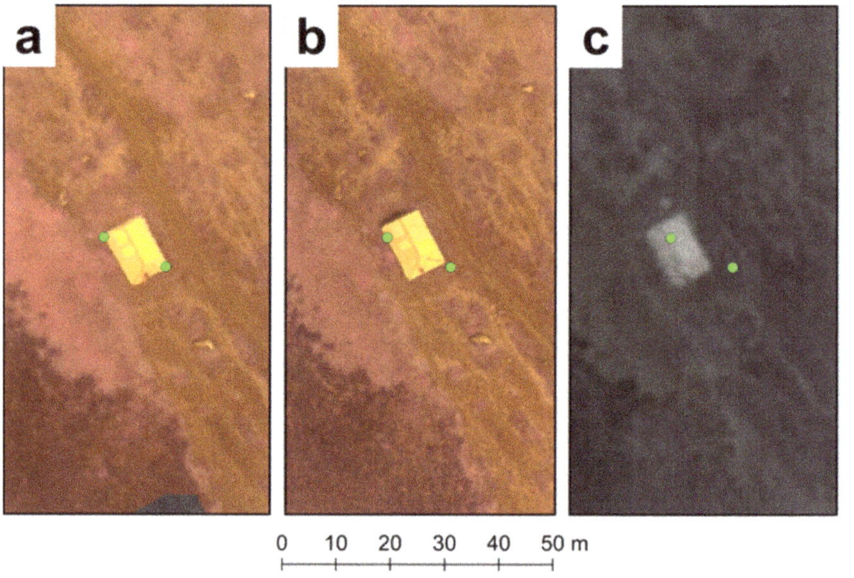

Figure 7. Assessment of geolocation obtained with the processed drone images (orthomosaics). The vector features (green circles) corresponding to opposite vertices of a rectangular ceiling as observed in one mission block (**a**), are overlayered on an adjacent overlapping mission block (**b**), and on the Google Earth imagery (**c**). When cartography is performed at 1:5000 scale, the differences among the imagery become negligible.

Given the constantly changing weather and sunlight conditions during image acquisition, the use of supervised classification approaches to extract the regions of interest from the imagery was impractical. Then, the final vector layers representing the 1877 deposits were elaborated by hand, through correlation and interpretation of the field data together with the drone-NDVI imagery. In some specific regions where drone imagery could not be obtained, the field data was interpreted using the freely available satellite images provided by the Google Earth server. Thus, the hand mapped contacts from the 1877 deposits were classified in four categories:

1. Identified in the image, when the contact was drawn from the NDVI imagery in correlation with field data;
2. Observed in the field, when the contact was obtained with a hand GPS track;
3. Identified in satellite, when the contact was drawn from Google Earth imagery in correlation with field data;
4. Inferred, when the contact was not identifiable in the images (NDVI or Google Earth) and was drawn by informally extrapolating and interpreting field data.

Consequently, the contacts belonging to the first three categories may be considered as most reliable, while those inferred are considered as less reliable. As a final product, two geospatial vector layers at 1:5000 scale were created from the mapped contacts. One layer contains line features, where information about the contact categories defined above was included as an attribute value. The second layer contains polygone-features, which represented a merger of the previous layer lines, with an attribute defining the deposit type (i.e., primary lahar or PDC) as a value. Both vector layers are provided as supplementary material to this study.

4. Results

During geological field surveys, a total of 165 control points were recorded in the whole study area, from which 105 corresponded to lateral contacts and 33 to outcrops of the 1877 deposits, while 27 corresponded to outcrops belonging to deposits from other eruptions (Table 2). Additionally, a total of 38 GPS track lines representing lateral contacts of the 1877 deposits were also acquired, with a total length of 10.96 km.

Table 2. Summary of the cartographic data acquired during geological field surveys of the study area. Locations are shown in Figures 8, 10, 12 and 14.

Plain	Control Points			Tracks	
	1877 Contact	1877 Outcrop	Other Deposit Outcrop	Number of Tracks	Total Track Length (km)
Sindipamba	12	15	6	5	2.33
Víctor Punina	10	6	2	17	2.96
North-eastern	57	1	14	9	2.81
Eastern	26	11	5	7	2.55

For the drone surveys, the study area was divided into 21 mission blocks, covering a total surface of 28.4 km^2. To cover this area, 62 drone flights were required with a total flight time of 18 h 12 m, during which a total of 8973 individual frames were obtained (Table 3).

Table 3. Summary of the drone surveys performed in the study area.

Plain	Number of Mission Blocks	Number of Flights	Frames Obtained	Area (km^2)	Duration (h)	Average Altitude (m above Ground)
Sindipamba	7	27	3291	8.32	7.25	146
Víctor Punina	3	7	1103	3.36	2.13	168
North-eastern	5	19	2420	10.29	5.4	156
Eastern	6	9	2159	6.4	3.42	203

The processing of the 21 mission blocks through SfM was performed by means of a commercial software and yielded RMS reprojection errors ranging between 0.54 and 1.21 pixels, for ground resolutions ranging from 5 to 7 cm/pixel. The number of tie points identified in the mission blocks for the initial image alignment varied from ~217 to ~697 thousand, and the processed dense clouds contained between ~88 and 473 million points. Thus, 21 orthomosaic images were created from the mission blocks covered by the drone surveys, containing NIR, red and green bands, and were all resampled to ground resolutions of 25 cm/pixel. Consequently, 21 single-band images with NDVI values were also obtained and were the main supply to perform the cartography together with the geological field data. Following, the cartographic results are presented for each plain.

4.1. Sindipamba Plain

Geological fieldwork in this plain was characterized by a relative difficulty in differentiating the 1877 deposits from those belonging to other preceding eruptions of Cotopaxi. Firstly, the 1877 lahar deposits in the Sindipamba plain contained few of the characteristic black cauliflower-shaped bombs, and consequently the field survey required special care for their identification (Figure 5). Secondly, deposits from other primary lahars with scarce vegetation covering, that occurred shortly before the 1877 eruption (i.e., 1854), were widespread in the plain, as already noticed by Pistolesi et al. [17]. Thus, from 33 control points obtained in this plain, 15 corresponded to the 1877 outcrops (i.e., locations where the 1877 black bomb was found in the surface) and only 12 to 1877 contacts (i.e., locations where the 1877 lateral limit could be identified), while 5 tracks of lateral contacts covering 2.3 km could be recorded (Figure 8).

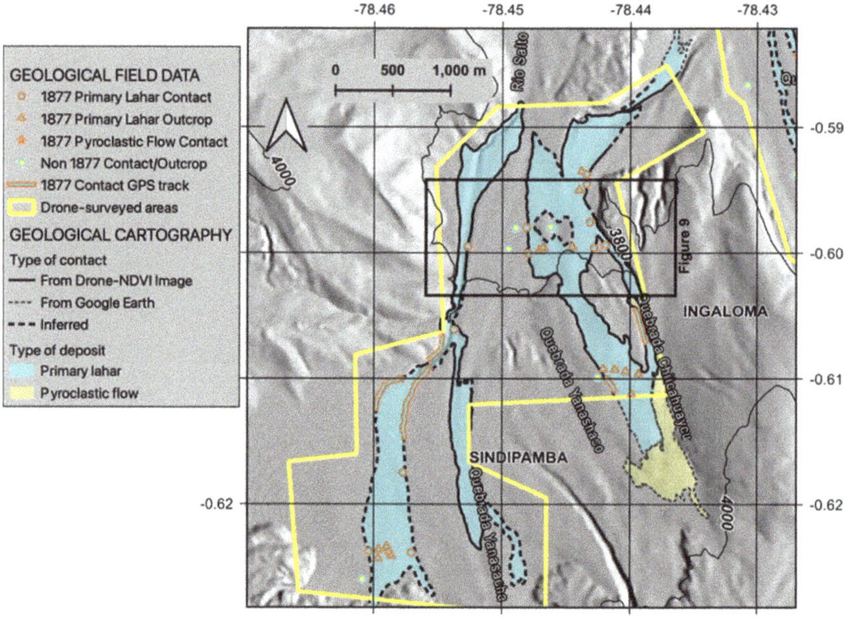

Figure 8. Results of cartography performed in the Sindipamba plain (see Figure 2 for location).

The difficulty in identifying lateral contacts of the 1877 deposits persisted in the NDVI images, given the presence of other deposits with similar vegetation covering, which produced reduced contrast in the images. Thus, significant interpretation and inference was necessary during contact drawing (Figures 8 and 9). A total of 39.84 km of 1877 lateral contacts were mapped in this plain, from which 10.72 km (27%) were inferred and 26.8 km (67.3%) were identified either in the NDVI or in the Google Earth images (Table 4).

Table 4. Summary of the contact length by types, obtained during the mapping performed for the 1877 deposits. All figures are in meters.

Plain	Drone Image	Satellite	Inferred	Field (GPS Track)	Total
Sindipamba	19,198	7595	10,720	2330	39,843
Víctor Punina	10,068	5288	11,597	2957	29,910
North-eastern	81,980	8677	397	2805	93,859
Eastern	25,646	15,488	3412	2554	47,100

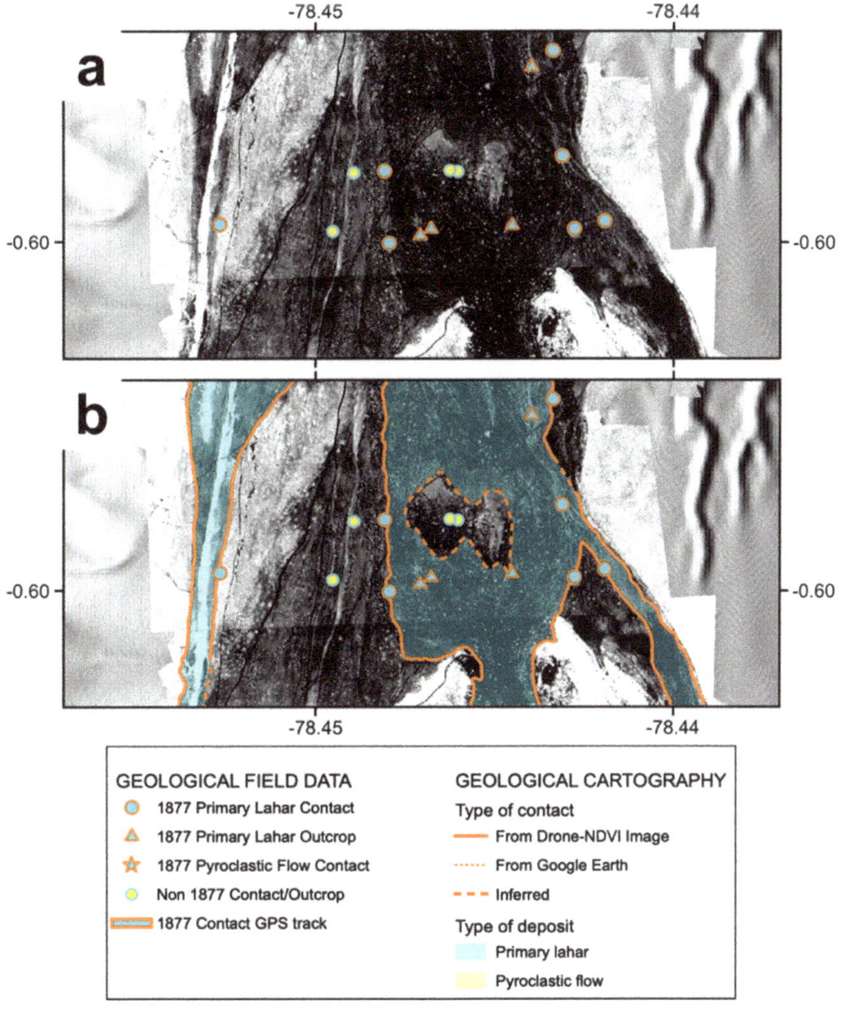

Figure 9. Detail of (**a**) geological control points observed in the field superposed on the drone-NDVI imagery, as compared with (**b**) the mapped lateral contacts in a segment of the Sindipamba plain. See Figure 8 for location.

The results of carthography show that the 1877 primary lahar flowed through three of the four main drainages of the Sindipamba plain. The eastern Chilcahuaycu drainage perhaps channelized the largest volumes and discharge rates, and fed flows into both the Rio Salto northwards and the Rio Pita northeastwards (Figure 8). Conversely, the Yanasacha and the unnamed (to the west) drainages diverted apparent smaller volumes directly to the Rio Salto. A pyroclastic fow deposit has been mapped in the upper Chilcahuaycu drainage, from both drone-NDVI and Google Earth imagery (Figures 6 and 8). The field observations show that the PDC deposit overlies the primary lahar deposit.

4.2. Victor Punina Plain

Similarly to the Sindipamba, the geological field survey in the Victor Punina plain was complicated due to the low abundance of the specific black bombs in the 1877 deposits and to the presence of other recent deposits with poor vegetation covering. Only 10 contact and

6 outcrop-control points could be obtained in the plain, although 17 tracks covering 2.96 km of 1877 lateral contacts could be recorded, most of them outside the drone-surveyed area, in the upper stream above the 3800 m level (Table 2, Figure 10).

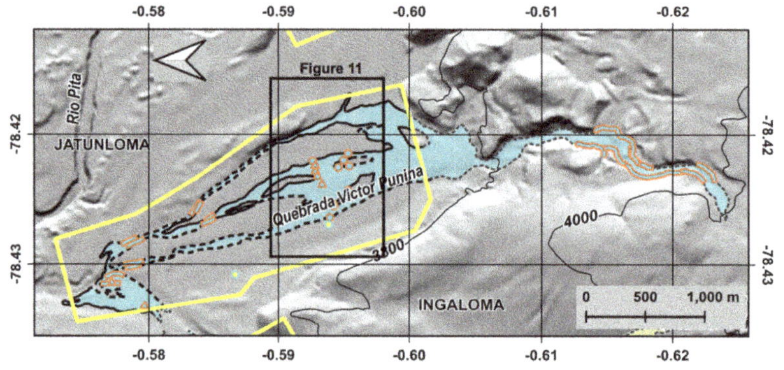

Figure 10. Results of cartography performed in the Victor Punina plain (see Figure 2 for location). The map has been rotated counterclockwise (north is pointing to the left) to optimize space. Map legend is the same shown in Figure 8.

Contact cartography thus also required significant interpretation in the Victor Punina plain where differences between deposits in the drone-NDVI images were slight due to the lack of contrast (Figure 11). A total of 29.91 km length of contacts were mapped, from which 11.6 km (38.8%) were inferred and 10.07 km (33.7%) were considered as identifiable from the drone-NDVI images (Table 4). Moreover, this plain is characterized by a significant amount of contacts mapped from the Google Earth imagery, with a total of 5.29 km (17.7%) (Figure 10). In this plain it was evident that even in close locations, there might be significant changes in the vegetal covering of the 1877 deposits, as is observed in the eastern zone of the Victor Punina plain where the drone-NDVI imagery shows lighter gray hues (Figure 11).

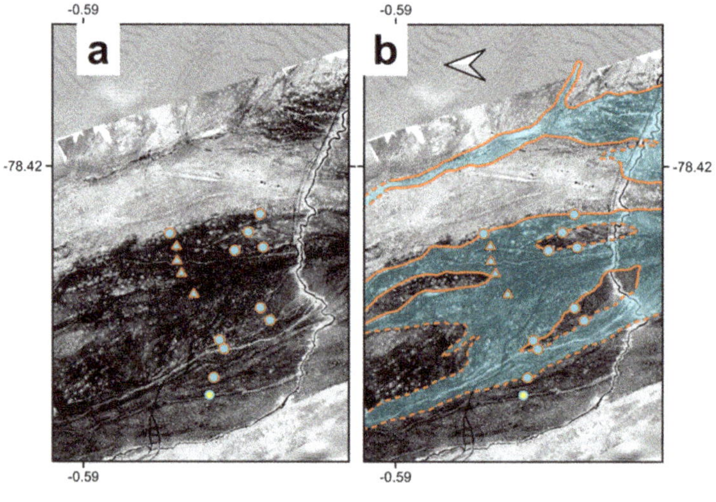

Figure 11. Detail of (**a**) geological control points observed in the field superposed on the drone-NDVI imagery, as compared with (**b**) the mapped lateral contacts, in a segment of the Victor Punia plain. See Figure 10 for location. The map has been rotated clockwise (north is pointing to the right) in order to fit with Figure 10. Map legend is the same as in Figure 9.

The cartography obtained in the Victor Punina plain shows that the 1877 flow was channelized in the deep valley of the upper drainage, between the 3800 and the 4000 m level (Figure 10). Upon arrival to the flatter and wider valley floor, the lahar experienced significant deposition and formed a fan, after which three narrow ravines contained the remaining flow towards the conjunction with the Rio Pita northwards (Figure 10). Given the amount of field data acquired and the difficulty in identifying contacts in the imagery, the cartography in the Victor Punina plain is considered to have the most uncertainty.

Similar to the case of the Sindipamba plain (Figure 9), other poorly vegetated deposits can be observed around the 1877 deposits in the Victor Punina plain (Figure 11) in the field and in the NDVI imagery. These deposits were probably emplaced shortly before 1877 and belong to the 1854 eruption of Cotopaxi volcano, as also suggested in [17].

4.3. North-Eastern Plain

The North-eastern plain was generally characterized by the occurrence of clearly distinguishable 1877 deposits. Throughout the plain, the deposits abundantly contained the specific 1877 black bombs, while displaying the least amount of vegetation covering with respect to the neighboring deposits, facilitating the identification of lateral contacts (Figures 4 and 12). Thus, a total of 57 contact and only 1 outcrop control points were recorded for the 1877 deposits, while 14 outcrops of other deposits were also identified (Table 2, Figure 12). Additionally, nine GPS tracks were recorded covering a total 2.81 km of lateral contacts.

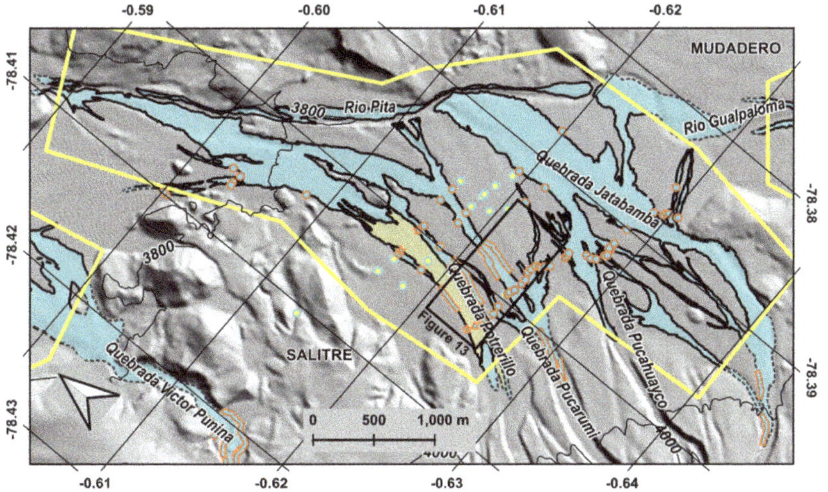

Figure 12. Results of cartography performed in the North-eastern plain (see Figure 3 for location). The map has been rotated counterclockwise (north is pointing to the left) in order to optimize space. Map legend is the same shown in Figure 8.

Identification and cartography of the 1877 deposits present in the North-eastern plain was also relatively simple in the drone-NDVI images, given their significant contrast and the clear correlations with the geological field data (Figure 13). A total of 93.86 km length contacts were mapped in the plain, from which 81.98 km (87.3%) were considered as identifiable and only 0.4 km (0.4%) were inferred in the NDVI imagery (Table 4). Additionally, 8.68 km were obtained from the Google Earth imagery, mainly in the upper flanks, close to the 4000 m level (Figure 12).

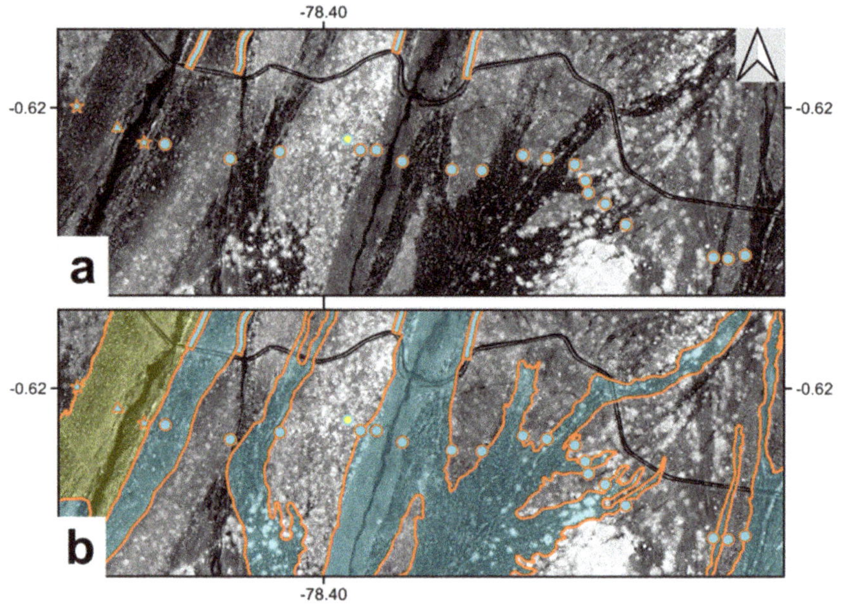

Figure 13. Detail of (**a**) geological control points observed in the field superposed on the drone-NDVI imagery, as compared with (**b**) the mapped lateral contacts, in a segment of the North-eastern plain. See Figure 12 for location. Map legend is the same shown in Figure 9.

The cartography in the North-eastern plain suggests that the total volume of 1877 primary lahars was significantly larger than in the previous plains, and certainly richer in juvenile clasts (specific black bombs). They descended through the Potrerillo, Pucarumi, Pucahuayco and Jatabamba drainages, all of which display wide and deep canyons in the middle and upper NE flanks of Cotopaxi. Once in the flatter zones, the flows displayed a complex behavior, producing an intricate network of contacts, specially between the Potrerillo and the Jatabamba drainages (Figures 10 and 11). Though significant deposition must have occurred in the plain, large volumes of the flows seem to have reached the Rio Pita, given the long intersections between the mapped deposits and the river course. The NDVI images also show the presence of other lahar deposits with changing amounts of vegetation (pale gray shades) surrounding the 1877 deposits (Figure 13).

Finally, a PDC deposit which descended the Potrerillo drainage was also observed in the plain, which was composed almost exclusively of the specific 1877 black bombs. From field observations, this deposit is unequivocally overlying the 1877 primary lahar deposit, which implies explosive activity at Cotopaxi after the June 26 eruption. This is in agreement with the report by Wolf [1], who mentions that the volcano produced loud explosions for several weeks after the main event.

4.4. Eastern Plain

Similarly to the North-eastern plain, the Eastern plain was characterized by the widespread occurrence of the 1877 deposits, with abundant presence of the specific black bombs and clear differences in vegetation covering with respect to the neighboring deposits. Therefore, also in this case the identification of lateral contacts was relatively simple in the field. A total of 26 contact- and 11 outcrop-control points for the 1877 deposits were obtained, together with 5 outcrops of older deposits in the plain (Figure 14, Table 2). Additionally, seven GPS tracks recorded lateral contacts in the Eastern plain for a total distance of 2.55 km.

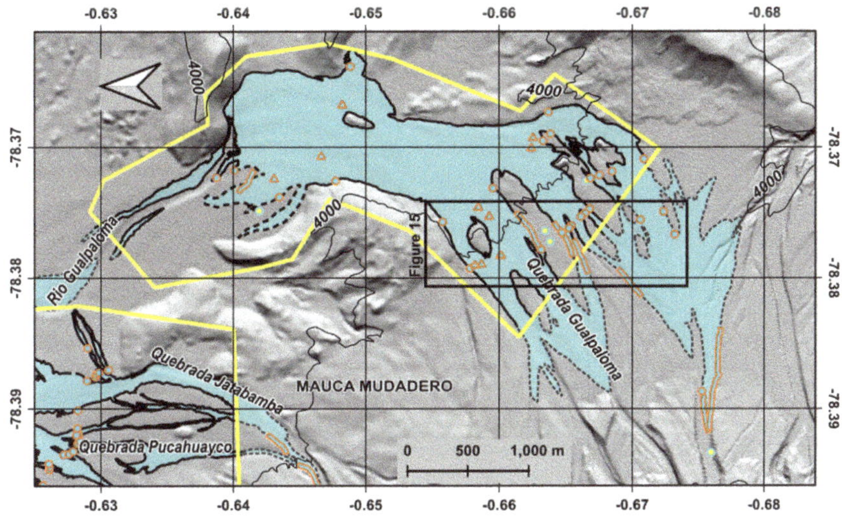

Figure 14. Results of cartography performed in the Eastern plain (see Figure 3 for location). The map has been rotated counterclockwise (north is pointing to the left) in order to optimize space. Map legend is the same shown in Figure 8.

The identification of the 1877 contacts in the drone-NDVI images was facilitated by the clear contrasts produced with other deposits and sharp correlations with the field data (Figure 15). A total of 47.1 km of contacts were cartographed in the Eastern plain, from which 25.65 km (54.5%) were considered identifiable and 3.41 km (7.2%) inferred from the NDVI imagery, while 15.49 km (32.9%) were obtained from Google Earth, mainly corresponding to the upper zones of the plain (Figures 14 and 15).

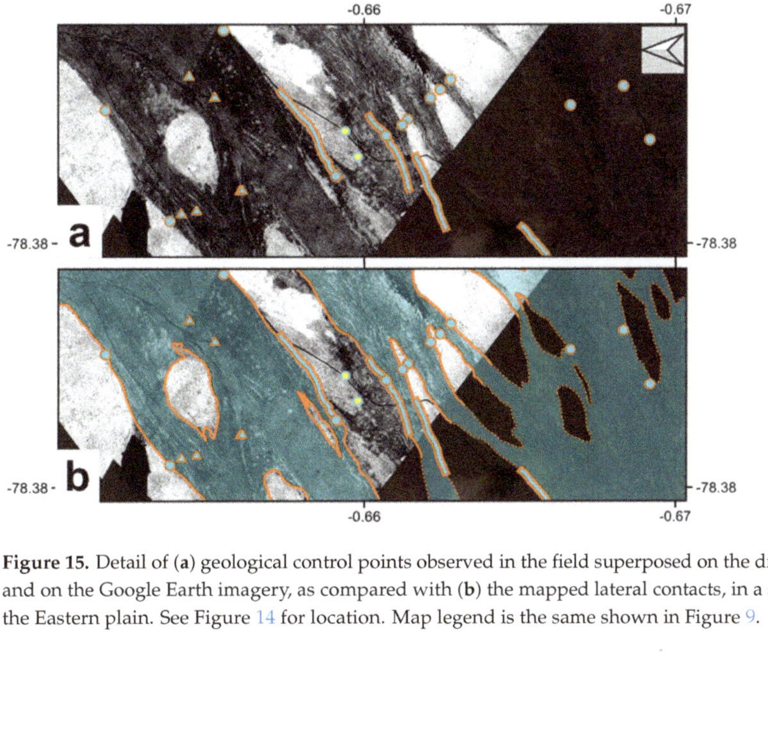

Figure 15. Detail of (**a**) geological control points observed in the field superposed on the drone-NDVI and on the Google Earth imagery, as compared with (**b**) the mapped lateral contacts, in a segment of the Eastern plain. See Figure 14 for location. Map legend is the same shown in Figure 9.

The results show that the 1877 primary lahar flows in the Eastern plain were possibly the largest of all in the study area and contained abundant specific black bombs. This suggests that, similar to the North-Eastern plain, the process of glacier melting by contact with PDCs during the eruption was somehow more widespread or more efficient than in the Sindipamba and Victor Punina plains to the north. Five deep and wide drainages channelized the flows descending the eastern higher flank of Cotopaxi, which eventually coalesced and flooded nearly the whole Eastern plain, before diverting most of the released volume into the Rio Gualpaloma which are the headwaters of Rio Pita (Figure 14). In the southernmost Eastern plain, a part of the primary lahar's discharge indeed flowed into the Rio Tamboyacu drainage, which runs towards the Amazonian basin (Figures 1 and 14). No evidence of PDC deposits belonging to the 1877 eruption were identified in this plain.

5. Discussion

5.1. Level of Detail and Inference during Contact Cartography

The present work has been an attempt to produce a detailed geological mapping of the primary lahars that occurred in the 26 June 1877 eruption of the Cotopaxi volcano. This means that the mapped contacts represent the zones where those deposits can currently be found in surface, at 1:5000 scale. However, it is important to recognize the various limitations and constraints of the cartography performed here.

One of the main flaws of the cartography presented here is that it ignores the actual existence of relatively extensive alluvial deposits corresponding to the erosion that has occurred at Cotopaxi since 1877, which partially cover the primary lahar. These alluvial deposits outcrop extensively in the study area, usually displaying light reddish hues and forming either fans or intricate branch systems, with sizes ranging from a few to tens of meters wide. In some places, deposition is actively going on, while in others it seems to be paused or have stopped completely. The identification of these deposits is easy in the field, but may be very challenging in the drone-imagery in the case of the thinner branches. Due to these difficulties, an initial attempt to map these deposits and extract them from the 1877 deposits was abandoned.

It is important to consider that the cartography presented here only partially represents the original deposits. Specially for the case of the primary lahars, the mapped contacts should not be considered as the originally flooded regions, but only the places where enough deposits have remained in place. For lahar flows, it is well known that the inundation extension is always larger than the observable deposits, even if the study area is relatively flat. Additionally, as mentioned above, the post-1877 deposition and erosion have modified the original deposit. Finally, during fieldwork it was evident that the 1877 deposits were always the least vegetated of any specific place, but the amount of vegetation was not constant all around the volcano. Thus, in some places, vegetation can be expected to have completely covered an originally thinner or finer grained lahar deposit.

5.2. Comparison with Previous Works

Though several investigations have been published about the 1877 primary lahars of Cotopaxi, only a few have presented geological maps of the deposits. The earliest of those is the report prepared by Theodore Wolf [1], who extensively visited the volcano shortly after the eruption, in early September, 1877. This report includes a detailed recount of the events that occurred right before and during the June 26 eruption, a description of the observations he made during the field visit, which included an ascension to the volcano summit, and surprisingly wise interpretations about the eruption and the processes that led to the formation of the primary lahars. A map of Cotopaxi is presented at the end of the report, where the main geomorphologic features of the volcano are displayed together with a cartography of the "water and mud avenues" which is indeed the way the author refers to the primary lahars in the text. The segment of that map covering the proximal northern drainage of Cotopaxi is shown in Figure 16a, together with the results of the present study in Figure 16b.

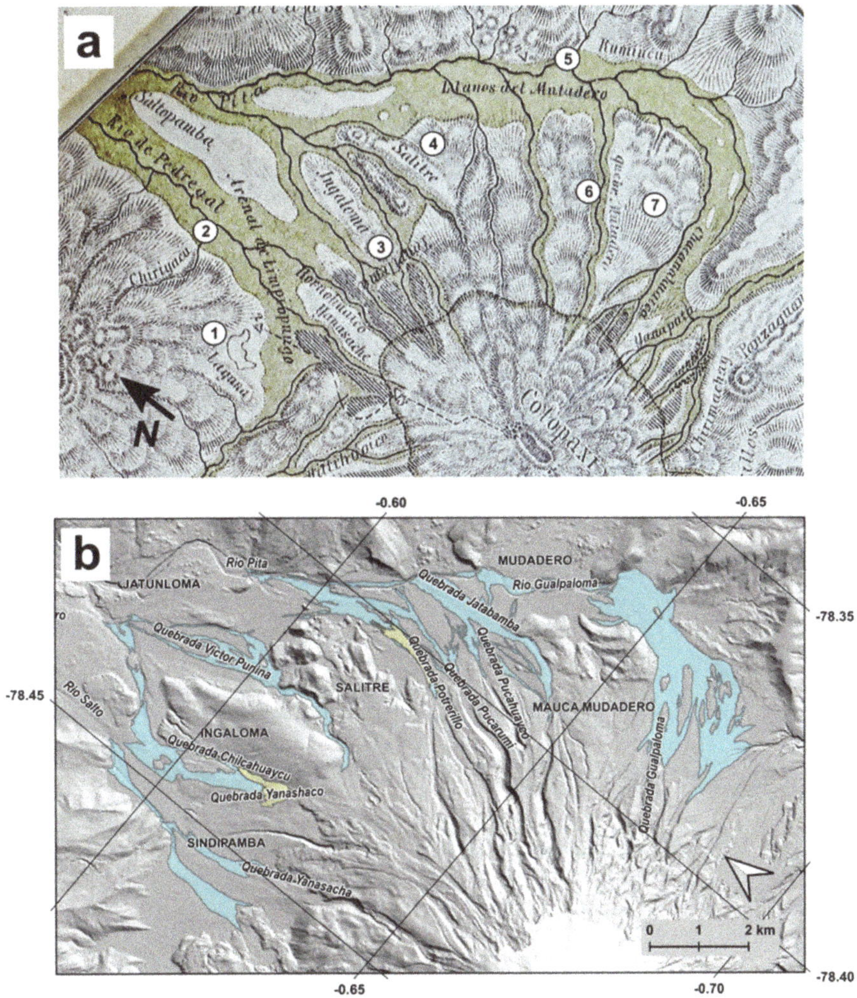

Figure 16. Comparison between the geological map published by Theodore Wolf [1] (**a**) and that obtained in the present study (**b**). Numbers in (**a**) denote changes in names or highlight localities for an improved comparison with (**b**). 1: Laguna Limpiopungo, 2: Rio Salto, 3: Ingaloma, 4: Salitre, 5: Mudadero, 6: Quebrada Jatabamba, 7: Mauca Mudadero. Map legend for deposits in (**b**) is the same shown in Figure 8. North arrow in (**a**) is approximate.

The cartography obtained in the present study is in significant agreement with Wolf's map, and although several location and drainage names are different, it is still possible to make a direct comparison between both maps. From Limpiopungo lake, eastwards until Ingaloma, both maps show three main lahar branches joining into the Rio Salto (*Rio de Pedregal* in Wolf's map). Just east of Ingaloma, both maps show the Victor Punina drainage and plain, but Wolf's map includes an additional lahar branch at Salitre where the present study could not find deposit evidence. East and South from Salitre, Wolf's map shows the whole North-eastern plain as covered by lahar deposits descending from only two drainages, the southernmost corresponding to the Quebrada Jatabamba (*Quebrada Mutadero* in Wolf's map) and the other most probably to Pucarumi. Interestingly, Wolf's map shows an unnamed drainage equivalent to Quebrada Potrerillo as not channeling the

1877 deposits, although most modern studies coincide to show primary lahar and/or PDC in that drainage [6,17,20,48]. Finally, south and east from Mauca Mudadero, Wolf's map shows a fully flooded valley corresponding to the Eastern plain in the present study.

A more recent cartographic effort has been performed by Hall and Mothes [20]. This survey was performed at a 1:50,000 scale, covering the whole volcano spatially and stratigraphically, and displays a map of the 1877 deposits. Figure 17 shows an overlying of the lahar deposits in that map with the results here, only for the purpose to highlight the level of improvement achieved. Indeed, it is difficult to compare both maps, given the differences in scale and base information used to obtain each one; i.e., it can be noticed that the topographic base used by Hall and Mothes [20] does not match the one available for the present study.

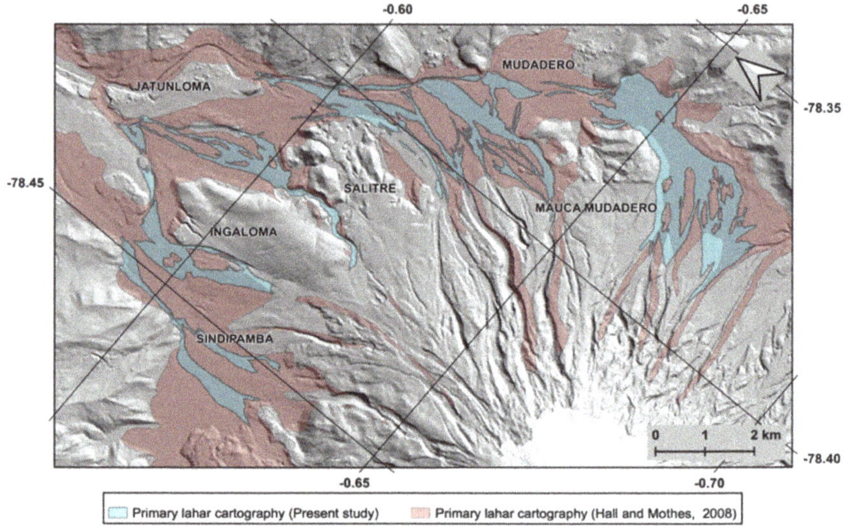

Figure 17. Comparison between the lahar deposits mapped in the present study and those presented by Hall and Mothes [20].

Finally, Pistolesi et al. [17] present a detailed geological map coincident with the area of the present study, based on an extensive stratigraphic study of the Cotopaxi primary lahars that have occurred during the last eight centuries. Unfortunately, the cartographical data (geospatial layers) are not available from the paper publisher, and a digitalization of the published map was unpractical due to the figure size and the lack of a coordinates grid in the map. Nonetheless, significant coincidences and differences may be highlighted with respect to the present results.

In general, the Pistolesi et al. [17] map displays the 1877 deposits outcropping in all the same four plains as the results here. The contacts mapped in the Sindipamba and Victor Punina plain seem generally coincident in distribution and size with respect to Figures 8 and 10, although the lahar branch of the Quebrada Chilcahuaycu is lacking in the former. However, clear differences arise in the North-eastern and Eastern plains. In both cases, the 1877 outcrops mapped by Pistolesi et al. [17] appear in much smaller zones with respect to Figures 12 and 14. In the North-eastern zone, only the PDC deposits at Quebrada Poterillo and a very small primary lahar branch of the Quebrada Jatabamba are mapped, while in the Eastern plain only the western half of the valley appears to contain the 1877 primary lahar deposits. This lack of deposits in the North-eastern and Eastern plains in the Pistolesi et al. [17] map, and is also in contradiction with the direct observations made by Wolf shortly after the eruption (Figure 16a).

The Pistolesi et al. map also shows that all the plains are largely covered by deposits belonging to older Cotopaxi eruptions. This agrees with observations made in the present study even if the cartography is not coincident. Older deposits could be observed both in the field and in the drone-NDVI imagery, some of them with apparent lateral extensions suggesting much larger volumes than the 1877 deposit, especially in the Sindipamba, Victor Punina and North-eastern plains. A detailed cartography of those deposits using the drone-NDVI imagery could also be possible if specific facies criteria were available to constrain them in the field, i.e., the case of the fragile cauliflower-shaped black-bombs present in the 1877 deposits.

5.3. Future Lahar Hazard Assessment and Modelling

The main findings of the present study should be taken into account when lahar hazard at Cotopaxi is re-assessed with improved numerical models in the future. Two of the most recent and developed numerical approaches tested for lahar hazard assessment at Cotopaxi are based on: the numerical solution of fluid-dynamics partial differential equations that represent lahars by a Voellmy-Salm reology (RAMMS software) [29]; and, the use of a multicomponent Cellular Automata of semiempirical models of surface flows (LLUMPIY model) [30]. Both are 3D models capable of simulating phenomena such as erosion and deposition of the moving flow, as well as providing fundamental quantitative information (i.e., for further risk assessment) such as basal, lateral and impact pressure, or flow height and average velocity at any point of a high-resolution drainage topography. In those studies, RAMMS and LLUMPIY have produced simulations starting in the higher Cotopaxi flanks, with the latter being even capable of simulating contrasting rates of water release from the glacier during the eruption. Finally, and as usual, in both studies the models were calibrated using geological information related to the 1877 eruption and primary lahars.

Thus, in the simplest approach, the results of the present study could be useful to improve the calibration of new generation models such as RAMMS or LLUMPIY [29,30]. This means that the adjustment of parameters governing any 1877 model should be such that the produced simulations are at least somewhat similar to the cartography presented here, considering that the current day topography might be significantly different from that prior to 1877 and, as already discussed, the cartography presented here only represents the current day lithic deposit left since the original event. In this comparative approach, the calibration simulations produced by RAMMS [29] seem closer to the results obtained here, although they ignore a significant part of the flows that descended the Eastern plain, which certainly has an impact on the final results.

Conversely, the 1877 calibration simulations obtained with LLUMPIY [30] are significantly different from the cartography of this study. One of the main reasons for such differences may be related to another finding highlighted here. The present survey has shown that the 1877 primary lahar deposits observed in the Sindipamba and Victor Punina plains were clearly depleted in the 1877-specific black bombs and were, in general, smaller in size with respect to those observed in the North-Eastern and Eastern plains. This suggests that important differences existed between the PDCs that melted the glacier in the North flank with respect to those in the East flank of Cotopaxi. Such differences are explained by the topography of the summit crater, whose northern rim is ~120 m higher than its eastern rim, which produced that only the more diluted bomb-depleted PDCs descended northwards, while the denser bomb-rich PDCs descended towards the East. However, one of the main assumptions made in [30] during LLUMPIY calibrations, is that PDCs (and consequently primary lahars) are evenly distributed all around the crater rim, which ultimately produces the differences between the calibration simulations and the cartography obtained here.

6. Conclusions

A detailed (1:5000 scale) geological cartography has been performed for the deposits belonging to primary lahars occurred at Cotopaxi volcano on 26 June 1877, as observed in the Sindipamba, Victor Punina, North-Eastern and Eastern plains, which compose the northern proximal drainage. The final cartography was possible through a combination of geological fieldwork with the analysis and interpretation of high-definition imagery obtained by drone surveys. During fieldwork, contacts and outcrops of the 1877 deposits were recorded in locations where a very specific cauliflower-shaped black bomb was identified. On the other hand, the drone imagery included near-infrared, red and green bands, which allowed obtaining 25 cm/pixel NDVI imagery with the help of SfM procedure, covering a total surface of 28.4 km^2, where the 1877 deposits were identified and cartographed at 1:5000 scale. Additionally, Google Earth imagery was used in some regions where the drone surveys could not be performed.

From those data, a total of 210.63 km length of contacts was drawn for the 1877 deposits, from which 136.84 km were obtained from the drone-NDVI imagery, 37.02 km from the Google Earth imagery, 10.64 km were GPS field tracks and 26.13 km were classed as inferred. The results show that the primary lahars that occurred in the Sindipamba and Victor Punina plains were clearly depleted in the specific 1877 black bomb and smaller in size with respect to those that flooded the North-Eastern and Eastern plains. This suggests that significant differences existed between the PDCs that melted the glacier in the North flank with respect to the East flank, during the formation of primary lahars which, in turn, might be a consequence of Cotopaxi's summit crater topography.

The obtained cartography is in general agreement with respect to previous studies, but it represents a significant advance in the level of detail achieved and in the methodological approach, which can be applied in other volcanic areas. However, it is important to mention that the cartography represents the currently identifiable outcrops, which are expected to be different from the originally flooded areas, due to the action of erosion/deposition processes and vegetation development that have occurred at Cotopaxi since 1877.

Finally, next generation lahar hazard assessments at Cotopaxi volcano might benefit from the results from this study. Specifically, the calibration of developed numerical models could be improved by either comparative approaches to the cartography presented here, or by the new insight obtained with respect to the 1877 eruption.

Author Contributions: Conceptualization, methodology, field-work, data acquisition and analysis and original draft preparation, S.D.A.; field-work, data acquisition, curation and analysis, E.S., S.C. and V.N.; resources and data acquisition G.L. and J.B. All authors have read and agreed to the published version of the manuscript.

Funding: E.S and S.C. were funded by Escuela Politécnica Nacional (Ecuador) through the grant number PIMI 1608 (for S.D.A.). Additionally, V.N. was funded by the UK United Kingdom Research and Innovation (UKRI) Collective Fund using Official Development Assistance (ODA) funding through the UKRI GCRF: Urban Disaster Risk Hub—Tomorrow's Cities (UKRI grant number NE/S00900-1).

Data Availability Statement: The following datasets are freely available online at https://zenodo.org/deposit/5866730 (accessed on 17 November 2021) (doi:10.5281/zenodo.5866730): (1) Drone-NDVI imagery at 25cm/pixel for the four surveyed plains; (2) Geospatial vector layer of the geological fieldwork control points; (3) Geospatial vector layer of the cartographed 1877 deposit contact lines; (4) Geospatial vector layer for the cartographed 1877 deposit polygons.

Acknowledgments: The authors would like to thank Elisa Sevilla and Karina Barragán for the high-quality digital historical information provided. Also, the invaluable support during geological and drone fieldwork provided by staff at Instituto Geofísico—Escuela Politécnica Nacional is deeply acknowledged.

Conflicts of Interest: The authors declare no conflict of interest. The funders had no role in the design of the study; in the collection, analyses, or interpretation of data; in the writing of the manuscript, or in the decision to publish the results.

References

1. Wolf, T. *Memoria sobre el Cotopaxi y su última Erupción acaecida el 26 de Junio de 1877*; Imprenta de El Comercio: Guayaquil, Ecuador, 1878; p. 48.
2. Sodiro, L. *Relación Sobre la Erupción del Cotopaxi acaecida el día 26 de Junio de 1877*; Imprenta Nacional: Quito, Ecuador, 1877; p. 40.
3. Miller, C.; Mullineaux, D.; Hall, M. *Reconnaissance Map of Potential Volcanic Hazards from Cotopaxi Volcano, Ecuador*; U.S. Department of the Interior: Washington, DC, USA, 1978.
4. Mothes, P. Lahars of Cotopaxi Volcano, Ecuador: Hazard and Risk Evaluation. In *Geohazards, Natural and Man-Made*; McCall, G., Laming, D., Scott, S., Eds.; Chapman and Hall: London, UK, 1992; pp. 53–64.
5. Mothes, P.A.; Hall, M.L.; Janda, R.J. The Enormous Chillos Valley Lahar: An Ash-Flow-Generated Debris Flow from Cotopaxi Volcano, Ecuador. *Bull. Volcanol.* **1998**, *59*, 233–244. [CrossRef]
6. Mothes, P.; Hall, M.L.; Andrade, D.; Samaniego, P.; Pierson, T.C.; Ruiz, A.G.; Yepes, H. Character, Stratigraphy and Magnitude of Historical Lahars of Cotopaxi Volcano (Ecuador). *Acta Vulcanol.* **2004**, *16*, 85–108.
7. Aguilera, E.; Pareschi, M.T.; Rosi, M.; Zanchetta, G. Risk from Lahars in the Northern Valleys of Cotopaxi Volcano (Ecuador). *Nat. Hazards* **2004**, *33*, 161–189. [CrossRef]
8. Pierson, T.C.; Janda, R.J.; Thouret, J.-C.; Borrero, C.A. Perturbation and Melting of Snow and Ice by the 13 November 1985 Eruption of Nevado Del Ruiz, Colombia, and Consequent Mobilization, Flow and Deposition of Lahars. *J. Volcanol. Geotherm. Res.* **1990**, *41*, 17–66. [CrossRef]
9. Hall, M.L.; von Hillebrandt, C. *Mapa de los Peligros Volcánicos Potenciales Asociados con el volcán Cotopaxi, zona Norte 1988*; Instituto Geofísico, Escuela Politécnica Nacional: Quito, Ecuador.
10. Hall, M.L.; von Hillebrandt, C. *Mapa de los Peligros Volcánicos Potenciales Asociados con el volcán Cotopaxi, zona sur 1988*; Instituto Geofísico, Escuela Politécnica Nacional: Quito, Ecuador.
11. Hall, M.L.; Mothes, P.A.; Samaniego, P.; Yepes, H.; Andrade, S.D. *Mapa Regional de los Peligros Volcánicos Potenciales del Volcán Cotopaxi—Zona Norte 2004*; Instituto Geofísico, Escuela Politecnica Nacional: Quito, Ecuador.
12. Hall, M.L.; Mothes, P.A.; Samaniego, P.; Yepes, H.; Andrade, S.D. *Mapa Regional de los Peligros Volcánicos Potenciales del Volcán Cotopaxi—Zona Sur 2004*; Instituto Geofísico, Escuela Politecnica Nacional: Quito, Ecuador.
13. Mothes, P.A.; Espín, P.A.; Hall, M.L.; Vasconez, F.; Sierra, D.; Andrade, S.D. *Mapa Regional de Amenazas Volcánicas Potenciales del Volcán Cotopaxi—Zona Norte 2016*; Instituto Geofísico, Escuela Politécnica Nacional: Quito, Ecuador.
14. Mothes, P.A.; Espín, P.A.; Hall, M.L.; Vasconez, F.; Sierra, D.; Córdova, M.; Santamaría, S. *Mapa Regional de Amenazas Volcánicas Potenciales del Volcán Cotopaxi—Zona Sur 2016*; Instituto Geofísico, Escuela Politécnica Nacional: Quito, Ecuador.
15. Vasconez, F.; Sierra, D.; Almeida, M.; Andrade, S.D.; Marrero, J.; Mothes, P.A.; Bernard, B.; Encalada, M. *Mapa Preliminar de Amenazas Potenciales del Volcán Cotopaxi—Zona Oriental 2017*; Instituto Geofísico, Escuela Politécnica Nacional: Quito, Ecuador.
16. Barberi, F.; Carusso, P.; Macedonio, G.; Pareschi, M.T.; Rosi, M. Reconstruction and Numerical Simulation of the Lahar of the 1877 Eruption of Cotopaxi Volcano (Ecuador). *Acta Vulcanol.* **1992**, *2*, 35–44.
17. Pistolesi, M.; Cioni, R.; Rosi, M.; Cashman, K.V.; Rossotti, A.; Aguilera, E. Evidence for Lahar-Triggering Mechanisms in Complex Stratigraphic Sequences: The Post-Twelfth Century Eruptive Activity of Cotopaxi Volcano, Ecuador. *Bull. Volcanol.* **2013**, *75*, 698. [CrossRef]
18. Pistolesi, M.; Cioni, R.; Rosi, M.; Aguilera, E. Lahar Hazard Assessment in the Southern Drainage System of Cotopaxi Volcano, Ecuador: Results from Multiscale Lahar Simulations. *Geomorphology* **2014**, *207*, 51–63. [CrossRef]
19. Ettinger, S.; Mothes, P.; Paris, R.; Schilling, S. The 1877 Lahar Deposits on the Eastern Flank of Cotopaxi Volcano. *Geomorphologie* **2007**, *13*, 271–280. [CrossRef]
20. Hall, M.; Mothes, P. The Rhyolitic–Andesitic Eruptive History of Cotopaxi Volcano, Ecuador. *Bull. Volcanol.* **2008**, *70*, 675–702. [CrossRef]
21. Sierra, D.; Vasconez, F.; Andrade, S.D.; Almeida, M.; Mothes, P. Historical Distal Lahar Deposits on the Remote Eastern-Drainage of Cotopaxi Volcano, Ecuador. *J. S. Am. Earth Sci.* **2019**, *95*, 102251. [CrossRef]
22. Christie, R.; Cooke, O.; Gottsmann, J. Fearing the Knock on the Door: Critical Security Studies Insights into Limited Cooperation with Disaster Management Regimes. *J. Appl. Volcanol.* **2015**, *4*. [CrossRef]
23. Ariyanti, V.; Gaafar, T.; De La Sala, S.; Edelenbos, J.; Scholten, P. Towards Liveable Volcanic Cities: A Look at the Governance of Lahars in Yogyakarta, Indonesia, and Latacunga, Ecuador. *Cities* **2020**, *107*, 102893. [CrossRef]
24. Rodriguez, F.; Toulkeridis, T.; Sandoval, W.; Padilla, O.; Mato, F. Economic Risk Assessment of Cotopaxi Volcano, Ecuador, in Case of a Future Lahar Emplacement. *Nat. Hazards* **2017**, *85*, 605–618. [CrossRef]
25. Mothes, P.A.; Ruiz, M.C.; Viracucha, E.G.; Ramón, P.A.; Hernández, S.; Hidalgo, S.; Bernard, B.; Gaunt, E.H.; Jarrín, P.; Yépez, M.A.; et al. Geophysical Footprints of Cotopaxi's Unrest and Minor Eruptions in 2015: An Opportunity to Test Scientific and Community Preparedness. In *Volcanic Unrest: From Science to Society*; Gottsmann, J., Neuberg, J., Scheu, B., Eds.; Springer: Cham, Switzerland, 2019; pp. 241–270. ISBN 978-3-319-58412-6.
26. Hidalgo, S.; Battaglia, J.; Arellano, S.; Sierra, D.; Bernard, B.; Parra, R.; Kelly, P.; Dinger, F.; Barrington, C.; Samaniego, P. Evolution of the 2015 Cotopaxi Eruption Revealed by Combined Geochemical and Seismic Observations. *Geochem. Geophys. Geosyst.* **2018**, *19*, 2087–2108. [CrossRef]
27. Toapaxi, J.A.; Ortega, P.; Casa, E.; Santamaría, J.; Hidalgo, X. Análisis de la Modelación Numérica del Flujo Producto de una Erupción del Volcán Cotopaxi—Flanco Norte. *Rev. Politéc.* **2019**, *44*, 7–14. [CrossRef]

28. Vera, P.; Ortega, P.; Casa, E.; Santamaría, J.; Hidalgo, X. Modelación Numérica y Mapas de Afectación por Flujo de Lahares Primarios en el Drenaje Sur del Volcán Cotopaxi. *Rev. Politéc.* **2019**, *43*, 61–72. [CrossRef]
29. Frimberger, T.; Andrade, S.D.; Weber, S.; Krautblatter, M. Modelling Future Lahars Controlled by Different Volcanic Eruption Scenarios at Cotopaxi (Ecuador) Calibrated with the Massively Destructive 1877 Lahar. *Earth Surf. Process. Landf.* **2021**, *46*, 680–700. [CrossRef]
30. Lupiano, V.; Catelan, P.; Calidonna, C.R.; Chidichimo, F.; Crisci, G.M.; Rago, V.; Straface, S.; Di Gregorio, S. LLUNPIY Simulations of the 1877 Northward Catastrophic Lahars of Cotopaxi Volcano (Ecuador) for a Contribution to Forecasting the Hazards. *Geosciences* **2021**, *11*, 81. [CrossRef]
31. Turner, D.; Lucieer, A.; Watson, C. An Automated Technique for Generating Georectified Mosaics from Ultra-High Resolution Unmanned Aerial Vehicle (UAV) Imagery, Based on Structure from Motion (SfM) Point Clouds. *Remote Sens.* **2012**, *4*. [CrossRef]
32. Cress, J.J.; Hutt, M.; Sloan, J.; Bauer, M.; Feller, M.; Goplen, S. *US Geological Survey Unmanned Aircraft Systems (UAS) Roadmap 2014*; US Department of the Interior, US Geological Survey: Reston, VA, USA, 2015.
33. Favalli, M.; Fornaciai, A.; Isola, I.; Tarquini, S.; Nannipieri, L. Multiview 3D Reconstruction in Geosciences. *Comput. Geosci.* **2012**, *44*, 168–176. [CrossRef]
34. Westoby, M.J.; Brasington, J.; Glasser, N.F.; Hambrey, M.J.; Reynolds, J.M. 'Structure-from-Motion' Photogrammetry: A Low-Cost, Effective Tool for Geoscience Applications. *Geomorphology* **2012**, *179*, 300–314. [CrossRef]
35. James, M.R.; Robson, S. Sequential Digital Elevation Models of Active Lava Flows from Ground-Based Stereo Time-Lapse Imagery. *ISPRS J. Photogramm. Remote Sens.* **2014**, *97*, 160–170. [CrossRef]
36. Jordan, B.R. Collecting Field Data in Volcanic Landscapes Using Small UAS (SUAS)/Drones. *J. Volcanol. Geotherm. Res.* **2019**, *385*, 231–241. [CrossRef]
37. Amici, S.; Turci, M.; Giulietti, F.; Giammanco, S.; Buongiorno, M.F.; La Spina, A.; Spampinato, L. Volcanic Environments Monitoring by Drones—Mud Volcano Case Study. *Int. Arch. Photogramm. Remote Sens. Spat. Inf. Sci.* **2013**, *XL-1/W2*, 5–10. [CrossRef]
38. Favalli, M.; Fornaciai, A.; Nannipieri, L.; Harris, A.; Calvari, S.; Lormand, C. UAV-Based Remote Sensing Surveys of Lava Flow Fields: A Case Study from Etna's 1974 Channel-Fed Lava Flows. *Bull. Volcanol.* **2018**, *80*, 29. [CrossRef]
39. De Beni, E.; Cantarero, M.; Messina, A. UAVs for Volcano Monitoring: A New Approach Applied on an Active Lava Flow on Mt. Etna (Italy), during the 27 February–02 March 2017 Eruption. *J. Volcanol. Geotherm. Res.* **2019**, *369*, 250–262. [CrossRef]
40. Turner, N.; Perroy, R.L.; Hon, K.A.; Rasgado, V. Kīlauea June 27th Lava Flow Hazard Mapping and Disaster Response with UAS; American Geophysical Union Fall Meeting Abstracts. 2015. Available online: https://agu.confex.com/agu/fm15/meetingapp.cgi/Paper/85504 (accessed on 17 November 2021).
41. Tsunetaka, H.; Shinohara, Y.; Hotta, N.; Gomez, C.; Sakai, Y. Multi-Decadal Changes in the Relationships between Rainfall Characteristics and Debris-Flow Occurrences in Response to Gully Evolution after the 1990–1995 Mount Unzen Eruptions. *Earth Surf. Processes Landf.* **2021**, *46*, 2141–2162. [CrossRef]
42. Walter, T.R.; Salzer, J.; Varley, N.; Navarro, C.; Arámbula-Mendoza, R.; Vargas-Bracamontes, D. Localized and Distributed Erosion Triggered by the 2015 Hurricane Patricia Investigated by Repeated Drone Surveys and Time Lapse Cameras at Volcán de Colima, Mexico. *Geomorphology* **2018**, *319*, 186–198. [CrossRef]
43. Gomez, C.; Shinohara, Y.; Tsunetaka, H.; Hotta, N.; Bradak, B.; Sakai, Y. Twenty-Five Years of Geomorphological Evolution in the Gokurakudani Gully (Unzen Volcano): Topography, Subsurface Geophysics and Sediment Analysis. *Geosciences* **2021**, *11*, 457. [CrossRef]
44. Bunds, M.; Toké, N.; DuRoss, C.; Gold, R.; Reitman, N.; Johnson, K.; LaJoie, L.; Personius, S.; Briggs, R.; Fletcher, A. High-Resolution Topographic Mapping for Geologic Hazard Studies Using Low-Altitude Aerial Photographs and Structure from Motion Software: Methods, Accuracy, and Examples. In Proceedings of the GSA Annual Meeting, Baltimore, MD, USA, 1–4 November 2015; p. 733.
45. Rodgers, M.; Dixon, T.H.; Gallant, E.; López, C.M.; Malservisi, R.; Ordoñez, M.; Richardson, J.A.; Voss, N.K.; Xie, S. Terrestrial Radar Interferometry and Structure-from-Motion Data from Nevado Del Ruiz, Colombia for Improved Hazard Assessment and Volcano Monitoring. In Proceedings of the AGU Fall Meeting Abstracts, San Francisco, CA, USA, 14–18 December 2015; p. G41A-1017.
46. Vezzoli, L.; Apuani, T.; Corazzato, C.; Uttini, A. Geological and Geotechnical Characterization of the Debris Avalanche and Pyroclastic Deposits of Cotopaxi Volcano (Ecuador). A Contribute to Instability-Related Hazard Studies. *J. Volcanol. Geotherm. Res.* **2017**, *332*, 51–70. [CrossRef]
47. Pistolesi, M.; Rosi, M.; Cioni, R.; Cashman, K.V.; Rossotti, A.; Aguilera, E. Physical Volcanology of the Post-Twelfth-Century Activity at Cotopaxi Volcano, Ecuador: Behavior of an Andesitic Central Volcano. *Geol. Soc. Am. Bull.* **2011**, *123*, 1193–1215. [CrossRef]
48. Sklenář, P.; Kovář, P.; Palice, Z.; Stančík, D.; Soldán, Z. Primary Succession of High-Altitude Andean Vegetation on Lahars of Volcán Cotopaxi, Ecuador. *Phytocoenologia* **2010**, *40*, 15–28. [CrossRef]
49. Piras, M.; Taddia, G.; Forno, M.G.; Gattiglio, M.; Aicardi, I.; Dabove, P.; Russo, S.L.; Lingua, A. Detailed Geological Mapping in Mountain Areas Using an Unmanned Aerial Vehicle: Application to the Rodoretto Valley, NW Italian Alps. *Geomat. Nat. Hazards Risk* **2017**, *8*, 137–149. [CrossRef]

50. Weiss, J.L.; Gutzler, D.S.; Coonrod, J.E.A.; Dahm, C.N. Long-Term Vegetation Monitoring with NDVI in a Diverse Semi-Arid Setting, Central New Mexico, USA. *J. Arid Environ.* **2004**, *58*, 249–272. [CrossRef]
51. Bhandari, A.K.; Kumar, A.; Singh, G.K. Feature Extraction Using Normalized Difference Vegetation Index (NDVI): A Case Study of Jabalpur City. *Procedia Technol.* **2012**, *6*, 612–621. [CrossRef]
52. Wahab, I.; Hall, O.; Jirström, M. Remote Sensing of Yields: Application of UAV Imagery-Derived NDVI for Estimating Maize Vigor and Yields in Complex Farming Systems in Sub-Saharan Africa. *Drones* **2018**, *2*, 28. [CrossRef]
53. Remondino, F.; Spera, M.G.; Nocerino, E.; Menna, F.; Nex, F. State of the Art in High Density Image Matching. *Photogram. Rec.* **2014**, *29*, 144–166. [CrossRef]
54. Motohka, T.; Nasahara, K.N.; Oguma, H.; Tsuchida, S. Applicability of Green-Red Vegetation Index for Remote Sensing of Vegetation Phenology. *Remote Sens.* **2010**, *2*, 2369–2387. [CrossRef]
55. Potere, D. Horizontal Positional Accuracy of Google Earth's High-Resolution Imagery Archive. *Sensors* **2008**, *8*, 7973. [CrossRef]

remote sensing

Article

Submarine and Subaerial Morphological Changes Associated with the 2014 Eruption at Stromboli Island

Daniele Casalbore [1,2,*], Federico Di Traglia [3], Alessandro Bosman [2], Claudia Romagnoli [4], Nicola Casagli [3,5] and Francesco Latino Chiocci [1,2]

1 Dipartimento Scienze della Terra, Università Sapienza di Roma, Piazzale Aldo Moro 5, 00185 Rome, Italy; francesco.chiocci@uniroma1.it
2 Istituto di Geologia Ambientale e Geoingegneria (IGAG), Consiglio Nazionale delle Ricerche, Struttura Congiunta DICEA, Via Eudossiana 18, 00184 Rome, Italy; alessandro.bosman@cnr.it
3 Dipartimento di Scienze della Terra, Università degli Studi di Firenze, Via La Pira 4, 50121 Firenze, Italy; federico.ditraglia@unifi.it (F.D.T.); nicola.casagli@unifi.it (N.C.)
4 Dipartimento Scienze Biologiche, Geologiche e Ambientali, Università di Bologna, P.za Porta S. Donato 1, Bologna 40126, Italy; claudia.romagnoli@unibo.it
5 National Institute of Oceanography and Applied Geophysics—OGS, 34010 Trieste, Italy
* Correspondence: daniele.casalbore@uniroma1.it

Abstract: Stromboli is an active insular volcano located in the Southern Tyrrhenian Sea and its recent volcanic activity is mostly confined within the Sciara del Fuoco (SdF, hereafter), a 2-km wide subaerial–submarine collapse scar, which morphologically dominates the NW flank of the edifice. In August-November 2014, an effusive eruption occurred along the steep SdF slope, with multiple lava flows reaching the sea. The integration of multisensor remote sensing data, including lidar, photogrammetric, bathymetric surveys coupled with SAR amplitude images collected before and after the 2014 eruption enabled to reconstruct the dynamics of the lava flows through the main morphological changes of the whole SdF slope. Well-defined and steep-sided ridges were created by lava flows during the early stages of the eruption, when effusion rates were high, favoring the penetration into the sea of lava flows as coherent bodies. Differently, fan-shaped features were emplaced during the declining stage of the eruption or in relation to lava overflows and associated gravel flows, suggesting the prevalence of volcaniclastic breccias with respect to coherent lava flows. The estimated volume of eruptive products emplaced on the SdF slope during the 2014 eruption, accounts for about 3.7×10^6 m^3, 18% of which is in the submarine setting. This figure is different with respect to the previous 2007 eruption at Stromboli, when a large lava submarine delta formed. This discrepancy can be mainly related to the different elevation of the main vents feeding lava flows during the 2007 eruption (around 400 m) and the 2014 eruption (around 650 m). Besides slope accretion, instability processes were detected both in the subaerial and submarine SdF slope. Submarine slope failure mobilized at least 6×10^5 m^3 of volcaniclastic material, representing the largest instability event detected since the 2007 lava delta emplacement.

Keywords: lava delta; slope failure; repeated bathymetric surveys; digital elevation models; LiDAR; PLÉIADES; morphological monitoring

Citation: Casalbore, D.; Di Traglia, F.; Bosman, A.; Romagnoli, C.; Casagli, N.; Chiocci, F.L. Submarine and Subaerial Morphological Changes Associated with the 2014 Eruption at Stromboli Island. *Remote Sens.* **2021**, *13*, 2043. https://doi.org/10.3390/rs13112043

Academic Editor: Balázs Székely

Received: 28 March 2021
Accepted: 20 May 2021
Published: 22 May 2021

1. Introduction

Stromboli is the north-easternmost island of the Aeolian Archipelago in the Southern Tyrrhenian Sea, representing the tip of a large, 3000 m-high volcanic edifice (Figure 1). Stromboli and Panarea volcanic edifices are part of a 45 km long volcanic area developed along a NE-SW regional extensional fault system (Figure 1b) above the thinned continental crust of the Calabrian Arc [1,2]. The evolution of Stromboli island occurred in the last 85 ka (apart from the older Strombolicchio edifice dated at about 200 ka) and can be divided in six main growth stages, with magmas ranging from calc-alkaline to potassic series, typical of an arc-volcanic setting, whose formation can be related to the subduction of the Ionian crust beneath the Calabrian Arc ([2] and references therein).

The Stromboli edifice is characterized by a quasi-bilateral symmetry with respect to a tectonically controlled NE-SW axis (Figure 1c), which corresponds to the main orientation and distribution of dykes, eruptive fissures, secondary vents and active summit craters during most of the volcano evolution [1–3]. Dyke intrusion along this axis also promoted the development of recurrent lateral collapses along the NW and SE flanks of the volcano [2,4,5]. Specifically, multiple sectors collapses affected the NW flank of the edifice in the last 13 ka, the last of them likely occurred during medieval times, leading to the development of the Sciara del Fuoco scar, SdF hereafter (Figure 1c) [2].

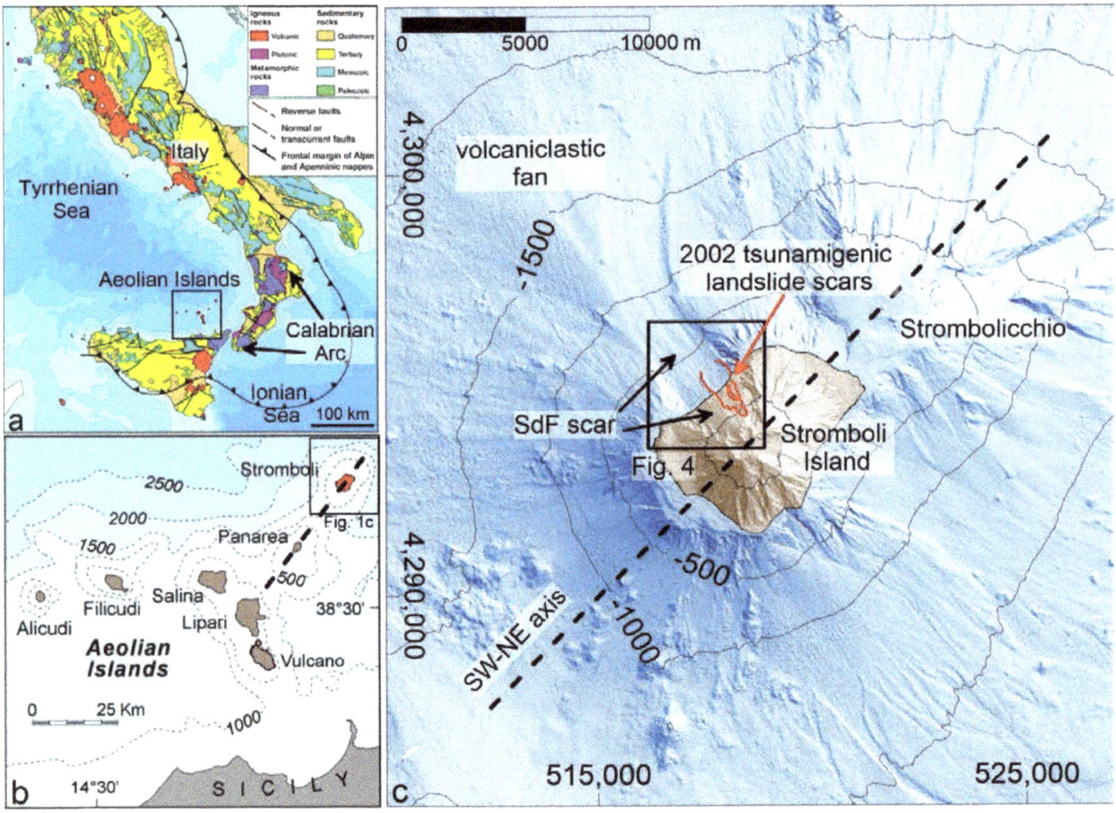

Figure 1. (**a**) Simplified geological map of Italy (modified from [6]), with the location of the Aeolian Islands (black box); (**b**) zoom of the Aeolian archipelago in the Southern Tyrrhenian Sea, with the geographical location of Stromboli Island (in red) and the main SW–NE regional system (black dashed lines) controlling the Stromboli–Panarea alignment; (**c**) shaded relief map and contour lines (equidistance 500 m) of the Stromboli edifice (location in Figure 1b), with the indication of the 2002 tsunamigenic landslide scars (see also Figure 4a) and the main rift axis oriented along the SW–NE direction.

The SdF is a very steep subaerial/submarine partially filled depression that funnels eruptive material produced at the summit craters into the sea, where a large volcaniclastic fan, composed by debris avalanche deposits and overlying turbidite deposits, is recognizable to over 3000 m water depth (mwd, hereafter) [7]. This volcaniclastic system is mostly fed by the persistent activity, occurring at the summit craters (South-Western, Central and North-Eastern Craters, SWC, CC and NEC, respectively, in Figure 1a). This activity, mostly consisting of low- to mid-energy Strombolian explosions, shows intensity and frequency fluctuations over time and is punctuated by: (a) high-energy explosive events, called paroxysms [8,9], often associated with pyroclastic flows as observed in 2019 [10], and (b) lava overflows from the summit crater area [11,12] and/or effusive flank eruptions, as recently occurred in 1985–1986, 2002–2003, 2007 and 2014 [13–16].

Dyke intrusions and effusive eruptions also induce stress changes on the volcanic flanks and can generate slope instability at different spatial scales, as observed during the initial phases of the last four flank eruptions [16–18]. Of these events, only the 2002–2003 eruption was associated with the occurrence of tsunamigenic submarine and subaerial landslides (red dashed line in Figure 1c; [19,20]). The tsunami waves related to the 2002 submarine and subaerial landslides severely impacted the Stromboli coastline with a maximum runup of 10 m [21], evidencing the high hazard associated with similar events that have repeatedly occurred also during the previous century [22,23]. Since then, SdF slopes have been carefully monitored through repeated topographic surveys for their submarine [24–26] and subaerial part [27]. These surveys are part of a more complex monitoring network including: ground-based thermal infrared data [3]; infrasonic monitoring [28], seismic network [29], borehole strainmeters [10], dilatometric sensors [30], ground tiltmeters [31], plume and soil gas geochemical monitoring [32,33], two ground-based interferometric synthetic aperture radar sites [34] and satellite thermal monitoring [35]. However, except for the study of the 2002-2003 eruption and related tsunamigenic landslides [36], no comprehensive subaerial and submarine studies of the eruptive events occurring at SdF were performed until now.

In this paper, we present an integrated analysis of multisensor remote sensing data for the characterization of the morphological changes associated with the 2014 effusive eruption occurred at Stromboli volcano, during which lava flows entered the sea in the NW shallow offshore of the island. In detail, the bathymetric comparison between multibeam surveys performed before and after the 2014 eruption have been integrated with lidar, photogrammetric and SAR amplitude images [16], aimed at fully understanding the morphological evolution of the SdF slope involved in this eruption. The integration of submarine and subaerial DEMs also allows one to estimate the volume of the eruptive products emplaced on the SdF slope during the 2014 eruption and compare it with the values previously estimated using satellite and ground based thermal observation [28,37], thus providing a crossvalidation of the two methodological approaches.

More generally, the results of this study are important to better understand the behavior and morphological evolution of lava flows penetrating into the sea, mainly in relation to the paucity of bathymetric monitoring of coastal areas during eruptive crisis [25,38], with observations mainly limited to the subaerial part of lava deltas [27,39,40] or scuba dives at water depth less than 50 m [41,42]. This is a very important issue considering the possible hazard associated with these events in active insular and coastal volcanoes.

2. The 2014 Eruption: Chronology and Subaerial Morphological Analysis

The 2014 effusive eruption began on the morning of 7 August 2014 and ended on 13 November 2014. The eruption was preceded by two months of more intense activity and anomalous values in geophysical and geochemical parameters [28,32,43], with stronger and more frequent Strombolian explosions, overflows from the crater terrace and small landslides along the subaerial SdF triggered by crater-rim collapse or by remobilization of volcaniclastic material by overflows (Figure 2a, [44]). The most frequent explosive activity and the lava overflows were generated mainly by the NEC, with the accumulation and remobilization of volcaniclastic material along the central part of the SdF [16,44]. On 6 August 2014, starting in the early afternoon, a series of overflows from the NEC area reached the sea. These events were associated with crater-rim collapse landslides that occurred in the same sector (Figure 2a; [16]). The actual onset of the eruption (between 3 and 5 a.m. UTC on 7 August 2014) was marked by the propagation of magma from the crater terrace towards the NEC, with the development of an eruptive fracture that fed a lava vent at an altitude of 650 m above sea level, asl hereafter (Figure 2d,e) located 100 m lower than the crater terrace altitude.

Figure 2. (**a**) Landslide associated with an overflow occurred on 6 August 2014; (**b**) initial phases of the 2014 eruption, with the entry into the sea of the various branches of the lava flow on 7 August 2014; (**c**) the Sciara del Fuoco during the 2014 effusion, with the extension of the 2014 lava field, the area affected by the different overflows of July-early August 2014 and by subsequent debris flows to November 2014; (**d**) single thermogram (approximately 0.67 m pixel resolution) showing the active lava flows, the various branches of the lava field, and the overflows of emplaced on 6 August 2014 and subsequent debris flows to November 2014; (**e**) proximal area of the upper SdF slope that highlights the emission zone of the July-early August 2014 overflows (F1 to F3) and the vent of the 2014 lava field (F4, in red).

The propagation of the fracture triggered the collapse of the volcaniclastic talus below the NEC [16]. The first days of the 2014 eruption were characterized by a high effusive rate, which decreased dramatically in the following days, and then remained low for the duration of the eruption [28,45]. During this initial phase, a series of lava flows reached the sea in different points, always located in the northernmost area of the SdF (Figure 2b–d; [46]). Subsequently, with the decrease of the effusive rate, the lava flows did not propagate far from the vent, with lava fronts that stood at around 400 m a.s.l.. This produced the typical morphology of Stromboli lavas with vent positioned at high altitudes [14], with the development of a proximal shield, a medial zone fed by small flows, and characterized by frequent lava crumbling down slope and producing a debris field, with debris moving downslope, to the coastline (Figure 2c,d; [46]).

The volume of lava emitted by the 2014 eruption has been estimated to be 7.4 $\times$ 10^6 m^3 by [37] and 5.5 $\times$ 10^6 m^3 by [28], with 2.697 $\pm$ 0.190 $\times$ 10^6 m^3 of lava emplaced on the subaerial slope (black dashed line in Figure 1b; [47]). After the 2014 eruption, at least for the period 2015-2016, the eruptive activity remained at very low levels [29], mainly characterized by sporadic, low intensity Strombolian explosions. In this period the volcaniclastic sedimentation from the crater area towards the SdF was significantly reduced, and erosion of the 2014 lava field mainly occurred, with small landslides mobilizing volumes in the order of 10^3–10^4 m^3 [34].

3. Data and Methods

Data used for this research mainly rely on the analysis of morphological changes detected by using high-resolution topo-bathymetric surveys collected before and after the 2014 eruption along the SdF collapse scar and time-lapse SAR (synthetic aperture radar) amplitude images collected between May and August 2014 (Figure 3 and Table 1).

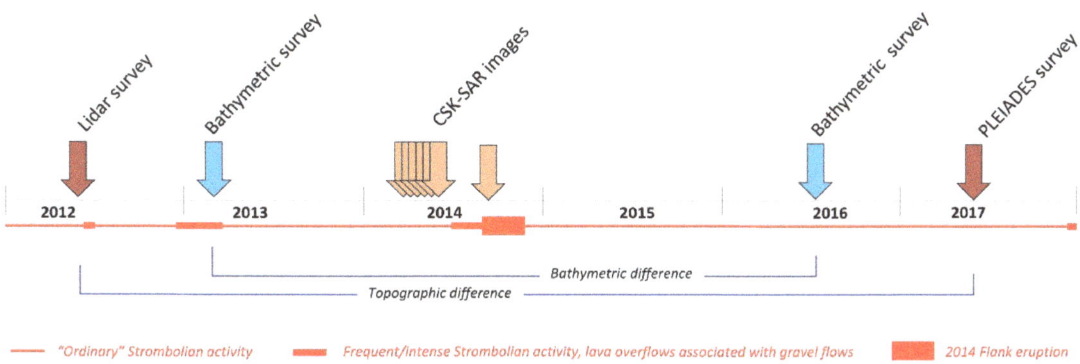

Figure 3. Timeline of the surveys performed for this study, with the indication of the 2014 flank eruption and period of "ordinary" and intense Strombolian activity.

Table 1. Timetable of the surveys used for monitoring the morphological evolution of the August-November 2014 effusive eruption at Stromboli. Time elapsed refers to months before (−) and after (+) the 2014 eruption.

Survey	Date	Time Elapsed	Elevation Range	Cell-Size
LiDAR	4-8/05/2012	−27 months	0–981 m a.s.l.	1 m
Bathymetry EM710 (70–100 kHz)	15/2/2013	−18 months	25–700 mwd	3 m
SAR amplitude COSMO-SkyMed	8/5/2014	−91 days	0–981 m a.s.l.	3 m
SAR amplitude	24/5/2014	−75 days	0–981 m a.s.l.	3 m

Table 1. *Cont.*

Survey	Date	Time Elapsed	Elevation Range	Cell-Size
SAR amplitude	9/6/2014	−59 days	0–981 m a.s.l.	3 m
SAR amplitude	25/6/2014	−43 days	0–981 m a.s.l.	3 m
SAR amplitude	11/7/2014	−27 days	0–981 m a.s.l.	3 m
SAR amplitude	27/7/2014	−11 days	0–981 m a.s.l.	3 m
SAR amplitude	12/8/2014	+5 days after the onset	0–981 m a.s.l.	3 m
Bathymetry Teledyne Reson 7125 (455 kHz) Teledyne Reson 8160 (45 kHz)	1/1/2016	+14 months	10–100 mwd 50–700 mwd	1–3 m
Photogrammetric PLÉIADES tri-stereo	26/05/2017	+30 months	0–981 m a.s.l.	1 m
CSK-SAR dataset description: COSMO-SKYMED				
Wavelength: 3.12 cm	Geometry: Descending	Sensor mode: H4-01		Track: 98
Line of sight angles	25.48°	11.19°		
Line of sight versors	V: 0.903	N: −0.084		E: 0.422

3.1. Lidar, Photogrammetric and Bathymetric Surveys

The difference map, obtained by comparing repeated digital elevation models (DEMs, hereafter), enable us to quantify the morphological change occurred on the SdF slope during the time interval between the surveys, with positive and negative values associated with slope accretion and erosion, respectively. The volumes associated with slope changes were obtained by integrating the difference in depth over the area of interest through the software Global Mapper. The reliability of computed volumes depends on the resolution and accuracy of the DEMs, as described below both for the submarine and subaerial slope.

For the submarine slope, seafloor changes were computed through the difference between multibeam bathymetries collected in 2013 and 2016 onboard the R.V. Urania and Minerva 1 (CNR). On 15 February 2013, bathymetric data were collected between 25 and 700 m water depths using the multibeam echosounder Kongsberg Simrad EM710 working at a frequency of 70–100 kHz. On 1 January 2016, bathymetric data were collected in the depth range 10-700 m using Teledyne Reson 7125 and 7160 multibeam echosounders working at a frequency of 400 kHz and 45 Hz in shallow- (<100 mwd) and deep-water (>100 mwd), respectively. In these surveys, data were DGPS-positioned and processed with hydrographic software (Caris Hips and Sips Professional), using daily sound speed profiles and patch test of transducers in the survey zone. Furthermore, a hull-mounted sound speed sensor was used to update in real-time the sound velocity values close to the flat face of the multibeam transducer. Tidal corrections were performed using data from the nearby tide gauge stations (www.mareografico.it). Random spikes and organized noise (multiple) produced by local very steep slopes were removed by applying geometrical and statistical filters. Cleaned multibeam data were gridded using a weighted averaging algorithm to produce the DEMs at variable resolution, from 1 m in shallow-water (<100 mwd) to 5 m at greater depths. DEMs used for bathymetric comparison were gridded with cell-size of 3 m and limited at depths <500 m, because at greater depths acoustic noise largely increased, making unreliable the computed changes. A vertical accuracy range of ±0.5 m was estimated for the difference map by comparing the difference in depth of stable benchmarks between pairs of successive bathymetric sets. Small depth changes below this error range in the difference map were not considered in volume computations.

For the subaerial slope, topographic changes were estimated by comparing two DEM reconstructed starting from a LiDAR airborne survey (4–8 May 2012) and PLÉIADES-1 tri-stereo satellite imagery (26 May 2017). The LiDAR-DEM was obtained through the processing of the 3D point cloud that was acquired by using the Leica ADS80 sensor, which has instrumental vertical and horizontal accuracy of 0.10–0.20 and 0.25 m, respectively [46,47]. The acquired point cloud has a mean point density of 8 pt/m^2. The 2017 DEM derived from the tri-stereo optical imagery acquired by the PLÉIADES-1 satellites, provided by Airbus with a nominal xy resolution of 1 m × 1 m. The PLÉIADES-1A (PHR1A) and PLÉIADES-1B (PHR1B) satellites can sense different synchronous images of the same area, with an angle variable between 6 and 28°. The tri-stereo approach consists of the acquisition of three nearly simultaneously images (one backward looking, one forward looking and a third near-nadir image) allowing the DEM reconstruction through the photogrammetric processing of the three stereo images [48]. Although the acquisitions used in this study are 100% cloud-free, the presence of the gas plume from the crater terrace prevented the reconstruction of the topography in this area. To assess the accuracy of the heights and their horizontal position in the PLÉIADES-1 DEM, ground control points (GCPs) were collected on the map database (cartographic XY standard deviation: 0.15 m). A block adjustment including all the satellite scenes was performed. The block adjustment was validated when the following accuracy was achieved: (i) an image's pixel xy bias smaller than 0.3 pixels, (ii) an image's pixel xy standard deviation smaller than 0.3 pixels and (iii) an image's pixel xy maximum residuals smaller than 2 pixels (for details see [16,47]). The z standard deviation was about 1.5 m (47).

3.2. Change Detection with SAR Amplitude Images

Data from SAR sensors have often been used to maps areas that have been affected by lithological and morphological changes, i.e., to identify areas that have been impacted by eruptive and post-eruptive (landslides or floods) phenomena [49–51]. The SAR backscattering is determined by several factors, as the local morphology and the surface microrelief related to the grain-size, and the dielectric constant of the material at the surface [52,53]. To define which roughness scale affects the backscatter properties, the Rayleigh criterion was applied. In this way, the root mean squared height (hrms) variation on horizontal surfaces has been evaluated following [52] and, for the COSMO-SkyMed SAR (CSK-SAR) images used in this study, the hrms is approximately 4 mm. In an area with volcaniclastic sedimentation, this indicates that it is possible to identify the variation of the sedimented material between fine-grained (i.e., ash-dominated deposits) and coarse-grained (i.e., blocks and bombs dominated deposits, such as grain flows or 'a'ā lava surfaces). Local morphology produces irregularities having wavelengths at least twice as large as the satellite resolution cell due to changes in the local incidence angles, and it is not directly possible to separate the local morphology effects from the grain-size influence. Therefore, the term "roughness" is used to represent a combination of both factors [52].

To detect and interpret changes in land cover in connection with the SdF slope, RGB color composites are used [44,52]. The composite is created using these combinations:

- RED: 8 May 2014 image, considered as a reference for all the others;
- GREEN: image that is analyzed;
- BLUE: ratio of the amplitudes between the analyzed image and that of 8 May 2014.

The ratio is used because it depends on the relative average radar backscattering changes between two images, and therefore it does not depend on the pixel's intensity level [54].

A dataset comprising 7 COSMO-SkyMed SAR (CSK-SAR) images, acquired in descending orbit between 8 May 2014 and 12 August 2014, was used for this study (Figure 3 and Table 1). The products were coregistered, using the offset refinement based on the shuttle radar topography mission (SRTM) digital elevation model (DEM) at 1 arc second forming one unique stack, which is cropped around the target area. The cropped stack is geocoded by correcting SAR geometric distortions using SRTM DEM, producing SAR orthorectified map-projected images. Backscattered intensity of each image is transformed in amplitude image and then decibel scaled, converting the data into a virtual band with the expression 10*log10 (amplitude). Finally, the quality of the images was enhanced using a multitemporal speckle filter that reduces the "salt and pepper-like" texturing (speckle) of the CSK-SAR data.

4. Results

4.1. Morphological Evolution of the Submarine SdF between 2013 and 2016

The DEMs used to constrain seafloor changes associated with the August–November 2014 eruption were collected in February 2013 and January 2016, respectively (Figure 3). Despite the time gap before and after the eruption, our 20 years' experience on submarine monitoring of the SdF has evidenced that the main morphological changes occur during eruptive crisis or increased period of Strombolian activity, while the "ordinary" evolution of the slope is limited to small readjustments of the shallowest part of the SdF [20,24–26].

The pre-2014 eruption submarine SdF slope can be morphologically divided in the SW and NE sectors, whose boundary roughly corresponds to the SW limit of the 2002 landslide scars (dashed red lines in Figure 4a). Considering also previous surveys, the SW sector in the time frame 2002–2017 was affected by minor morphological changes, whereas significant changes occurred in the NE sector both due to the morphological readjustment of the 2002 landslide scar, and to the emplacement of the 2007 lava delta (dotted blue lines in Figure 4a). These events left a depressed area in the southern part of the NE sector of the SdF slope, hereafter defined as the central part of the SdF.

The difference between pre- and post-2014 bathymetries shows that the main morphological changes on the NE sector of the SdF occur within the first 250 mwd, even if they locally extend down to 500 mwd (seaward limit of the difference map, Figure 4b). Seafloor accretion largely overwhelms erosion, accounting for a total estimated volume of $+1.75 \times 10^6$ m^3 and -3.5×10^5 m^3, respectively (Table 2).

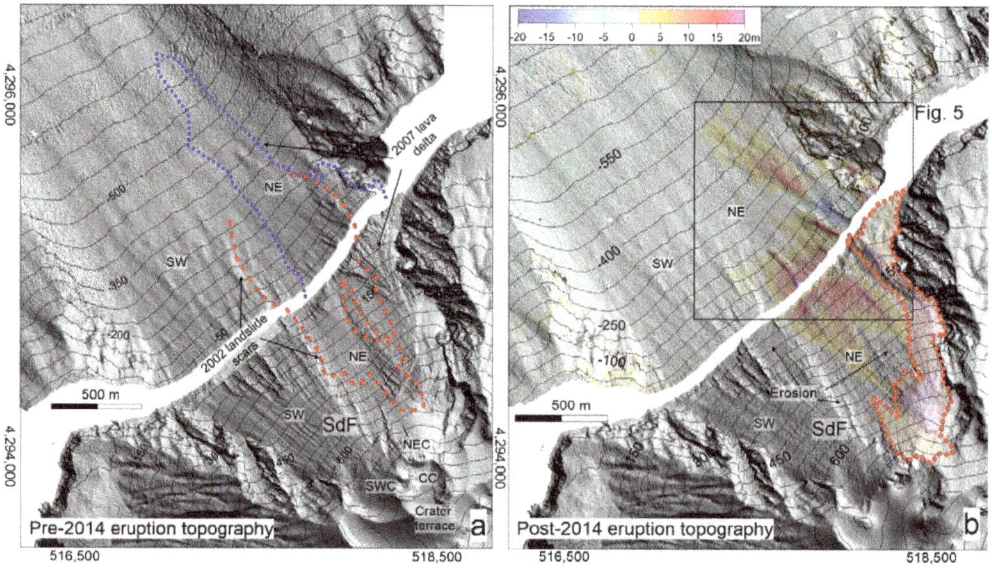

Figure 4. Shaded relief and contours (equidistance of 50 m) of the SdF in the NW Stromboli flank (location in Figure 1c), using (**a**) pre-2014 eruption topography, with the limits of the 2002 tsunamigenic landslides scars (dashed red lines) and the 2007 submarine lava delta (dotted blue line); (**b**) difference map (50% of transparency) between the pre- and post-2014 eruption topography, draped over the shaded relief and contours (equidistance of 50 m) of the SdF using post-2014 eruption DEM. The pre-/post-eruption comparison for the subaerial part of the SdF shows only accretion that largely overwhelms erosion (limited to a few sectors here indicated by arrows, see also (adapted from [47]). The dashed red line delimits the area with main morphological changes associated with the 2014 lava flows entering the sea.

Table 2. Volumetric estimation related to (a) submarine morphological changes (depth range 10–500 mwd) occurred between 2013 and 2106, (b) accretion and erosion of the entire SdF slope associated with the 2014 effusive eruption, (c) volcaniclastic material emplaced in the central part of the SdF, mainly before the 2014 eruption (see text for detail). The red and light-blue colors are referred to the subaerial and submarine flanks, respectively; the black color is used for the total volume. * Volume inferred in the gap area between the coastward limit of the bathymetric survey (around 10 mwd) and the coastline. Note that the error on z regard to the submarine and subaerial DEMs is on the order of 0.5 m and 1.5 m, respectively. The symbol ≈ is used for approximately.

2013–2016 submarine morphological changes	Volume (m^3)
Seafloor accretion (10–500 mwd)	$\approx +1.75 \times 106$ m^3
Seafloor erosion (10–500 mwd)	$\approx -3.5 \times 105$ m^3
2014 Eruption (accretion and erosion)	**Volume (m^3)**
Main subaerial lava flows in the NE part of the SdF	$\approx 2.7 \times 10^6$
Subaerial lava flows in the central part of the SdF	$\approx 3.5 \times 10^5$
Total subaerial volume related to the 2014 eruption	$\approx 3.05 \times 10^6$
Seafloor accretion off the main lava flows (A2 and A3)	$\approx 3.3 \times 10^5$
Seafloor accretion in the central part of the SdF (A1)	$\approx 1.5 \times 10^5$
* Seafloor accretion inferred within the first 10 mwd	$\approx 2 \times 10^5$
Total submarine volume related to the 2014 eruption	$\approx 6.8 \times 10^5$
Total volume related to the 2014 eruption	$\approx 3.73 \times 10^6$
Main submarine landslide related to the 2014 eruption	$\approx -3.5 \times 10^5$
Submarine landslide deposits related to the 2014 eruption (A4)	$\approx 6 \times 10^5$
Volcaniclastic material in the central part of the SdF	**Volume (m^3)**
Volcaniclastic in the subaerial slope (first 500 m a.s.l.)	$\approx 1.35 \times 10^6$
Volcaniclastic in the submarine slope (10–250 mwd)	$\approx 4.9 \times 105$
* Volcaniclastic inferred in the submarine slope (<−10 mwd)	$\approx 2.1 \times 105$
Total volcaniclastic in the submarine slope	$\approx 7 \times 105$
Total volcaniclastic material in the SdF slope	$\approx 1.84 \times 10^6$

Seafloor erosion (light-blue area in Figures 4b and 5) was almost confined in an area of about 50.000 m^2 facing the central part of the 2007 lava delta between 20 (coastward limit of the bathymetric survey) and 200 mwd: here erosion affected the seabed up to 20 m of thickness (E1 area in Figure 5).

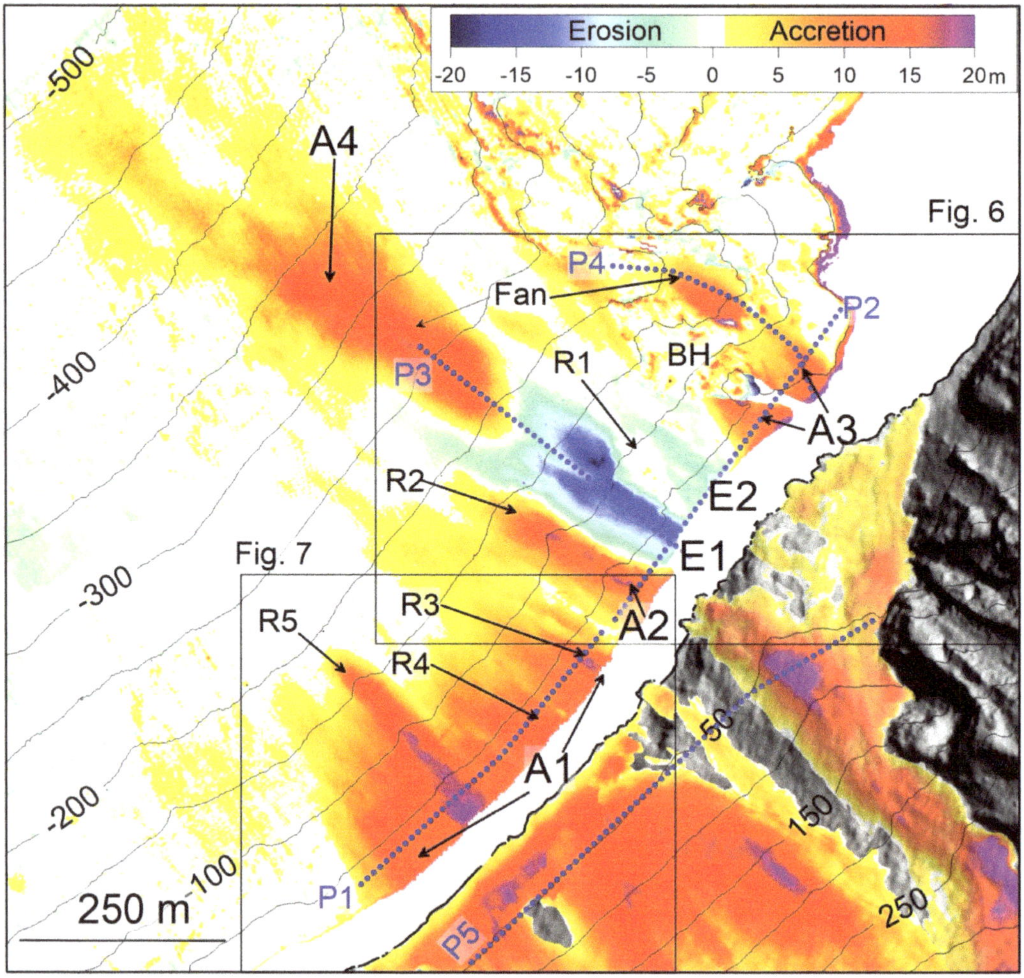

Figure 5. Zoom of the difference map between pre- and post-2014 eruption topography (location in Figure 4b, contours are derived from post-2014 eruption DEMs), with the indication of the main areas characterized by seafloor accretion (A1 to A4) and erosion (E1 and E2). Morphological ridges (R1 to R5) elongated downslope recognizable on the post-2014 bathymetry (Figures 6 and 7) and a previously existing basement high (BH) are also indicated, along with the trace of the bathymetric profiles (P1 to P5, blue dotted lines) shown in Figure 8.

A smaller erosive area (E2 in Figure 5), with thickness lower than 5 m, was recognizable just to the north of E1. These erosive areas correspond to the formation of channelized features on the 2016 DEM, bounded by steep-sided ridges (R1 and R2 in Figures 5 and 6b).

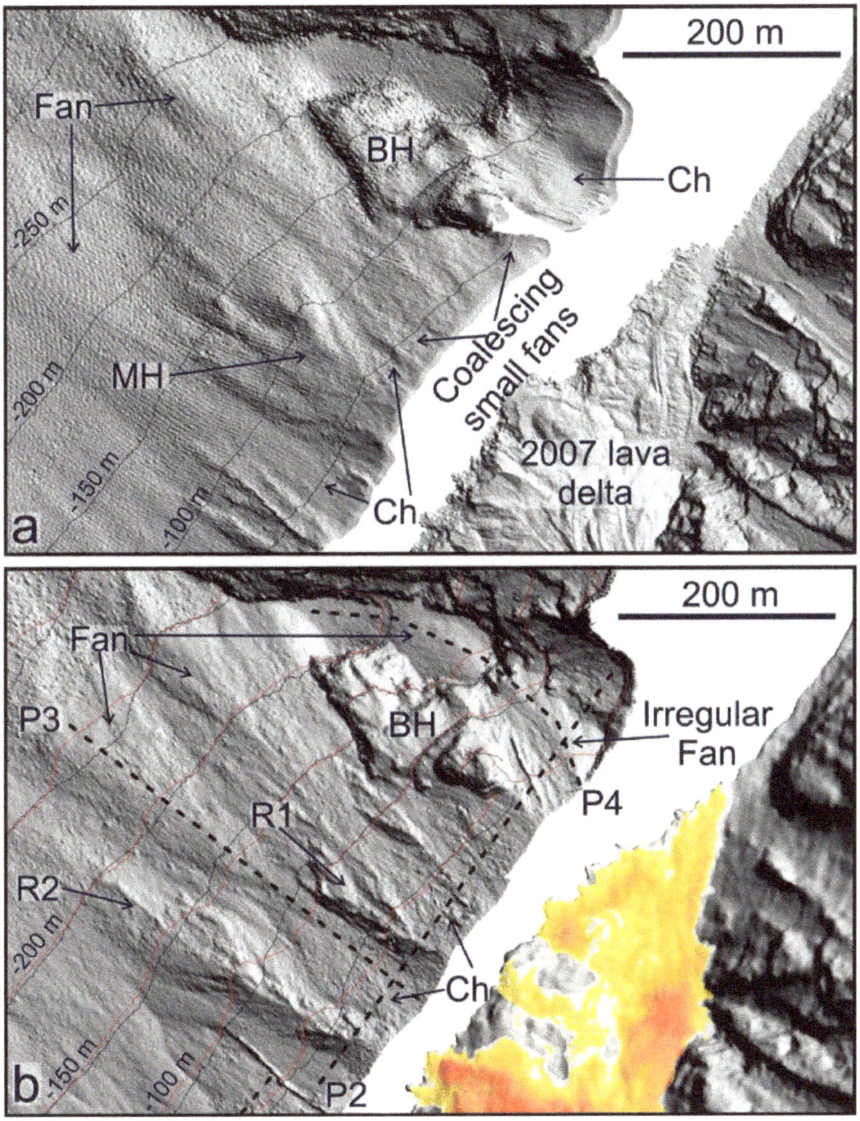

Figure 6. Zoom showing the (**a**) pre- and (**b**) post-2014 shaded relief and contours (black and red lines, respectively; equidistance 50 m) in the north-easternmost sector of the SdF collapse scar (location in Figure 5). In (**b**), the subaerial difference map between pre- and post-2014 eruption is also draped over the coastal topography (for the color scale refer to Figure 5) The dashed black lines (P2 to P4) are the trace of the bathymetric profiles shown in Figure 8. MH: morphological high; BH: basement high; R: ridge; Ch: channelized area.

The E1 channelized area (Figure 5 and Ch in Figure 6b) roughly matches the location of a large morphological high (MH in Figure 6a) on the 2013 DEM; the smaller E2 erosive area corresponds, instead, to an area previously characterized by small-scale coalescing fan-shaped features (Figure 6a).

Seafloor accretion, as resulting from the comparison between pre- and post-2014 eruption, can be divided in four main areas (A1–A4 in Figure 5): A1–A3 were mainly located in the first 200 mwd, while A4 was located at depths greater than 200 mwd.

The A4 area covers a surface of 1.8 × 10^5 m^2 down to 500 mwd (seaward limit of the difference map) for a total estimated volume of +6 × 10^5 m^3; it morphologically corresponds to the development of a large fan-shaped feature, as visible on the 2016 DEM (Figure 6b), with its apex located within the E1 channelized area (see P3 in Figure 8).

The A1 area covers a surface of 1.25 × 10^5 m^2, with a thickness of few tens of meters (Figure 5 and P1 in Figure 8) for a total estimated volume of +8.35 × 10^5 m^3; it morphologically corresponds to the development of small coalescing fan-shaped features alternated with downslope elongated ridges (R3, R4 and R5 in Figures 5 and 7b), characterized by a rough morphology and steep lateral gradients (over 60°, P1 in Figure 8) on the 2016 DEM.

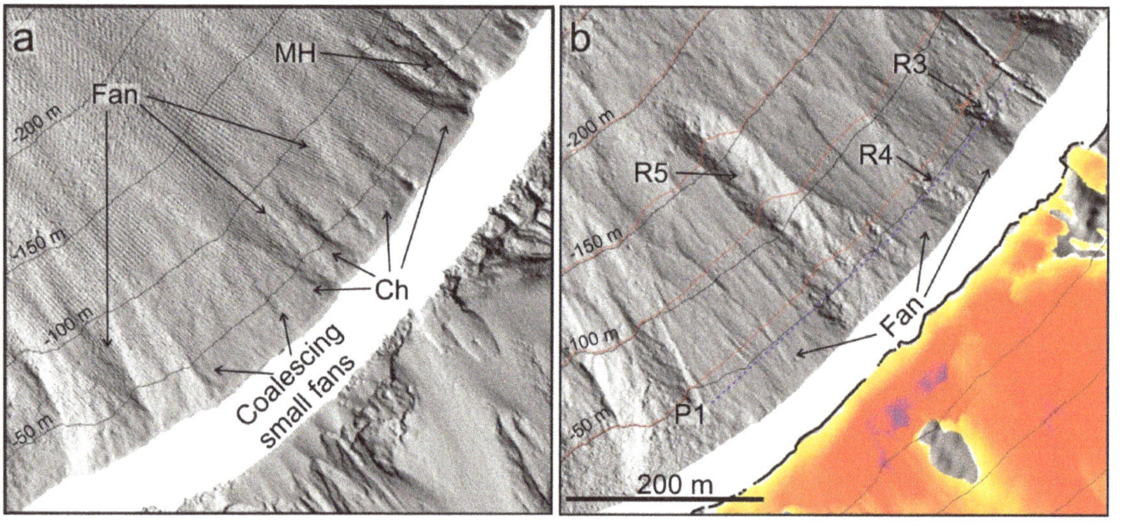

Figure 7. Zoom showing the (**a**) pre- and (**b**) post-2014 shaded relief and contours (black and red lines, respectively; equidistance 50 m) in the central sector of the SdF collapse scar (location in Figure 5). In (**b**), the subaerial difference map between pre- and post-2014 eruption is also draped over the coastal topography (for the color scale refer to Figure 5) The dashed blue line (P1) is the trace of the bathymetric profile shown in Figure 8. MH: morphological high; Ch: channelized area; R: ridge.

These ridges coincide, in fact, with the areas characterized by the highest values (up to 20 m) of seafloor accretion in A1 (Figure 5), mostly matching the location of previous channelized features (Ch in Figure 7a) recognizable on the 2013 DEM. The westernmost, main ridge (R5 in Figures 5 and 7b) extended down to 200 mwd for a length of 275 m, a width of 60 m and height of 5–6 m with respect to the surrounding seafloor (P1 in Figure 8).

The A2 accretion area also corresponded to a well-defined ridge (R2 in Figure 6b) on the 2016 DEM, having steep flanks (slope gradients up to 60°, P2 in Figure 8); this ridge extended over an area of 2.6 × 10^4 m^2 down to 200 mwd, with a maximum thickness of 20 m (Figures 5 and 6b) for an estimated volume of +1.7 × 10^5 m^3. R2, together with smaller ridge (R1 in Figures 5 and 6b), bound the E1 erosive area. R2 was not present on the 2013 DEM, roughly matching the location of a previous channelized feature (Ch in Figure 6a), while R1 was already present before the eruption as part of a morphological high (MH in Figure 6a) and did not correspond to an accreted area in the difference map of Figure 5.

Finally, A3 was located in the NE-most part of the study area, covering a surface of 30.000 m^2 down to 200 mwd, with an average seafloor accretion of approximately 10 m (Figure 5 and P2 in Figure 8) for an estimated volume of about +1.5 × 10^5 m^3. Seafloor accretion here morphologically corresponds to the development of irregular and smooth fan-shaped features on the 2016 DEM, developed above and below of a marked slope break (Figure 6b and P4 in Figure 8). These fan-shaped features were emplaced above a previous channelized feature recognizable on the 2013 DEM (Ch in Figure 6a), carving the NW flank of a basement high (BH in Figures 5 and 6) recognizable along the northern shoulder of the SdF collapse scar.

In order to compare submarine and subaerial morphological changes associated with the 2014 eruption, a topographic profile (P5 in Figure 8) crossing the entire NE sector of the SdF (location in Figure 5) is reported, showing the slope accretion occurred between 2012 and 2017 surveys.

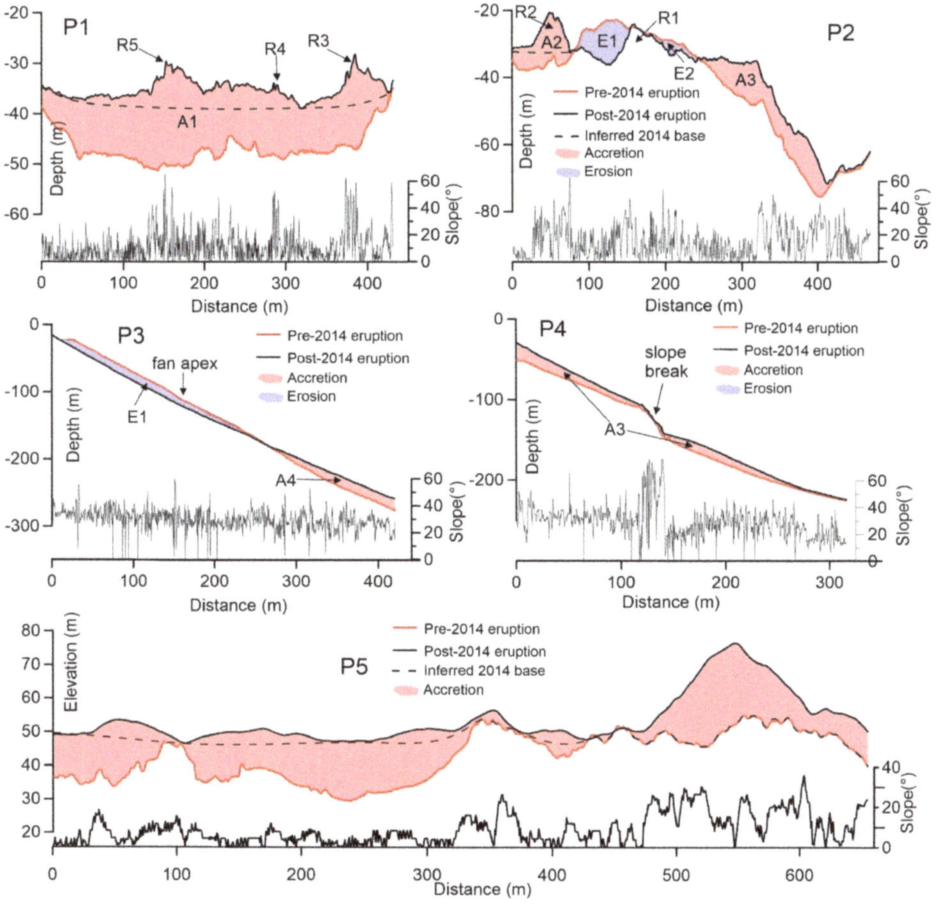

Figure 8. Bathymetric (**P1–P4**) and topographic (**P5**) profiles (location in Figures 5–7) showing seafloor accretion and erosion (red and light blue polygons, respectively) between pre- and post-2014 eruption topography. Acronyms as in the previous figures. Below the profiles, the graph of slope gradients is referred to the post-eruption topography. The location of the apex fan recognizable on the hillshade bathymetry of Figure 6b is reported on profile P3, indicating that the scar is partially filled by landslide deposits.

4.2. CSK-SAR Data Collected between May and August 2014

CSK-SAR images, collected between 8 May 2014 and 24 May 2014, show a decrease in the amplitude ratio (i.e., loss in SAR backscattering between the two consecutive images) in the central portion of the SdF (Figure 9a), testifying the erosion of volcaniclastic deposits. Afterwards, the progressive increase in the explosion rate (16), and the emplacement of lava overflows and rock-falls evolving in gravel flows, produced the volcaniclastic sedimentation in the central part of the SdF (Figure 9b,c).

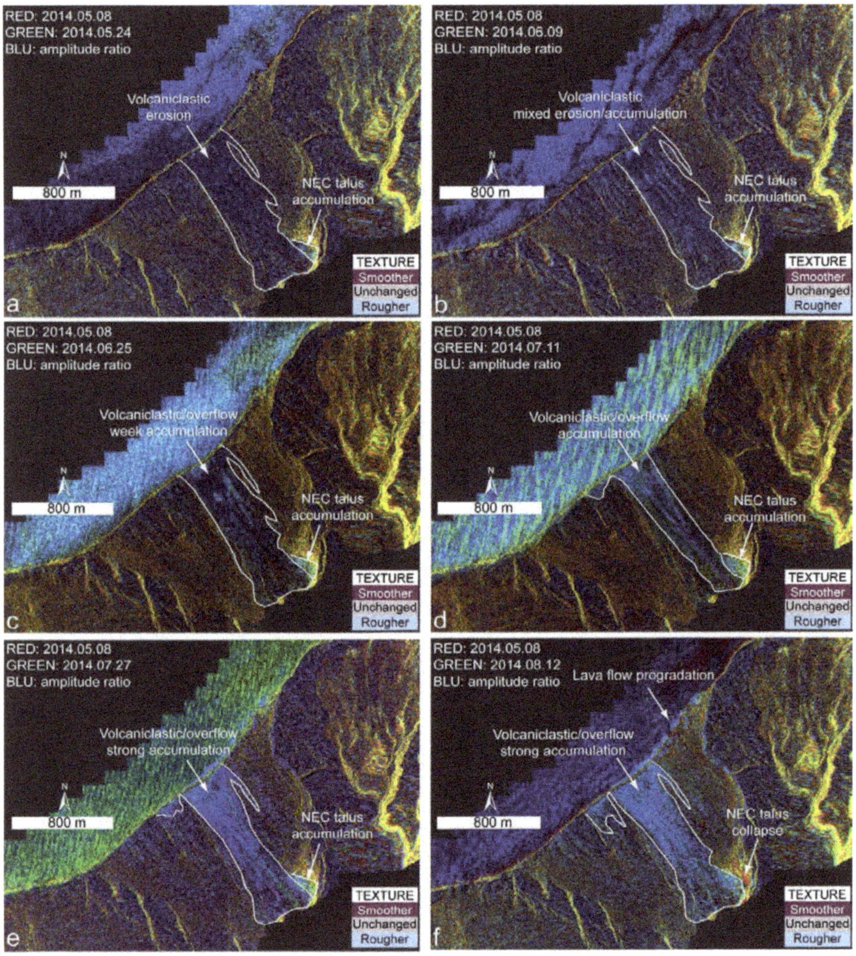

Figure 9. CSK amplitude RGB color composites (RED: 8 May 2014 amplitude image; GREEN: analyzed image; BLUE: ratio between the analyzed image and the 8 May 2014 image). For the significance given to "roughness" see the text. (**a**) comparison image between 8 May 2014 and 24 May 2014, with a reduction in the SAR amplitude in the central part of the SdF; (**b**) comparison image between 8 May 2014 and 9 June 2014, showing areas of reduced and increased SAR amplitude; in (**c**–**e**) it is possible to observe how the areas characterized by increased SAR amplitude increase over time (up to 27 July 2014). (**f**) The central area of the SdF is completely in shades of light blue, reflecting the increase of volcaniclastic sedimentation and overflows that occurred before the onset of the 2014 eruption. Note also the strong reduction of the SAR amplitude in the NEC talus area due to its collapse on 7 August 2014, while the 2014 lava field is characterized by little or no variation in the surface roughness.

This sedimentation (and related superficial roughness) clearly increased in the images collected before the onset of the eruption (11 and 27 July 2014; Figure 9d,e). The events of 6 August 2014, with a series of overflows associated with frequent landslides, significantly increased the roughness in the central part of the SdF (Figure 9f). In the same image, the variation in amplitude on the coastline is due to the entrance into the sea of the lava flows. The lava effusion that began on 7 August 2014 instead produced no remarkable changes in SAR backscattered amplitude, probably due to the similar texture between the pre-effusive and the early-lava flows surfaces. Instead, it is possible to note how the area of proximal volcaniclastic accumulation (i.e., NEC talus) is characterized by accumulations during the whole period considered, except for the last image where an erosive process is highlighted (reduction in the backscattered amplitude in Figure 9f), consistent with the landslide occurred in the early stages of the eruption.

5. Discussion

5.1. Seafloor Accretion Associated with the 2014 Eruption

The seafloor accretion observed in areas A2 and A3 (Figure 5) fits well the entry points of the main lava flows reaching the sea on the NE-most sector of the SdF during the 2014 eruption (outlined by the dashed black lines in Figure 1b), emplaced above the remnants of the 2007 lava delta (Figure 6a; [16]). Once entered the sea, these lava flows built 40–60 m wide and some hundreds m long lobes (for instance R2 in Figure 6b), similarly to what was observed during the emplacement of the 2007 lava flows at Stromboli [25]. However, some morphological differences can be noted between the lobes emplaced in A2 and A3: while the former one corresponds to a well-defined and steep-sided ridge (R2 in Figure 6b), a more irregular, fan-shaped feature is present in the area A3 (see P2 in Figure 8 for comparison). This difference in morphology could be ascribed to the different amount of coherent lava flows with respect to chaotic breccias, using as reference the scuba observations made for the 2007 lava delta, where this kind of products were recognized [25]. Similarly, ancient analogous of 'a'ā lava-fed deltas outcropping in Antarctica were found to be made up by the coalescence of "hyaloclastic lobes", with variable percentages of coherent lavas and chaotic breccias [57].

The higher slope gradients (locally over 60°, P2 in Figure 8) observed along the flanks of R2 can be related to the high percentage of coherent lava flows forming this ridge. Lava flows feeding R2 were, in fact, active along the slope during the first days of the eruption, when the effusion rates were high (up to 20 m^3/s according to the model of [28]). The high effusion rates can also justify the thicker accretion recognizable on the difference map in the coastal sector of the subaerial slope (red and violet colors in Figure 5 just above A2). Such conditions would have favored the penetration of lava flows into the sea as coherent bodies, as suggested by several laboratory and field studies [55,56]. On the meantime, the lower thickness observed for the 2014 lava flows in the coastal sector facing A3 (yellow colors in Figure 5) would justify the development of the irregular fan-shaped feature in this area, due to the prevalence of volcaniclastic breccias on the slope. The volcaniclastic contribution appeared predominant also moving downslope, where the emplacement of a smoothed fan is observed in A3, down to almost 200 mwd (Figure 6b). This composite fan likely formed during the declining stage of the 2014 eruption, when effusion rates reached a steady value of 0.4 m^3/s, and the entrance of well-fed, coherent lava flows into the sea was no longer observed. Moreover, the steep slope break at around 100 mwd could have favored the brecciation of the less sustained submarine lava flows that reached this area, as testified by the thinned seafloor accretion over the slope break (see P4 in Figure 8).

The overall comparison between the pre- and post-bathymetries thus suggests a key role played by the paleomorphology of the SdF slope in controlling the location of the different lava flows and volcaniclastic lobes, which were emplaced within previous channelized features (Figure 6; P2 in Figure 8). Beyond the SW limit of the main 2014 lava flows (as reported by [16], indicated with the dashed line in Figure 4b), a series of steep-sided ridges (R3, R4 and R5 in Figures 5 and 7b) form the accretion area A1. From the comparison with onshore data (Figure 10) they can be interpreted as the submarine extension of lava flows that were mainly emitted during the first days of the 2014 eruption, when effusion rates were very high, and multiple flows reached this coastal sector (Figure 2b,d).

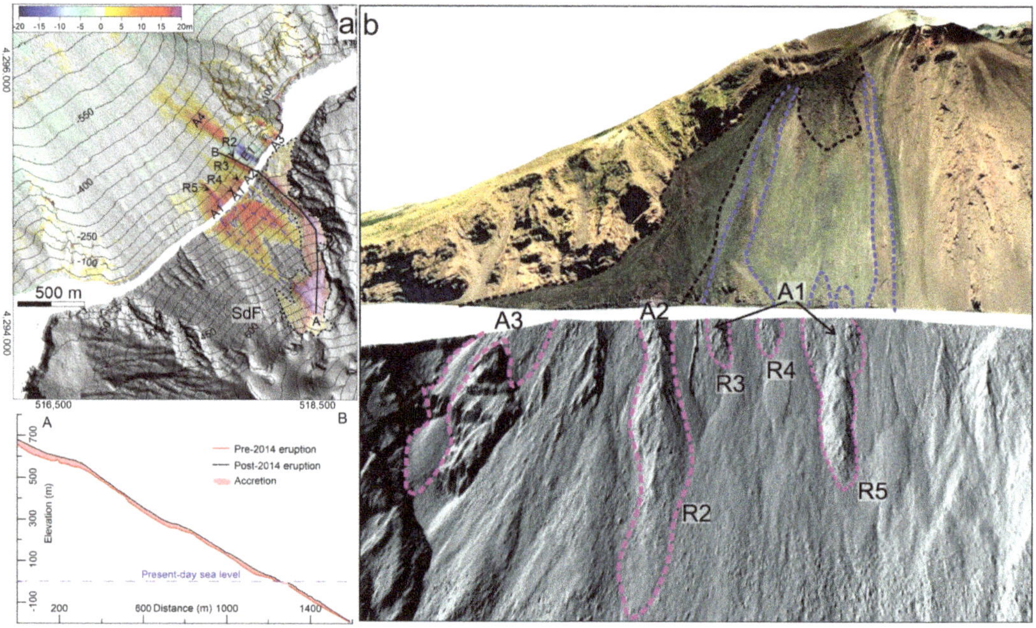

Figure 10. (**a**) Correlation between submarine and subaerial topographic changes, where A1–A4 are areas with seafloor accretion between 2012–2013 and 2016–2017; E1 is the main area with seafloor erosion during the same period. The profile A-B shows the slope accretion associated with the 2014 eruption along the main lava flows that characterized this eruption; (**b**) Main morphologies identified on the subaerial and submarine slope of the SdF; ridges R2 to R5 in magenta are compared with subaerial lava flows drawn as dashed blue lines associated with the 2014 eruption; see text for further details. Dashed black line on the subaerial SdF delimits the area with main morphological changes associated with the 2014 lava flows entering the sea from (adapted from [16]), whereas the blue dashed lines show the other subaerial ridges that in this study have been related to the 2014 lava flows through the integration of submarine/subaerial data.

Besides these ridges, the seafloor accretion in A1 is also made up of coalescing and overlapping small fan-shaped features (Figure 7b) that can be interpreted as the result of erosive-depositional processes associated with gravel flows entering the sea, fed by small-scale instabilities on the steep (>30°) subaerial and submarine slope of the SdF. These small fan-shaped features are a common, but ephemeral feature, often observed in many bathymetric surveys performed since the 2002. They are associated with the morphological adjustment of the SdF slope that commonly occurs due to severe winter storms affecting the NW flank of the island. A good match is observed between subaerial and submarine accretion in the central sector of the SdF (Figures 5 and 10a), due to the progressive infilling of this depressed area (P1 and P5 in Figure 8). This infilling can be mainly related to the abundant volcaniclastic material derived from (a) overflow-induced landslides, (b) the NEC-talus collapse (Figure 2a) and (c) erosion of the upper part of the SdF (arrow in Figure 4b), occurred before, during and after the 2014 eruption.

CKS-SAR data show a relevant accumulation of volcaniclastic material within the depressed area encompassed between the 2007 lava delta and the SW limit of the 2002 landslide scar some months before the starting on the 2014 eruption (Figure 9). Differently, relatively mild slope dynamics, mainly characterized by weak erosive-depositional processes, occurred in the SdF in the period late 2014–early 2017 (Figure 3, [34]), when the post-2014 submarine/subaerial surveys were realized. A possible limit between the volcanic material emplaced in the submarine slope before and during the 2014 effusive eruption has been inferred by through the analysis of the bathymetric cross-sections by reconstructing a hypothetical surface that approximately matches the base of the ridges before mentioned (R2 to R5, dashed black line indicated as "inferred 2014 base" in P1 and P2 of Figure 9).

5.2. Seafloor Erosion Associated with the 2014 Eruption

During the phase preceding the onset of the eruption, there were frequent gravel flows both from the edge of the crater terrace and along the SdF, mainly triggered by overflows. The volume of these gravel flows never reached 10^6 m^3, remaining in the order of magnitude 10^4–10^5 m^3 of remobilized material [16]. In the submarine slope, two failure events (E1 and E2 in Figures 5 and 10a) occurred where the main 2014 lava flows entered the sea, indicating a strong link between slope failures and 2014 eruptive dynamics. It is noteworthy that E1 is the larger submarine instability event recorded through the bathymetric monitoring along the SdF since the 2007 lava delta emplacement.

The computed volume of this event is around -3.5×10^5 m^3 (Table 2), but this figure is very likely underestimated because: (a) bathymetric data are limited to water depth of 10 m, and the difference map indicates that the scar likely cuts back coastward and (b) the scar is partially filled both by landslides deposits in its distal part (see Figure 6b and P3 in Figure 8) and by small-scale fan-shaped features related to successive erosive-depositional process in its proximal part (Figure 6b). A more realistic estimate of the mobilized volume can be obtained by taking into account the A4 seafloor accretion area (Figures 4b, 5 and 10a), interpreted as the landslide deposits recognizable at the base of the scar and accounting for approximately $+6 \times 10^5$ m^3 (Table 2). The low runout of the mobilized material is likely due to a rapid dissipation of pore pressures during the slide, similarly to what was reported for small submarine slope failures affecting the 2018 'a'ā lava delta off Hawaii [38]. Again, this figure is likely underestimated because (a) the mapping of the landslide deposits is limited down to 500 mwd (Figure 5), and (b) the distal part of the moving landslide could be evolved downslope as sedimentary gravity flow, due to the steep gradients characterizing the SdF slope.

Regardless, the estimated volume of the E1 slope failure is two order magnitude larger than any instability processes occurred since the emplacement of the 2007 lava delta, and it also affected the more stable part of the delta according to the previous bathymetric surveys [26]. Such findings support our inference that triggering mechanisms for this larger slope failure are likely different from those common involved in the "ordinary" readjustment of the submarine slope, such as the loading of the 2014 lava flows on the submarine slope or the seismic activity associated with the 2014 eruptive event. However, the combined action of storm-waves cannot be excluded among the triggering mechanisms due to the shallow depths.

5.3. Volumetric Considerations on the 2014 Eruption and SdF Infilling

The estimation of the submarine volume emplaced during the 2014 eruption is difficult, because the subaerial and submarine surveys were not performed just after and before the eruption, as for instance realized during the 2007 eruption [15,25]. Data were collected one year before and one after the eruption for the submarine part and a slightly longer time frame (2012–2017) for the subaerial part. Despite such limits, some volumetric constraints on the overall material emplaced in the SdF during the 2014 eruption can be derived.

According to [16] the main lava flows emitted during the 2014 were emplaced in the NE-most sector of the SdF, encompassed by the dashed black line in Figures 1 and 9a, with a volume of approximately $+2.7 \times 10^6$ m^3 (Table 2). Seafloor accretion in A2 and A3 (Figures 4b and 10) can be confidently associated with these lava flows, accounting for approximately $+3.3 \times 10^5$ m^3 (Table 2). However, taking into account a southwestern, further extension of lava flows in the central part of the SdF on the base of submarine evidence (from R3 to R5 in Figures 5, 7, 8 and 10) we proposed in this work that further $+1.5 \times 10^5$ m^3 can be added as products of the 7 August–13 November 2014 eruption (Table 2). Moreover, this figure is surely underestimated because the bathymetric comparison pre-/post-eruption is limited at depths greater than 10 m. In order to compute the volume not considered, an average thickness of 5 and 10 m has been assumed for the SW and NE sectors, respectively, based on the trend derived from the difference map and integrated over the coast. The estimated volume is +2 × 10^5 m^3 of material that added to the previous two figures accounts for a total estimated volume of ≈6.8 × 10^5 m^3 in the submarine part of the SdF (Table 2).

A similar approach has been also applied for the central part of the subaerial SdF facing A1, where the volume possibly related to the 2014 eruption was distinguished with respect to the underlying volcaniclastic material by reconstructing a surface interpolated through the base of the subaerial ridges facing the submarine ones (P5 in Figure 8). This volume was estimated to about +3.5 × 10^5 m^3, which summed to the previous +2.7 × 10^6 m^3 accounts for a total volume of +3.05 × 10^6 m^3 emplaced on the subaerial slope (Table 2). It is noteworthy that the volume emplaced in the surveyed submarine area was markedly lower (about 5 times) than the volume emplaced on the subaerial SdF (Figure 11 and Table 2). This is different from what observed during the previous 2007 eruption, where the volume of the erupted material estimated for the submarine part (where a large lava delta in the order of 7 × 10^6 m^3 of volume was formed, see [25]) was approximately three times larger than the subaerial one (Figure 11 and Table 2).

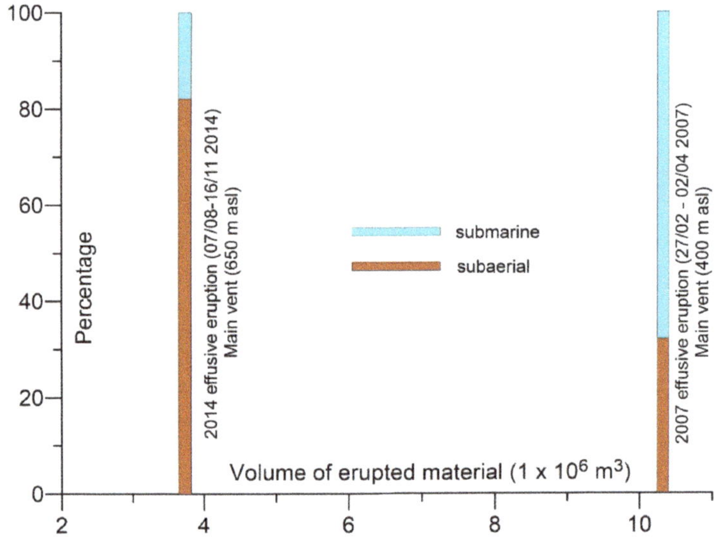

Figure 11. Bar graph showing the difference in volumes (see also Table 2) and percentages between volcanic material accreted on the subaerial (light brown) and submarine (light blue) slope of the SdF in correspondence of different eruptive phases.

This discrepancy can be mainly associated with the different elevation of the main vents feeding lava flows during the 2007 eruption (around 400 m a.s.l.) and the 2014 eruption (around 650 m a.s.l). Considering that, during the 2014 eruption, effusion rates drastically fell from 20 to 0.5 m^3/s just two days after the eruption [28], lava flows erupted by a high-elevation vent during the declining stage of the eruption could have remained mostly confined to the median part of the SdF slope, without reaching the sea. Another interesting finding is that the total volume estimated for the 2014 eruption through the above-described computations (approximately +3.7 × 10^6 m^3, Table 2) is markedly lower than the total bulk volume of +7.4 × 10^6 m^3 estimated using data derived by the new satellite Technology Experiment Carrier-1 (TET-1) [37], whereas it is more comparable to the total bulk volume of +5.5 × 10^6 m^3 obtained by thermal images from satellites using the moderate resolution imaging spectroradiometer (MODIS) sensor [28]. This large difference can be mostly explained by the different methods used for constraining the total volume, as in the case of MODIS approach, where the error is usually estimated at 50% [58].

Nevertheless, our approach relies on topographic differences realized not contextually at the start and end of eruption, so hindering a clear discrimination between material the emplaced during the eruption with respect to the volcaniclastic material emplaced (mostly) before and after the eruption, especially in the central part of the SdF. This area, in fact, is a morphological low with respect to the surrounding NE and SW sectors, representing a main pathway for the funneling to the sea of volcaniclastic material produced by persistent strombolian activity and erosive processes acting in the upper SdF slope (Figures 4b and 10), and this makes difficult to discriminate between such deposits and the previous 2014 overflows. As a whole, our results show a good matching between the subaerial and submarine slope accretion in the central part of the SdF (Figures 4b, 5, 9 and 10a). As regards the subaerial slope, this infilling is mostly limited between 500 m and the coastline, resulting in a volume of approximately +1.35 $\times$ 10^6 m^3 (Table 2). For the submarine slope, the infilling is mostly limited down to -250 m, with an average accretion of 10 m for an estimated volume of approximately +7 $\times$ 10^5 m^3, 4.9 $\times$ 10^5 m^3 of which were directly measured through the difference map (limited to depths greater than 10–15 m). The cumulative value between submarine and subaerial slope accounts for approximately 1.84 $\times$ 10^6 m^3 accumulated during few years, representing not only eruptive dynamics but also gravity instability processes reworking the SdF slope. It is noteworthy that this infilling is also responsible for the development of a steep submarine volcaniclastic apron in the proximal sector of the submarine SdF that was one of the main factors controlling the geometry of the 2002 tsunamigenic landslide [20].

6. Conclusions

The integration of repeated bathymetric, lidar and photogrammetric surveys along with SAR amplitude images was a fundament tool to understand the morphological evolution of the SdF slope induced by the 2014 eruption and its pre-eruptive stage, which was characterized by two months of intense Strombolian activity. Difference maps showed that slope accretion largely overwhelmed erosion along most of the SdF slope both before and during the 2014 eruption. The total volume of erupted material during the 2014 eruption was approximately estimated in +3.7 $\times$ 10^6 m^3, $\approx$80% of which emplaced in the subaerial slope. This figure is different from what depicted for the previous 2007 eruption, where a large submarine lava delta formed, representing $\approx$75% of the total volume. This difference can be explained by the location of the main vents feeding the 2007 (around 400 m a.s.l.) and 2014 eruption (around 650 m). Before the 2014 eruption, the total volume estimated is about +1.84 $\times$ 10^6 m^3, the 60% of which emplaced in the subaerial slope between 500 m and the coastline. The remaining 40% was emplaced in the submarine slope down to 250 mwd, promoting the formation of a steep volcaniclastic apron, a feature that increase the tsunamigenic potential of the SdF, as it was considered to have controlled the geometry of the submarine tsunamigenic landslide occurred in 2002.

The 2014 eruption also promoted significant slope instabilities along the SdF slope. In the subaerial slope, landslides mobilized volumes in the order of $-10^4/10^5$ m^3, mostly affecting the upper part of the SdF during the early stages of the eruption. In the submarine slope, a main landslide scar was detected in the NE part of the SdF within the first 200 mwd, just off the entry points of the main 2014 lava flows. Landslide deposits are recognizable at the base of the scar, extending down to 500 mwd, corresponding to the limit of the bathymetric survey. This landslide mobilized a volume of at least -6 $\times$ 10^5 m^3, representing the largest slope failure affecting the submarine part of the 2007 lava delta since its emplacement.

DEM analysis also showed a good correlation between the subaerial and submarine slope and a variability in submarine morphologies associated with the entrance into the sea of lava flows (i.e., steep-sided ridges and fan-shaped features), reflecting a different ratio between coherent lava flows and volcaniclastic breccias that, in turn, was likely controlled by changes in effusion rates and paleo-topography. This observation is very important for the interpretation of seafloor volcanic morphologies around insular and coastal volcanoes, mainly in relation to the paucity of marine studies monitoring the behavior of lava flows penetrating into the sea.

The presented results highlight the importance of integrated submarine and subaerial studies to monitor active volcanoes, providing a comprehensive view of the main processes (constructive vs destructive) associated with eruptive dynamics, with significant implications also for the related geohazard assessment.

Author Contributions: D.C. conceived and designed the study; A.B. and D.C. acquired and processed the multibeam data, D.C., F.L.C., C.R. and A.B. analyzed and interpreted the multibeam data; F.D.T. and N.C. analyzed and interpreted the topographic and CKS-SAR data; D.C., F.D.T. and C.R. led the writing of the main text, with the contribution of F.L.C. and A.B.; D.C., F.D.T. and F.L.C. prepared the figures. All authors have read and agreed to the published version of the manuscript.

Funding: This research was funded by the National Research Council and "Presidenza del Consiglio dei Ministri–Dipartimento della Protezione Civile" (Presidency of the Council of Ministers–Department of Civil Protection), through CNR-DPC and UniFi-DPC respective agreements, however, does not reflect the position and the official policies of the Department.

Data Availability Statement: The data presented in this study are available on request from the corresponding author. The data are not publicly available due to their acquisitional for institutional purposes.

Acknowledgments: Crew of R/V Urania and Minerva1 are gratefully acknowledged along with the people who took part in the surveys. This work was carried out using CSK® Products© ASI (Italian Space Agency), delivered under an ASI license to use in the framework of COSMO-SkyMed Open Call for Science (Project Id: 191; Project title: Volcano Instability—VITA; Scientific Coordinator: F. Di Traglia)". The COSMO-SkyMed images were exploited from the COSMO-SkyMed data hub, managed by E-GEOS for ASI. The authors are grateful to E. Di Cuia (E-GEOS) for the technical support during the exploitation phase. X-band Cosmo-SkyMED images were processed using the freely delivered Sentinel Application Platform (SNAP) software. Three anonymous reviewers are gratefully acknowledged for their suggestions that improved the quality of the paper.

Conflicts of Interest: The authors declare no conflict of interest.

References

1. Bosman, A.; Chiocci, F.L.; Romagnoli, C. Morpho-structural setting of Stromboli volcano revealed by high-resolution bathymetry and backscatter data of its submarine portions. *Bull. Volcanol.* **2009**, *71*, 1007–1019. [CrossRef]
2. Francalanci, L.; Lucchi, F.; Keller, J.; De Astis, G.; Tranne, C.A. Chapter 13 Eruptive, volcano-tectonic and magmatic history of the Stromboli volcano (north-eastern Aeolian archipelago). *Geol. Soc. Lond. Memoirs* **2013**, *37*, 397–471. [CrossRef]
3. Calvari, S.; Bonaccorso, A.; Madonia, P.; Neri, M.; Liuzzo, M.; Salerno, G.G.; Behncke, B.; Caltabiano, T.; Cristaldi, A.; Giuffrida, G.; et al. Major eruptive style changes induced by structural modifications of a shallow conduit system: The 2007–2012 Stromboli case. *Bull. Volcanol.* **2014**, *76*, 1–15. [CrossRef]
4. Kokelaar, P.; Romagnoli, C. Sector collapse, sedimentation and clast population evolution at an active island-arc volcano: Stromboli, Italy. *Bull. Volcanol.* **1995**, *57*, 240–262. [CrossRef]
5. Romagnoli, C.; Casalbore, D.; Chiocci, F.L.; Bosman, A. Offshore evidence of large-scale lateral collapses on the eastern flank of Stromboli, Italy, due to structurally-controlled, bilateral flank instability. *Mar. Geol.* **2009**, *262*, 1–13. [CrossRef]
6. Bosellini, A. Outline of the Geology of Italy. In *Landscapes and Landforms of Italy*; Springer: Cham, Switzerland, 2017; pp. 21–27.
7. Romagnoli, C.; Kokelaar, P.; Casalbore, D.; Chiocci, F.L. Lateral collapses and active sedimentary processes on the northwestern flank of Stromboli volcano, Italy. *Mar. Geol.* **2009**, *265*, 101–119. [CrossRef]
8. Barberi, F.; Rosi, M.; Sodi, A. Volcanic hazard assessment at Stromboli based on review of historical data. *Acta Vulcanol.* **1993**, *3*, 173–187.
9. Bevilacqua, A.; Bertagnini, A.; Pompilio, M.; Landi, P.; Del Carlo, P.; Di Roberto, A.; Aspinall, W.; Neri, A. Major explosions and paroxysms at Stromboli (Italy): A new historical catalog and temporal models of occurrence with uncertainty quantification. *Sci. Rep.* **2020**, *10*, 17357. [CrossRef] [PubMed]
10. Giudicepietro, F.; López, C.; Macedonio, G.; Alparone, S.; Bianco, F.; Calvari, S.; De Cesare, W.; Donne, D.D.; Di Lieto, B.; Esposito, A.M.; et al. Geophysical precursors of the July-August 2019 paroxysmal eruptive phase and their implications for Stromboli volcano (Italy) monitoring. *Sci. Rep.* **2020**, *10*, 10296. [CrossRef]

11. Calvari, S.; Intrieri, E.; Di Traglia, F.; Bonaccorso, A.; Casagli, N.; Cristaldi, A. Monitoring crater-wall collapse at active volcanoes: A study of the 12 January 2013 event at Stromboli. *Bull. Volcanol.* **2016**, *78*, 39. [CrossRef]
12. Calvari, S.; Di Traglia, F.; Ganci, G.; Giudicepietro, F.; Macedonio, G.; Cappello, A.; Nolesini, T.; Pecora, E.; Bilotta, G.; Centorrino, V.; et al. Overflows and Pyroclastic Density Currents in March-April 2020 at Stromboli Volcano Detected by Remote Sensing and Seismic Monitoring Data. *Remote Sens.* **2020**, *12*, 3010. [CrossRef]
13. De Fino, M. The Stromboli eruption of 6 December 1985–25 April 1986: Volcanological, petrological and seismological data. *Rend. Soc. Ital. Mineral. Petrol.* **1988**, *43*, 1021–1038.
14. Lodato, L.; Spampinato, L.; Harris, A.; Calvari, S.; Dehn, J.; Patrick, M. The morphology and evolution of the Stromboli 2002–2003 lava flow field: An example of a basaltic flow field emplaced on a steep slope. *Bull. Volcanol.* **2007**, *69*, 661–679. [CrossRef]
15. Marsella, M.; Baldi, P.; Coltelli, M.; Fabris, M. The morphological evolution of the Sciara del Fuoco since 1868: Reconstructing the effusive activity at Stromboli volcano. *Bull. Volcanol.* **2011**, *74*, 231–248. [CrossRef]
16. Di Traglia, F.; Calvari, S.; D'Auria, L.; Nolesini, T.; Bonaccorso, A.; Fornaciai, A.; Esposito, A.; Cristaldi, A.; Favalli, M.; Casagli, N. The 2014 Effusive Eruption at Stromboli: New Insights from In Situ and Remote-Sensing Measurements. *Remote Sens.* **2018**, *10*, 2035. [CrossRef]
17. Bonaccorso, A.; Calvari, S.; Linde, A.; Sacks, S.; Boschi, E. Dynamics of the shallow plumbing system investigated from borehole strainmeters and cameras during the 15 March, 2007 Vulcanian paroxysm at Stromboli volcano. *Earth Planet. Sci. Lett.* **2012**, *357–358*, 249–256. [CrossRef]
18. Di Traglia, F.; Nolesini, T.; Intrieri, E.; Mugnai, F.; Leva, D.; Rosi, M.; Casagli, N. Review of ten years of volcano de-formations recorded by the ground-based InSAR monitoring system at Stromboli volcano: A tool to mitigate volcano flank dynamics and intense volcanic activity. *Earth Sci. Rev.* **2014**, *139*, 317–335. [CrossRef]
19. Tommasi, P.; Baldi, P.; Chiocci, F.L.; Coltelli, M.; Marsella, M.; Pompilio, M.; Romagnoli, C. The Landslide Sequence Induced by the 2002 Eruption at Stromboli Volcano. In *Landslides*; Springer: Berlin/Heidelberg, Germany, 2005; pp. 251–258.
20. Chiocci, F.L.; Romagnoli, C.; Tommasi, P.; Bosman, A. The Stromboli 2002 tsunamigenic submarine slide: Characteristics and possible failure mechanisms. *J. Geophys. Res. Solid Earth* **2008**, *113*. [CrossRef]
21. Tinti, S.; Pagnoni, G.; Zaniboni, F. The landslides and tsunamis of the 30th of December 2002 in Stromboli analysed through numerical simulations. *Bull. Volcanol.* **2006**, *68*, 462–479. [CrossRef]
22. Maramai, A.; Graziani, L.; Tinti, S. Tsunamis in the Aeolian Islands (southern Italy): A review. *Mar. Geol.* **2005**, *215*, 11–21. [CrossRef]
23. Fornaciai, A.; Favalli, M.; Nannipieri, L. Numerical simulation of the tsunamis generated by the Sciara del Fuoco landslides (Stromboli Island, Italy). *Sci. Rep.* **2019**, *9*, 1–12. [CrossRef] [PubMed]
24. Chiocci, F.L.; Romagnoli, C.; Bosman, A. Morphologic resilience and depositional processes due to the rapid evolution of the submerged Sciara del Fuoco (Stromboli Island) after the December 2002 submarine slide and tsunami. *Geomorphology* **2008**, *100*, 356–365. [CrossRef]
25. Bosman, A.; Casalbore, D.; Romagnoli, C.; Chiocci, F.L. Formation of an 'a'ā lava delta: Insights from time-lapse multibeam bathymetry and direct observations during the Stromboli 2007 eruption. *Bull. Volcanol.* **2014**, *76*, 1–12. [CrossRef]
26. Casalbore, D.; Passeri, F.; Tommasi, P.; Verrucci, L.; Bosman, A.; Romagnoli, C.; Chiocci, F.L. Small-scale slope instability on the submarine flanks of insular volcanoes: The case-study of the Sciara del Fuoco slope (Stromboli). *Int. J. Earth Sci.* **2020**, *109*, 2643–2658. [CrossRef]
27. Marsella, M.; Proietti, C.; Sonnessa, A.; Coltelli, M.; Tommasi, P.; Bernardo, E. The evolution of the Sciara del Fuoco subaerial slope during the 2007 Stromboli eruption: Relation between deformation processes and effusive activity. *J. Volcanol. Geotherm. Res.* **2009**, *182*, 201–213. [CrossRef]
28. Valade, S.; Lacanna, G.; Coppola, D.; Laiolo, M.; Pistolesi, M.; Donne, D.D.; Genco, R.; Marchetti, E.; Ulivieri, G.; Allocca, C.; et al. Tracking dynamics of magma migration in open-conduit systems. *Bull. Volcanol.* **2016**, *78*, 78. [CrossRef]
29. Giudicepietro, F.; Calvari, S.; Alparone, S.; Bianco, F.; Bonaccorso, A.; Bruno, V.; Caputo, T.; Cristaldi, A.; D'Auria, L.; De Cesare, W.; et al. Integration of Ground-Based Remote-Sensing and In Situ Multidisciplinary Monitoring Data to Analyze the Eruptive Activity of Stromboli Volcano in 2017–2018. *Remote Sens.* **2019**, *11*, 1813. [CrossRef]
30. Di Lieto, B.; Romano, P.; Scarpa, R.; Linde, A.T. Strain Signals Before and During Paroxysmal Activity at Stromboli Volcano, Italy. *Geophys. Res. Lett.* **2020**, *47*. [CrossRef]
31. Ripepe, M.; Donne, D.D.; Genco, R.; Maggio, G.; Pistolesi, M.; Marchetti, E.; Lacanna, G.; Ulivieri, G.; Poggi, P. Volcano seismicity and ground deformation unveil the gravity-driven magma discharge dynamics of a volcanic eruption. *Nat. Commun.* **2015**, *6*, 6998. [CrossRef]
32. Rizzo, A.L.; Federico, C.; Inguaggiato, S.; Sollami, A.; Tantillo, M.; Vita, F.; Bellomo, S.; Longo, M.; Grassa, F.; Liuzzo, M. The 2014 effusive eruption at Stromboli volcano (Italy): Inferences from soil CO2 flux and 3 He/4 He ratio in thermal waters. *Geophys. Res. Lett.* **2015**, *42*, 2235–2243. [CrossRef]
33. Inguaggiato, S.; Vita, F.; Cangemi, M.; Calderone, L. Increasing Summit Degassing at the Stromboli Volcano and Relationships with Volcanic Activity (2016–2018). *Geoscience* **2019**, *9*, 176. [CrossRef]
34. Schaefer, L.N.; Di Traglia, F.; Chaussard, E.; Lu, Z.; Nolesini, T.; Casagli, N. Monitoring volcano slope instability with Synthetic Aperture Radar: A review and new data from Pacaya (Guatemala) and Stromboli (Italy) volcanoes. *Earth Sci. Rev.* **2019**, *192*, 236–257. [CrossRef]

35. Massimetti, F.; Coppola, D.; Laiolo, M.; Valade, S.; Cigolini, C.; Ripepe, M. Volcanic Hot-Spot Detection Using SENTINEL-2: A Comparison with MODIS–MIROVA Thermal Data Series. *Remote Sens.* **2020**, *12*, 820. [CrossRef]
36. Baldi, P.; Bosman, A.; Chiocci, F.L.; Marsella, M.; Romagnoli, C.; Sonnessa, A. Integrated subaerial-submarine morphological evolution of the Sciara del Fuoco after the 2002 landslide. In *The Stromboli Volcano: An Integrated Study of the 2002–2003 Eruption*. Available online: https://agupubs.onlinelibrary.wiley.com/doi/10.1029/182GM15 (accessed on 21 May 2021).
37. Zakšek, K.; Hort, M.; Lorenz, E. Satellite and Ground Based Thermal Observation of the 2014 Effusive Eruption at Stromboli Volcano. *Remote Sens.* **2015**, *7*, 17190–17211. [CrossRef]
38. Soule, S.A.; Zoeller, M.; Parcheta, C. Submarine lava deltas of the 2018 eruption of Kīlauea volcano. *Bull. Volcanol.* **2021**, *83*, 1–16. [CrossRef]
39. Mattox, T.N.; Mangan, M.T. Littoral hydrovolcanic explosions: A case study of lava–seawater interaction at Kilauea Volcano. *J. Volcanol. Geotherm. Res.* **1997**, *75*, 1–17. [CrossRef]
40. Poland, M.P.; Orr, T.R. Identifying hazards associated with lava deltas. *Bull. Volcanol.* **2014**, *76*, 1–10. [CrossRef]
41. Moore, J.; Phillips, R.L.; Grigg, R.W.; Peterson, D.W.; Swanson, D.A. Flow of lava into the sea, 1969–1971, Kilauea volcano, Hawaii. *GSA Bull.* **1973**, *84*, 537–546. [CrossRef]
42. Tribble, G.W. Underwater observations of active lava flows from Kilauea volcano, Hawaii. *Geology* **1991**, *19*, 633–636. [CrossRef]
43. Di Traglia, F.; Battaglia, M.; Nolesini, T.; Lagomarsino, D.; Casagli, N. Shifts in the eruptive styles at Stromboli in 2010–2014 revealed by ground-based InSAR data. *Sci. Rep.* **2015**, *5*, 13569. [CrossRef]
44. Di Traglia, F.; Nolesini, T.; Ciampalini, A.; Solari, L.; Frodella, W.; Bellotti, F.; Fumagalli, A.; De Rosa, G.; Casagli, N. Tracking morphological changes and slope instability using spaceborne and ground-based SAR data. *Geomorphology* **2018**, *300*, 95–112. [CrossRef]
45. Ripepe, M.; Pistolesi, M.; Coppola, D.; Donne, D.D.; Genco, R.; Lacanna, G.; Laiolo, M.; Marchetti, E.; Ulivieri, G.; Valade, S. Forecasting Effusive Dynamics and Decompression Rates by Magmastatic Model at Open-vent Volcanoes. *Sci. Rep.* **2017**, *7*, 1–9. [CrossRef] [PubMed]
46. Di Traglia, F.; Nolesini, T.; Solari, L.; Ciampalini, A.; Frodella, W.; Steri, D.; Allotta, B.; Rindi, A.; Marini, L.; Monni, N.; et al. Lava delta deformation as a proxy for submarine slope instability. *Earth Planet. Sci. Lett.* **2018**, *488*, 46–58. [CrossRef]
47. Di Traglia, F.; Fornaciai, A.; Favalli, M.; Nolesini, T.; Casagli, N. Catching geomorphological response to volcanic activity on steep slope volcanoes using multi-platform remote sensing. *Remote Sens.* **2020**, *12*, 438. [CrossRef]
48. Bagnardi, M.; Gonzalez, P.J.; Hooper, A. High-resolution digital elevation model from tri-stereo Pleiades-1 satellite imagery for lava flow volume estimates at Fogo Volcano. *Geophys. Res. Lett.* **2016**, *43*, 6267–6275. [CrossRef]
49. Arnold, D.; Biggs, J.; Wadge, G.; Ebmeier, S.; Odbert, H.; Poland, M. Dome growth, collapse, and valley fill at Soufrière Hills Volcano, Montserrat, from 1995 to 2013: Contributions from satellite radar measurements of topographic change. *Geosphere* **2016**, *12*, 1300–1315. [CrossRef]
50. Arnold, D.W.D.; Biggs, J.; Anderson, K.; Vargas, S.V.; Wadge, G.; Ebmeier, S.K.; Naranjo, M.F.; Mothes, P. Decaying Lava Extrusion Rate at El Reventador Volcano, Ecuador, Measured Using High-Resolution Satellite Radar. *J. Geophys. Res. Solid Earth* **2017**, *122*, 9966–9988. [CrossRef]
51. Chaussard, E. A low-cost method applicable worldwide for remotely mapping lava dome growth. *J. Volcanol. Geotherm. Res.* **2017**, *341*, 33–41. [CrossRef]
52. Wadge, G.; Cole, P.; Stinton, A.; Komorowski, J.-C.; Stewart, R.; Toombs, A.C.; Legendre, Y. Rapid topographic change measured by high-resolution satellite radar at Soufriere Hills Volcano, Montserrat, 2008–2010. *J. Volcanol. Geotherm. Res.* **2011**, *199*, 142–152. [CrossRef]
53. Ebmeier, S.K.; Biggs, J.; Muller, C.; Avard, G. Thin-skinned mass-wasting responsible for widespread deformation at Arenal volcano. *Front. Earth Sci.* **2014**, *2*, 35. [CrossRef]
54. Rignot, E.J.; Van Zyl, J.J. Change detection techniques for ERS-1 SAR data. *IEEE Trans. Geosci. Remote Sens.* **1993**, *31*, 896–906. [CrossRef]
55. Carr, P.F.; Jones, B.G. The influence of palaeoenvironment and lava flux on the emplacement of submarine, near-shore Late Permian basalt lavas, Sydney Basin (Australia). *J. Volcanol. Geotherm. Res.* **2001**, *112*, 247–266. [CrossRef]
56. Gregg, T.K.; Fink, J.H. A laboratory investigation into the effects of slope on lava flow morphology. *J. Volcanol. Geotherm. Res.* **2000**, *96*, 145–159. [CrossRef]
57. Smellie, J.L.; Wilch, T.I.; Rocchi, S. 'A'ā lava-fed deltas: A new reference tool in paleoenvironmental studies. *Geology* **2013**, *41*, 403–406. [CrossRef]
58. Coppola, D.; Laiolo, M.; Piscopo, D.; Cigolini, C. Rheological control on the radiant density of active lava flows and domes. *J. Volcanol. Geotherm. Res.* **2013**, *249*, 39–48. [CrossRef]

MDPI
St. Alban-Anlage 66
4052 Basel
Switzerland
Tel. +41 61 683 77 34
Fax +41 61 302 89 18
www.mdpi.com

Remote Sensing Editorial Office
E-mail: remotesensing@mdpi.com
www.mdpi.com/journal/remotesensing

www.ingramcontent.com/pod-product-compliance
Ingram Content Group UK Ltd.
Pitfield, Milton Keynes, MK11 3LW, UK
UKHW060122010826
14089UKWH00006B/109

9 783036 551197